P9-AGV-171

HANDBOOK OF PETROLEUM EXPLORATION AND PRODUCTION

6

Series Editor

JOHN CUBITT

Previous volumes in this series:

Volume 1	Operational Aspects of Oil and Gas Well Testing
Volume 2	Statistics for Petroleum Engineers and Geoscientists
Volume 3	Well Test Analysis
Volume 4	A Generalized Approach to Primary Hydrocarbon Recovery of Petroleum Exploration and Production
Volume 5	Deep-Water Processes and Facies Models: Implications for Sandstone Petroleum Reservoirs

STRATIGRAPHIC RESERVOIR CHARACTERIZATION FOR PETROLEUM GEOLOGISTS, GEOPHYSICISTS, AND ENGINEERS

STRATIGRAPHIC RESERVOIR CHARACTERIZATION FOR PETROLEUM GEOLOGISTS, GEOPHYSICISTS, AND ENGINEERS

HANDBOOK OF PETROLEUM EXPLORATION AND PRODUCTION
6

Roger M. SLATT
University of Oklahoma
Norman, Oklahoma 73019
U.S.A.

ELSEVIER
Amsterdam • Boston • Heidelberg • London • New York • Oxford
Paris • San Diego • San Francisco • Singapore • Sydney • Tokyo

Elsevier
Radarweg 29, PO Box 211, 1000 AE Amsterdam, The Netherlands
The Boulevard, Langford Lane, Kidlington, Oxford OX5 1GB, UK

First edition 2006

Library of Congress Cataloging-in-Publication Data
A catalog record for this book is available from the Library of Congress

British Library Cataloguing in Publication Data
A catalogue record for this book is available from the British Library

ISBN-13: 978-0-444-52818-6
978-0-444-52817-9 (CD ROM)
ISBN-10: 0-444-52818-0
0-444-52817-2 (CD ROM)
Series ISSN 1567-8032

Printed and bound in The Netherlands

06 07 08 09 10   10 9 8 7 6 5 4 3 2 1

Contents

Preface

I have been very fortunate in having experienced a dual career spanning fourteen years in the petroleum industry and sixteen years in academia. Thus, this book carries with it my experience in the scientific and technical aspects of reservoir characterization as well as my understanding of the necessity for explaining a concept or practice in a manner that is understandable to technical people from a variety of experience levels and interests. To accomplish this, I have drawn upon personal experiences in reservoir characterization and also upon both classic and recently published comprehensive literature. The course from which this book evolved is actually a flexible series of topics presented to both undergraduate and graduate-level students, to domestic and international petroleum companies, to petroleum societies and organizations, and even to the interested public as an online web (distance-learning) course.

Reservoir characterization as a discipline grew out of the recognition that more oil and gas could be extracted from reservoirs if the geology of the reservoir was understood. Prior to that awakening, reservoir development and production were the realm of the petroleum engineer. In fact, geologists of that time would have felt slighted if asked by corporate management to move from an exciting exploration assignment to a more mundane assignment working with an engineer to improve a reservoir's performance.

Slowly, reservoir characterization came into its own as a quantitative, multidisciplinary endeavor requiring a vast array of skills and knowledge sets. Perhaps the biggest attractor to becoming a reservoir geologist was the advent of fast computing, followed by visualization programs and theaters, all of which allow young geoscientists to practice their computing skills in a highly technical work environment. Also, the discipline grew in parallel with the evolution of data integration and the advent of asset teams in the petroleum industry. Finally, reservoir characterization flourished with the quantum improvements that have occurred in geophysical acquisition and processing techniques and that allow geophysicists to image internal reservoir complexities.

Unfortunately, universities have lagged behind this discipline's growth, so that young geoscience students typically do not get the opportunity to mingle with their engineering counterparts. Certain university faculties still take the naïve view that something called "characterization" cannot be science-based and should be confined to vocational education.

However, some universities, particularly several in Europe and a few in the US, have picked up the ball and developed solid, rigid educational programs. Students at those universities now are reaping the benefits of frantic industry recruitment that aims to avoid the demographic trap of workers collectively approaching retirement age at a time when energy demand is increasing globally.

Thus, this book is intended as a primer for geologists and geophysicists whose education and careers have taken them to this fascinating, multifaceted discipline. The book is also for petroleum engineers who seek to understand what geologists and geophysicists do and to explore how all three groups can help improve reservoir performance in a team setting.

The book focuses on stratigraphic aspects of characterization, with particular emphasis on understanding the primary control that depositional processes and systems exert on reservoir performance, and the extent (or sometimes the limits) to which stratigraphic features can be predicted away from the wellbore. Students and professionals have told me that this primer has helped them understand the stratigraphic intricacies of reservoirs and how such intricacies affect reservoir performance.

I have purposefully avoided discussing structural aspects of reservoirs much, because that is a vast field in itself. For the same reason, I apply engineering principles in only a peripheral manner. Finally, I have omitted a chapter on reservoir geologic modeling and its application to simulation. That discipline has evolved so rapidly that it is difficult to keep up with new algorithms, programs, and approaches to quantitative characterization. Suffice it to say that as geologists and geophysicists become increasingly involved in reservoir characterization, more complex and more-quantitative modeling programs will continue to evolve, often specific to a particular reservoir problem.

Yes, the discipline of reservoir characterization is complex, comprehensive, multidisciplinary, and exciting. It promises many careers for young people entering the petroleum industry, and for more experienced individuals seeking to broaden their horizons.

I would like to acknowledge individuals with whom I have had the honor and pleasure to work with, and learn from, during my years in petroleum geoscience. These people include, but are not limited to Hamid Al-Hakeem, Al Barnes, Greg Browne, Mike Burnett, John Castagna, Bob Davis, Marlan Downey, Jim Ebanks Jr., Eric Eslinger, Camilo Goyeneche, Neil Hurley, Cretis Jenkins, Doug Jordan, John Kaldi, T.K. Kan, Marcus Milling, Shankar Mitra, Clyde Moore, Dave Pyles, Mark Scheihing, Bob Sneider, Charles Stone, Rod Tillman, John Warme, Bob and Paul Weimer, and Alan Witten. The patience and encouragement of my wife Linda Gay during the writing of this book is also greatly acknowledged. My son Andrew Slatt completed most of the excellent graphics used in the book, and Anne Thomas patiently edited each chapter to make them more readable than I could have accomplished. Carol Drayton spent many long, and sometimes frustrating hours securing permissions to publish figures. My other son Tom Slatt frequently provided food for thought and nourishment. Finally, I am especially indebted to Robert Stephenson for providing the incentive for me to take up residence at the University of Oklahoma to enhance its energy program, and for providing the financial support through a start-up donation that allowed me to complete this book.

Roger M. Slatt
Gungoll Family Chair Professor of
Petroleum Geology and Geophysics
University of Oklahoma

Chapter 1

Basic principles and applications of reservoir characterization

1.1 Introduction

The volume of information that is being generated and made publicly available about oil and gas reservoirs is increasing at an exponential rate, as is most "knowledge". The "information age" applies equally to oil and gas exploration and development as to other global issues.

Partly because of the volume and the nature of available information, and the lessons learned and discussed from specific projects, the field of reservoir characterization is approaching a healthy level of maturity. Not many years ago, being assigned to evaluate a reservoir was considered a less desirable job for an exploration geologist. As exploration declined during the 1980s, stratigraphers, who had been accustomed to doing exploration evaluations, turned to describing reservoirs in order to enhance their employment capabilities in their chosen field. Exploration geophysicists also found a niche in reservoir development. Biostratigraphers and geochemists, among others, later found that their skills were applicable to reservoir characterization.

Even the American Association of Petroleum Geologists (AAPG) has now recognized the need for better balance between exploration and production. President P.J.F. Gratton (2004) stated: "... the growth of hydrocarbon recovery technology as a supplement and/or substitute for our traditional focus on discovery technology requires our attention and response".

1.2 Integrating expertise for reservoir characterization

Today, the field of reservoir characterization routinely involves disciplines of geology, geophysics, petrophysics, petroleum engineering, geochemistry, biostratigraphy, geostatistics, and computer science. Even behavioral science must be included in this list, because people in the different disciplines do not think or act similarly and sometimes must be encouraged to work together in a team setting. A popular quote (of unknown origin) is very appropriate to the oil and gas industry: Two stonecutters were asked what they were doing. The first said, "I'm cutting this stone into blocks". The second replied, "I'm on a team that's building a cathedral".

Different disciplines even use their own technical languages, so that communication is sometimes lacking and costly mistakes are made. An example is the term "deep water", which a geologist uses in the context of the deposition of sediments in water depths beneath storm wave base (slope and basinal depths). A drilling engineer refers to deep water in the context of drilling an offshore well in present-day water depths greater than 500 m (1,500 ft) above the mudline (seafloor).

Visualization technologies and equipment, introduced in earnest in the mid-1990s (Slatt et al., 1996) and now routinely used in all large and many midsize companies, have provided an effective means of breaking down the communication barriers among the disciplines. In part, this is due to the greater willingness of young entrants into the petroleum industry – geoscientists who were raised in an age of home and school computers and video arcades – to seek out, be comfortable with, and use computers for most tasks. Although we all acknowledge the advances that have been made in oil and gas exploration and development as a result of computers, there are a number of occasions of computer overuse (i.e., using the computer instead of knowledge to attempt to solve a problem). University professors are sometimes chided for producing "nintendo geologists" (a term first introduced to me by W. Camp, a longtime petroleum-industry geologist).

Thus, the field of reservoir characterization is quite comprehensive and challenging. In fact, definitions of reservoir characterization now vary according to the technologies available for characterization and the skills of the technologists. A rather vague definition that I like (perhaps because it is vague) has been provided by Halderson and Damsleth (1993): "The principal goal of reservoir characterization is to outsmart nature to obtain higher recoveries with fewer wells in better positions at minimum cost through optimization".

1.3 Oil and gas: the main sources of global energy

1.3.1 Resources and reserves

The volume of oil and gas that is present beneath the earth's surface can be classified in several ways. Figure 1.1 illustrates one classification, from the resource (undiscovered, total estimated) to the reserves (either proven or probable at a given price). Determining the resource and reserves at any one time is a difficult task, and the values will change over time. As shown in Fig. 1.1, undoubtedly

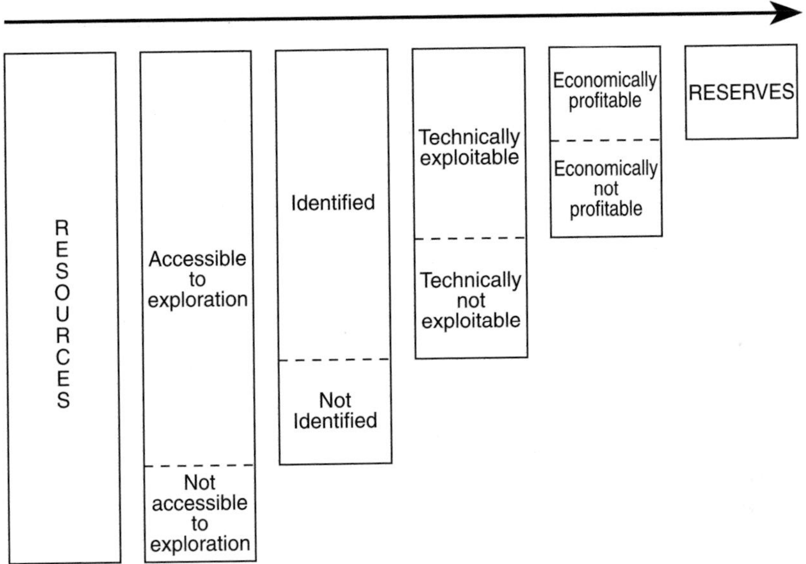

Fig. 1.1. A visual definition of resources and reserves. Modified from Favennec (2002). (Reprinted with permission of Institut Francais du Petrole.)

the total anticipated resource is not accessible to exploration, for geographic, political, and economic reasons. However, what is not accessible today may become accessible in the future, so that the total resource can change with time. Then, once the resource is identified and discovered, only a percentage of it may be technically recoverable. However, future technologies (some of which will focus on improved reservoir characterization) will modify this percentage from time to time. Finally, although the resource might be technically recoverable, economics will dictate how much is produced at a given time and price. Because global and local economics change more or less continually, the amount of resource that is actually extracted or extractable will also vary with time. It is this amount of the resource that can be considered the reserves, both proven and probable.

1.3.2 Predicting the remaining resource

Over the years, many estimates and calculations have been made regarding when the global supply of oil and gas will be exhausted. Perhaps the most widely cited set of predictions are those of M.K. Hubbert. The "Hubbert Curve", which attracted significant attention, was published in 1957, when Hubbert predicted that annual US oil production would peak as early as 1965 and no later than 1970, based on reserve estimates at that time of 150–200 billion barrels (bbl) of oil. In fact, US oil production did peak in 1970, at 3.44 billion bbl (Deming, 2001).

Table 1.1 Some estimates of ultimate recovery (EUR) and peak production, made in different years. Modified from Edwards (2001). (Reprinted with permission of AAPG, whose permission is required for further use)

Year of estimate	EUR (billion bbl)	Year of peak production
1969	2,100	2000
1978	3,200	2004
1983	3,000	2025
1989	2,000	2010
1993	3,000	2010
1994	1,650	1997
1994	1,750	2000
1996	2,600	2007–2019
1997	2,836	2020
1998	2,800	2010–2020
1998	4,700	2030
1999	2,700	2010
2000	2,659	
2001	3,670	2020–2030

Hubbert's prediction was based on the assumptions of: (1) a finite supply of oil and (2) exponential growth, peak, and decay of the production rate. Such predictions can be fraught with difficulty because of the inaccuracy or unavailability of data from which to estimate the finite supply. Aside from Hubbert's two assumptions, his production predictions were based solely upon time. However, other factors that can affect supply and production rates include ever-changing economics, global politics, complex reporting procedures, and technological advancements for both exploration and production. Some of the estimates made during the past 30 years, using different philosophies and information, are summarized in Table 1.1.

These data show that the estimated ultimate recovery (EUR) predictions vary by a factor of 2×, without a systematic increase or decrease over the 30-year time span of predictions. Most predictions place peak production within the next 5–25 years. If this proves to be true, there is still uncertainty as to how long production would continue on a downward exponential trend, because that would depend on the various nongeologic factors mentioned above, in addition to the size of the resource. The more pessimistic predictions of the end of significant oil and gas production have fueled the debate over conservation, industrialization, environmentalism, alternate fuels, hydrogen economy, and fusion, to name a few. They have also stimulated research, development, and advancements in these areas.

1.3.3 The US Geological Survey assessment

Table 1.2 examines, in more detail, the year 2000 global prediction listed in Table 1.1.

Definitions of the descriptors in Table 1.2 are:

- *Conventional deposit*: A discrete accumulation, commonly bounded by a downdip water contact, that is significantly affected by the buoyancy of petroleum in water. This is opposed to unconventional deposits, such as basin-center tight-gas sands, coalbed methane, tar sands, and oil shales.
- *Reserve growth*: The increases in known petroleum volume that commonly occur as oil and gas fields are developed and produced. It refers to oil and gas that are recoverable by current technology, through improved reservoir management.
- *Undiscovered resources*: Resources postulated from geologic information and theory to exist outside of known oil and gas fields.
- *Remaining reserves*: The volume of oil and gas that exists in discovered fields, but that has not yet been produced. Remaining reserves are calculated by subtracting cumulative production from known (predicted) volumes.
- *Cumulative production*: Reported cumulative volume of oil and gas that has been produced.

Table 1.2 Predicted global resource of oil and gas by the US Geological Survey, 2000. (Reprinted with permission of US Geological Survey)

		Mean value of gas	
	Mean value of oil (billions of barrels)	Trillion cubic feet (TCF)	Billion barrels oil equivalent (BBOE)
World (excluding US)			
Undiscovered conventional	649	4,669	778
Reserve growth (conventional)	612	3,305	551
Remaining reserves	859	4,621	770
Cumulative production	539	898	150
Total	2,659	13,493	2,249
United States			
Undiscovered conventional	83	527	88
Reserve growth (conventional)	76	355	59
Remaining reserves	32	172	29
Cumulative production	171	854	142
Total	362	1,908	318
World total	3,021	15,401	2,567

1.3.4 Some significant comparisons

Table 1.2 reveals some significant points, including:

- On a global basis, more reserves (both oil and gas) remain than have been produced, but US cumulative production has far exceeded remaining reserves.
- On a global basis, more undiscovered conventional oil and gas exists than has been produced, but US cumulative production has exceeded what is considered to be the undiscovered conventional resource.
- On a global basis, reservoir growth exceeds, or is about equal to, cumulative production for oil and is about $4\times$ that of gas production, but US reserve growth is $1/2$ to $1/3$ that of cumulative production for both oil and gas.
- On both a global and a US basis, reserve growth is about equal to the volumes of undiscovered conventional oil and gas.

This last point is particularly significant for three reasons. First, it raises the issue of the economics of exploration versus development. Because approximately the same volume of reserves is thought to remain undiscovered as is presently left in existing fields, is it better to place more emphasis on exploring for new resources or on exploiting the existing reserves? Recently, some companies have reduced exploration budgets in favor of acquiring properties with known reserves. However, the larger answer to this question is not simple and varies according to the geography and the remoteness of exploration and field locations, the potential for infrastructure to move hydrocarbons once they have been discovered, and the value of oil and gas over the short term. Second, an incremental increase in production, above what was expected from the remaining reserves in a mature field, can add significantly to the field's livelihood. Third, in the US, for example, about twice the volume of reserves of both oil and gas is derived from reserve growth, compared with remaining reserves (Table 1.2), which suggests that initial estimates of field reserves are often too conservative. As one moves from exploration to discovery, then to appraisal and development/production, more and more data are generated that provide a better understanding of the reservoir. Improved understanding improves the accuracy of resource estimates and reduces uncertainty about reserve calculations and ultimate recovery (Fig. 1.2).

1.3.5 Energy consumption

Figure 1.3 illustrates the past and the predicted future trends of global consumption of various energy sources, to the year 2020 (Durham, 2003). Consumption trends, which are directly related to population growth, suggest that, at least for the foreseeable future, oil and natural gas will provide the bulk of global energy. Thus, the need exists for continued global exploration and exploitation of this

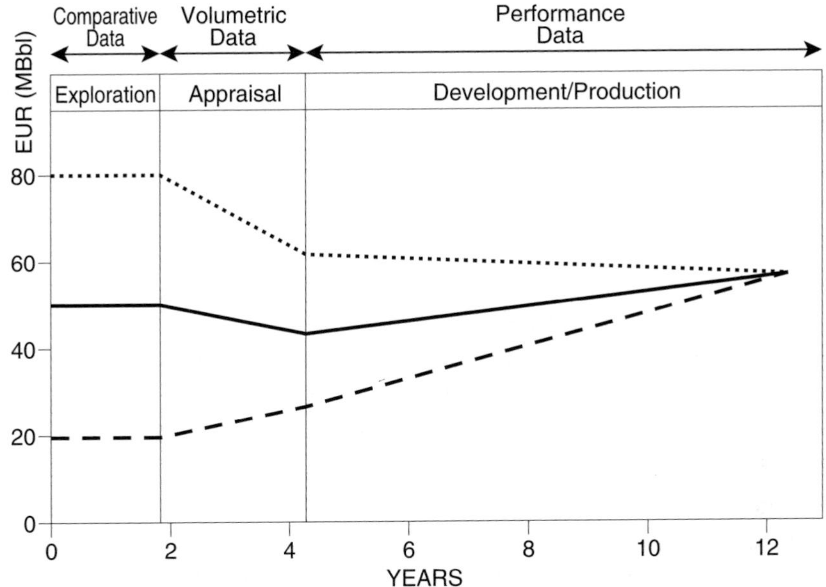

Fig. 1.2. Evolution of uncertainty in resource estimation with time; the vertical axis represents the Estimated Ultimate Recovery (EUR). Modified from Ross (1997). (Reprinted with permission of Society of Petroleum Engineers.)

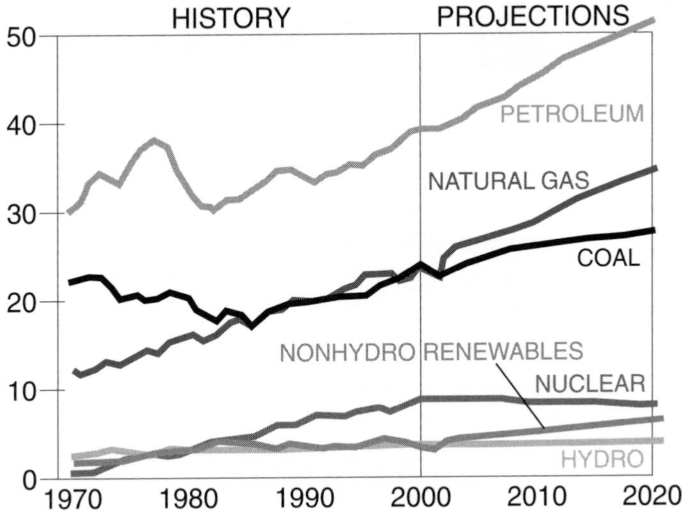

Fig. 1.3. Energy consumption by fuel, 1970–2020 (quadrillion Btu). After Durham (2003). (Reprinted with permission of AAPG, whose permission is required for further use.)

valuable resource. Irrespective of predictions, it is a certainty that oil and natural gas are finite resources that must be exploited efficiently to give maximum benefit to a global population whose growth rate is also exponential (Edwards, 2001).

1.4 The added value of reservoir characterization

The preceding discussion illustrates the value of reservoir characterization. If a proper reservoir characterization is conducted for a field and it leads to an incremental improvement in production beyond what was anticipated, then there is economic value to the characterization. For example, if the characterization of a field that was originally estimated to contain 100 MMBO recoverable improves that field's recovery by an additional 5%, an extra 5 MMBO is produced. Production improvements can come about through a better understanding of the geologic complexities of the field (Fig. 1.4), which may result either from sound geologic evaluation and/or new technologies applied to the field (i.e., from improved reservoir characterization). Recovery in many mature fields has improved by using 3D seismic to image fine-scale stratigraphic and structural

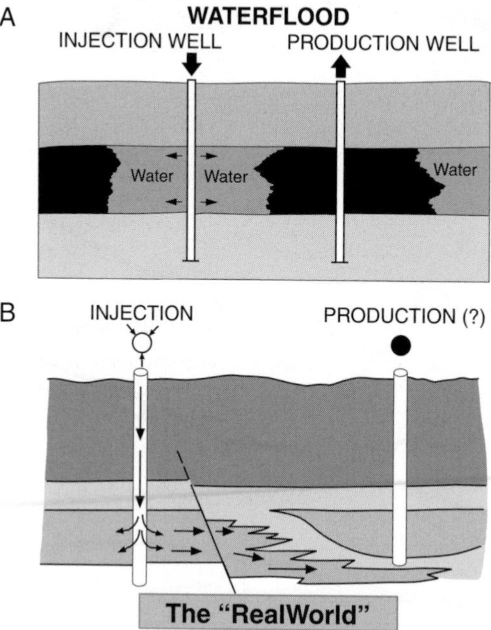

Fig. 1.4. (A) Simplistic perception of a continuous reservoir sandstone undergoing waterflood. (B) Stratigraphic and structural complexities between wells that can affect the waterflood. (Figure provided by W.J. Ebanks Jr.)

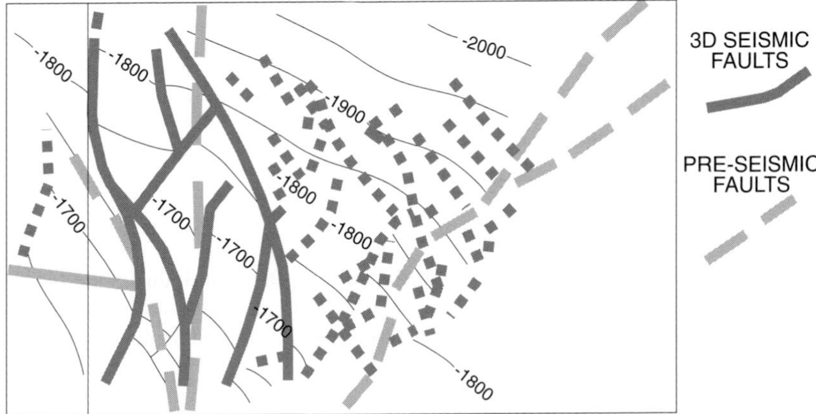

Fig. 1.5. Faults mapped within a sandstone reservoir before (orange*/light) and after (green/solid and dashed) a 3D seismic survey was acquired. This field was to be subjected to an expensive tertiary recovery project and the positions, orientations and number of faults were quite important. (Figure provided by W.J. Ebanks Jr.)

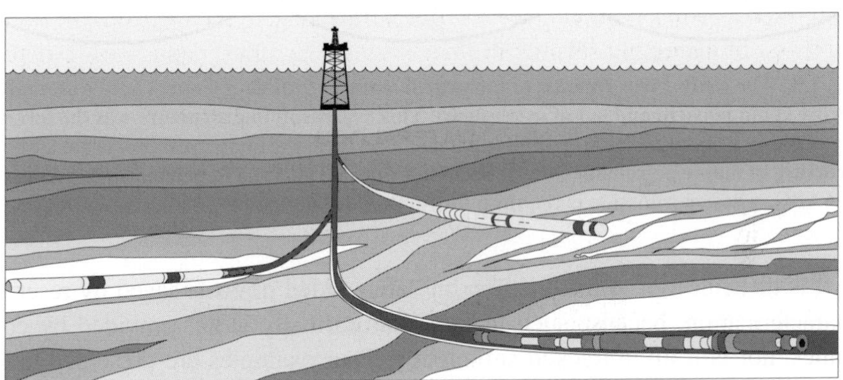

Fig. 1.6. Schematic illustration of multilateral horizontal wells drilling into sandstones (yellow/white), exhibiting stratigraphic pinchout, offset stacking and compartmentalization by shales, and good lateral and vertical continuity and connectivity. Horizontal drilling is ideally suited for stratigraphically and structurally complex reservoirs. Modified from AAPG Explorer.

features that were previously unnoticed (Fig. 1.5) or by employing horizontal drilling when the orientation, geometry, and compartmentalization of a reservoir are understood (Fig. 1.6). For this reason, the characterization of reservoirs has evolved, during the past 15 or so years, from a simple engineering evaluation, to multidisciplinary teams of geologists, geophysicists, petrophysicists, and petroleum engineers working together.

---

*The indicated color is for a CD which contains all of the figures in color.

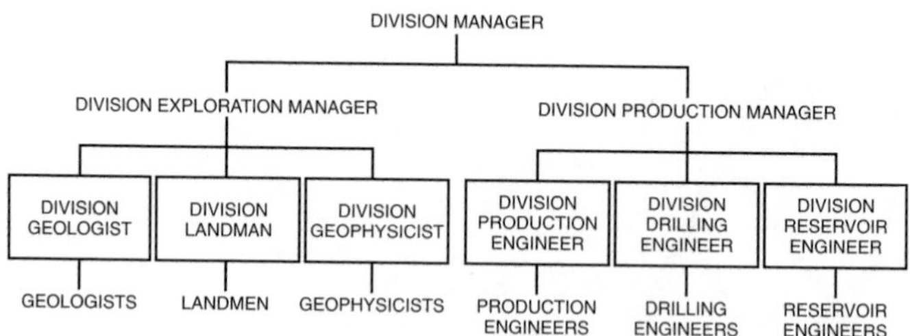

Fig. 1.7. Traditional discipline organization within the petroleum industry up to the late 1980s, when companies began to form more integrated organizations. After Sneider (1999). (Reprinted with permission of AAPG, whose permission is required for further use.)

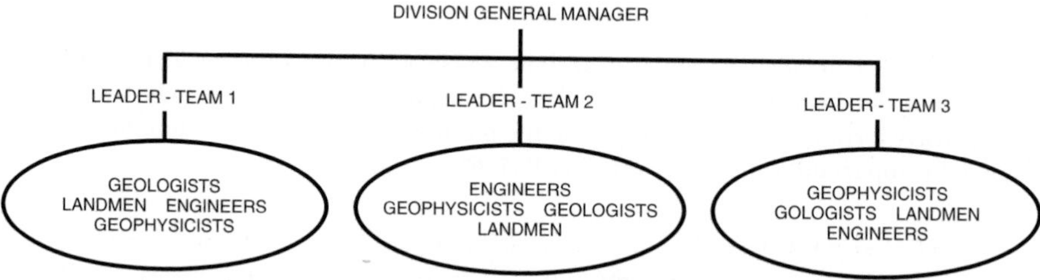

Fig. 1.8. The early 1990s concept of integrated teams, of which members with different expertise could move in and out of as required. This organizational structure was the forerunner of today's "asset teams" (or similar names), that form the main upstream organizational structure of major petroleum companies. After Sneider (1999). (Reprinted with permission of AAPG, whose permission is required for further use.)

It is difficult to find information on value added or costs saved by reservoir characterization, because such information normally is not provided by companies, nor is it often tracked sufficiently. Four examples are presented below that provide some insights into the economic value of integration and use of technologies that were new at the time.

The first example (Sneider, 1999) concerns a major oil and gas company that, for many years, was organized in a traditional manner (Fig. 1.7). As an experiment in the value of integration, the company formed a small subsidiary organization composed of personnel with different skills and expertise who worked in synergistic teams (Fig. 1.8). At the end of a 5-year trial period, the finding costs and proven reserves of the major company and the small subsidiary were compared, with startling results (Fig. 1.9). Not only did the synergistic company find 2.8 times the reserves of the large company, but they did so at less than half the finding cost! Experiments like this paved the way for the modern organizational unit within the petroleum industry – often called the "asset team" (or similar titles). This example demonstrates the positive value of teamwork, in both an exploration and production setting.

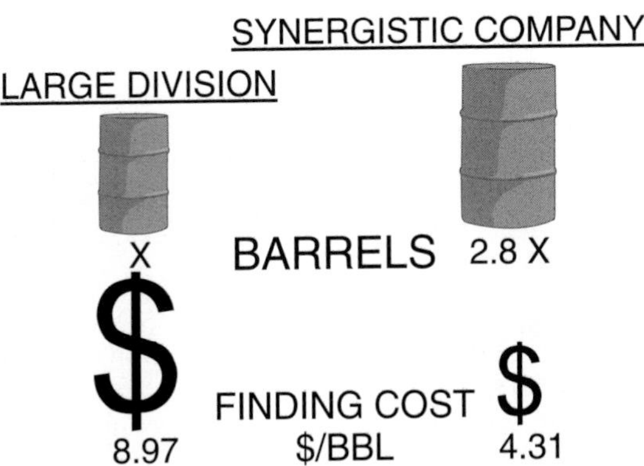

Fig. 1.9. Proved reserves and finding costs after the five year experimental term of the subsidiary company. Almost three times as much oil was found by the synergistic company as was discovered by the large exploration division of the company, and at a finding cost of less than half that of the large division. After Sneider (1999). (Reprinted with permission of AAPG, whose permission is required for further use.)

The second example is of a small company that went from exploration to property acquisition during the industry downturn in the 1980s (Durham, 2001). The following criteria were established as being necessary for the company to consider acquiring a property:

(1) the field was a possible waterflood candidate;
(2) the field had an existing waterflood that had yielded poor results with existing well spacing and position;
(3) the field had unrecognized pay in low-resistivity rocks (i.e., "low-resistivity pay");
(4) the field had unrecognized structural or stratigraphic compartments; or
(5) there were unrecognized field extensions that might be identified from seismic.

The company's acquisition program had excellent results, with the purchase of 46 mature fields that had more than 625 MMBOE added reserves at a cost of $2.69/BOE! The criteria that this company used were based simply on good reservoir characterization, applying integrated knowledge and existing technologies.

The third example demonstrates the value of 3D seismic in field development. In 1995, a 3D seismic survey was shot over a small field in the Rocky Mountains (Sippel, 1996; Montgomery, 1997). Prior to the survey, the field was mapped from well logs and production information as a continuous sandstone body (Fig. 1.10). After the 3D seismic shoot, analysts recognized that the field was subdivided into a number of "functional compartments" that were mutually isolated. This finding prompted realignment of the initial waterflood design and selective emplacement of additional infill wells, which led to a greater

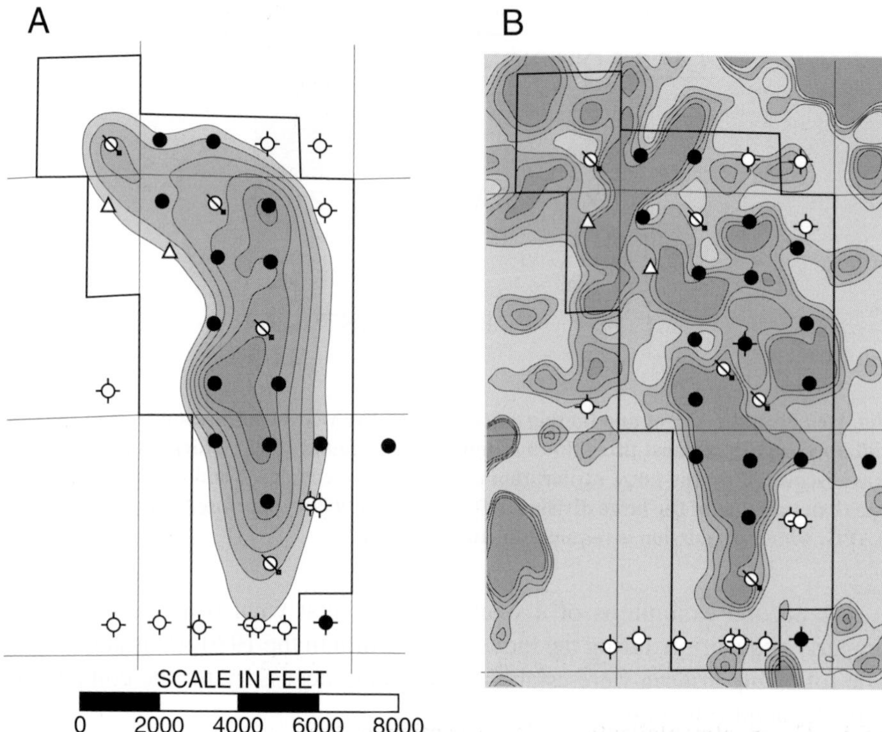

Fig. 1.10. (A) Net pay thickness determined from well control only and (B) from 3D seismic and well control. Contour interval is 5 ft, from 0 to 25 ft. The 3D seismic clearly shows the high degree of compartmentalization of the reservoir sandstone, unlike the more continuous nature of the sandstone as mapped from only the well control. Note that some sandstone thicks have not been penetrated by wells (black dots), so represent untapped parts of the total reservoir. After Sippel (1996) and Montgomery (1997). (Reprinted with permission of American Association of Petroleum Geologists and European Association of Geoscientists and Engineers.)

than 100% increase in daily unit production, an increase in estimated OOIP from 5.9 to 6.9 MMBO, and an increase in projected total recovery to 32.6% of OOIP. Also, costs were reduced and expenditures became more efficient. This example provides clear proof of economic success through the use of existing technology.

The fourth example provides statistics on the application of 3D seismic to production in south Louisiana gas fields (Fig. 1.11). Gas production there began to decline in the mid-1970s. By the mid-1990s, 3D seismic was being used more extensively, and gas production increased dramatically in a number of fields after 3D seismic imaging was used to improve reservoir characterization and well emplacement.

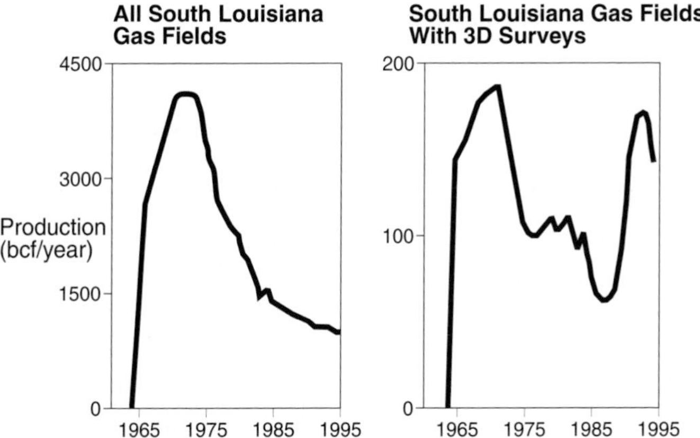

Fig. 1.11. Production in bcf/year for all south Louisiana gas fields and those fields with 3D seismic surveys. Although the vertical scales are different between the graphs, it is apparent that there is a steady production decline for all of the fields, which is reversed, with improved production in those fields in which 3D surveys were acquired. (Provided by Scotia Group, Inc.)

1.5 Compartmentalization of oil and gas reservoirs

1.5.1 Compartmentalization – The exception, or the rule?

Integration of technical disciplines over the past two decades has changed our perception of the characteristics of oil and gas reservoirs. Whereas it used to be commonly perceived that oil and gas reservoirs were relatively simple geologic features, the reality is that they are quite complex, and they can be subdivided into architectural elements or compartments on the basis of several structural and stratigraphic features (Fig. 1.4). Part of the misconception comes from the fact that one cannot actually see a reservoir, because it is beneath ground level, in the subsurface. Slatt (1998) has claimed that in Rocky Mountain basins, compartmentalized reservoirs are the rule, rather than the exception. With increased flow of information into the public domain concerning reservoirs worldwide, this claim appears to be valid beyond the Rockies. Thus, in the initial through final stages of characterization, investigators should assume that the field will be compartmentalized and segmented, even at scales too small to recognize by normal subsurface technologies. By integrating the various disciplines mentioned above, it is possible to accurately quantify the characterization process.

Often, the most limiting factor to a proper reservoir characterization is time. For a variety of reasons, the integrated team may not have sufficient time to complete the required work before drilling or other production steps are initiated (i.e., the cart is placed before the horse).

1.5.2 The significance of compartmentalization

A compartmentalized shoreface reservoir is illustrated in Fig. 1.12. Various isolated components that together comprise the entire sandstone body are illustrated. Hypothetical wells have been equally spaced.

Because of this equal well spacing, some of the isolated components will not be drilled or produced, and the field will not live up to its production potential. Even if barriers to fluid communication – such as the shales in this example – break as a result of production pressure drawdown, production of the hydrocarbons within the isolated compartment will have been delayed.

Lack of understanding of compartmentalization can have a very profound effect on waterflood performance. For example, the upper sandstone shown in Fig. 1.13 was mapped originally as a single sandstone body, on the basis of well control. A waterflood was designed on this basis, and it failed to provide the anticipated results. When a geologist reexamined the log and core characteristics, he recognized that notches in the log response (Fig. 1.14) represented

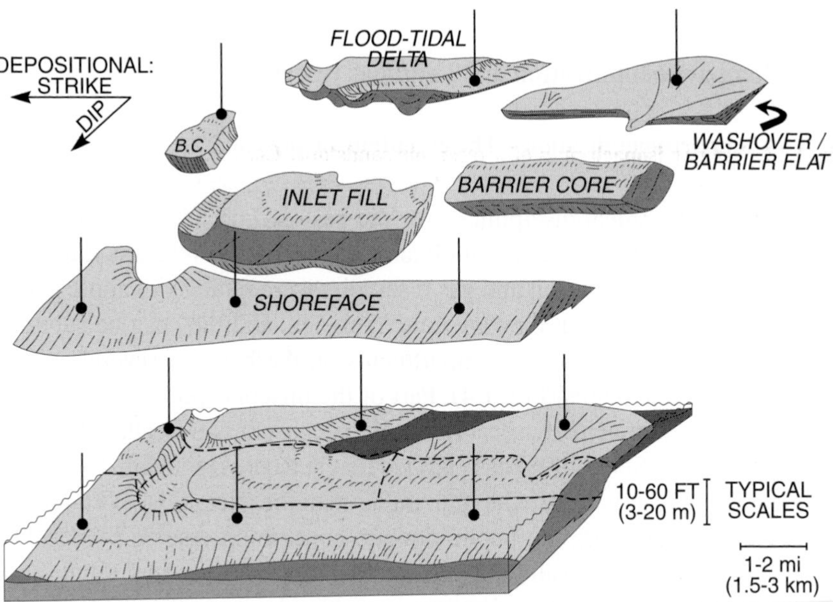

Fig. 1.12. A coastal sandstone sequence that might be considered to be a continuous deposit. However, it is broken into component parts, deposited within different depositional environments of the coastal zone. Some of these sandstone (yellow/bright-gray) deposits are encased in shale (orange/gray), and are thus isolated or compartmentalized from other sandstone deposits. Six equally spaced wells illustrate that, in this hypothetical example, two deposits, the inlet fill and the barrier core, would not have been penetrated by a well, and would thus contain untapped reserves. Modified from Galloway and Hobday (1983). (Reprinted with permission of Springer Science and Business Media.)

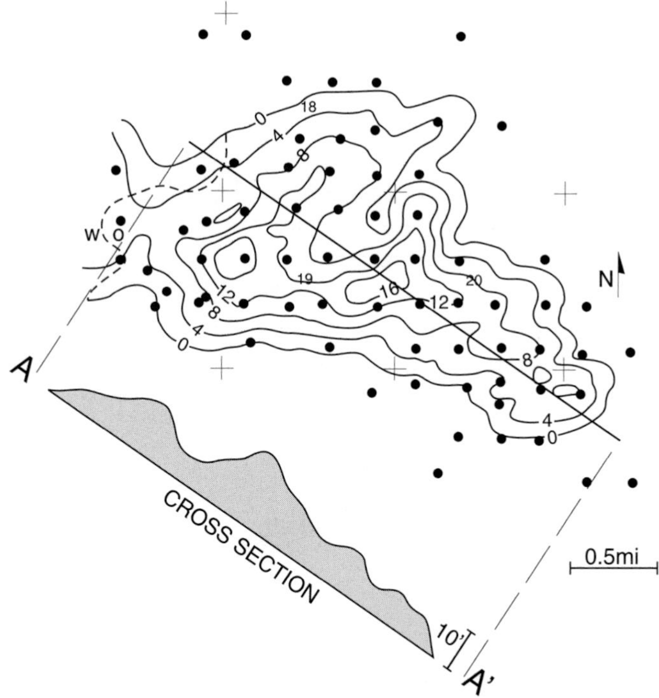

Fig. 1.13. Net pay isopach map of a reservoir sandstone. Contour interval is in feet. This map, and the cross-section assume that the sand is one uniform and continuous body. (Figure courtesy of W.J. Ebanks Jr.)

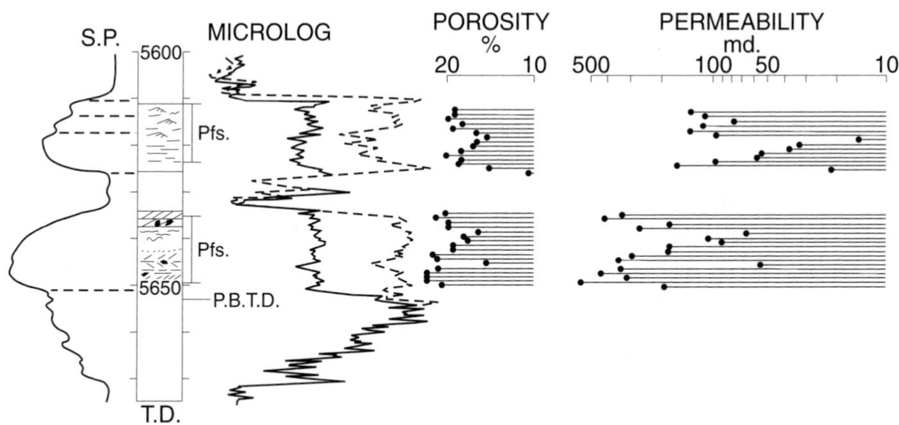

Fig. 1.14. Electric log and core-derived porosity and permeability values for two reservoir sandstones. The upper sandstone is the one that is mapped in Fig. 1.13. Note the notches on the SP and micrlog curves (highlighted by horizontal dashes) and low porosity and permeability zones. (Figure courtesy of W.J. Ebanks Jr.)

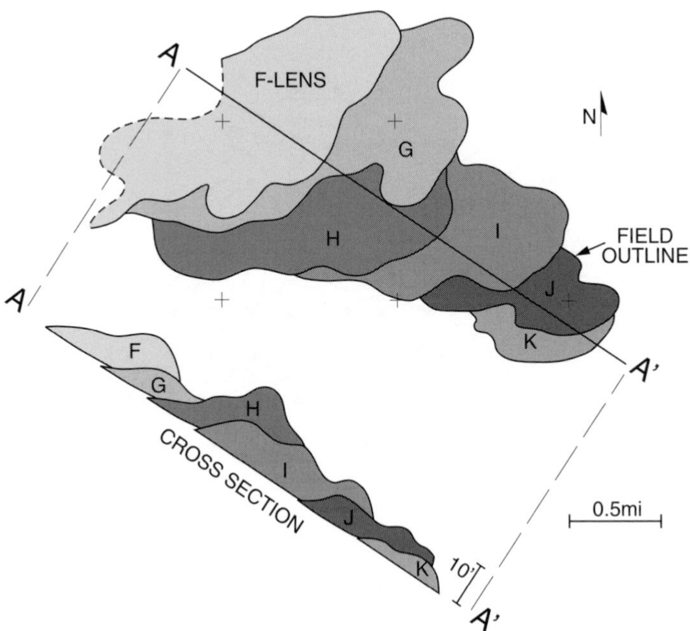

Fig. 1.15. Isopach map showing the six (F–K) isolated (by shales) sandstone lenses that comprise the sandstone reservoir. (Figure provided by W.J. Ebanks Jr.)

thin shales that could be mapped across the field, and that mutually separated the sandstone into a series of isolated sandstone lenses (Fig. 1.15). The shales prevented the injected water from reaching the targeted sandstones.

1.5.3 The nature of compartmentalization

A good analogy for reservoirs is a pocket knife with a number of attachments or component parts (Fig. 1.16). The total volume and shape of the knife is analogous to a new-field discovery, in which the data are only sufficient to estimate the volume, external shape, and gross internal properties. Once the field is discovered, the different components of the knife are analogous to the different "architectural elements" and internal properties of the reservoir, because each is characterized by a different size, shape, and performance capability.

Even within a mature field that contains a large number of wells and production information, the undrilled areas between the wells (the "interwell areas") – even if they are small – represent areas of geologic uncertainty. The interwell area can offer surprises that may hinder or enhance oil and gas production. For example, faults and stratigraphic pinchouts can prevent fluid communication between a waterflood injection well and a producing well, thus reducing the effectiveness of the waterflood plan (Fig. 1.4). Many failed waterfloods are a result of such common but undetected features inside a reservoir.

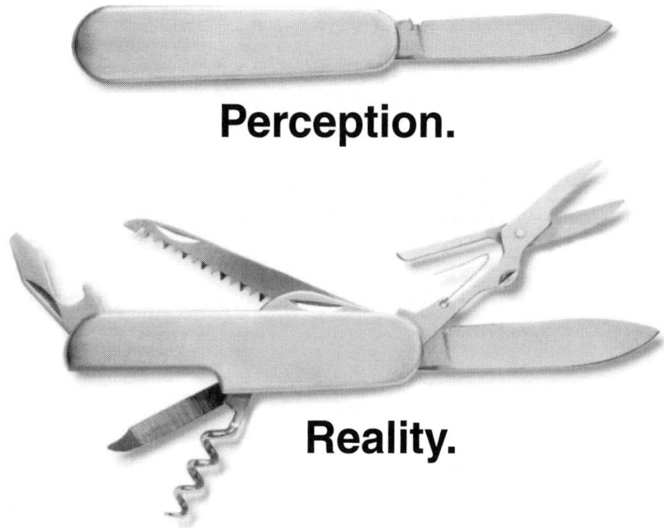

Fig. 1.16. Perception of a relatively simple, single-bladed pocket knife, and the reality of a multi-component, multi-functional knife. Modified from Oklahoma Independent Petroleum Association (2001).

Even 3D seismic reflection surveys, which now are commonly acquired for reservoir characterization, do not image all of the features inside a reservoir that may control performance. Such features have been termed "subseismic-scale" features (Slatt and Weimer, 1999).

Thus, reservoirs typically consist of a number of component parts or architectural elements that, together, comprise the reservoir, but that individually control the volume of hydrocarbons present and the production behavior of the reservoir. These architectural elements are defined by the size, geometry, orientation, internal continuity, and vertical connectivity of reservoir and seal beds, as well as by their reservoir quality.

1.6 Depositional environments and types of deposits

Many environments on the earth's surface experience sediment deposition in the form of detrital and biological particles (shells, etc.) and chemical precipitates. In this book, we are concerned with sediments that are derived by weathering and erosion of preexisting rock and are transported to a final depositional site by wind, water, and/or ice, that is, the clastic sediments. Sediments from different depositional environments can be grouped broadly into continental, transitional, and marine deposits (Fig. 1.17).

Each of the different depositional systems exhibits different architectural elements, with different features resulting from the processes of sediment transport and subsequent deposition into the various environments. Shapes, sizes,

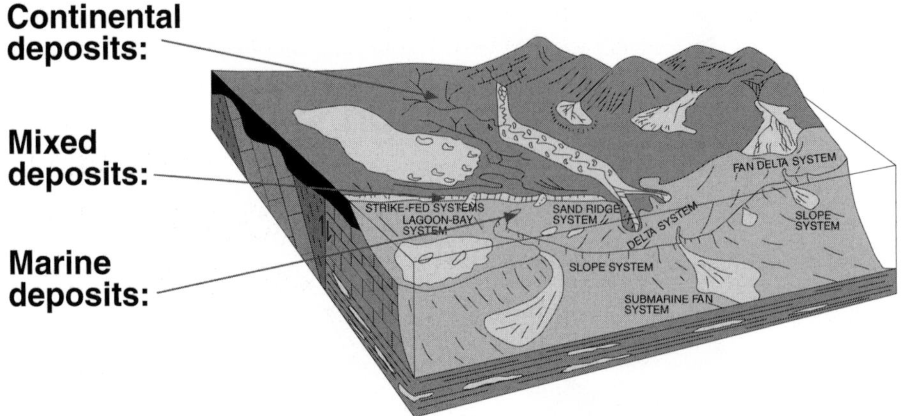

Fig. 1.17. Major clastic sedimentary environments (yellow/bright-gray). The three major depositional environments (Level 1) are continental (non-marine), mixed (marine to non-marine or coastal setting), and marine (submarine). Each of these deposits has a characteristic set of processes and resulting deposits which form different reservoir types. Further subdivision is shown in Figs. 1.20, 1.21 and 1.22. Modified from Fisher and Brown (1984).

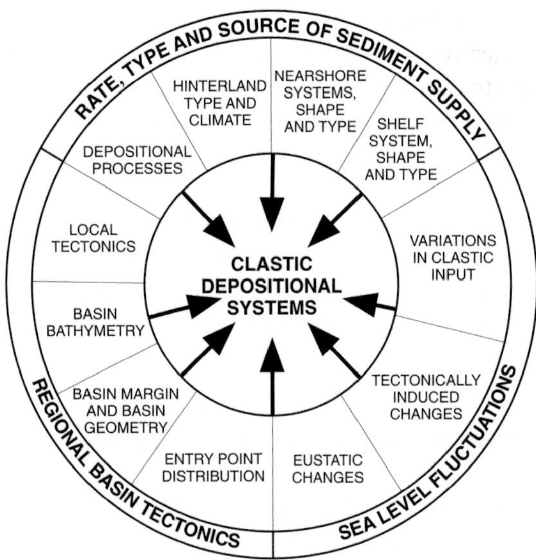

Fig. 1.18. Factors which influence clastic depositional systems. After Richards et al. (1998). (Reprinted with permission of Geological Society of London.)

net:gross values, continuity, orientation, and other reservoir characteristics are all a function of the nature of transport and depositional processes, basin configuration, climate, tectonic and eustatic sea-level fluctuations, and the like (Fig. 1.18). Petroleum engineers must understand the geologic complexities that control reservoir performance if they wish to maximize the placement, orientation, and number of delineation and development wells to improve ultimate production and reservoir management.

1.6.1 Scales and styles of geologic reservoir heterogeneity

A variety of types and scales of heterogeneity are found in most reservoirs. Figure 1.19 classifies heterogeneities according to scale; from the smallest to the largest scale, they are microscopic, mesoscopic, macroscopic, and megascopic heterogeneities (Krause et al., 1987).

Microscopic or pore/grain-scale heterogeneities are related to pores and arrangement of grains, including pore volume (porosity), pore sizes and shapes, grain-to-grain contacts that control permeability, and grain types.

Mesoscopic or well-scale heterogeneities can be recognized in the vertical dimension, such as in cores or well logs. Such heterogeneities include bedding and lithologic types, stratification styles, and the nature of bedding contacts.

Macroscopic or interwell scale heterogeneities occur at the scale of well spacing. Such heterogeneities include lateral bed continuity or discontinuity as a result of stratigraphic pinchout, erosional cutout, or faulting. This is the most difficult scale of heterogeneity to quantify, because the technologies required to image interwell-scale heterogeneities often exhibit resolutions that are too coarse for one to observe the feature (subseismic). Cross-hole tomography, 4D (time-lapse) seismic, and well tests can provide direct information on the presence or absence of such heterogeneities, but the inherent resolution of definable features with 2D or 3D seismic often is too high to be able to resolve important subseismic scale, interwell heterogeneities.

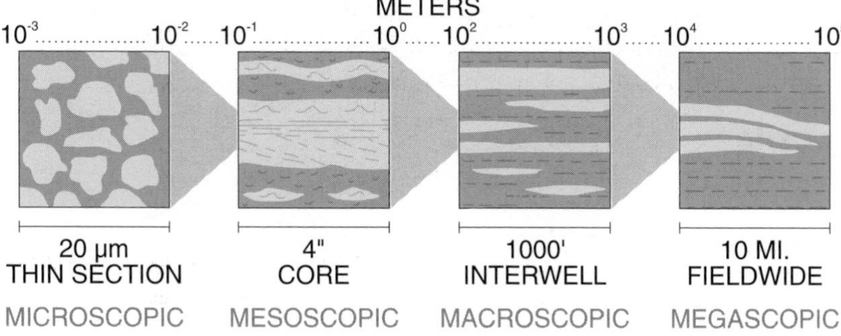

Fig. 1.19. Classification of heterogeneities in reservoirs according to scale. From the smallest to the largest, these are microscopic, mesoscopic, macroscopic, and megascopic heterogeneities. After Krause et al. (1987). (Reprinted with the permission of AAPG, whose permission is required for further use.)

Megascopic, or fieldwide heterogeneities, such as overall geometry and large-scale reservoir architecture (related to structure and/or depositional environment), normally can be delineated by 2D or 3D seismic, well tests, production information, and field-wide well log correlation. However, it is important to note that the size of the depositional system that comprises a field normally exceeds the size of the field itself. For this reason, regional mapping and field correlations should be extended beyond the geographic confines of the field.

1.6.2 Hierarchical scales of geologic heterogeneity (levels)

Mesoscopic, macroscopic, and megascopic heterogeneities of sandstone reservoirs can be subdivided further, according to the scale of the feature (Slatt and Mark, 2004). For fluvial systems (as an example), these subdivisions, or levels, are: Level 1: regional environments of deposition (i.e., continental, mixed or marine, Fig. 1.17); Level 2: major type of deposit (continental: fluvial, eolian, lacustrine or alluvial deposit, Fig. 1.20); Level 3: more specific types of deposit (continental, fluvial: meandering river, braided river, or incised valley fill) (Fig. 1.21); Level 4: architectural elements of specific reservoir types comprising a continental (Level 1), fluvial (Level 2), meandering river deposit (Level 3) composed of floodplain, point bar, cut bank, mud plug, fining-upward, and cross-bedded elements (Level 4) (Fig. 1.22). As will be discussed in subsequent chapters, all depositional systems can be subdivided at these levels.

Microscopic heterogeneities also can be subdivided according to features such as grain-size distribution, porosity, permeability, capillarity, grain-packing arrangements, and well log signature, as is also discussed in subsequent chapters.

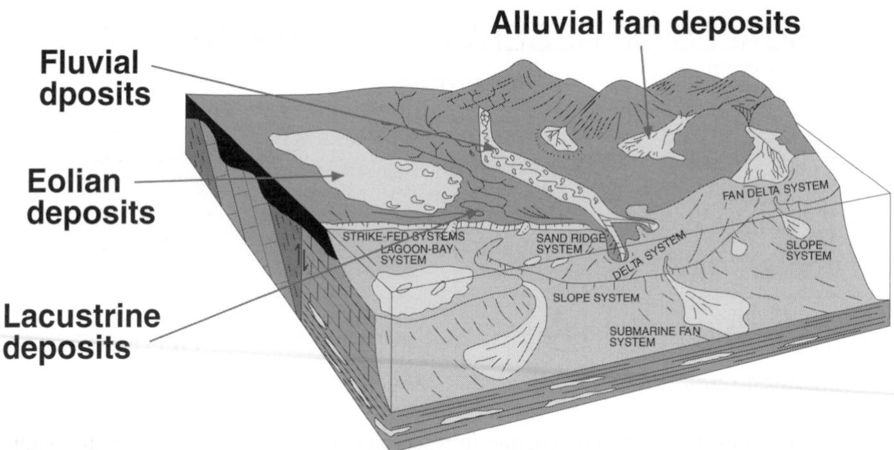

Fig. 1.20. Level 2 environments include all of those within the continental (as this example shows), mixed, or marine environments (Level 1). Modified from Fisher and Brown (1984).

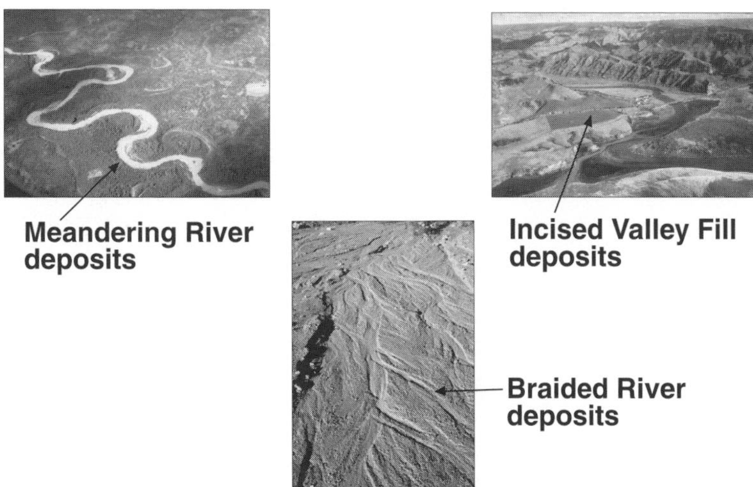

Fig. 1.21. Level 3 environments that may occur within each Level 2 environment. In this example, Level 2 fluvial environments and deposits occur as meandering river, incised valley fill or braided river systems. Each system has its own unique characteristics and trends. The three photos are of modern surficial deposits.

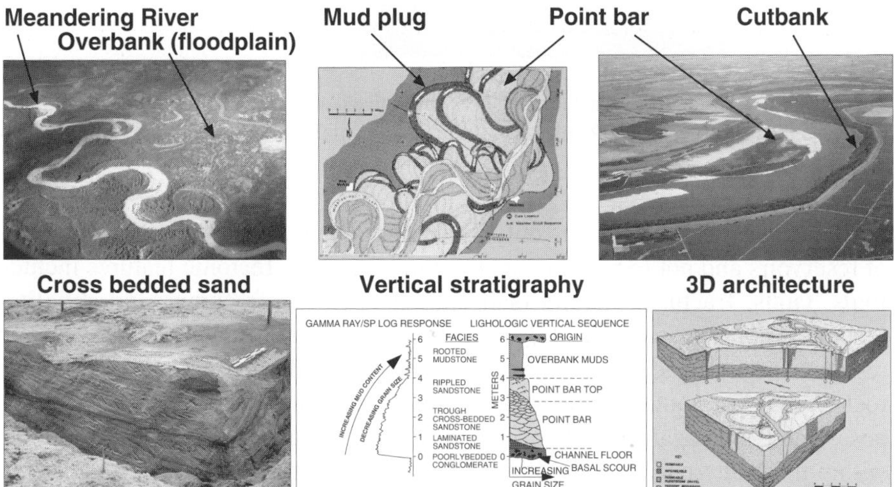

Fig. 1.22. Level 4 environments and deposits are composed of smaller scale features which are part of Level 3 deposits. In this example, the meandering river (Level 3) is composed of a series of features. From upper left to lower right, these features are a modern meandering river and floodplain, a map reconstruction of part of the modern Mississippi River showing point bar (reservoir) sands isolated by mud plugs, the point bar and cutbank sides of a meander bend, cross bedded, point bar sands along a trench wall, the ideal vertical stratigraphy of a point bar deposit, and a 3D model showing the complexities of the modern Mississippi River example. (Mississippi River examples were provided by D. Jordan.)

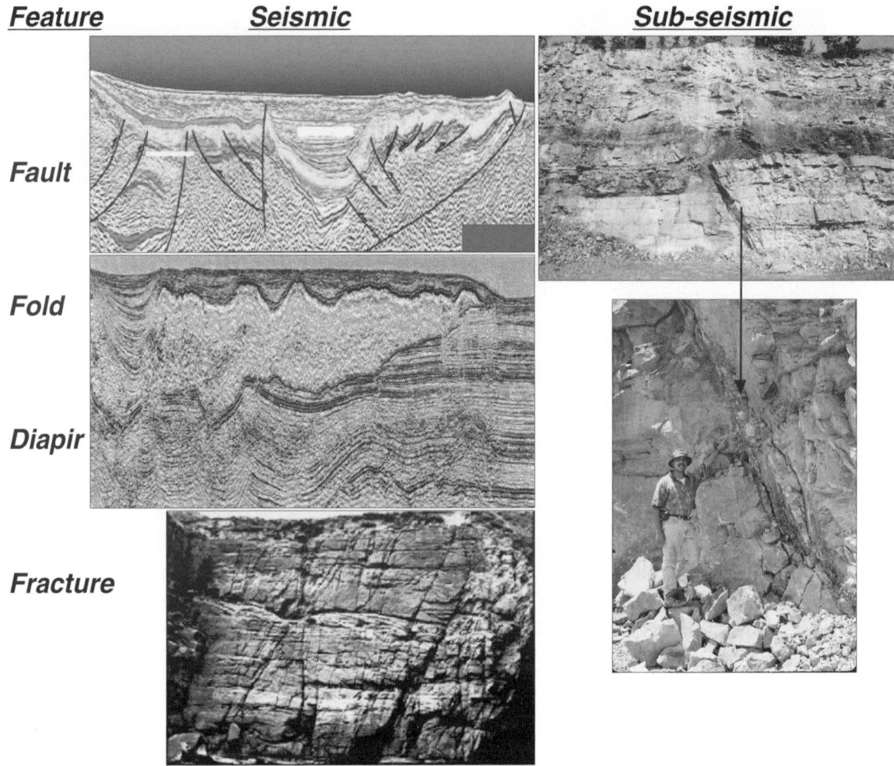

Fig. 1.23. Tectonic features at both seismic and subseismic scales, including faults, folds, diapirs and fractures. These features, both large and small, can influence reservoir performance.

The above discussion refers only to stratigraphic and sedimentologic features of reservoirs and not to tectonic or structural features. Tectonic features include folds, faults, fractures, diapirs (salt and shale), microfractures, and stylolites (chemical compaction) (Fig. 1.23). Such features can have a profound effect on reservoir performance. For example, at one end of the scale, faults can isolate rock bodies horizontally into compartments, particularly if a fault offset juxtaposes sandstone against shale or a significant amount of fault gouge builds up within the fault zone (Fig. 1.24). At the other end of the scale, open faults can facilitate hydrocarbon migration into reservoir intervals and can enhance fluid communication across the faulted intervals. Reservoirs that produce hydrocarbons directly from fractures are common but often may be short-lived (e.g., "fractured reservoirs").

In addition to structural and pore-level characteristics, Level 4 is probably the most important in reservoir characterization because properties at this scale often control or influence the reservoir's performance (see the various complexities in Fig. 1.22, for example) and because they are often subseismic in

Fig. 1.24. Closer view of the sub-seismic scale fault shown in Fig. 1.23. This is the west wall of Hollywood Quarry, Arkansas (Fig. 30; Slatt et al., 2000) showing thick bedded sandstones on the lower half of the wall (light colored) and interbedded sandstones and shales (darker colors) on the upper half. A strike-slip fault with a component of throw has juxtaposed thinner bedded sandstones against shales on the upper half of the wall (horizontal compartment). Sandstone is juxtaposed against sandstone on the lower half, but a thick gouge zone of fine-grained, crushed sandstone separates the sandstones on both sides of the fault. Blue and red dots refer to the same beds on opposite sides of the fault, to illustrate the vertical offset. The curve on the left side of the wall is an outcrop gamma ray log superimposed on the quarry wall.

scale. Therefore, geologic knowledge must replace direct imaging. Unfortunately, even with geologic information, many reservoirs are only described at Levels 1–3 and are understood structurally only in the broadest sense.

1.7 When is reservoir characterization important in the life cycle of a field?

1.7.1 The life cycle of a field

Figure 1.25 shows the phases of an oil (or gas) field's life cycle. Initially, mapping and reconnaissance are conducted by exploration geologists and geophysicists. They gather data from outcrops, old wells, seismic, and any other

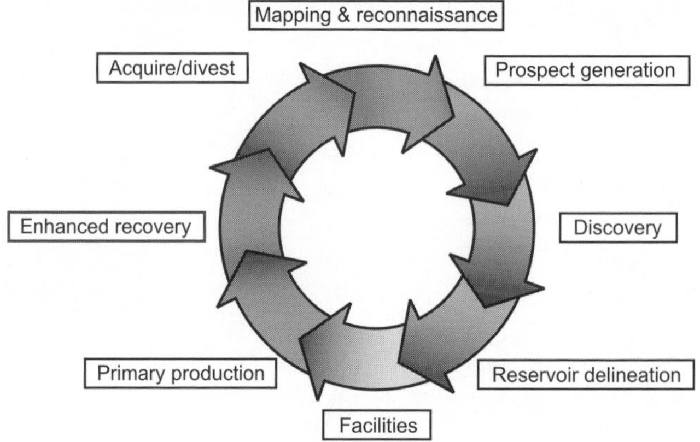

Fig. 1.25. Phases of a typical oil (or gas) field life cycle. Modified from the Oil & Gas Journal.

information that is available, and they use it to develop a regional understanding of the area's geology. Of prime concern at this stage are the structure of the basin and the subregional features (i.e., are there fault and/or fold traps for hydrocarbons?), the stratigraphy (are there reservoir rocks with porosity and permeability, and are there organic shales that can generate hydrocarbons in the basin? Are there shale seals?), and the burial history of the basin (have the source rocks been buried sufficiently for hydrocarbon generation, and have the reservoir rocks been cemented during burial?). By addressing these questions, investigators may identify and select parts of the larger area for further study and ultimately may generate a prospect evaluation (prospect-generation phase).

To evaluate the prospect further, seismic is shot if none is available, or if the vintage of existing seismic is too old. The seismic is reviewed and a decision is made regarding whether to drill a well. If the seismic further suggests the potential for hydrocarbon accumulation, a well is drilled. About 10–30% of the exploratory wells that are drilled actually discover oil and/or gas. Once a well discovers hydrocarbons (in the discovery phase), the well is logged with logging tools (most wells, even those without hydrocarbons, normally are logged) to evaluate the fluids and rocks and to determine the reservoir interval.

Assuming that the initial tests are positive and a decision is made to spend more money on the prospect, the reservoir delineation phase begins, which involves drilling delineation wells and perhaps shooting 3D seismic. This new information allows reservoir engineers and geologists to calculate the volume of oil and/or gas that is present in the reservoir. Once the volume and flow rates can be determined, facilities engineers design and build appropriate facilities for collecting, separating, refining, and delivering the hydrocarbon products. By this time, a reservoir management plan has been developed, production wells are put in place, and primary production begins.

As the hydrocarbons are extracted from the reservoir, the reservoir pressure drops, making it increasingly difficult to extract primary hydrocarbons. Usually, the field then is waterflooded (water is injected into the reservoir interval to push more hydrocarbons toward a production wellbore). Waterfloods (in this enhanced-recovery phase) can extend the life of a field for many years. At some point, when the waterflood production declines significantly, a decision must be made either to apply a tertiary recovery process (such as carbon dioxide injection, fire flood, etc.) or to abandon the field.

Abandonment does not necessarily mean closure of the field. More typically, the operating company will try to sell the reservoir to another company that has a better enhanced-recovery process or that has finances for additional production (this is the acquire/divest phase).

As might be expected, individuals and teams with different backgrounds are responsible for the different phases. Entrepreneurial geologists tend to be in charge of mapping through prospect generation and discovery. When hydrocarbons are discovered, reservoir engineers play the major role in delineating the reservoir, but geologists and geophysicists still should have a major impact in these phases. Facilities engineers are responsible for building the facilities into which the hydrocarbons are collected and processed. Production is managed by production engineers. Divestiture of a field is usually handled by the financial and legal branches of an organization. Large companies have specialists in each area. However, smaller companies, particularly independent operators, often require their people to wear several hats and conduct several, if not all of the activities.

1.7.2 Applying reservoir characterization

Thus, with all of these activities, when is reservoir characterization most important? Many professionals will answer this question by stating that characterization begins as soon as the discovery is made and the first data become available (usually from seismic and the discovery well and perhaps from earlier dry holes in the area). As more wells are drilled (Fig. 1.26), more information becomes available. Therefore, reservoir characterization is an ongoing process, and the characterization is (or should be) updated as new data are acquired. It is important to reiterate that, even with a large number of wells drilled in the field, the majority of the field acreage is still undrilled, and there can be many surprises at the interwell scale. The smaller the well spacing, the fewer the surprises (but the greater the expense) and the greater degree of reservoir connectivity (Fig. 1.27).

The goal of many reservoir characterization studies is to provide a 2D or 3D geologic model to petroleum engineers for reservoir performance simulation and for well planning. The three stages of stratigraphic model-building are

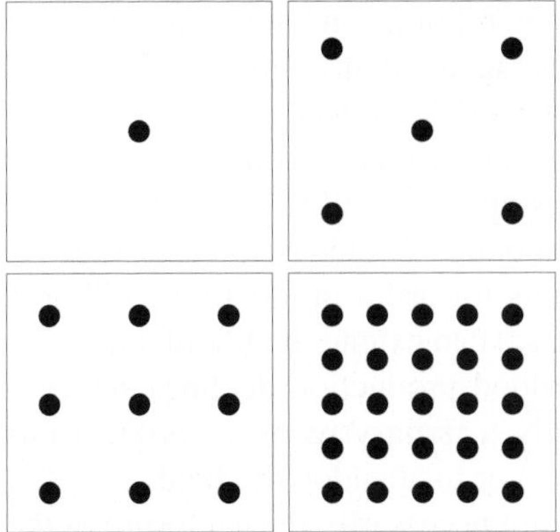

From top to bottom and left to right: field discovery,
primary recovery, secondary recovery and infill drilling.

Fig. 1.26. Well locations (black dots) at discovery (upper left), appraisal/primary production (upper right), primary/secondary production (lower left) and tertiary recovery or post-infill drilling. Modified from Al-Quahtani and Ershaghi (1999). (Reprinted with permission of AAPG, whose permission is required for further use.)

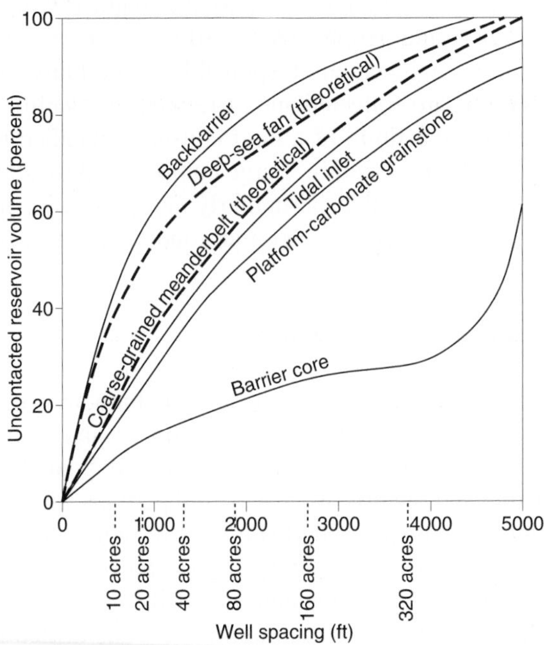

Fig. 1.27. Curves depicting the percent of uncontacted reservoir volumes in various types of reservoirs. Those reservoir types thought to have the highest degree of heterogeneity and compartmentalization exhibit the highest proportion of uncontacted reservoir volume. These reservoir types will have poor reservoir sweep efficiency. Modified from Ambrose et al. (1991). (Reprinted with permission of Society for Sedimentary Geology (SEPM).)

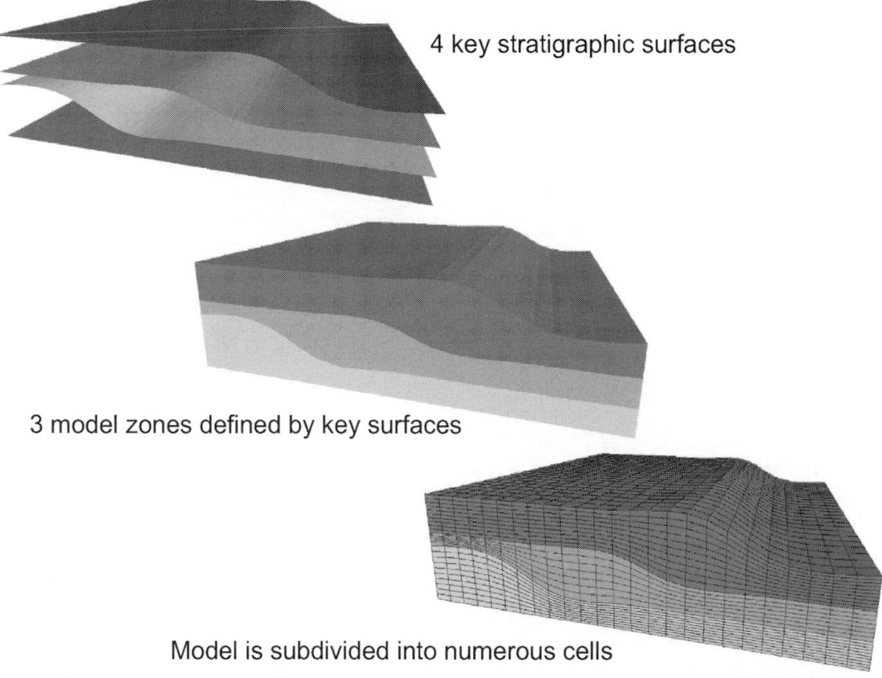
4 key stratigraphic surfaces
3 model zones defined by key surfaces
Model is subdivided into numerous cells

Fig. 1.28. Three stages in development of a 3D stratigraphic model for reservoir performance simulation. Stage 1 is to define key stratigraphic surfaces to subdivide the model into layers. The second stage is to fill the space between the layers with facies or architectural elements. The third stage is to grid the model and fill the grid blocks with reservoir parameters. Input of structural properties to the model is not illustrated here, but is equally important to building a final geologic model. (Source of figure is unknown.)

shown in Fig. 1.28, and include defining layer boundaries, filling the volume between the boundaries with strata, and gridding the model and inputting numerical values of reservoir parameters into the grid blocks. Not shown in the figure is the addition of structural attributes such as faults, fractures, and folds, which is essential to complete the geologic model. The chapters in this book are geared toward these three steps in model-building, although not in the order presented in Fig. 1.28. Chapters 4 and 11 deal with the identification of boundaries (sequence boundaries and maximum flooding surfaces) that can define stratigraphic layers for model-building. Chapters 6–10 deal with the strata that are placed between the boundaries. Chapters 3 and 5 deal with reservoir petrophysical properties that are input into the grid blocks of a model. Multiple random or constrained iterations of input parameters are generated until a desirable output model is produced (Fig. 1.29). The resulting geologic model may take many forms depending upon the input information and the characteristics of the reservoir. Some examples of 3D geologic models are presented in

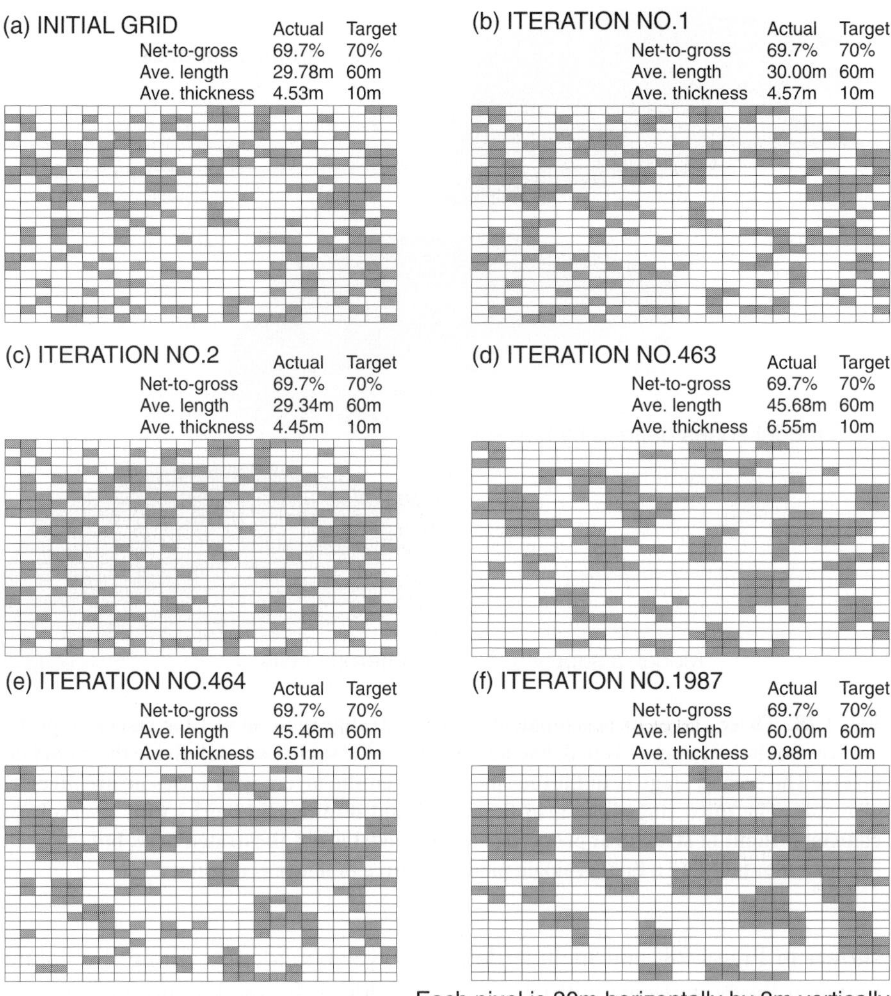

Fig. 1.29. Example of multiple iterations of input parameter to a 2D gridded model. The input parameter into grid blocks is sand (brown/dark). In this case the desired output (target) is a net-to-gross value of 70%, with an average sandstone length of 60 m and average thickness of 10 m. Sand blocks are randomly input into grid blocks through multiple iterations. After 1,987 iterations, the actual output values matched the target values. After Srivastava (1994).

Fig. 1.30. These examples are of facies distribution and permeability distribution. Many other parameters may be substituted for these two parameters within a model. In addition, the shapes and orientations of sand bodies may be modeled for simulation and well planning purposes (Fig. 1.31).

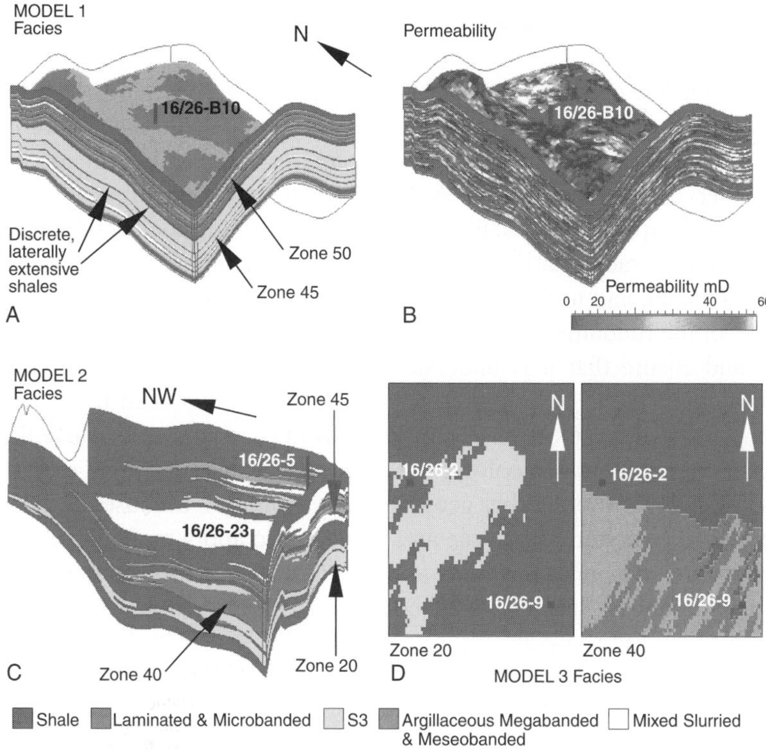

Fig. 1.30. Examples of 3D models. A and C are two iterations of a 3D facies model. B is a model showing the distribution of permeability. D shows plan views of the models shown in A and C. After Chapin et al. (2000).

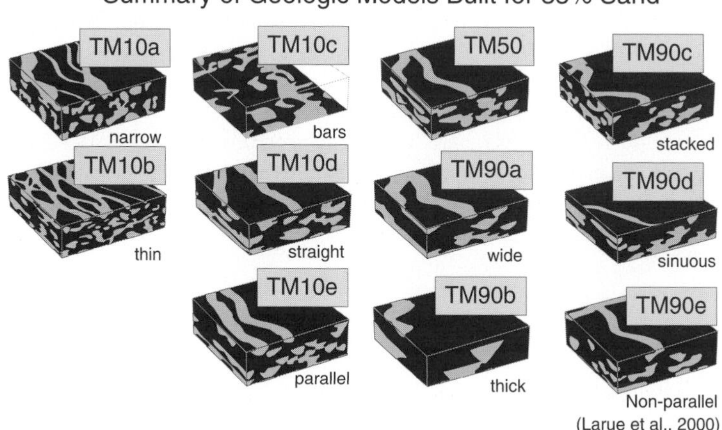

Fig. 1.31. Several iterations of channel sandstone distribution using the common constraint of 35% sand within the model volume. After Larue and Friedmann (2000).

1.8 The value of case studies

Because reservoir characterization is a comprehensive field, with many disciplines involved, it is difficult to compile all its various components into a single book. This book is intended to provide an overview, or introduction, to the field of reservoir characterization. Although the theme is heavily geologic, other aspects and disciplines are discussed. The strategy of this book is to present various reservoir characterization procedures, tools, and knowledge, in brief fundamental segments, and to follow these segments with case histories that provide examples of the fundamentals. In my personal experience, the best way to make a point and ensure that it is understood is to present a case study. However, there is a downside to this approach, because it has also been my experience that people too often reject or do not learn from case studies because they are "not in my geographic or geologic area". It is a mistake to reject case studies because of their location (either geographic or geologic), because the principles that apply in a case study from one area usually also apply more globally. Thus, I encourage the readers of this book to be open to more widespread applications of the individual case studies presented here.

Chapter 2

Tools and techniques for characterizing oil and gas reservoirs

2.1 Introduction

The oil and gas industry is a technology-driven industry. Our ability to locate and extract hydrocarbons from beneath the ground's surface is tied directly to the evolution of technologies, concepts, and interpretative sciences. Figure 2.1 illustrates how petroleum-recovery efficiency has improved since 1950. Interpretation techniques and concepts are listed by year beneath the black curve, and related new technologies are listed above the curve. The technologies are mainly seismic-based methods for imaging features beneath the ground's surface. Other technologies that also have vastly improved our ability to extract hydrocarbons include advances in well logging techniques, improvements in our ability to drill in deep water beyond the continental shelf, and the advent of horizontal drilling, to name a few.

This chapter provides an overview of the common techniques for characterizing oil and gas reservoirs. The techniques can be subdivided into those that measure static reservoir properties and those that measure dynamic reservoir properties.

2.1.1 Static and dynamic properties of reservoirs

Static reservoir properties are those rock and fluid properties that normally do not change during the life of a field. They are the result of primary depositional processes coupled with postdepositional burial, diagenesis, and tectonics (Fig. 1.18). Static properties include:

- stratigraphy
- geometry
- size
- lithologies

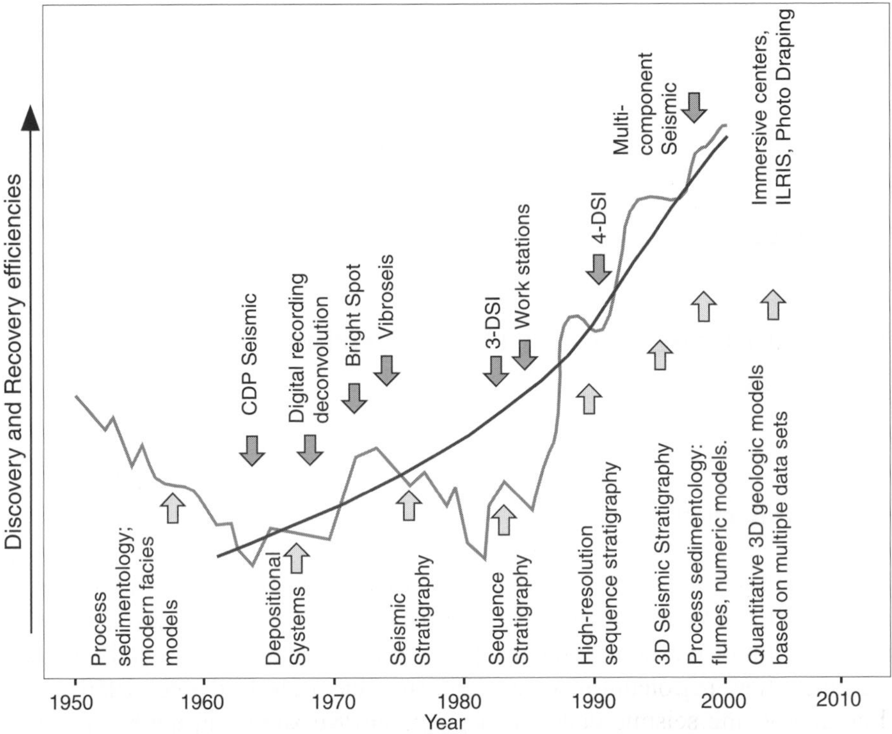

Fig. 2.1. Graph showing the US discovery and recovery efficiencies since 1950. Blue*/black line is a normalized curve. Also plotted are the timing of major technology developments (green/dark arrows) and new interpretive disciplines (yellow/bright arrows). Figure is modified significantly from Fisher (1991). (Reprinted with permission of The Leading Edge.)

- structure
- initial porosity and permeability
- temperature.

Dynamic properties are those that do change significantly during the life of a field. For example, fluid saturations, compositions, and contacts, as well as reservoir pressure, change as the field is produced. Porosity and permeability can change as the reservoir pressure changes over time or as injected fluids react with formation minerals (either to precipitate new minerals that fill pore spaces or to dissolve minerals and thereby provide new pore spaces). Dynamic properties include:

- fluid saturations
- fluid contacts
- production and fluid-flow rates

*The indicated color is for a CD which contains all of the figures in color.

- pressure
- fluid compositions, including gas-to-oil ratio (GOR) and water-to-oil ratio (WOR)
- acoustic (seismic) properties.

Acoustic properties, which are measured and documented as seismic attributes, are dependent upon porosity, fluid type and content, and the nature of the reservoir rock. Seismic attributes are dynamic, because fluid type and content change during oil and gas production. By comparing seismic attributes at different times in the life of a field, it is possible to indirectly measure fluid movement in the reservoir.

2.2 Measuring properties at different scales

The three main stages of field development are exploration (through to discovery), appraisal, and production. In exploration, one starts with a conceptual geologic model (Fig. 2.2), which may be based on geologic knowledge of the area, including basin evolution, structure, and stratigraphy (labeled 1 in Fig. 2.3). Conventional 2D or 3D seismic-reflection analysis (Fig. 2.2) normally is the next phase of exploration, because subsurface seismic profiles and 3D volumes can provide a regional-scale image of the area being explored (labeled 2 in Fig. 2.3). If the seismic data reveal potential drill sites, and a well is drilled, the well is logged with conventional logging tools (Fig. 2.2) to determine rock and fluid properties in the wellbore (labeled 3 in Fig. 2.3).

If potentially economic accumulations of hydrocarbons are found in the exploratory well (labeled 4 in Fig. 2.3), the decision may be made to drill more wells and/or to shoot more seismic (particularly 3D seismic) to appraise the field (labeled 5 in Fig. 2.3). Some of these appraisal wells may be cored through selected intervals, and/or borehole-image logs might be acquired to evaluate the downhole structure and stratigraphy (Fig. 2.2).

If this delineation phase is favorable, more-detailed knowledge of the reservoir is required. Outcrop analog studies (Fig. 2.2) to determine the depositional environment and the reservoir's stratigraphy are particularly valuable at this point in order to build a 3D geologic model of the reservoir's stratigraphic continuity and connectivity (labeled 6 in Fig. 2.3). Of course, to appropriately compare outcrop features with a reservoir, the outcrop must be of the same depositional system as is the reservoir. This will be discussed more fully in subsequent chapters.

Also at this time, well tests, such as initial potential flow rates and pressure tests, are acquired to aid in the characterization process (labeled 7 in Fig. 2.3). With addition of hard data from wells and outcrops, and perhaps with additional 3D seismic-reflection analysis, the model can be quantified and a scaled, graphical, 3D reservoir geologic model (Fig. 2.2) can be built that petroleum engineers can use to simulate the reservoir's performance (labeled 8 in Fig. 2.3).

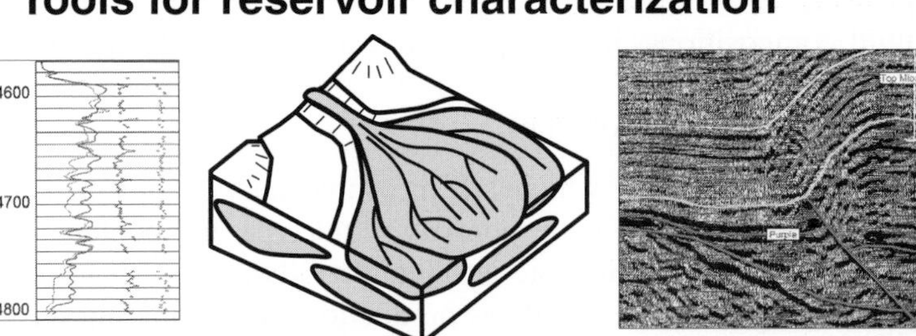

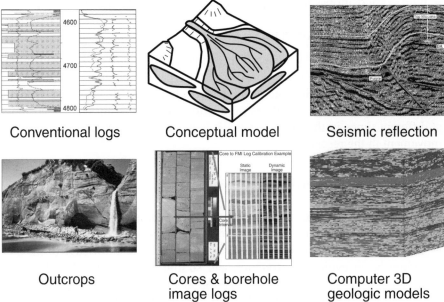

Fig. 2.2. Diagram showing some of the different data types used in the study of reservoirs. Clockwise from upper left are conventional well logs, a conceptual geologic model, 2D and 3D seismic reflection data (shallow and deep), outcrops, cores and borehole image logs, and geologic reservoir models. Not shown are geochemical and biostratigraphic data, which also are important in reservoir characterization.

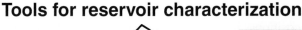

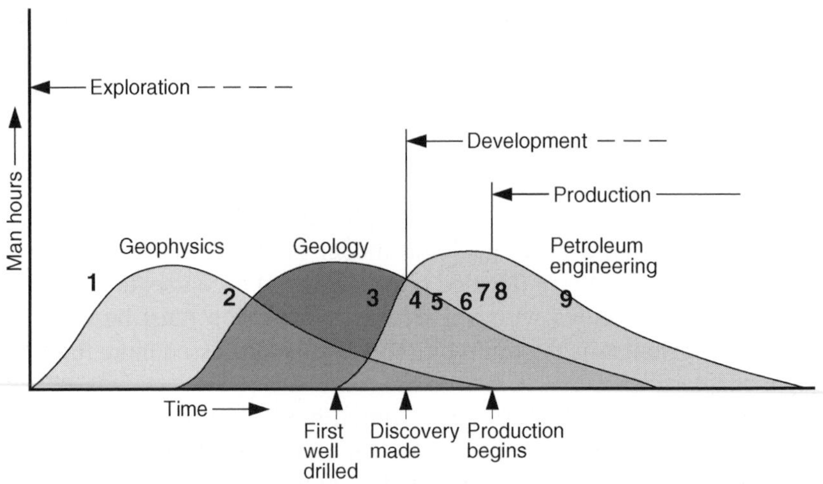

Fig. 2.3. Exploration, discovery, development, and production stages in the evolution of a reservoir over time (steps 1–9). Geophysics, geology, and petroleum engineering all play dominant roles at different times in the life of a field.

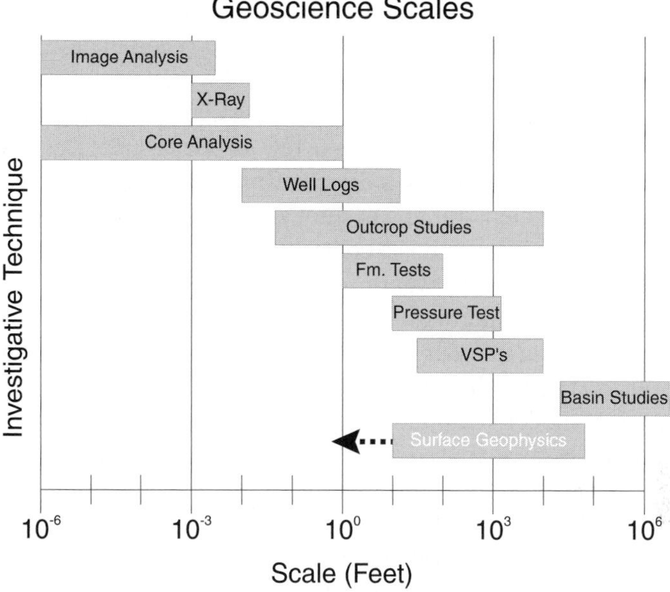

Fig. 2.4. Scales of measurement for different investigative techniques. The scales range from basin scale to pore scale. Different types of instruments are used to measure properties at these different scales. (Diagram courtesy of D. Minken.)

This workflow, from exploration to reservoir characterization to geologic modeling and field development, progresses from examining features at a large scale to examining very small features, using a variety of tools (Fig. 2.4). Once production begins, production information and new well data should be updated continually to refine the reservoir characterization (labeled 9 in Fig. 2.3).

2.3 Computers and the computing environment

Computers are essential tools for reservoir characterization. Most organizations provide adequate computing environments for their staff, from secretaries to geoscientists and engineers. However, some individuals and very small companies still prefer to hang cross sections on walls with a piece of string as a datum, or to interpret paper copies of seismic lines and well logs.

Progressively more geoscientists and engineers who are entering the petroleum industry grew up in a world in which, as toddlers, they had an early exposure to toy computers that helped them learn to visualize images on a screen. In grade school, they will have used computers in their classrooms and school libraries, and many of their families will have home computers. By the time they reach college and the work world, they are computer-literate and very comfortable in the computing environment.

Computers are ubiquitous in the upstream and downstream operations of the oil and gas industry. They are used to collect and manipulate seismic data for generating images, to generate well logs, to make maps, to evaluate numerical variables, to develop mathematical formulations, and to simulate fluid flow in a numerical reservoir model, to name a few applications. "Data mining" is a relatively new discipline that is a direct result of the ability of computers to search for data in efficient, rapid ways not previously possible or practical.

Improving computing speed is a constant challenge for hardware developers, because geoscientists always want to see the results of their studies as soon as possible. Even more importantly, geoscientists want to be able to input ever-larger volumes of data in order to build more-sophisticated and detailed (realistic) reservoir models for fluid-flow simulation.

A key asset of modern computing is its ability to provide visual 3D images of features and processes that used to be displayed either in 2D space, or verbally or numerically (Slatt et al., 1996). Advances in computer storage capacity and speed of data manipulation now allow rapid analysis of vast amounts of information for exploration and production (however, there is still the need for more and faster data gathering and manipulation). The old adage "A picture is worth a thousand words" is certainly true for animation and visualization technology. "Seeing is believing" is another expression that applies to these advancements.

Visualization is a form of communication that bridges the gap between verbal and technical languages (Fig. 2.5). This communication gap is one reason why

Fig. 2.5. Picture showing one of the principal values of computing in the petroleum industry: that of improving communication among disciplines. The picture illustrates a geologist (left) examining a computer geologic model built from well control, and a geophysicist (right) examining a seismic section on the computer screen. Both are attempting to develop a realistic geologic model of the petroleum reservoir, and they can be aided in this endeavor by iterating their visions of the reservoir on their computer screens. (Diagram courtesy of P. Romig.)

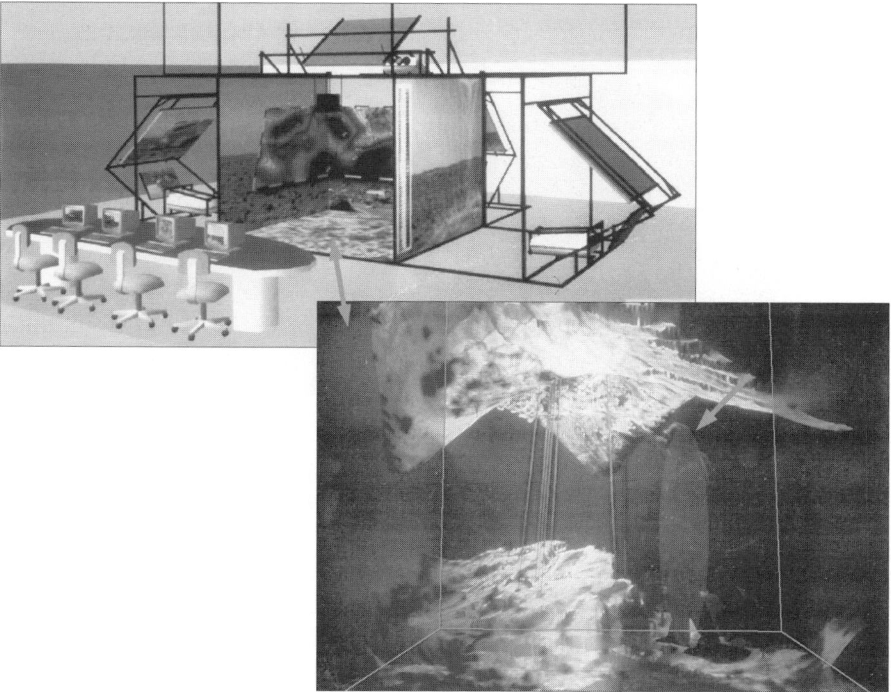

Fig. 2.6. Visualization theater (CAVE). Images, such as from a 3D seismic survey, are projected onto three walls, the floor, and the ceiling of the theater so that individuals and teams can appear to stand inside the seismic volume. The lower-right figure shows a person inside the theater evaluating potential drilling sites (near-vertical lines) through high-amplitude seismic intervals. (Diagrams courtesy of G. Dorn.)

many companies have invested large sums of money to develop visualization theaters, including those in which a team can stand in a semi-enclosed area that provides an "insider's" look at a reservoir (Fig. 2.6). In this way, a seismically-defined reservoir can be examined from the inside, and drilling scenarios can be simulated and calculated. These "CAVES" require high-speed computing capabilities, so they are relatively expensive. However, their high cost can be offset by the improvement in integrating disciplines and in the speed of solving reservoir problems (Shiralkar et al., 1996). The same is true in education. Online internet distance learning is now global and, in many cases, more convenient and less expensive than traditional university education (Fig. 2.7). Most companies also now have some sort of internal "intranet" for rapid global transfer of information among employees.

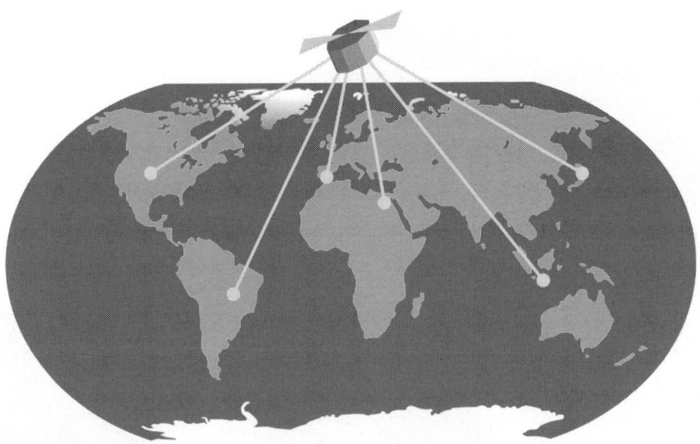

Fig. 2.7. Illustration pointing to the ease with which information can be transferred electronically to different locations, sometimes in different parts of the world, through satellite systems. (Diagram courtesy of P. Romig.)

2.4 Seismic-reflection and subsurface imaging

2.4.1 Two-dimensional (2D) seismic

Figure 2.8A shows a flat ground surface. By looking at the surface, there is no way that someone can truly know the geologic structure and stratigraphy that lie beneath it. The Grand Canyon (Fig. 2.8B) provides one rare example of being able to see the geology in a vertical section to a depth of about 1.6 km (1 mi) beneath the ground's surface. Seismic-reflection acquisition, or "shooting", provides an image of the subsurface that is not as detailed as the true geology, but that is adequate for imaging large- to medium-scale geologic structures and stratigraphy. Seismic-reflection analysis has become the dominant tool used in hydrocarbon exploration, and with some resolution limitations, it is becoming widely used for characterizing reservoirs.

The seismic-reflection method is based on the principle that an energy source, such as dynamite, generates sound waves that travel through the earth (Fig. 2.9A). When the sound waves reach an interface between two rocks that have different acoustic properties, some of the waves' energy will continue to penetrate through the rock beneath that interface, but some of the energy will be reflected off of the interface and will travel back toward the earth's surface (Fig. 2.10A). This reflected energy is recorded at the earth's surface by electronic receivers called "geophones" (Figs. 2.9B, C). The receivers are wired to a computer, inside a vehicle, that collects the reflected wave energy (Fig. 2.9D). The large amount of data that is collected during a seismic shoot can be processed either on site or at a facility that has more powerful computing capabilities. The resulting image shows seismic reflections that mimic features

Fig. 2.8. (A) A typical ground surface, with no indication of the geology that lies beneath the surface. (B) The Grand Canyon, which shows the internal structure and stratigraphy that is present beneath the ground surface at this locale. The roles of the geologist and the geophysicist are to image and evaluate the subsurface geology when it is not readily observable.

in the subsurface (Fig. 2.10B). The vertical axis is recorded not in depth beneath the ground surface, but rather, as two-way traveltime (TWT). Two-way traveltime is the amount of time (in seconds) that a sound wave takes to travel from its point of generation by the source at the ground's surface, to the subsurface interface that reflects it, and then back to the ground's surface where it is recorded by the geophones.

On land, other techniques are also available for generating the energy necessary to send waves traveling deep below the ground's surface. For example,

Fig. 2.9. Seismic reflection analysis has become the dominant tool for characterizing reservoirs as well as for exploring for subsurface hydrocarbon accumulations. It is based upon the principle that (A) an energy source such as dynamite generates sound waves that travel through the earth. When the sound waves reach a boundary in which acoustic rock properties vary significantly (such as off the interface of two different rock types), some of the wave energy will continue to penetrate rock beneath the interface, but some of the energy will be reflected back toward the earth's surface. This reflected energy is recorded at the earth's surface by electronic receivers called "geophones" (B and C), which are wired to a (D) receiving truck.

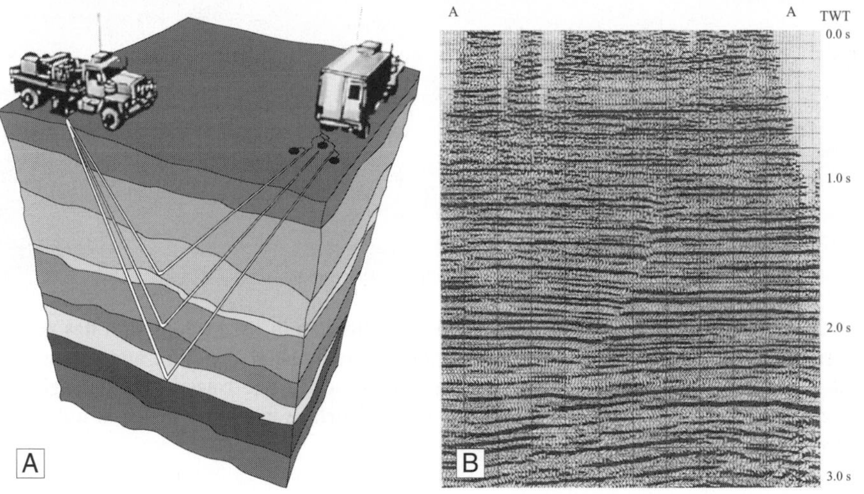

Fig. 2.10. (A) Cartoon showing the travel path of seismic energy down to rock interfaces, then back to the ground surface. After Brown (1988). (Reprinted with permission of AAPG, whose permission is required for further use.) (B) A 2D seismic line from the Alberta Basin of western Canada. The seismic amplitudes are representations of energy reflected off stratal boundaries. The vertical axis on this seismic line is the typical "2-way traveltime". Notice the high-amplitude (dark) reflections at about 1.3 seconds two-way traveltime. About two-thirds the distance from the left edge of the seismic line on the horizontal axis (which represents distance), these reflections are offset, indicating the presence of a thrust fault that cuts through the strata and dips toward the left.

Fig. 2.11. (A) Seismic energy can also be generated by hitting a metal plate with a sledgehammer. Behind the person is a geophone attached to a cable. (B) The reflected energy captured by the geophone is recorded on a laptop computer.

the source truck shown in Fig. 2.10A uses a heavy metal plate that is repeatedly dropped to the ground's surface, creating vibrations that travel through the earth. This energy source is called a "vibroseis". In Fig. 2.10A, the vibroseis truck generates wave energy that travels downward until it is reflected off of the rock interfaces and back up to the geophones, which are wired to the receiving truck on the right. If just the shallow subsurface is to be imaged, a sledgehammer can be used to hit a plate on the ground (Fig. 2.11A), thereby generating waves that reflect off of a shallow underground interface and travel back to be recorded by geophones wired to a laptop computer (Fig. 2.11B).

Seismic reflection is also the primary tool used for characterizing the structure and stratigraphy beneath the seafloor in the marine environment (Fig. 2.12). A large volume of seismic data has been collected in the nearshore and continental shelf areas of the world's ocean basins. In places like the US Gulf of Mexico and offshore West Africa, seismic shooting has gone beyond the shelf edge as the search for oil and gas has extended into deeper waters. Energy sources include generating electrical charges between metal plates (a "sparker") and using an air gun to generate vibrating air bubbles (a "pressure gun"). A boat tows the sound source via a cable attached to the stern, while another cable holds the hydrophones (the marine equivalent of a geophone) that receive the reflected waves.

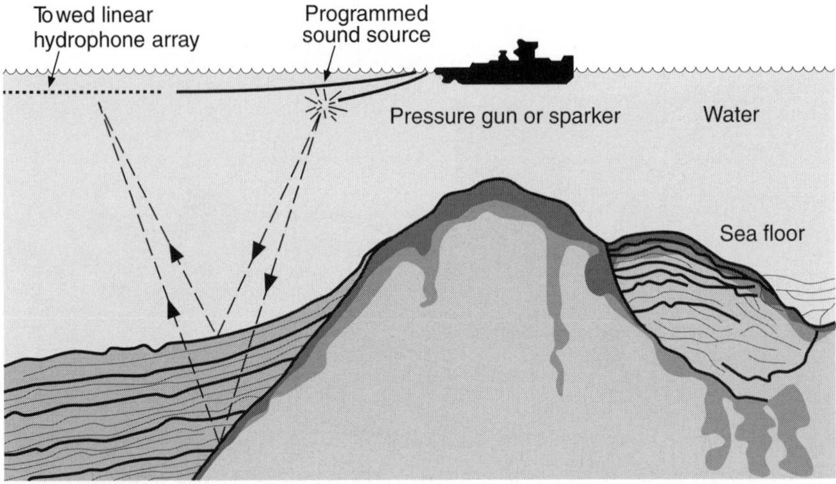

Fig. 2.12. In bodies of water, a seismic boat tows a cable that pulls a sound source, which may be an electrical charge from what is called a "sparker" or an air bubble from a pressure gun. Another cable, which contains "hydrophones", trails behind the boat at a set distance from the sound source and captures the waves that are reflected off the seafloor and subsea interfaces. (Source of figure is unknown.)

2.4.2 Three-dimensional (3D) seismic

In the early 1980s, the technology for acquiring and processing seismic data had improved enough that the costs were reduced and it became practical and economic to shoot 3D, rather than 2D seismic surveys (Fig. 2.1). The advantage of a 3D survey is obvious – a three-dimensional image of the subsurface is much more useful for exploration and field development than is one or more 2D vertical images. Three-dimensional seismic is designed to image a large area of the subsurface, including up to and beyond the size of a reservoir, both areally (horizontally) and stratigraphically (vertically) (Fig. 2.13).

Shooting a 3D seismic survey on land or in the marine environment requires rigorous planning, particularly in positioning the source and receivers. For land surveys, a 3D grid of geophones is placed on the ground surface (Fig. 2.14). Sound waves reflecting off of individual stratal boundaries will be picked up by this 3D array of geophones, and a 3D image of the subsurface will be generated as a series of seismic reflections. Generating the image requires a significant number of "processing" steps to reduce the effects of noise, topography, depth, and so forth. Color is now a common processing and display tool for enhancing the 3D seismic image (Fig. 2.14). In addition to imaging the 3D volume of data, it is possible to extract a 2D image in any orientation: a vertical profile (called either "crossline" or "inline", depending on the orientation), a horizontal plane (called a "seiscrop" or "horizon slice"), or a diagonal image (Fig. 2.14).

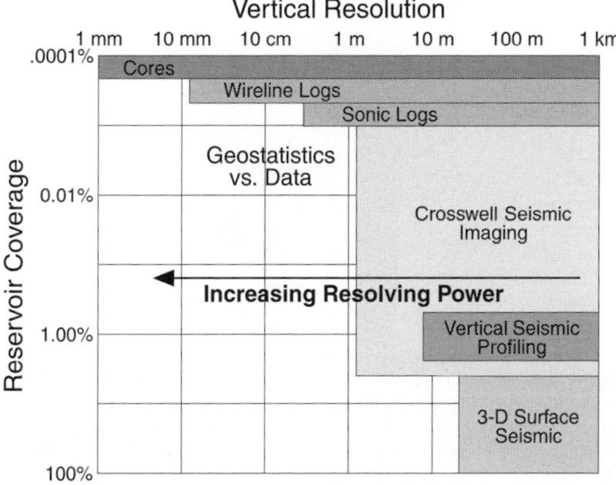

Fig. 2.13. Graph showing the vertical resolution of a reservoir on the horizontal axis and the horizontal coverage of the reservoir on the vertical axis. Cores can exhibit sedimentary features down to the scale of a millimeter or less, but the areal coverage is very small (15 cm, or 6 in, diameter). At the other extreme, 3D surface seismic covers a large area of a reservoir, but the features must be on the order of tens of meters to be fully resolved and imaged. Various other tools measure properties between these two end members. (Diagram courtesy of B. Marion and Z-Seis Corporation.)

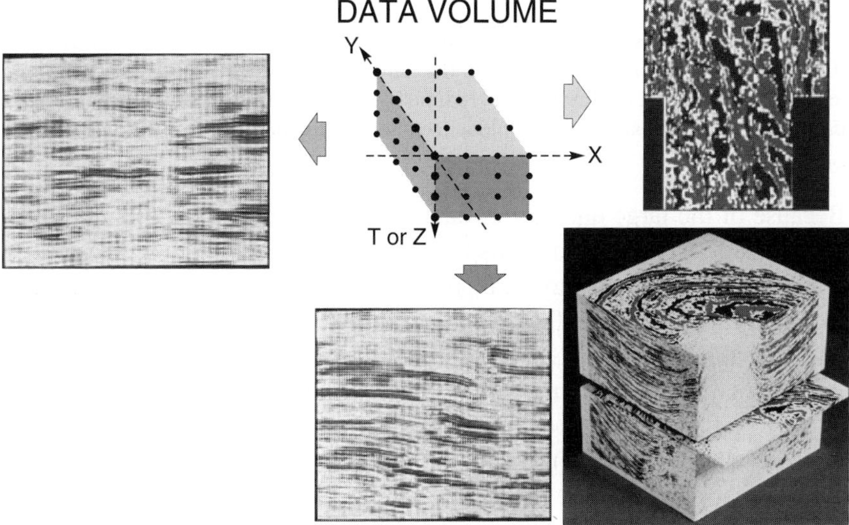

Fig. 2.14. Diagram showing three of many possible 2D images that can be generated from a 3D seismic survey. The two "crosslines" are vertical sections at different orientations within the 3D data volume. The "seiscrop" section is a horizontal section through the 3D data volume. In addition to these orientations, 2D sections can be generated at any orientation within the 3D data grid (diagonal, tilted, etc.) and a 3D image can also be generated. After Brown (1988). (Reprinted with permission of AAPG, whose permission is required for further use.)

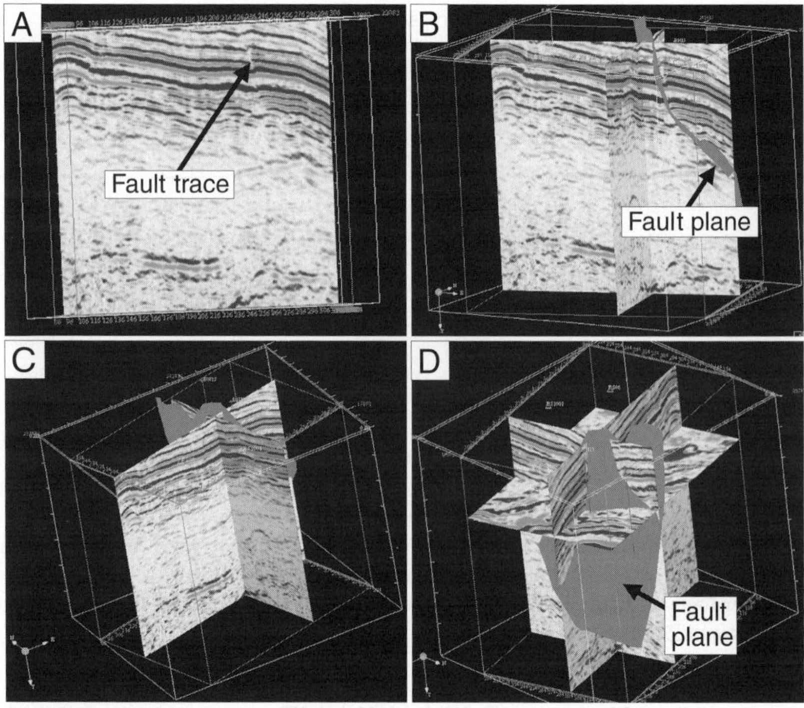

Fig. 2.15. (A) A single seismic profile from a 3D seismic survey. A fault trace is shown. (B) Crossing profiles from the survey with the fault plane highlighted. (C) Another view of the crossing profiles and the fault plane (red/dark-gray). (D) A horizontal plane or seiscrop section has been added, and the fault plane (red/dark-gray) mapped into the 3D volume.

Because of the large quantity of data collected during a 3D seismic survey, computers are essential for data display, as well as for acquisition and processing procedures. Many displays are possible, including 2D crossline and inline displays for imaging planar features in 3D space (Fig. 2.15) and full 3D volume and "chair" displays (Fig. 2.16). Figures 2.17 and 2.18 show examples of 3D seismic displays along with modern environmental analogs.

2.4.3 Four-dimensional (4D) seismic

Three-dimensional seismic data represent static reservoir properties, because the data are collected during one instant in time. Four-dimensional, or "time-lapse" seismic data, offer the opportunity to monitor the movement of reservoir fluids while they are being produced. Thus, 4D seismic measures dynamic reservoir properties. The underlying principle of 4D seismic is that acoustic properties of reservoir strata will change as a function of change in fluid content and type within the rock's pore spaces. Thus, 3D seismic surveys that are shot in the

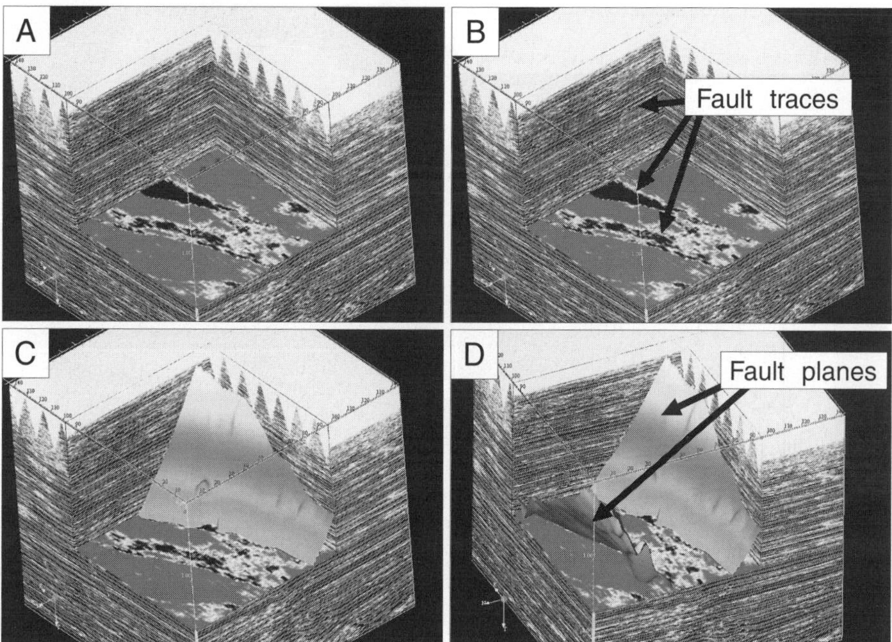

Fig. 2.16. Cutaway image of part of a 3D seismic reflection volume showing fault traces in 2D space (A and B) and fault planes in 3D space (C and D).

same (or similar) manner, at the same location, and at successive times during the life of a field, record rock-fluid changes that result from fluid movement.

ARCO Oil and Gas Company performed one of the earliest, if not the first, test of 4D seismic – in the Holt Sand Unit of north Texas (Greaves and Fulp, 1988) (Fig. 2.19). ARCO's goal was to conduct a combustion pilot test by igniting downhole air and gas to increase fluid (reservoir) temperature and thereby reduce the oil's viscosity so that it could flow more easily from injector to producer wells. Figure 2.19 shows seismic horizon slices for three separate 3D seismic surveys: one shot before the fireflood ("burn") was initiated, one shot during midburn (after several months), and one shot after the fireflood was completed. Air and gas were injected into the injection well and ignited (red/dark-gray triangle in the three figures). This injection well was surrounded by four production wells (red/dark-gray dots in the three figures). The concept was that the heated reservoir fluids would move outward in all directions from the well (arrows in the preburn seismic horizon). The bar scale shows the relative scale of seismic amplitude. At preburn conditions, there was little variation in seismic amplitude throughout the field. By midburn time, amplitude had changed significantly near the western production well, indicating fluid movement in that direction (arrow). By postburn time, amplitude had changed even more, and a new amplitude anomaly had developed near the eastern producer well, indicating

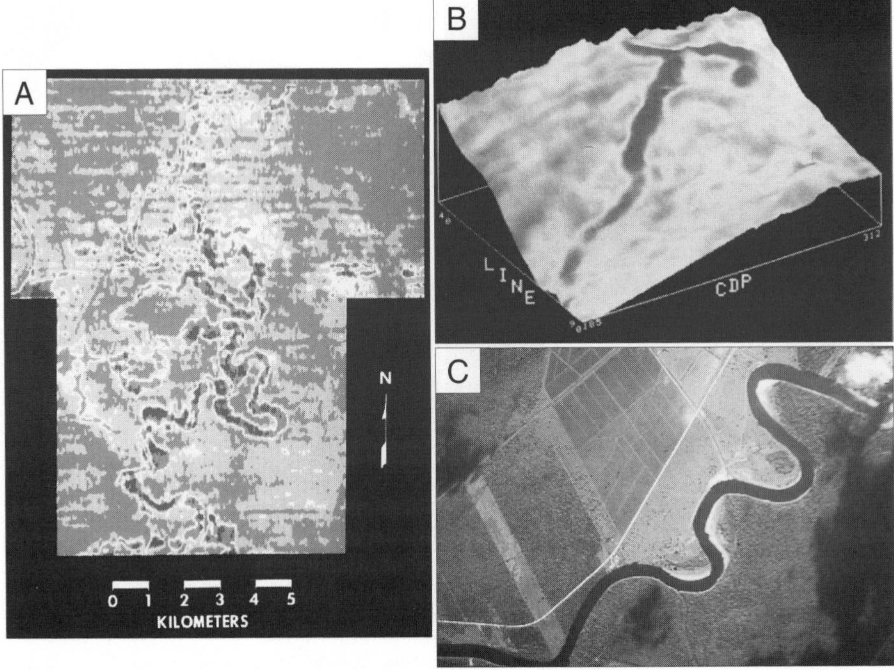

Fig. 2.17. (A) A 3D seiscrop or horizon slice showing a buried meandering river channel. The quality of the seismic image is so good that it gives almost the same appearance of the subsurface feature as if one were looking out of an airplane window at a similar feature on the ground surface; (B) shows a gas-charged fluvial sandstone in its true structural position – structural dip is toward the lower left. The red/dark coloration (amplitude strength) increases updip because of the presence of more gas, and consequently there is greater acoustic contrast between the gas-filled channel sandstone and adjacent strata. (A) and (B) after Brown (1988). (C) Areal photograph of a meandering river, similar in shape to that imaged in (A). The light colored areas on (C) are accumulations of sand. By knowing where sand accumulates in modern meandering channels, one can predict the occurrence of sandstone accumulations on the horizon slice of a similar subsurface feature (A). (Reprinted with permission of AAPG, whose permission is required for further use.)

some flow in that direction (arrow). The fact that amplitude changes recorded nonuniform fluid movement over the field indicates reservoir-rock heterogeneity or compartmentalization that preferentially "channeled" reservoir fluids in certain directions.

A popularized 4D seismic survey was completed in the Eugene Island area of offshore Gulf of Mexico (He et al., 1996) (Fig. 2.20). Here, three companies had run independent 3D seismic surveys at different times, using reasonably similar acquisition parameters. The earliest survey was shot by Pennzoil/Mobil/BP in 1985, the next was done by Texaco/Chevron in 1988, and the third was done by Shell/Exxon in 1992. Comparison of the 3D surveys, in the area where all three surveys overlapped, formed the basis for a 4D seismic survey.

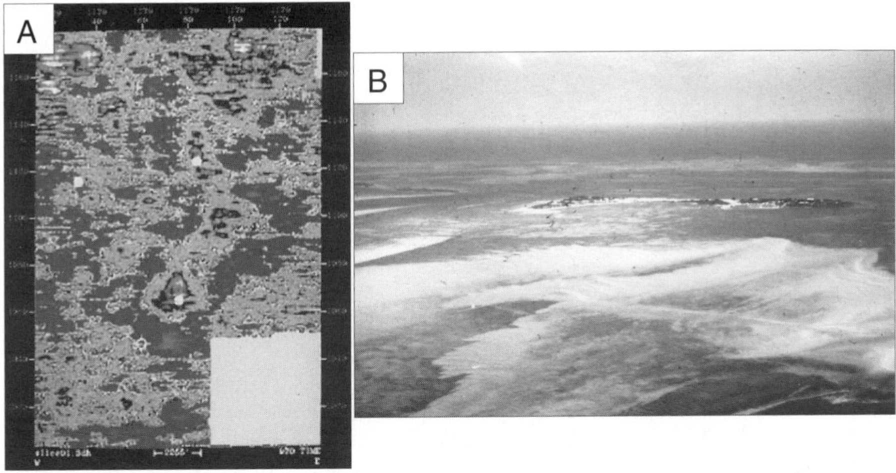

Fig. 2.18. (A) 3D seismic horizon slice showing a series of discontinuous high-amplitude (red, orange, and yellow/bright) areas that are interpreted to be pinnacle reefs. After Brown (1988). (B) Modern-day pinnacle reef for comparison. (Reprinted with permission of AAPG, whose permission is required for further use.)

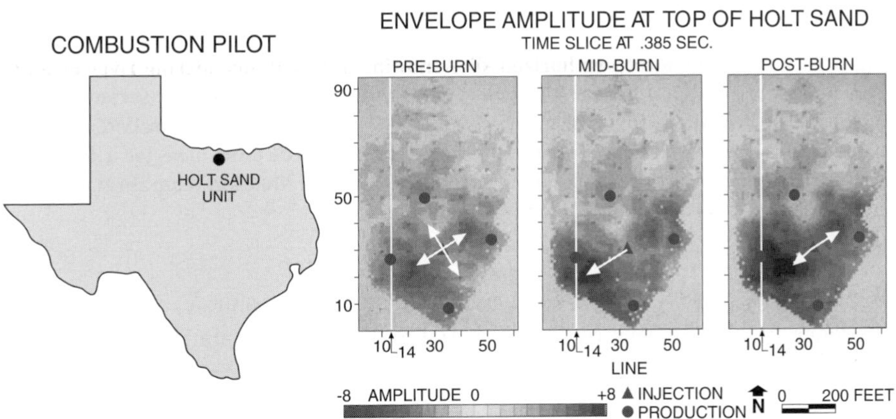

Fig. 2.19. Diagram showing the same seismic horizon slice for three separate 3D seismic surveys in the Holt Sand unit, northern Texas: one shot before a fireflood ("burn") was initiated, one shot during midburn (after several months), and one shot after the fireflood was completed. The relative measures of amplitude are shown on the bar scale. Variations in amplitude from pre- to post-burn time indicate migration of reservoir fluids as a result of the fireflood. After Greaves and Fulp (1988). (Reprinted with permission of AAPG, whose permission is required for further use.)

The results of the analysis were quite dramatic. Figure 2.21A shows the 3D survey area of the reservoir interval. Different colors represent seismic-amplitude change and lack thereof among the three 3D surveys. The data high-

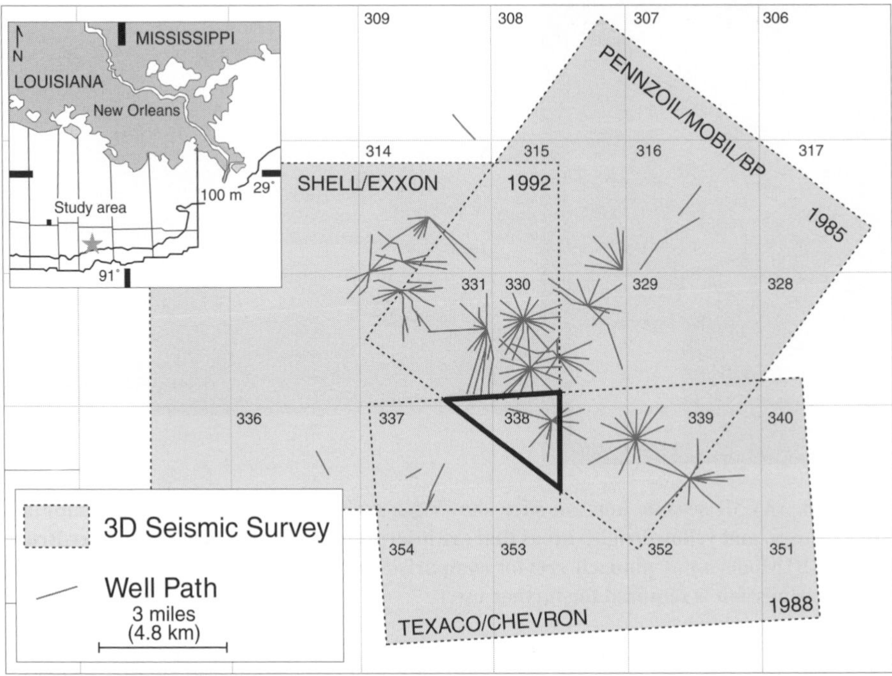

Fig. 2.20. One of the most popularized 4D seismic surveys was done in the Eugene Island area of offshore Gulf of Mexico. After He et al. (1996). An early 3D seismic survey was shot in 1985, followed by another survey nearby in 1988, then a third survey in 1992. The area where all three surveys overlap (red/black-solid triangle) formed the basis for a 4D seismic survey that was done by comparison of the three 3D surveys. Modified from He et al. (1996). (Reprinted with permission of Oil and Gas Journal.)

light a fairway in which there was no change in amplitude, suggesting the presence of oil that was bypassed or untapped. A horizontal well was drilled into the bypassed area. During the period 1972–1994 (prior to the seismic survey), 1.2 million barrels of oil had been produced from the interval, but during the post-survey years of 1994–1996, an additional 1.0 million barrels were produced from the horizontal well (Fig. 2.21B). The combination of 4D seismic and horizontal drilling paid off in handsome dividends to the operators and demonstrated a new technique that is becoming more widely used.

2.4.4 Other seismic imaging techniques

Industry and academia are conducting ongoing research to refine seismic imaging, particularly with regard to sharpening images and reducing the resolution and detection limits of imaged features. The addition of color to seismic sections has provided a simple but major improvement in our ability to mentally

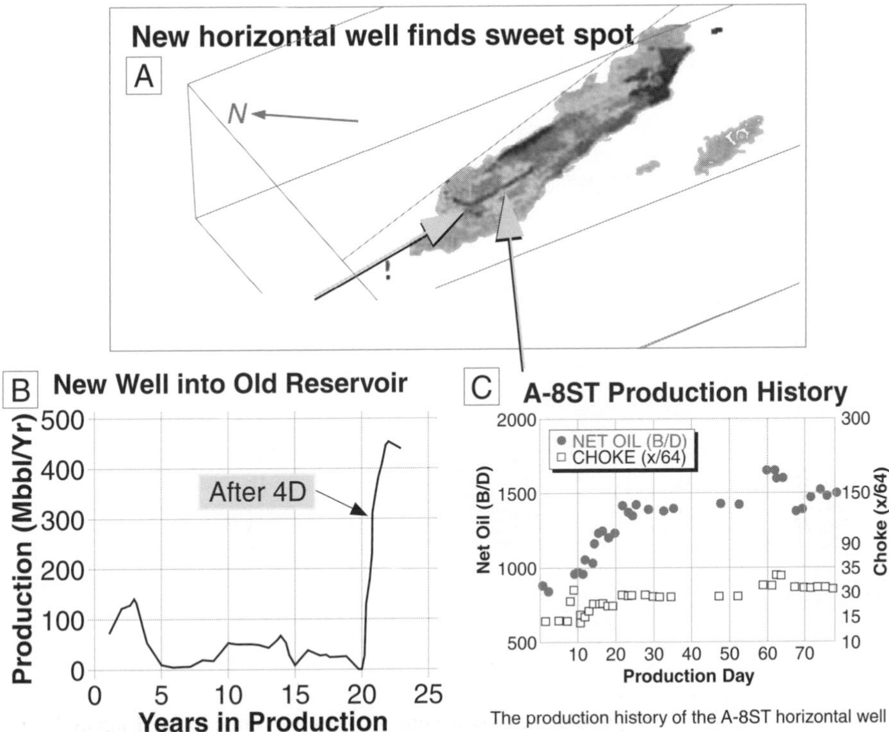

Fig. 2.21. The results of the 4D Eugene Island seismic survey. The 3D block shows the reservoir area, and the different colors represent areas in which seismic amplitudes changed or remained the same among the three 3D surveys. A fairway is highlighted in which there was no change in amplitude, which suggests the presence of bypassed or untapped oil. Red/dark denotes areas in which acoustic impedance decreased from 1985 to 1992. Blue denotes areas in which impedances increased over time. Green and yellow/bright denote areas with unchanged (>10%) impedances (where bypassed oil is most likely to occur). In this case, a horizontal well was drilled into the zone that appeared to contain bypassed oil and, as is shown on the lower left graph, production improved dramatically. During the period 1972–1994 (prior to the seismic survey), 1.2 MMBO were produced from the interval, but during just the two postsurvey years of 1994–1996, an additional 1.0 MMBO were produced. Modified from He et al. (1996). (Reprinted with permission of Oil and Gas Journal.)

visualize subsurface geology from seismic records (compare the seismic profile in Fig. 2.10B with those in Figs. 2.15, and 2.17).

Since the mid-1990s, "coherence-cube technology" has become a popular means of sharpening seismic images of geologic features (Taylor, 1995). This technology essentially reduces seismic noise, thus enhancing seismic signals from real geologic features. Figure 2.22 compares a 3D horizon slice without (left) and with (right) coherence-cube image processing (Bahorich and Farmer, 1995). The sharper image of faults on the right-hand diagram is obvious.

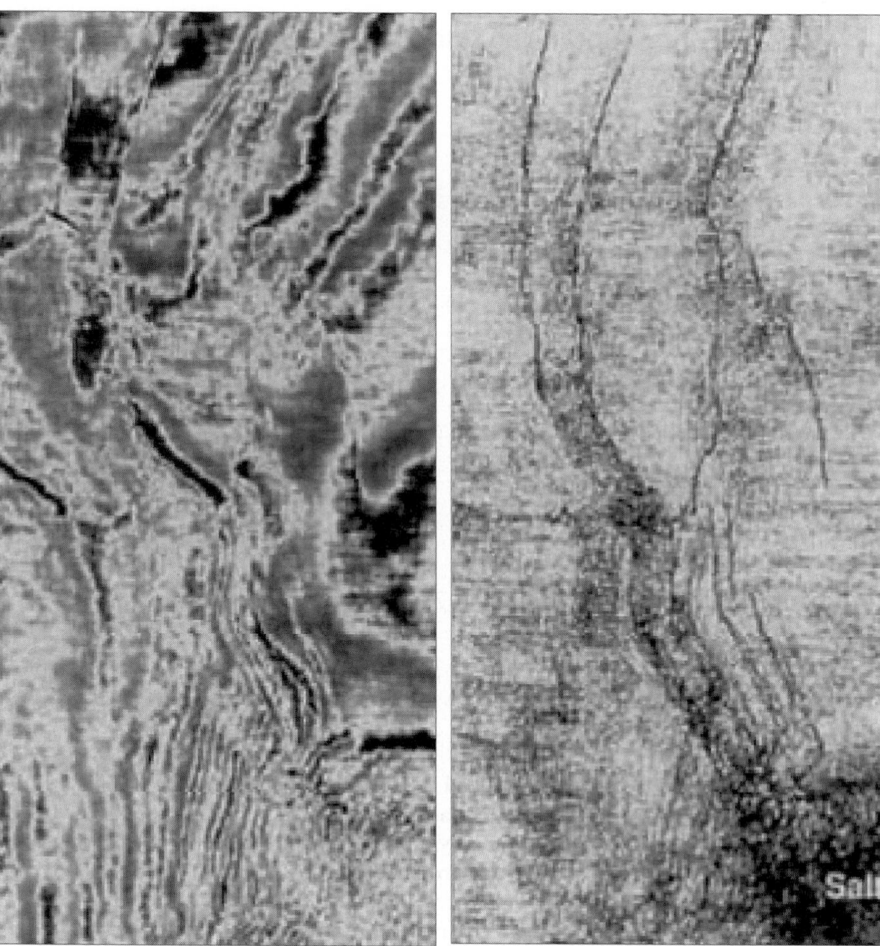

Fig. 2.22. Illustration from the cover of the October 1995 issue of The Leading Edge technical journal (Bahorich and Farmer, 1995). A 3D horizon slice without (left) coherence-cube image processing is compared with a slice with (right) coherence-cube image processing. This technology enhances seismic signals that represent real geologic features, at the expense of seismic "noise", thus providing sharper, more coherent images. The clear outlines of faults on the right-hand diagram are much more difficult to interpret on the standard 3D seismic horizon slice on the left. (Reprinted with permission of The Leading Edge.)

Spectral decomposition is another technique that is proving to be quite valuable in improving the clarity and resolution of seismic images of reservoirs. A seismic waveform is composed of a series of sinusoidal components that have variable frequencies and amplitudes (Fig. 2.23). The spectral decomposition process separates the seismic signal very precisely into individual frequencies, so that the associated amplitudes can be observed one at a time. Acoustic properties of rock and the fluids contained in that rock also vary with frequency.

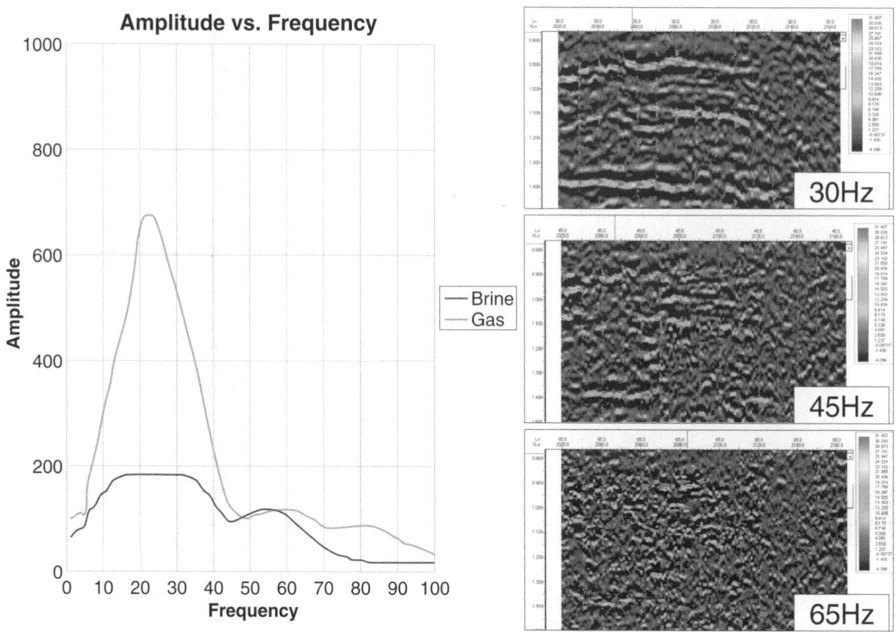

Fig. 2.23. Graph that shows the frequency dependence of seismic amplitude. The three seismic profiles on the right show amplitudes at 30-, 45-, and 65-Hz frequencies. Different features are highlighted at each frequency. After Sanchez (2004).

Thus, by decomposing the seismic signal into individual spectral components, associated amplitudes can be observed that vary with frequency as a function of the reservoir-rock and reservoir-fluid properties. Figure 2.23 shows the amplitude spectrum of gas and brine at different frequencies. Also shown are three seismic profiles of differing frequency. The amplitude variations with frequency represent the response of specific rock and fluid properties to the seismic signal. Figure 2.24 shows the same frequency-dependent amplitudes for horizon slices of a particular subsurface interval.

An offshoot of the spectral decomposition technique is multi-attribute inversion, a technique that provides excellent lithology resolution when it is compared with extracted seismic amplitudes. Figure 2.25 shows a good result obtained by comparing predicted versus actual gross sand thickness of a sandstone interval, and by comparing images of conventional horizon amplitudes and multi-attribute inversion amplitudes. Evolving technologies like these will continue to improve our ability to image and understand subsurface reservoirs.

2.4.5 Cross-well seismic investigation

Cross-well seismic is one of the only methods that can image subseismic-scale lateral and vertical attributes at interwell spacings. The method involves placing

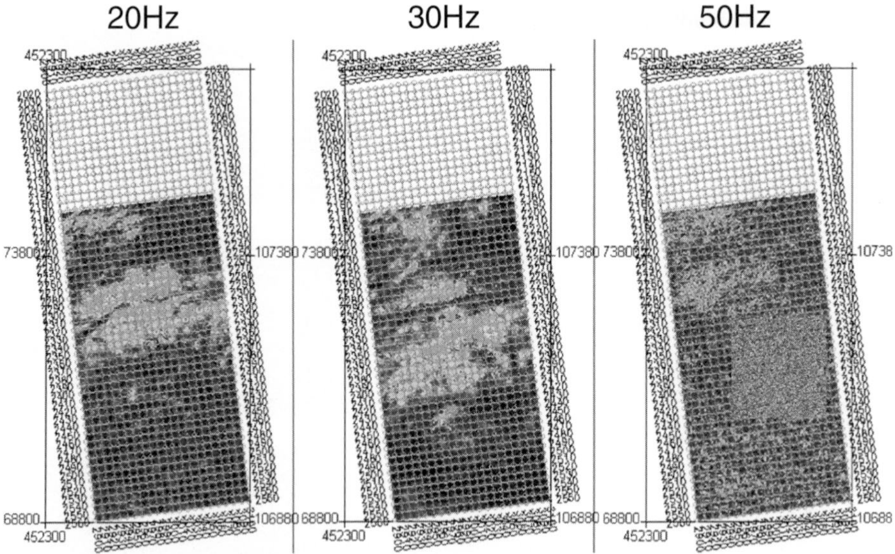

Fig. 2.24. The same seismic horizon slice, displayed at 20-, 30-, and 50-Hz frequencies. Different features are highlighted at each of the three frequencies. After Sanchez (2004).

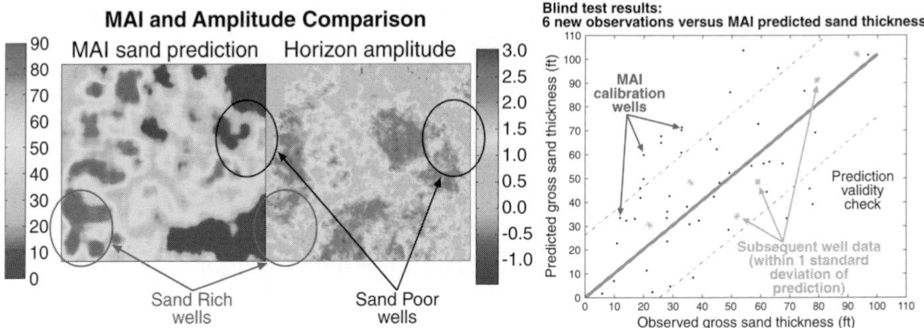

Fig. 2.25. Example showing that properly combined multiple attributes provide superior lithology resolution, compared with that from extracted seismic amplitude. The seismic amplitude exhibits poor correlation with the presence of sand in an area, whereas the multi-attribute inversion provides details of sand deposition within a meandering channel. (Figures courtesy of Fusion Petroleum Technologies, Inc.)

a seismic source string down one well and a set of geophone receivers down another well, and then shooting the seismic to obtain images between the wells (Fig. 2.26). Because the source and receivers are beneath the surficial weathering zone and are closely spaced, the resolution is an order of magnitude higher than that from surface seismic in the vertical dimension, though the aerial coverage is less (Figs. 2.13 and 2.27). Small faults and stratigraphic features, as well as porous (and nonporous) zones, can be imaged clearly with cross-well

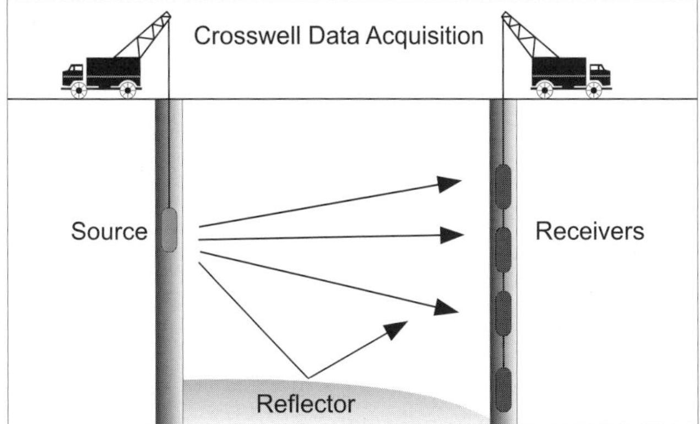

Fig. 2.26. Cross-well seismic provides an energy source that generates sound waves within a borehole. Receivers arranged in a series are placed down another well to detect the sound waves. Behavior of the waves from source to receiver can be related back to rock and fluid properties, such as lateral and vertical variations in porosity, at much higher resolutions than can be acquired from conventional surface seismic reflection.

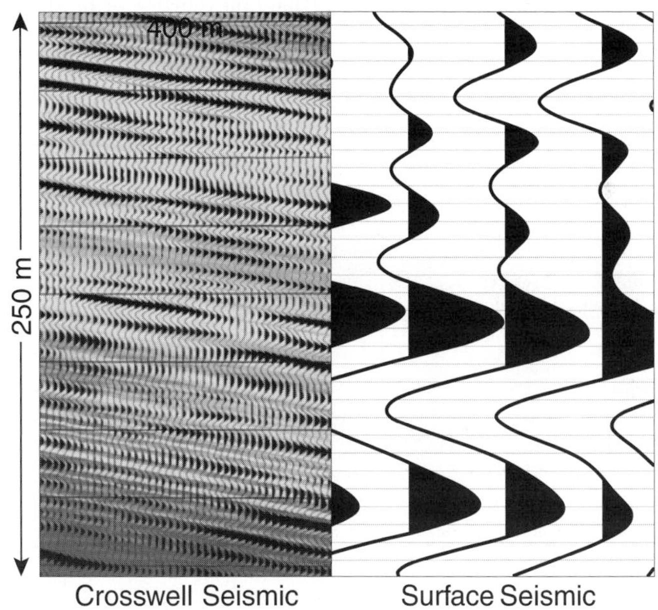

Fig. 2.27. A comparison of the resolution of cross-well seismic to that of surface seismic. The cross-well seismic provides much better resolution of features that can be related back to properties of the reservoir. (Diagram courtesy of B. Marion and Z-Seis Corporation.)

seismic (Figs. 2.28 and 2.29). Drawbacks to the method include its expense and the small areal extent of its coverage.

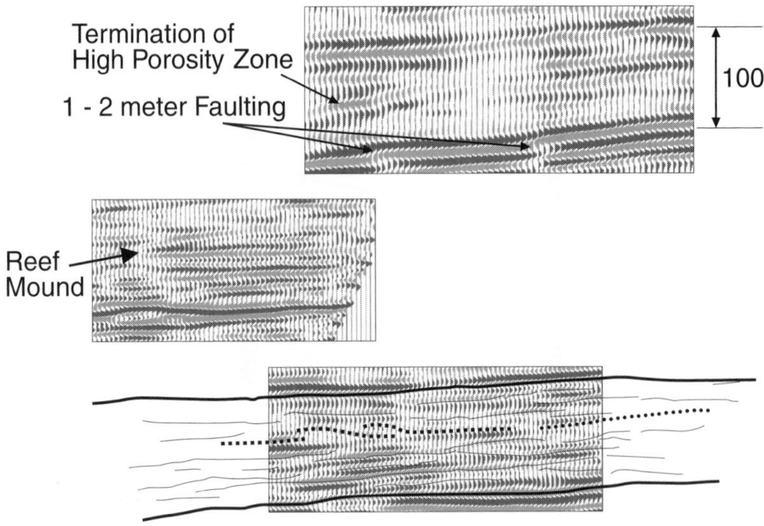

Fig. 2.28. Reservoir heterogeneities imaged from carbonate structures. The vertical and lateral extent of individual buildups requires cross-well seismic resolution to detect the buildups and to help target horizontal wells. (Diagram courtesy of B. Marion and Z-Seis Corporation.)

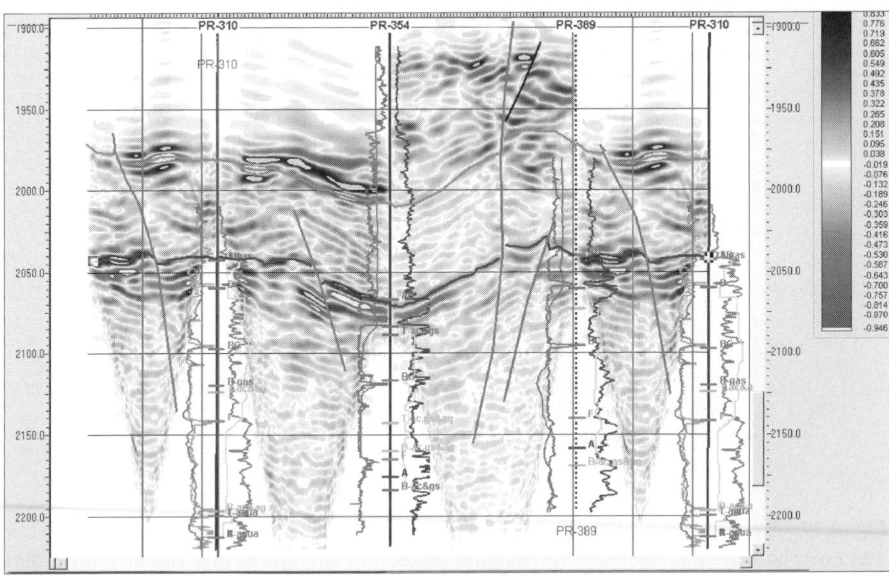

Fig. 2.29. Cross-well seismic showing small offset faults between wells and amplitude discontinuities representing stratigraphic discontinuities. These features are all beneath the resolution of conventional seismic data. (Diagram courtesy of B. Marion and Z-Seis Corporation.)

2.4.6 Multicomponent seismic investigation

Multicomponent seismic combines P (compressional)-wave and S (shear)-wave seismic energy. These two very different types of waveforms travel at different velocities through media, so that Vp (P-wave velocity) and Vs (S-wave velocity) can be measured and compared. Different Vp/Vs ratios are diagnostic of different rock/fluid combinations (Fig. 2.30).

2.4.7 Some pitfalls in seismic analysis

Although seismic-reflection analysis provides a means of imaging subsurface features, one must remember that the technique, though mature, is not without its flaws, and nonunique features are sometimes imaged. Dipping surfaces, such as faults and fold limbs, disperse seismic energy such that considerable post-acquisition processing of data is required, and even then, an uncertain image emerges. Because the seismic energy is in the form of wavefronts, the velocity of the waves and the medium through which they travel in a surveyed area must

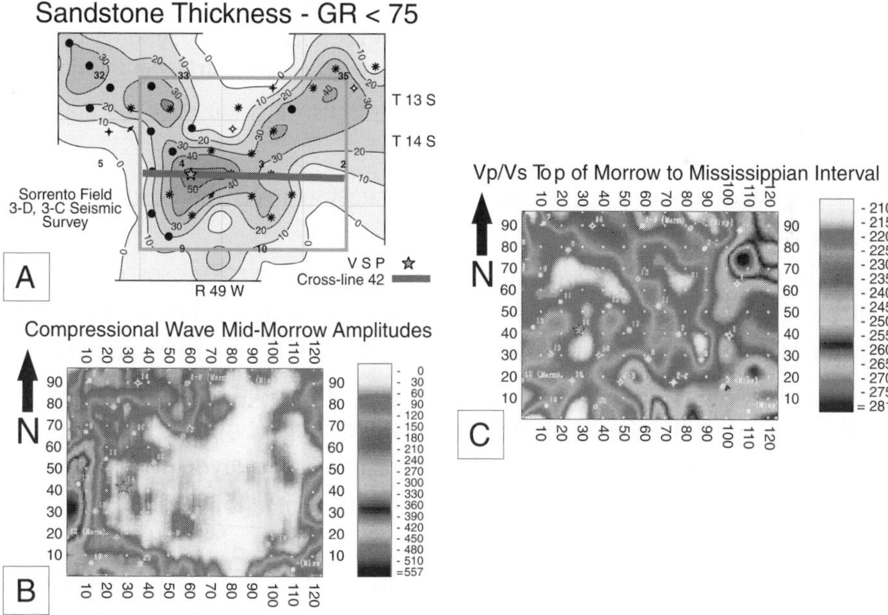

Fig. 2.30. (A) A sandstone-thickness isopach map based on well control (after Mark, 1998). (B) P-wave seismic horizon slice of the interval, showing some variability across the smaller area outlined in (A) by a green/gray square. (C) A horizon map of Vp/Vs, which images a greater level of heterogeneity than was mapped from either the well control or the horizontal seismic horizon slice. After Blott et al. (1999). (Reprinted with permission of The Leading Edge.)

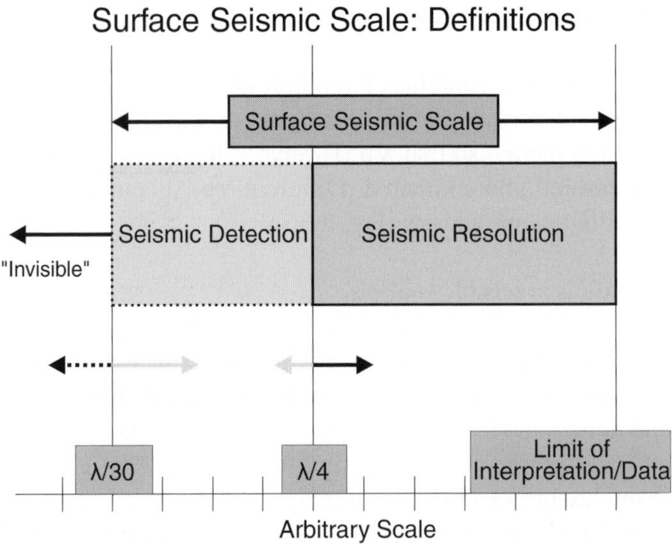

Fig. 2.31. Seismic resolution and detection are determined as a function of the wavelength of the seismic waveform. A complete image that includes the top and base of a sedimentary bed requires the bed to be a minimum of 1/4 the wavelength of the seismic waveform. Beds thinner than this will not be fully imaged, but if they are thicker than 1/30 wavelength, the seismic reflection will exhibit some amplitude response. (Diagram courtesy of D. Minken.)

be determined if one is to be able to accurately convert the vertical seismic interval from two-way traveltime to depth.

In detailed reservoir characterization, the greatest weaknesses of seismic-reflection analysis are the limits on the data's vertical resolution and on detection of small-scale features. Resolution is the ability of a seismic wave to identify and image both the top and base of a layer, and it usually is considered to be 1/4 the thickness of the seismic wavelength (Fig. 2.31). Detection is the ability of a seismic wave to "feel" and "respond" to a layer, even though the layer boundaries cannot be imaged (Fig. 2.31). The detection limit is considered to be 1/30 the thickness of the wavelength.

Many features that control, or at least affect, fluid flow in a reservoir are smaller than the seismic-reflection resolution and detection limits, and thus, are subseismic in scale. No single definition of the term subseismic exists, because resolution varies with wavelength and frequency of the waveform, subsurface depth of the reflecting surface, and other factors. But, if the wavelength is much longer than the thickness of the interval of interest, that interval will not be imaged (Fig. 2.32).

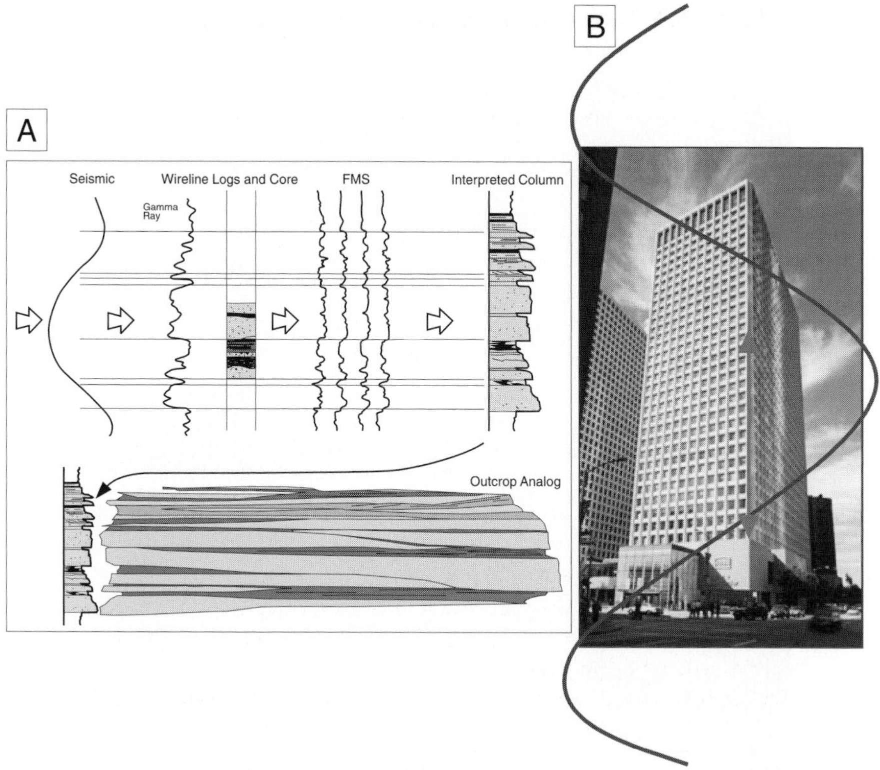

Fig. 2.32. (A) Conventional well log and a section of core. The response of the log can be calibrated to the different rock types comprising the core, so that the remainder of the well can be interpreted in terms of rock type. An FMS (Formation Micro-Scanner borehole-image log, which will be discussed later in this chapter) is also added. From these data, the geologist works to predict the rock types and their geometry away from the wellbore, where there are no real data. This is difficult and often is not assisted by seismic data if the rock stratification is beneath the resolution of the seismic wavelet, as is the case on the left. (B) This common situation is analogous to a seismic wavelet being superimposed upon the building. Internally the building consists of a series of rooms separated by floors, ceilings, and walls, but none of these compartments can be imaged because they are beneath the resolution (and detection) of the seismic wavelet. (Source of picture (A) is unknown. Picture (B) provided by D. Minken.)

2.5 Drilling and sampling a well

The only real way to identify subsurface rock types and their contained fluids is to drill a well and sample what is in the wellbore. Wells can be drilled on land with a land rig, or at sea from a floating platform or drillship (Fig. 2.33). The basic parts of a drill rig are shown in Fig. 2.34A. The drill bit is attached to a drill string of pipe that rotates through a turntable on the rig floor. At the end of the drill string is a drill bit that cuts through the rock. A water–mud mixture

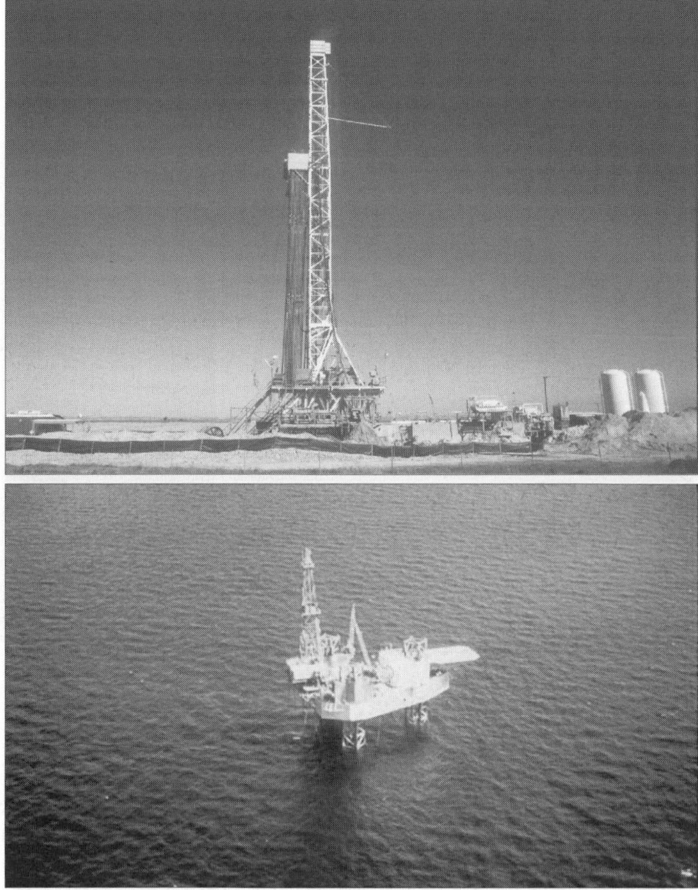

Fig. 2.33. The only real way to determine the types of rocks in the subsurface is to drill a well. Except in very structurally complex areas, seismic reflection data normally are acquired before an expensive well is drilled. Wells are drilled both on land and in the ocean, as is shown in these pictures.

(slurry) is continuously pumped down the hole to provide lubrication for the drilling and also to catch pieces of the borehole as it is cut. These pieces, called "cuttings", are carried upward to the ground surface by the circulating mud and are caught in a trap. Normally, a mud logger examines the cuttings on the drill floor, records the lithology, and measures the contained reservoir fluids, then bags the cuttings for future, more-detailed examination. To facilitate a thorough examination of cuttings samples for their lithology, composition, and for the presence of microflora and microfauna, the drilling mud must be washed from the cuttings to provide a clean sample (Fig. 2.35). Cuttings analysis may include relative proportions of rock types, biostratigraphy, and mineralogy (Fig. 2.36).

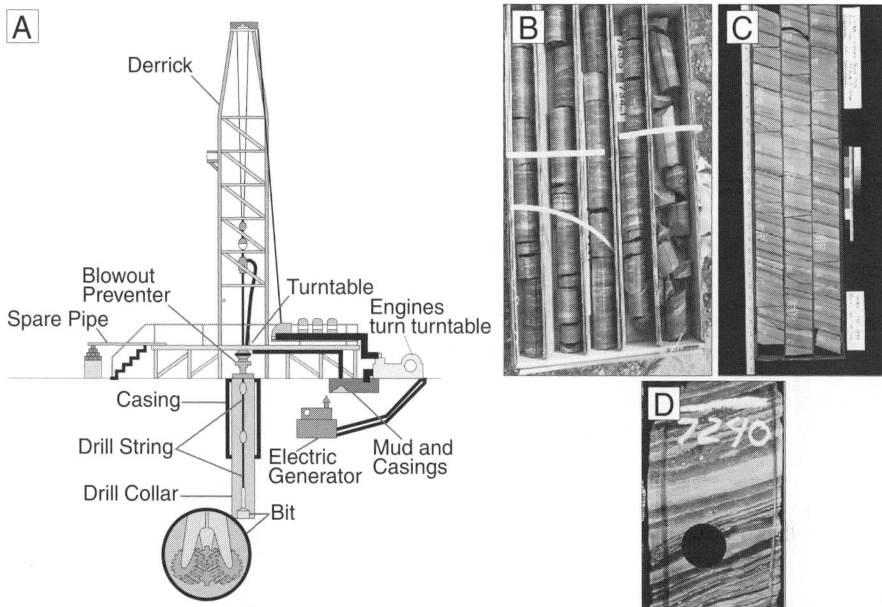

Fig. 2.34. (A) Basic parts of a drilling rig, including the derrick, the turntable that turns the drill string, and the bit that rotates and cuts through the rock. (B) Whole core that is boxed and ready to be shipped to a core analysis facility after being washed. (C) Slabbed core that has been cut lengthwise in two, with the side shown having been polished. (D) Close-up of a core piece from which a core plug has been obtained for porosity and permeability measurement. These plugs are obtained while the core is still whole (B).

Fig. 2.35. The least expensive sample of subsurface rock that can be obtained from the borehole is a collection of "cuttings" or chips of the rock formations that were drilled through. This figure shows five trays of cuttings from a well. The cuttings have been thoroughly washed and cleaned and are ready for examination by a geologist, who will determine the proportions of different rock types comprising the cuttings. The red/dark-gray numbers on the sides of the trays are the depth in feet at which the cuttings were obtained. In this case, the cuttings represent a composite of rocks over 3 m (10 ft) intervals. After Garich (2004).

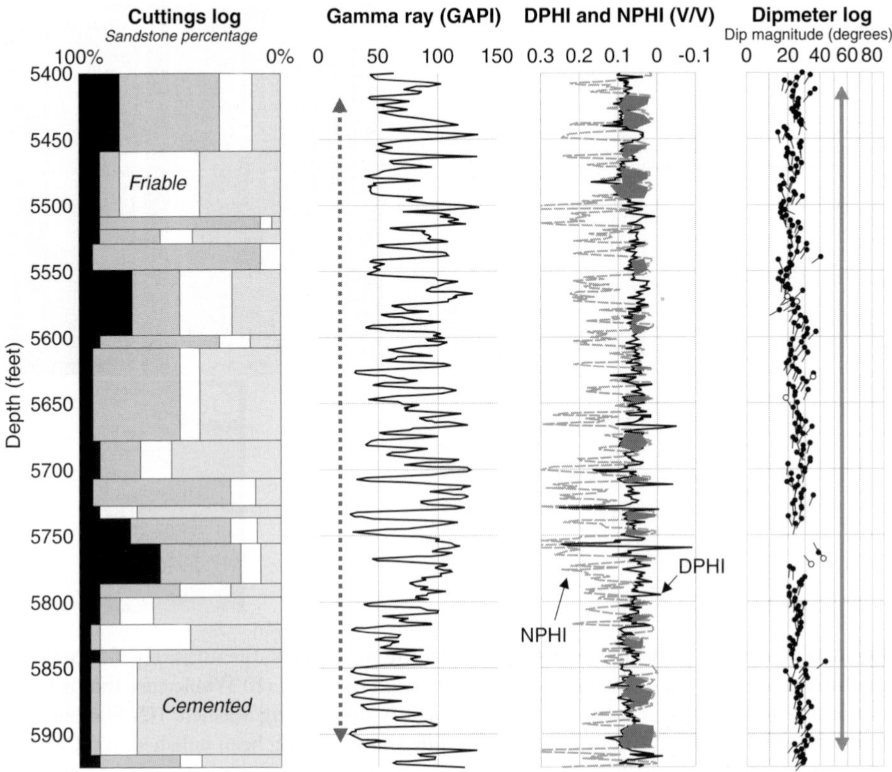

Fig. 2.36. Example showing different types of sandstones (friable and cemented; shades of yellow/white and green/bright-gray), mudstone (gray/dark-gray), and shale (black) that were determined from cuttings and whose proportions were counted through 3 m (10 ft) intervals. The cuttings log shows the proportions of each rock type throughout the entire 500-ft interval in this well. Various well logs are shown to the right. Red/dark-gray areas on the DPHI and NPHI logs show intervals of gas crossover effect. After Romero (2004).

If one wishes to obtain a core from a subsurface formation, a different assembly is attached to the drill string and a whole core of the subsurface rock formations is cut through a predetermined depth interval. When a core is obtained, it too is described at the wellsite. However, the cylindrical shape of the core's perimeter (Fig. 2.34B), coupled with the grooves and furrows produced by the coring operation, often make an adequate description of the geologic features of the core difficult. After the core is described at the wellsite, normally it is stored for later shipping to a laboratory for further treatment and examination. Core plugs, which generally are 2.54 cm (1 in) in diameter, are routinely obtained (Fig. 2.34D), normally at 0.3 m (1 ft) intervals along the length of the core, for analysis of porosity, permeability, and a measure of fluid saturation. If one desires, the core can then be cut into two vertical slabs. One slab is used for

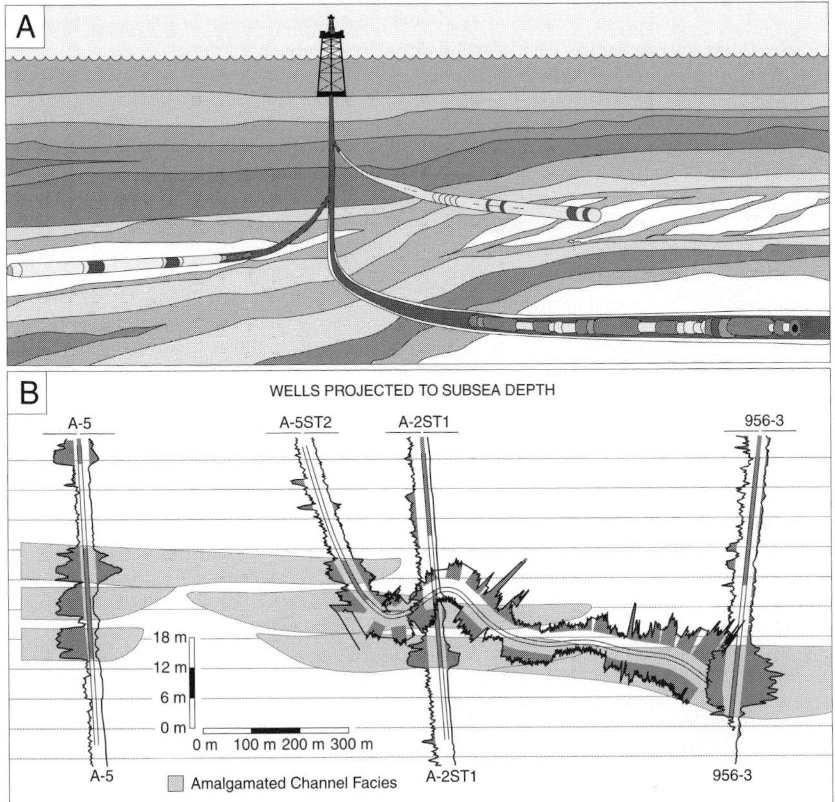

Fig. 2.37. Horizontal drilling of wells, such as is illustrated in these figures, is ideally suited for the type of situation shown here. (A) Cartoon of horizontal well drilling. (B) Reservoir in the Gulf of Mexico, with three vertical wells that penetrated some of the different lenticular sandstones and a horizontal well that also penetrated multiple channel sandstones. After Craig et al. (2003). (Reprinted with permission of AAPG, whose permission is required for further use.)

sampling, and the surface of the other slab is ground to a smooth finish and is used for detailed geologic description and photographing (Fig. 2.34C).

Horizontal-well drilling has expanded considerably in recent years because of improved technology and reduced costs (Fig. 2.37A). Although it is more expensive to drill a horizontal well than a vertical well, horizontal wells are particularly efficient for reaching compartmentalized reservoirs, such as sandstone lenses that are separated by impermeable shales (Fig. 2.37A,B) or sandstones that are cross-cut by impermeable fault zones (Fig. 2.38A,B). Horizontal wells can also be used for cross-well monitoring of enhanced-recovery projects (Fig. 2.39).

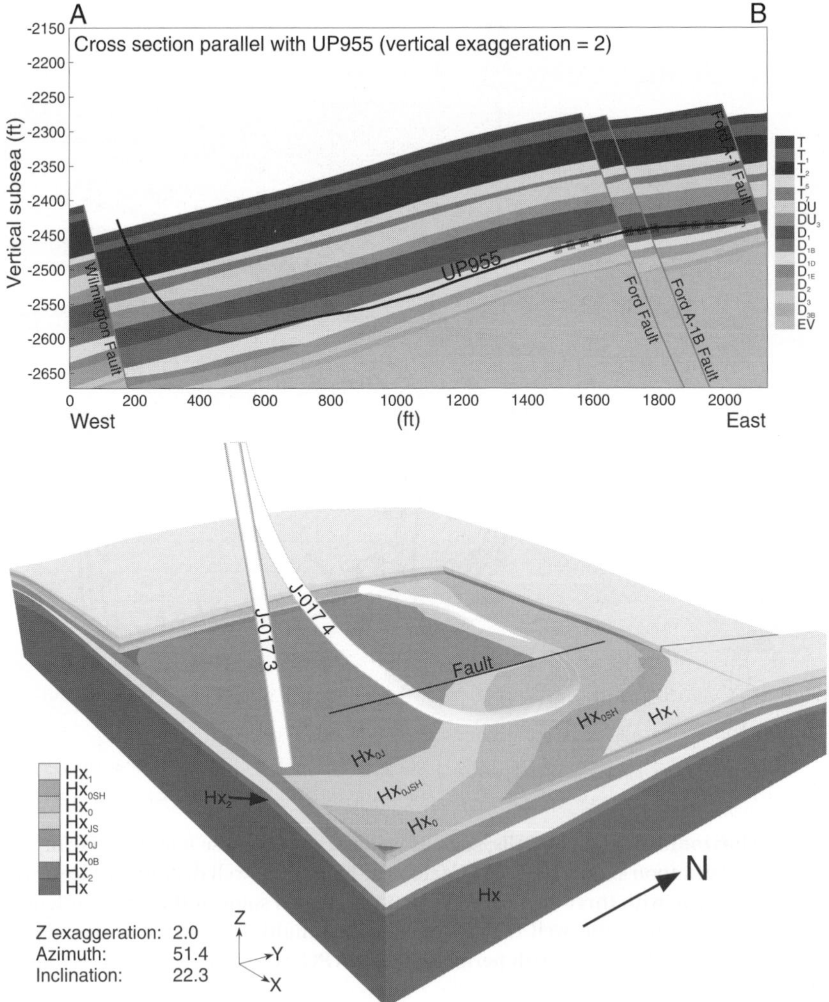

Fig. 2.38. (A) Schematic cross-section, showing the horizontal well course of well UP 955 through the thin D1 sand, and across two faults, Long Beach Unit, Wilmington field, California. (B) Horizontal well trace in the Hxo sandstone of the Long Beach Unit. Note the 180° curved well path across a fault. After Clarke and Phillips (2003). (Reprinted with permission of AAPG, whose permission is required for further use.)

2.5.1 Conventional logs

It is expensive and time-consuming to obtain cores, so drilling engineers and exploration/field managers prefer to avoid that cost, if possible. The most common method of determining subsurface rock and fluid properties is from conventional wireline logs. A variety of tools that measure different properties of

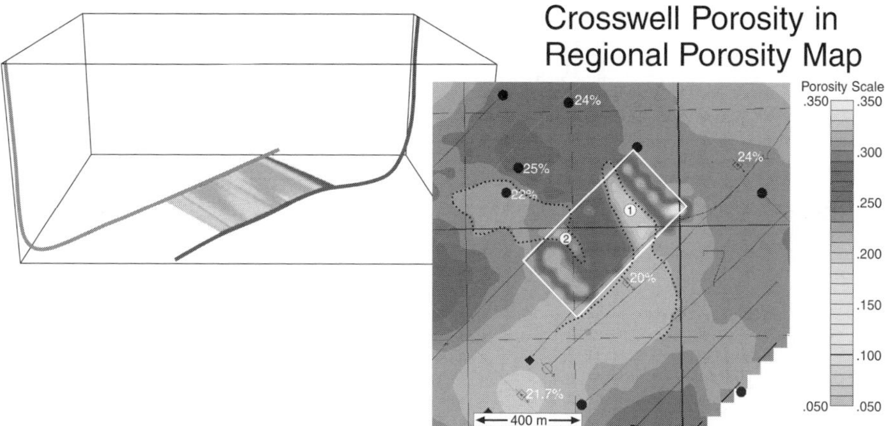

Fig. 2.39. A carbonate reservoir example that is the first stage, or baseline, of a CO_2 injection monitoring project using horizontal wells. Crosswell seismic attributes are shown in (A) and a crosswell porosity map superimposed on a regional porosity map is shown in (B). (Diagram courtesy of B. Marion and Z-Seis Corporation.)

the rock and fluid can be attached to the end of the drill string, lowered to the bottom of the hole, and retrieved at a steady rate, thereby obtaining a continuous record of the various properties being measured (Fig. 2.40A). Table 2.1 provides a list of the conventional types of well logs and the sorts of measurements they make. These tools measure static or dynamic reservoir properties at an instant in time. On a well log (Fig. 2.36), the vertical axis is the driller's depth (i.e., the depth below the ground's elevation, usually measured from the surveyed elevation of the Kelley Bushing on the drill-rig floor). The log shown in Fig. 2.36 exhibits a gamma-ray curve on the left track, density–porosity and neutron–porosity curves on the middle track, and a dipmeter (tadpole) plot on the right track.

Gamma-ray and SP logs provide indicators of lithology (Fig. 2.41A). On a typical gamma-ray log, the logging tool measures natural gamma radiation of the rock formation. Higher gamma-ray counts indicate the presence of shale, because shale constituents, including clay minerals, K-spar, and organic material, emit natural gamma radiation. With the exception of arkoses (K-feldspar rich), sandstones contain fewer, if any, of these components and therefore emit less radiation and have a lower gamma-ray count. Similar trends hold for SP logs (Fig. 2.41A).

Density logs measure the density of the formation, which is generally in the range of 2.00–3.00 g/cm^3 (Fig. 2.41B). Variations in formation density reflect the mineral composition of the rock, the porosity of the rock, and the fluids contained within the rock's pore spaces. Usually, a mineral density, such as that

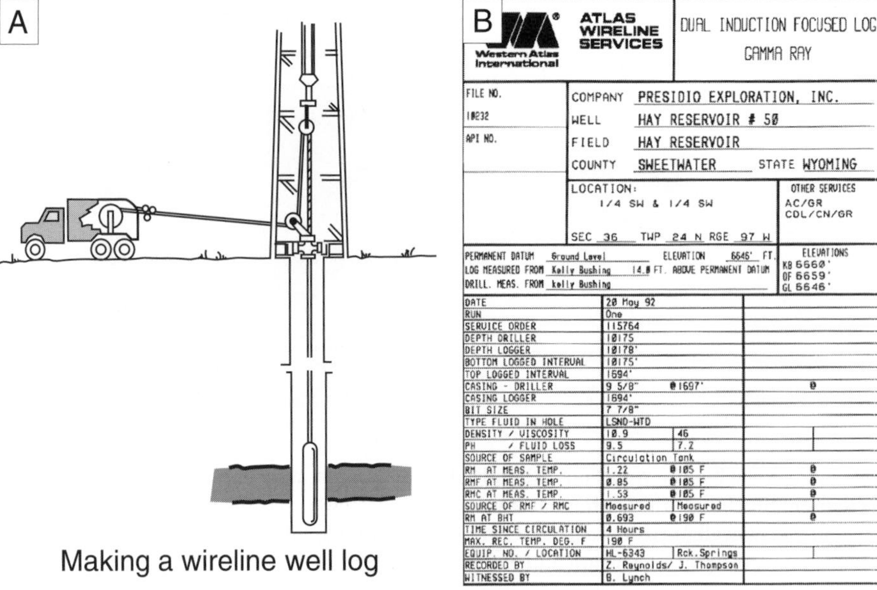

Fig. 2.40. (A) Electrical recording tools are sent down the wellbore to the bottom and then returned up the hole at a slow, constant rate on a wireline–winch assembly. (B) At the top of each well log (whether it be a paper or electronic copy) is a "Log Header" that provides basic information about the well, including the company that drilled and logged the well, the location of the well, the types of logs that were run, logging conditions, and depths of the logging runs.

Table 2.1 Conventional well logs and their applications

Type	Measures	Vertical resolution (m)	Uses
Electrical log-spontaneous potential, resistivity	Electrical properties of rocks and fluids	1.5–2.0	Calculation of S_W, fluids; correlation
Gamma-ray log	Natural radioactivity in rocks	0.2–0.3	Lithology indicator; shaliness; correlation
Neutron log	Hydrogen atom density	0.4	Porosity, gas
Density log	Rock density (includes pore space)	0.4	Porosity; sometimes lithology
Sonic log	Velocity of sound waves through rock; measured as transit time	0.6	Porosity; sometimes lithology
Caliper log	Size of well bore		Calibration of other logs; sometimes lithology; stress orientations
Dipmeter log	Orientation of subsurface rocks	0.01	Structure and depositional environment

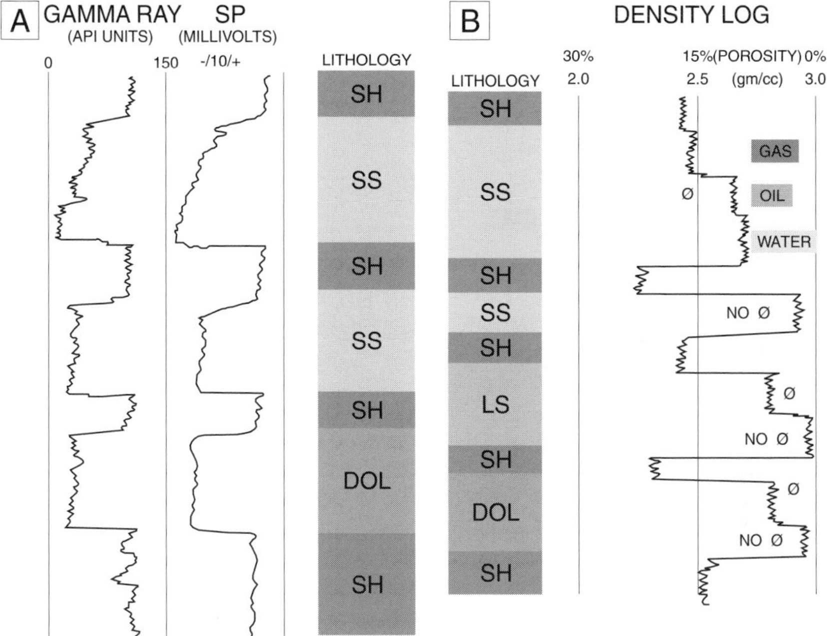

Fig. 2.41. (A) Figure showing gamma-ray and SP (self-potential or spontaneous potential) logs through an interbedded interval of sandstone (SS), shale (SH), limestone (LS), and dolomite (DOL). The effects of lithology are seen on the two logs in the picture. Gamma-ray counts (in API units of measure) are high for shales and low for sandstones and dolomite. The same is true for the SP log curve. Thus, both logs are lithology indicators. The gamma-ray log has a better resolving power than the SP log for detecting beds of a certain thickness, so the gamma-ray log provides a better characterization of lithology and bed thickness than does the SP log. (B) The density log measures the density of the rock and its contained fluids. Thus, the density log is sometimes referred to as a porosity log. Different fluids, particularly gas, can have a pronounced effect on the density measurement, as is shown on the diagram. Limestones and dolomites tend to have a higher density than do sandstones of the same porosity.

of quartz, with a density of 2.67 g/cm^3, is chosen to be the standard matrix or rock density, and variations from that value are attributed to porosity and fluid content (Fig. 2.41B).

Sonic velocities are a function of the rock's mineralogy, the amount of pore space in the rock, and the types of fluids in the pore spaces. The log measurement is called "sonic transit time", and it is measured in seconds per foot (i.e., the sound wave takes X seconds to travel 1 foot). The reciprocal of sonic transit time is rock velocity, which is a measure of the amount of rock, in feet or meters, through which the sound wave travels over a 1-second time span. Sonic logs measure the velocity of waves that pass through the rock (Fig. 2.42A). The sonic characteristics of a rock respond to the same components as does the den-

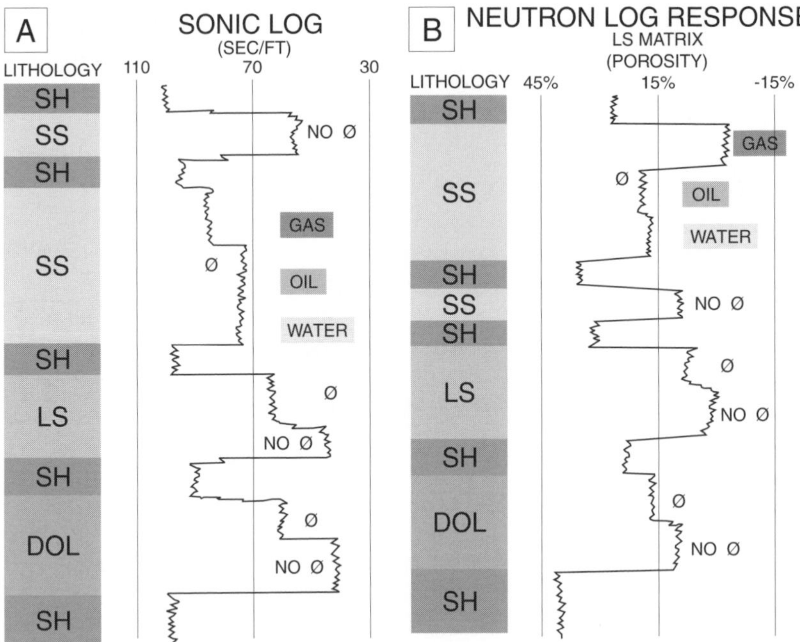

Fig. 2.42. (A) Sonic logs measure the velocity of sound waves as they travel through a rock. Note again the influence of gas on the sonic-log properties of the sandstone (SS) in the figure. Shales (SH) generally exhibit a large sonic transit time because it takes a long time for sonic waves to travel through them (i.e., they have a low sonic velocity). (B) The neutron log measures concentrations of hydrogen (water and hydrocarbons) in pores, expressed in units of percent porosity, and is also an indicator of gas content.

sity log; thus, denser minerals normally exhibit higher sonic velocities, whereas pore spaces and contained fluids reduce the velocities (Fig. 2.42A).

Neutron logs measure the concentrations of hydrogen atoms (held in water and hydrocarbons) that are present in formation pores, and records them in units of percent porosity (Fig. 2.42B). A combination neutron–porosity density–porosity log plots both curves together (Fig. 2.36). Such information is useful in determining lithology and also the presence of gas-bearing rocks. The gas effect is represented by an increase in density porosity and a decrease in neutron porosity, such that the two curves cross each other (Asquith, 1982). The twofold reason for this "crossover" is that: (1) gas is lighter than oil or water, so the density log records an anomalously high porosity, and (2) gas has a lower concentration of hydrogen atoms than does water or oil, so the neutron log's reading is anomalously low.

Subsurface geologic analysis relies heavily on well logs, and in particular on the gamma-ray log, because it indicates lithology. Not only can sandstones and shales be differentiated, but upward variations in "shaliness" or "sandi-

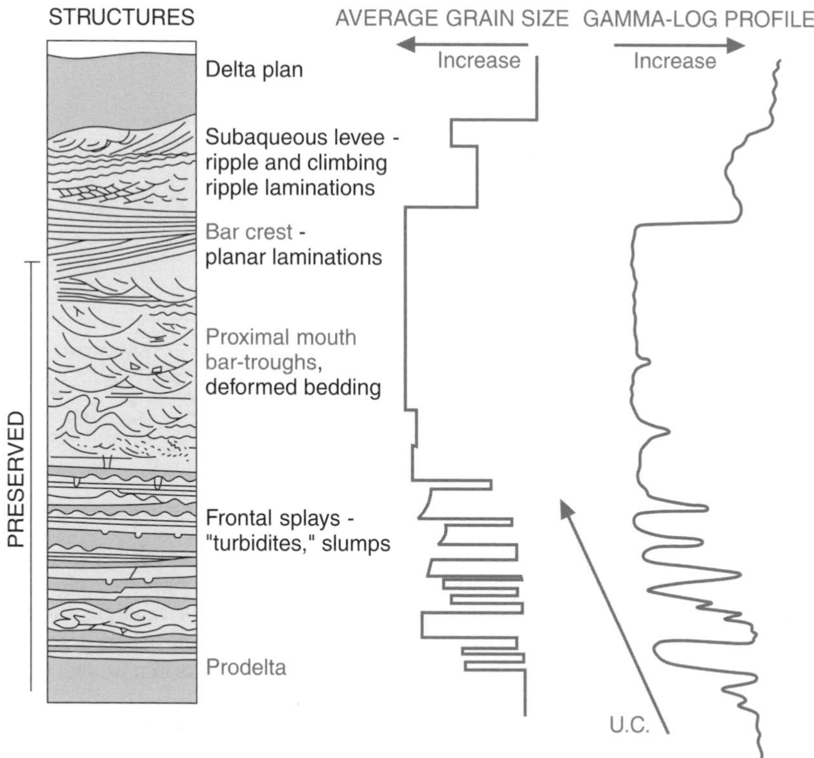

Fig. 2.43. Vertical profiles from well logs or from core or outcrop descriptions indicate changes in sedimentary facies that resulted from different depositional processes and environments. In this example, extending upward from the prodelta through the distributary mouth bar, grain size and the abundance of sand relative to shale (clay) both increase upward as a result of increased energy of the sedimentary environment. Upward-coarsening (cleaning) sequences are common in prograding shorelines. Various sedimentary structures are shown by different symbols. Modified from Galloway and Hobday (1996).

ness" can also be determined by the shape of the gamma-ray curve (Fig. 2.43). Specific shapes of the gamma-ray curve (and to a lesser extent, the SP curve) have been classified (Fig. 2.44) and often are related to the deposits of different sedimentary environments. For example, bell-shaped (fining-upward) intervals often are considered to represent deposition within a channel, where the lower fill is relatively coarse-grained and the upper fill is finer-grained. Funnel-shaped (coarsening/cleaning-upward) intervals are interpreted to represent upward shallowing of water and an increase in energy levels in the depositional environment. However, one must be cautious when using well logs to make genetic interpretations of vertical stratigraphy, because rarely is there a single depositional process that results in a unique stratigraphy. Also, the same type

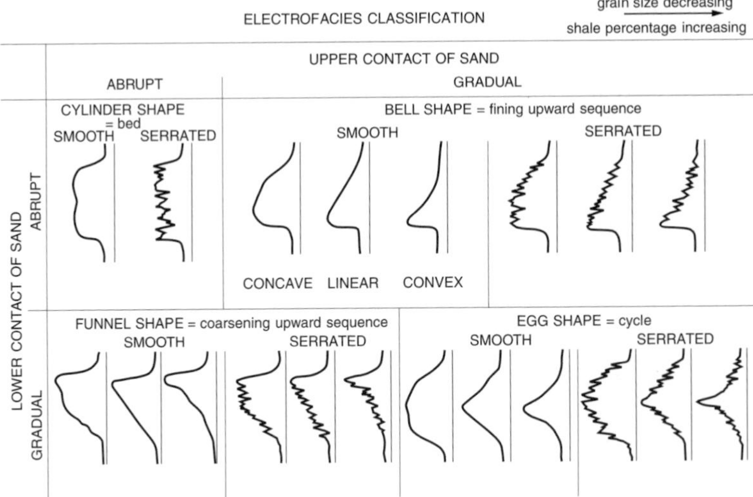

Fig. 2.44. Diagram showing the generalized range of gamma-ray or SP log responses. It represents an "electrofacies" classification of curve shapes. These responses often are interpreted in terms of rock types and depositional environments. For example, the vertical rock sequence shown in the previous figure would be classed as a funnel-shaped, serrated to smooth shape, and could be representative of deposition of the rocks in a prodelta-to-delta environment. (Source of figure is unknown.)

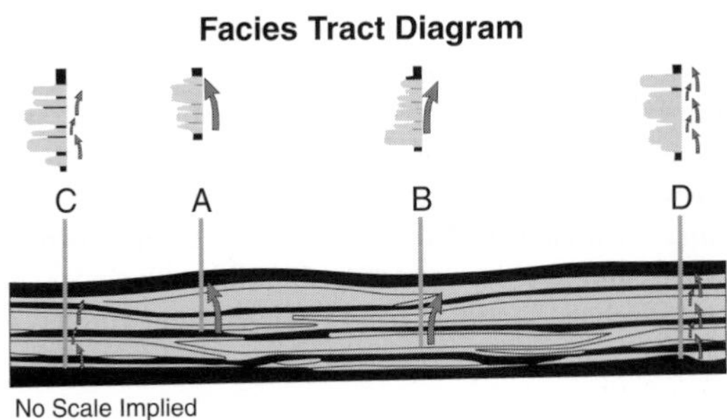

Fig. 2.45. Schematic illustration of compensation-style bedding. When sand is deposited on the seafloor, it creates a topographic high on the seafloor such that younger sand flows will seek the adjacent lower areas on the seafloor and be deposited there. The result is a series of lens-shaped sand bodies (yellow/bright-gray) separated by shale beds (black) that exhibit this compensation style of bedding. Four pseudo-well logs (A, B, C, and D) show the vertical sequence at each of the four locations. This example points to the difficulty in interpreting one-dimensional data (a well log vertical section) of a 3D object (a sediment body). Arrows point to the thinning- or thickening-upward nature of the bedding. Modified from Mutti (1985).

of deposit can present different well-log signatures, depending on where in the deposit the well is drilled and the log is run (Fig. 2.45).

2.5.2 Unconventional logs

2.5.2.1 Borehole-image logs

In addition to conventional well logs, more-advanced types of logs provide specialized information. Because these logs normally are more expensive to run than conventional log suites, they are not used as frequently.

One such logging tool is the borehole-imaging tool. This logging tool (Fig. 2.46) creates an image of the borehole wall by mapping the wall's resistivity using an array of small, pad-mounted button electrodes (Bourke et al., 1989). These buttons examine successive lateral increments of the formation and also small vertical increments (every 2.5 mm or 0.1 in), while the tool is pulled uphole at a constant rate. Figure 2.47 illustrates an inclined plane as it would appear in 3D space (the cylinder represents the borehole wall). The image is viewed in 2D by flattening it onto a plane such as paper or a computer screen, so that the inclined plane appears as a sinusoidal waveform. The steeper the dip of the inclined plane, the higher the amplitude of the sine wave representing that surface. The inclined planes that normally are imaged are faults, fractures, and dips of bedding planes (Fig. 2.47). The angle of dip and the strike orientation of the dip of an inclined plane also are routinely measured and recorded on the borehole-image log. Measurement of orientation information is a big advantage that borehole-image logs provide over conventional logs or even cores (unless a more expensive, oriented core is obtained).

Borehole-image logs image beds at a vertical resolution of a few mm, which is much better than the resolution of conventional logs (Table 2.1). Thus, many sedimentary features can be identified in addition to faults (Fig. 2.48), such as

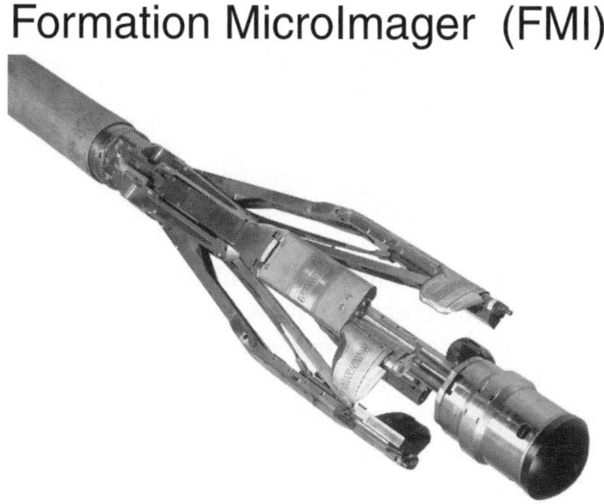

Fig. 2.46. Photograph of the FMI™ tool, showing the pads of 192 electrodes that provide a resistivity image of the wellbore wall. (Figure provided by Elizabeth Witten-Barnes. FMI is a Schlumberger trademark.)

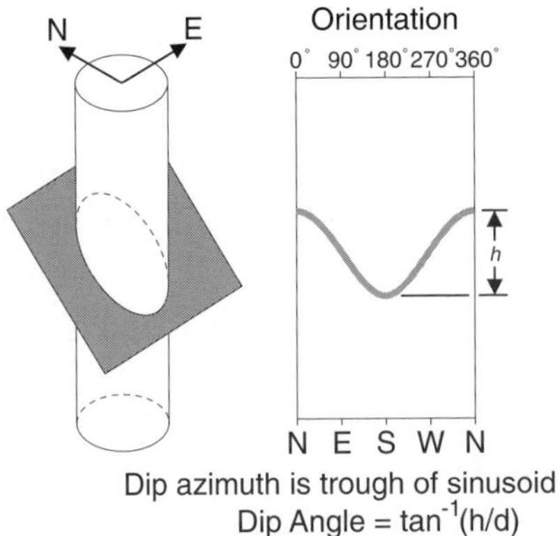

Fig. 2.47. Diagram illustrating an inclined plane as it would appear in 3D space, cutting through a cylinder (representing the borehole wall). The image is flattened for viewing on a 2D workstation screen or on paper, so the inclined plane appears as a sinusoidal surface. The steeper the dip of the inclined plane, the higher the amplitude of the sine curve (*h*) representing that surface. The orientation of the dipping plane with respect to north (azimuth) is also measured and recorded on the borehole-image log.

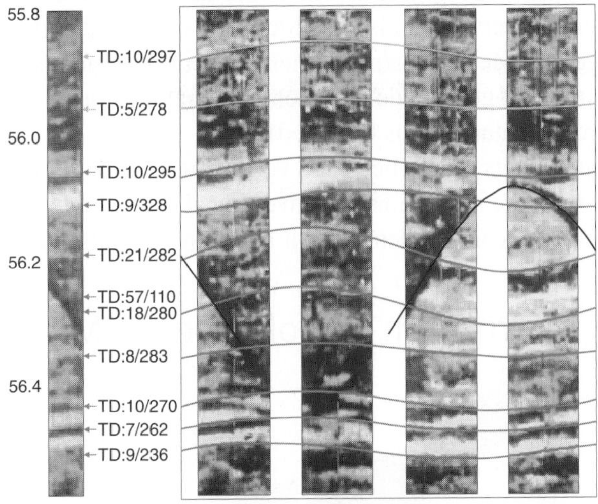

Fig. 2.48. This borehole-image log shows thin sandstones (light color) interbedded with mudstones (dark color). Tops and bases of bedding planes have been highlighted manually on a workstation and appear as sinusoidal curves (green/dark-gray and blue/black). These curves are matched to a standard set of curves from which dip magnitudes are determined. Orientations of the bedding planes are measured with a downhole compass. The notations on the far left (56.0, 56.2, etc.) are depths in meters beneath the ground surface. The notations TD (10/297, 5/278, etc.) represent the dip magnitude and the azimuth of the dipping bed, respectively. In this example, a fault, shown by the higher-amplitude sine wave with dimensions of 57° dip and 110° azimuth, crosscuts the more gently dipping sedimentary beds. (This example is from the Miocene Mt. Messenger Formation, New Zealand; Browne and Slatt, 2002.)

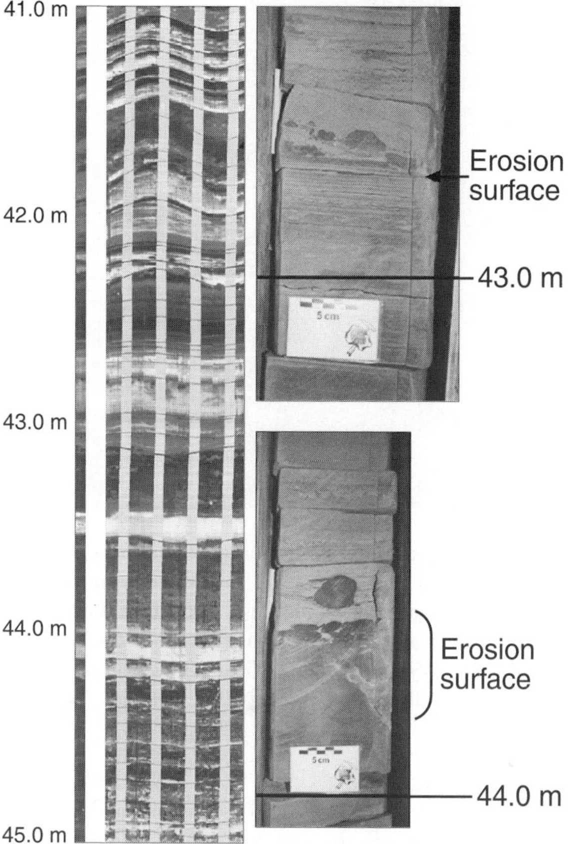

Fig. 2.49. Borehole-image log showing an unconformity (erosion) surface that separates thin-bedded strata of relatively steep dip (19–27°) from underlying, less-steeply dipping (6–8°) thin-bedded strata. A basal conglomerate bed immediately overlies the unconformity surface as shown on the core photographs. The black blobs just above the unconformity surface on the log are conglomerate clasts that can be seen in the core. The core depths differ from the log depths because no core-log correction has been applied. (This example is from the Miocene Mt. Messenger Formation, New Zealand; Browne and Slatt, 2002.)

erosional surfaces (Fig. 2.49), specific lithofacies (Fig. 2.50), internal physical and biogenic sedimentary structures (Fig. 2.51), bedding (Fig. 2.52), and stratification styles (Fig. 2.53). These features have proven useful in identifying depositional processes and environments (Fig. 2.54), from which we can predict sandstone trends, geometries, and reservoir quality at a distance from the borehole. Also, the ability to resolve thin beds using a borehole-image log has, in several instances, resulted in an upward recalculation of reserves. This has occurred with reservoir intervals that conventional-log analysis had indicated to be shaly, prior to a borehole-image log revealing the presence of low-resistivity, low-contrast, thin-bedded pay (Fig. 2.55).

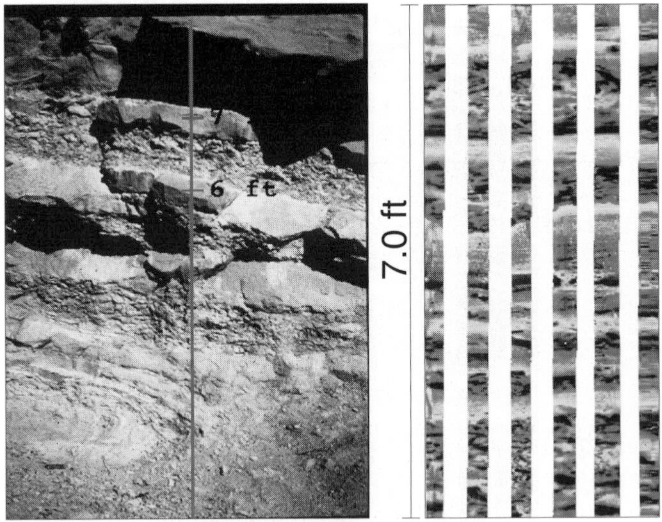

Fig. 2.50. Comparison of an outcrop of alternating turbidite sandstones and shale-clast (debrite) conglomerates with a borehole-image log of the same general stratigraphic interval from a well in a nearby gas field. Shale clasts appear on the image log as black blobs. Some erosional scour surfaces are also observable on the image log. This example is from the Cretaceous Lewis Shale, Wyoming. After Witton-Barnes et al. (2000). (Reprinted with permission of the Society of Sedimentary Geology (SEPM).)

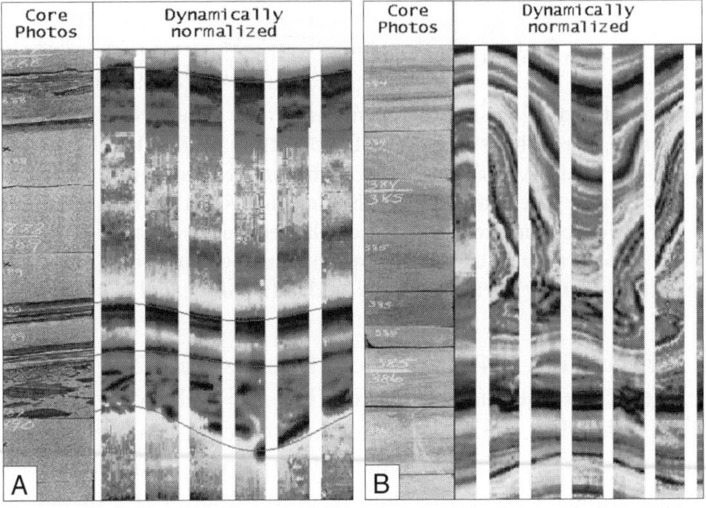

Fig. 2.51. (A) Borehole-image log illustrating an erosional surface lined with shale clasts, as calibrated to a core in the same interval. (B) Borehole-image log illustrating a contorted (slumped) bedset; the slumped beds are more easily recognized on the borehole-image log than in the core! This example is from the Cretaceous Lewis Shale, Wyoming. (Images and photographs provided by S. Goolsby.)

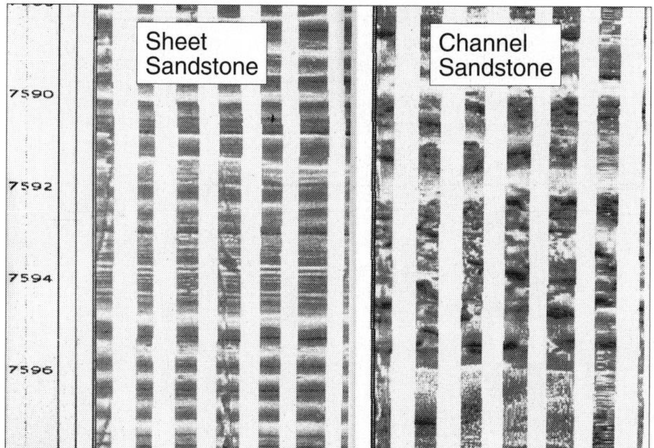

Fig. 2.52. Image (on left) showing what Lewis Shale deepwater sheet sandstones look like on a borehole-image log. The image on the right shows what Lewis Shale deepwater channel sandstones look like on a borehole-image log. The channel sandstones contain shale-clast conglomerates (clasts appear as black blobs; Fig. 2.50), which are not present in the Lewis Shale sheet sandstones. These differences can be used to differentiate deepwater sheet sandstones from channel sandstones in the Lewis Shale and elsewhere. After Witton-Barnes et al. (2000). (Reprinted with permission of the Society of Sedimentary Geology (SEPM).)

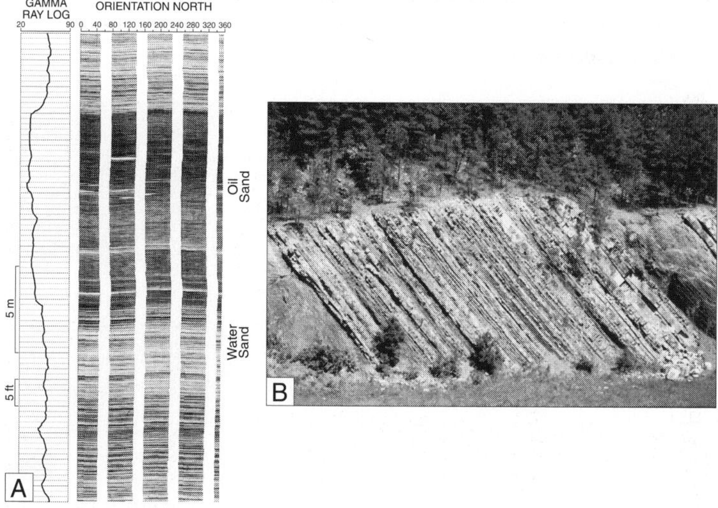

Fig. 2.53. (A) Borehole-image log from Pliocene deepwater deposits in the northern Gulf of Mexico. Beds become progressively thicker and cleaner upward, as also seen on the associated gamma-ray log (orange/black curve). The thick, dark-appearing sandstone near the top of the sequence is oil-filled. (B) An interpreted outcrop analog to this sequence is the Pennsylvanian Jackfork Sandstone, at DeGray Lake Spillway, Arkansas. This is a 27 m (80 ft) thick, thickening- and cleaning-upward stratigraphic interval (stratigraphic top is to the right). After Slatt et al. (1994).

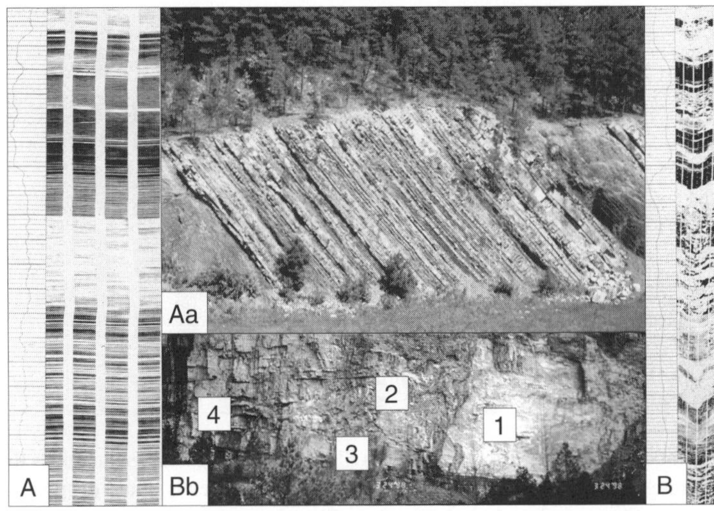

Fig. 2.54. (A) Borehole-image log of deepwater Pliocene sands from the northern Gulf of Mexico. Note the upward thickening of the beds and the uniform bedding. (Aa) Sheet sandstone outcrop analog of (A) from Jackfork Group turbidites, DeGray Lake Spillway, Arkansas. Stratigraphic top is toward the right. (B) Borehole-image log of deepwater Pliocene turbidites in the Gulf of Mexico showing irregular bedding indicative of channel strata. (Bb) Channel sandstone outcrop analog of (B) from Jackfork Group turbidites, Big Rock Quarry, Arkansas. 1 – massive sandstone of a channel thalweg; 2 – slumped and contorted channel-margin strata; 3 – erosional base of slumped channel margin; 4 – bedded levees. After Slatt et al. (1994).

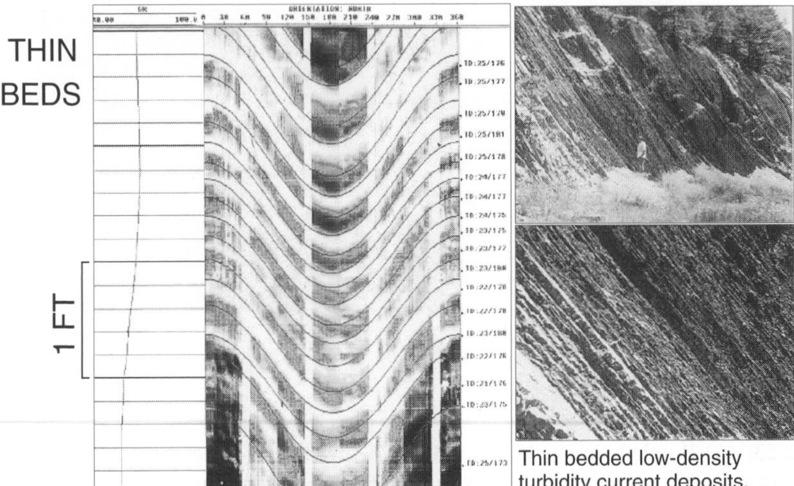

Fig. 2.55. Borehole-image log of thin-bedded, deepwater Pliocene strata from the northern Gulf of Mexico. Note the relatively smooth gamma-ray log curve on the left. Outcrop photographs are of thin-bedded turbidite strata from the Jackfork Group, DeGray Lake Spillway, Arkansas, analogous to the strata imaged on the log. After Slatt et al. (1994).

2.5.2.2 Dipmeter logs

Although dipmeter logs, or tadpole plots as they are sometimes called, are not uncommon or unconventional, they usually are obtained to gather information on subsurface structure (Fig. 2.56). A dipmeter log is a microresistivity measurement that records bed boundaries around a borehole wall, so that dip magnitudes and directions can be calculated from them. However, dip patterns also can be used to determine depositional trends and environments. Such determinations require calculating the structural trends of dip magnitude and orientation within the wellbore (usually measured from thick shale intervals) and then removing the structural dip from each measurement to determine residual dip magnitude and orientation (Fig. 2.57). Once residual dips are plotted, different trends and styles of dip patterns can be recognized and interpreted in terms of their sedimentary environments (Fig. 2.57, Fig. 2.58, Fig. 2.59 and Fig. 2.60). In one published example, dipmeter (and borehole-image) log patterns obtained from a shallow logged and cored borehole about 100 m (300 ft) inland from a coastal cliff section of Miocene turbidites revealed distinctive dip patterns for different facies. Those patterns in turn led to identification of hydrocarbon-productive facies from similar dip patterns

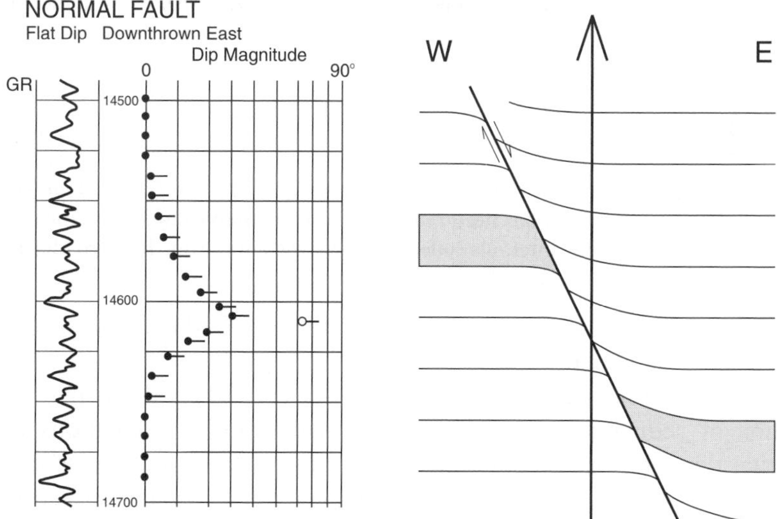

Fig. 2.56. Gamma-ray and dipmeter log from a well that penetrated a normal fault. Each "tadpole" denotes the strike and dip of strata at that particular footage in the well. The black dot or "head" of the tadpole provides the dip magnitude (see horizontal axis) from 0–90°, and the straight line or "tail" of the tadpole represents the orientation of the plane of bedding from 0–360°, with 0° being oriented to the north. Note how the dip magnitude increases as the fault is approached. The orientation of the beds does not change from a 90° azimuth.

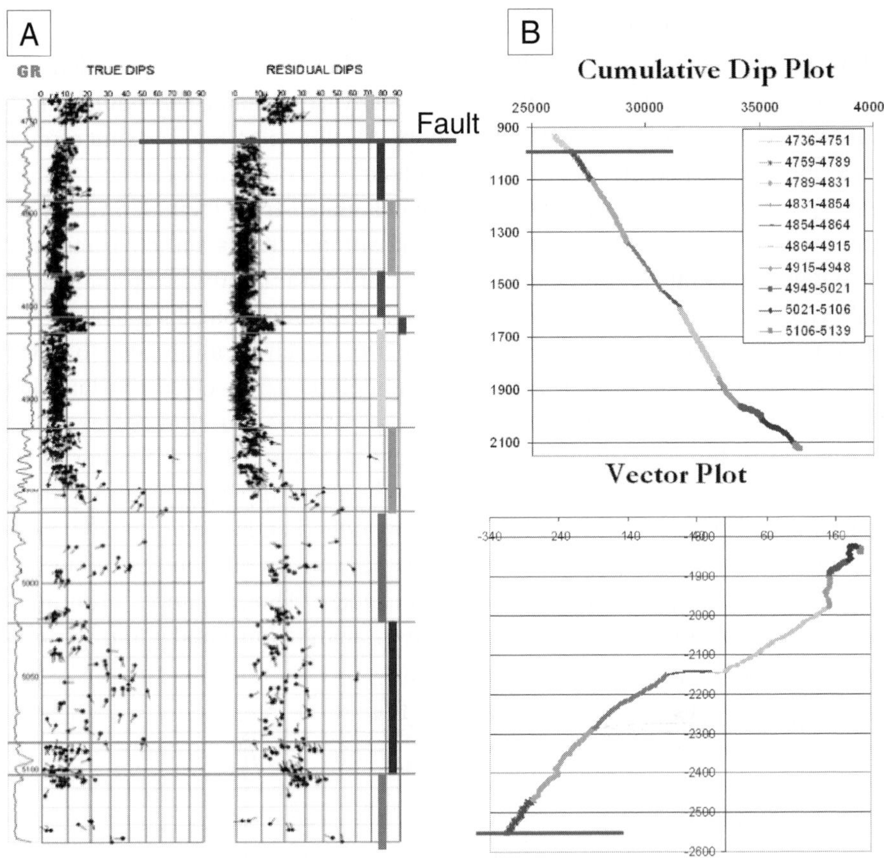

Fig. 2.57. (A) Dipmeter logs with the true measured dips (left) and the residual (sedimentary) dips after structural dip has been removed. (B) Cumulative dip and vector plots with different depths and genetic intervals coded to different colors. After Romero (2004).

in a Gulf of Mexico reservoir (Fig. 2.60 and Fig. 2.61) (Slatt et al., 1998; Clemenceau et al., 2000).

Hurley (1994) has developed techniques for correlating dip magnitude and direction of sedimentary beds, based on mathematical calculations applied to digital dipmeter or borehole-image data (Table 2.2). Cumulative dip plots (Hurley, 1994) are crossplots of dip magnitude versus evenly spaced sample numbers, which are a function of depth. Depths are converted to sample numbers for plotting, in order to account for unequally spaced depth measurements (Table 2.2). The shallowest depth is designated as number 1, and each subsequent depth is given the next highest integer. Each sample number then has its own dip magnitude associated with it. To obtain cumulative dip, each dip magnitude is added to the sum of the preceding dip magnitudes. The rationale behind this plot is that beds within a genetically-related depositional sequence

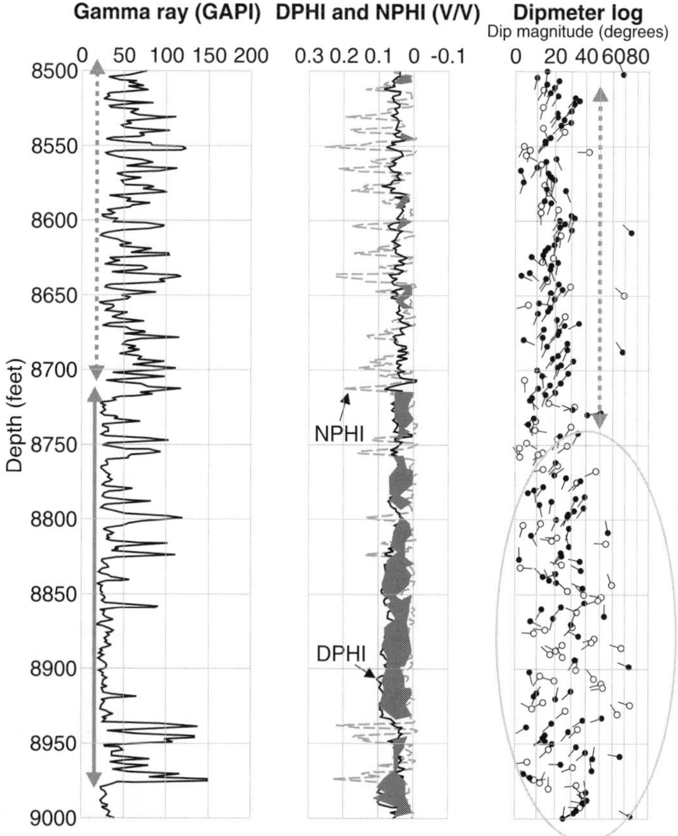

Fig. 2.58. Gamma-ray log on the left track, density–porosity (DPHI) and neutron–porosity (NPHI) logs in the middle track, and a dipmeter log on the right track. The red/dark-gray areas in the middle track represent the gas-bearing interval in the formation (density–neutron–porosity crossover). The dipmeter log shows two patterns. The upper half of the well has a relatively consistent dip pattern, suggestive of a deepwater sheet–sandstone interval. The lower (circled) interval exhibits more diversity in its dip pattern, suggestive of a deepwater channel–sandstone interval. After Romero (2004).

will exhibit uniform dip magnitudes. Therefore, genetically related strata can be identified as straight-line segments on a cumulative dip plot. Abrupt or subtle changes in dip magnitude indicate the presence of a fault, unconformity, or genetic difference in stratigraphic intervals (Fig. 2.57).

Vector-azimuth plots (Hurley, 1994) use the azimuthal dip orientation or directional data (Table 2.2; Fig. 2.57). A vector-azimuth plot is a crossplot of the cosine and sine of the dip direction (azimuth). These plots are designed to mimic the borehole trajectory as it would appear in plan view. To create such a plot, the azimuth data are converted into radians, and the sine and cosine functions are applied to each azimuth reading. Next, cumulative functions are assigned to

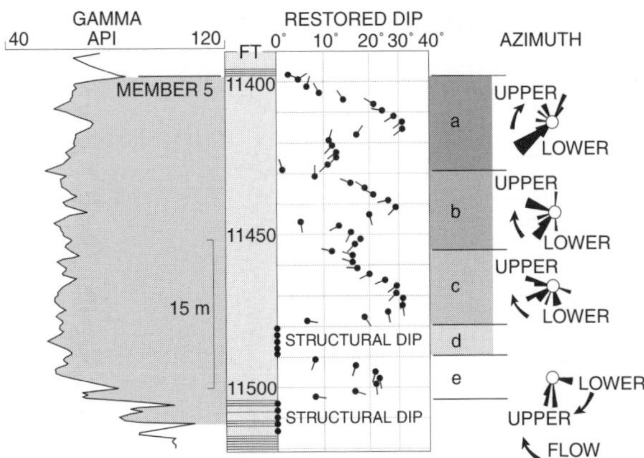

Fig. 2.59. Gamma-ray and dipmeter logs of Member 5 of the A3 sandstone, Indian Draw field, New Mexico, showing the variability in dip patterns, which provided a means of subdividing the sandstone into five separate intervals (a–e). Reservoir fluid flow will be very complex in this sandstone because of the variable dip magnitudes and orientations of beds. After Phillips (1987). (Reprinted with permission of the SEPM.)

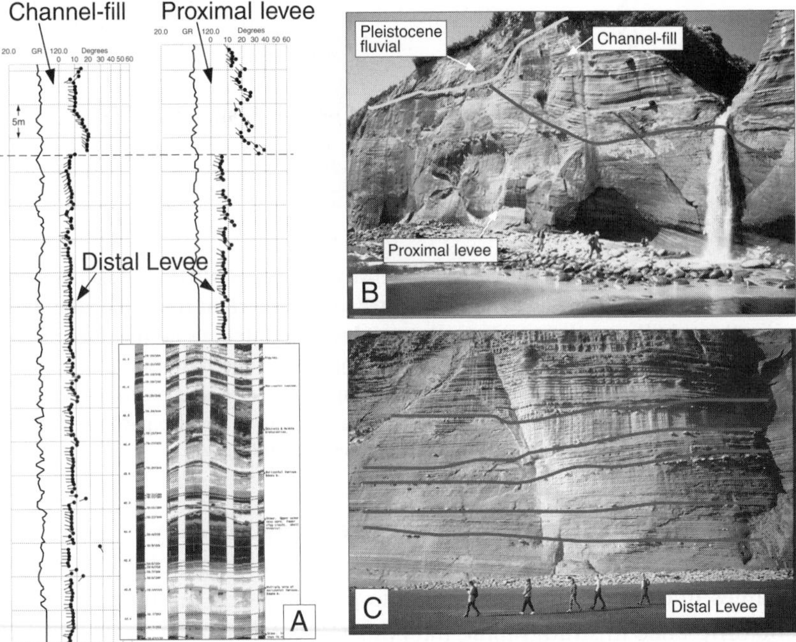

Fig. 2.60. (A) Two dipmeter logs taken in wellbores drilled behind the outcrop shown in (B). Three dip patterns are illustrated. Dips of channel-fill strata decrease from the base toward the top, as is shown in (B). Proximal-levee dip patterns are relatively high angle, as is seen in outcrop (B). Distal-levee dip patterns have a lower dip magnitude and more uniform orientation, as shown in outcrop (C). The inset is a borehole-image log of the contact between channel fill and the distal levee. After Browne and Slatt (2002).

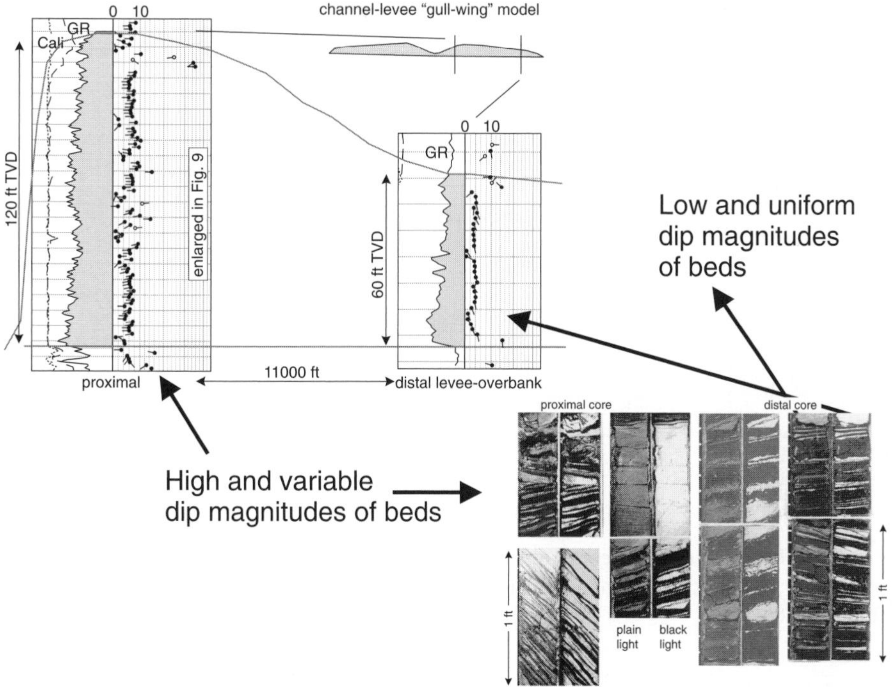

Fig. 2.61. Gamma-ray and borehole-image logs of proximal- and distal-levee beds in a well from the Ram-Powell L Sand field, Gulf of Mexico. The proximal-levee beds exhibit a relatively high angle of dip, as also is seen in the proximal-levee core (inset). Distal-levee dip patterns have a lower magnitude, as also is seen in the core. These patterns are the same as those observed in the New Zealand outcrop (Fig. 2.60). After Clemenceau et al. (2000). (Reprinted with permission of the Society for Sedimentary Geology (SEPM).)

Table 2.2 Calculations of cumulative dip magnitude and direction

Sample	Depth (ft)	Dip (°)n	Cumulative dip (°)	Dip direction (°)
1	3,767	2	2	257
2	3,775	6	8	221
3	3,776	5	13	240
4	3,782	4	17	247
5	3,791	4	21	234
6	3,793	3	24	226
7	3,797	5	29	230

both the "sine" and the "cosine" columns, and the resulting digits are plotted as a scatter plot, with the cosine on the x-axis and sine on the y-axis.

Cumulative-dip and vector-azimuth plots aid in the identification of "dip domains", which are genetically related stratigraphic intervals defined on the ba-

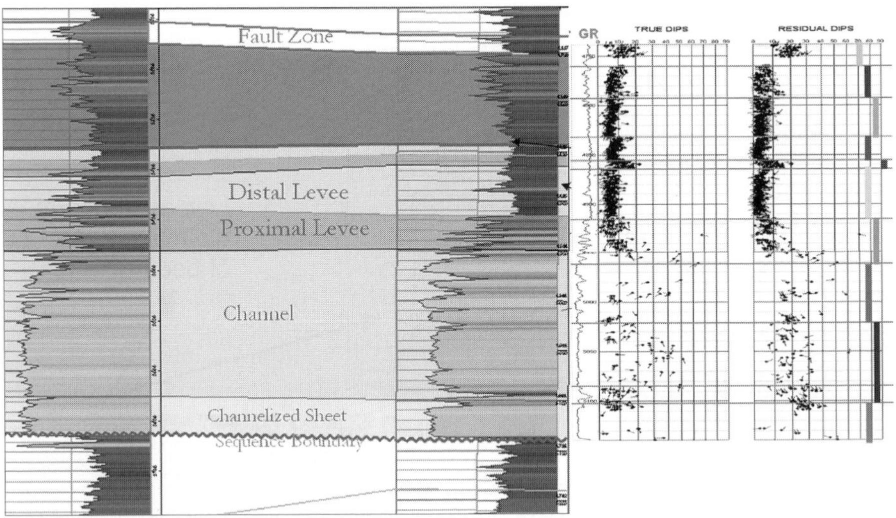

Fig. 2.62. Interpretation of components of a lowstand-to-highstand depositional sequence within the Jackfork Group of southeastern Oklahoma, based on dipmeter patterns. After Romero (2004).

sis of their dip characteristics (Fig. 2.57). Inflection points in the plots indicate boundaries of dip domains, and represent faults, unconformities, or major facies changes. The boundaries are determined by anomalies in both of the dip plots and by examining the lithology associated with the inflection point. Figure 2.57 illustrates how subtle changes in dip domains can be recognized by a combination of the two plots. These dip domains could then be interpreted in terms of sedimentary facies and depositional environments and processes (Fig. 2.62).

2.5.2.3 Nuclear magnetic resonance (NMR) logs

The basic theory behind nuclear magnetic resonance logging is as follows. Protons in hydrogen atoms within pore spaces of a rock react to an induced magnetic field by precessing at characteristic frequencies and by aligning themselves with the induced field. When the induced magnetic field is switched off, the protons respond by realigning themselves with the Earth's natural magnetic field. The process of realignment or relaxation is controlled by pulse sequences and amplitudes generated in the logging tool. The strength and rate of alignment are proportional to the volume of hydrogen atoms, the amount of water, and the effective size of the pore in which the protons are located. The rate of relaxation is modeled as a sum of exponentials whose characteristic time is inversely proportional to the pore size. The amplitude at any relaxation time, therefore, corresponds to the contribution made by the volume of protons existing in that pore size. The sum of all contributions is a measure of porosity. Thus, this distribution of relaxation times, referred to as a T2 distribution, can be interpreted as

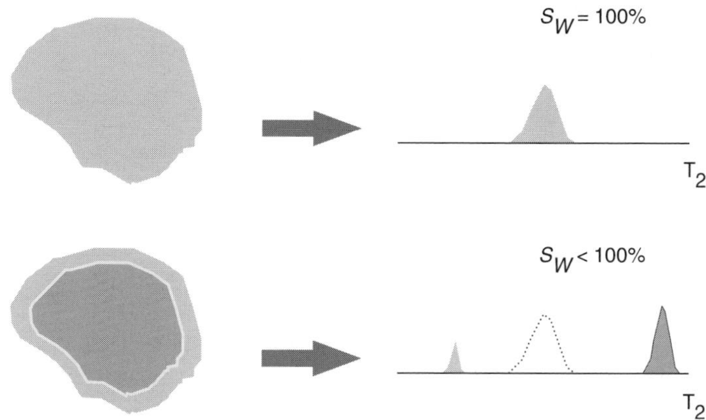

Fig. 2.63. Relaxation times T2 for oil and water. Water in a pore space is shown in blue/bright-gray, and a water-wet pore, with oil filling the interior of the pore space, is shown in green/dark-gray. The amplitudes for T2 show that T2 is much greater (longer) for oil than for water. (Diagram provided by C. Sondergeld.)

a pore-size frequency distribution, which yields insights into reservoir quality and producibility.

Interpretations of these distributions in terms of free and bound water have helped geoscientists develop empirical methods for predicting permeability from NMR logs (Coates et al., 1999). Additionally, changing the fluids to oil or gas alters the relaxation times significantly.

For example, for two equal-size pores, one filled with water and the other filled with oil in a water-wet pore, T2 for oil is much greater than T2 for water (Fig. 2.63). Fluids within a higher-permeability rock exhibit a longer T2 than do fluids in a lower-permeability rock, thereby suggesting the presence of hydrocarbons (Fig. 2.64). T2 distributions differ among clay-bound and capillary-bound (irreducible) water, and between producible water and hydrocarbons, thereby allowing us to differentiate them in a reservoir rock (Fig. 2.65).

Nuclear magnetic resonance can be measured both by a downhole logging tool and in the laboratory. In fact, to improve the accuracy of determinations of rock and fluid properties in a wellbore, it is best to calibrate the log response by first measuring the same properties on core samples.

Figure 2.66 shows an example of a well log suite that includes an NMR track. Producible and nonproducible fluids can be differentiated on the basis of T2 distribution. This figure's example is from a thin-bedded, low-resistivity, low-contrast pay interval. NMR logging has proven to be particularly useful in differentiating fluids within such thin-bedded strata.

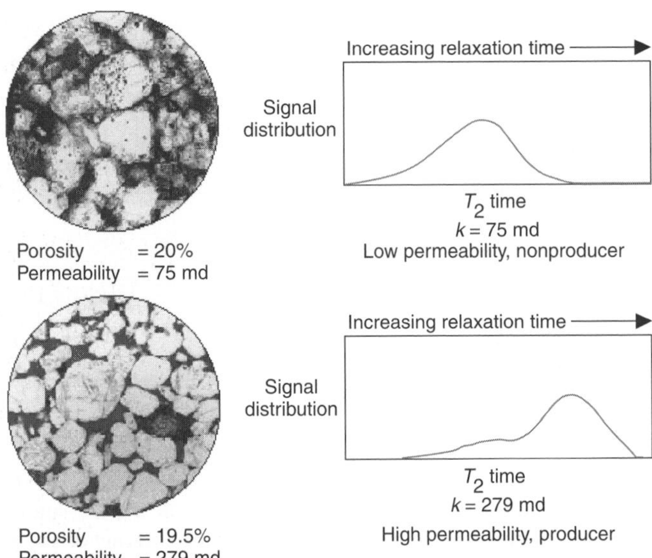

Fig. 2.64. Figure illustrating the contrasting magnetic-resonance data for two rock samples, each with about 20% porosity. The bottom sample, with permeability of 279 md, shows most of the magnetic-resonance signal occurring with a large pore diameter and free fluid that is producible. The top sample, with permeability of 7.5 md, shows almost all the magnetic resonance signal occurring as bound water. (Figure provided by C. Sondergeld.)

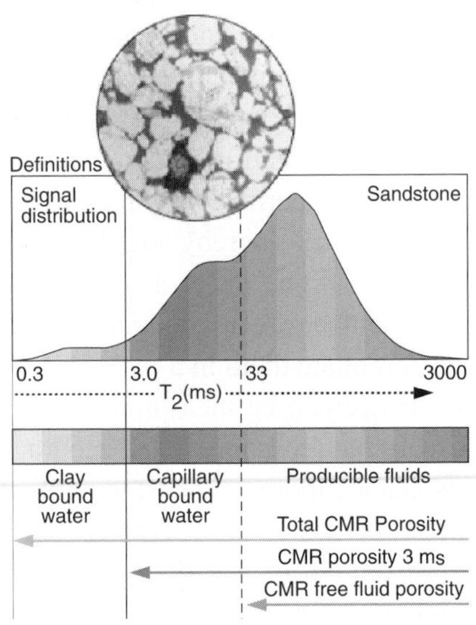

Fig. 2.65. Frequency distribution of T2 for clay-bound and capillary-bound water and producible fluids. The different fluids can be identified by this T2 distribution. (Figure courtesy of C. Sondergeld.)

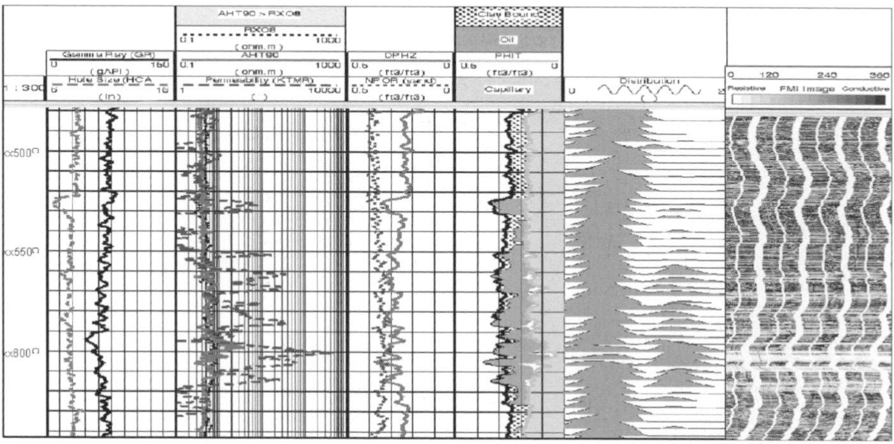

Fig. 2.66. Figure showing a log presentation for a very thin-bedded sand-mud interval, as shown by the borehole-image log on Track 6. Track 5 is the magnetic-resonance log. The continuous green band to the left of Track 5 indicates bound water. The segments of green/dark-gray to the right in the track represent producible fluid. Track 4 shows in green/dark-gray the intervals that would be expected to produce oil, based on this magnetic-resonance-tool interpretation. The thin-bedded interval was perforated and tested at more than 1000 BOPD!

2.6 Summary

There are many tools and techniques for characterizing oil and gas reservoirs. Seismic reflection techniques include conventional 2D and 3D seismic, 4D time-lapse seismic, multi-component seismic, crosswell seismic, and advanced processing techniques to enhance resolution and bed detection. These techniques are constantly being improved. Drilling and coring a well provides the "ground truth" for seismic interpretation. Rock formations are directly sampled by cuttings and by core, and indirectly characterized with a variety of well logs. More sophisticated logs are also available, including borehole image logs for imaging the borehole wall and measuring stratigraphic and structural features, and nuclear magnetic resonance logs for evaluating fluids and rock permeability. To maximize characterization, as many of these tools as possible should be employed. It is often less expensive to utilize a wide variety of tools, which measure reservoir properties at different scales, than to drill one or two dry holes.

Chapter 3

Basic sedimentary rock properties

3.1 Introduction

In this chapter, the basic properties of sedimentary rocks are discussed: their texture (grain size), sedimentary structures, and composition. We need to be able to recognize and characterize these features in cores and borehole-image logs of reservoir formations, so that we can use them to predict the reservoir's external geometry and internal architecture, the reservoir rock's orientation and trend, and potential interactions between reservoir fluids and reservoir rock. The subject of rock properties is very comprehensive, and this chapter will simply introduce the subject. For a more in-depth treatment, the reader is referred to numerous textbooks on the subject.

3.2 Classification and properties of sediments and sedimentary rocks

The Glossary of Geology (Bates and Jackson, 1980) defines sediment as: Solid fragmental material that originates from weathering of rocks and is transported or deposited by air, water, or ice, OR that accumulates by other natural agents, such as chemical precipitation from solution or secretion by organisms, and that forms in layers on the Earth's surface at ordinary temperatures in a loose, unconsolidated form. Sedimentary rock is defined as: A rock resulting from the consolidation of loose sediment that has accumulated in layers.

From this classification, three groups of sediments and sedimentary rocks can be defined: siliciclastic material (fragmental materials that originate from weathering), biogenic material (material originating from secretion by organisms), and chemical material (material produced by chemical precipitation).

The origins, formative processes, and depositional environments of sediments are shown schematically in Fig. 3.1. Sandstones and shales (and conglomerates) are collectively called siliciclastic sedimentary rocks because they are derived by weathering of pre-existing rocks into fragments of various sizes. Weathering normally occurs in mountainous regions where rain, snow, and ice

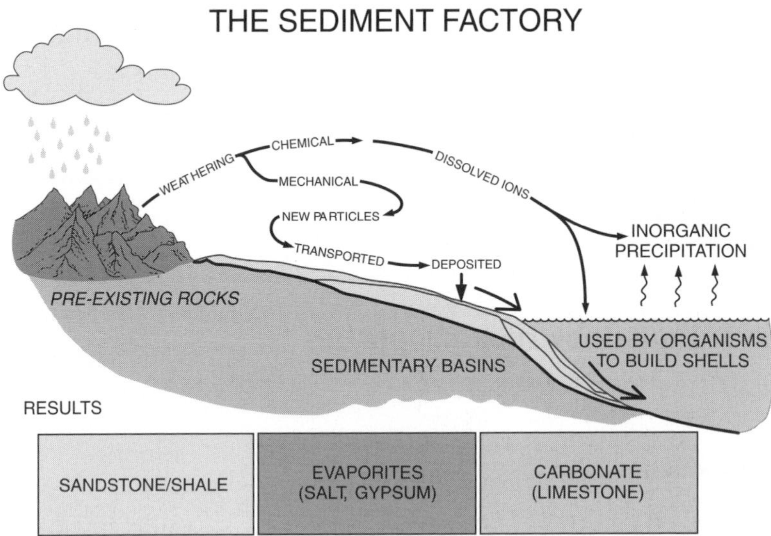

Fig. 3.1. The three common groups of sedimentary rocks and how they form. Details are provided in the text. (Figure provided by T. Cross.)

can react with the rock to physically fragment it. Fragments are transported and deposited in various environments. As new deposits occur, older material becomes buried under the new sediments. Upon burial, sediments become lithified into a conglomerate, sandstone, or shale, depending upon the size of the individual fragments or grains.

Evaporites form by inorganic precipitation of elements from solution, such as might occur in arid basins or along shorelines where saltwater evaporates and minerals such as salt (halite) and gypsum precipitate. Organisms in the oceans form calcium carbonate shells, or tests, that ultimately go into making biogenic sediments. Most organisms secrete the minerals aragonite or calcite to form their shells, and when the organisms die, their shells fall to the ocean floor and accumulate. With time, the shell-containing sediments lithify, and the resulting rock is limestone.

More details of the transport and depositional processes of sediment are presented below and in the chapters devoted to the different reservoir types.

3.2.1 Siliciclastic sediments and sedimentary rocks

Siliciclastic sediments are produced by weathering processes that break down preexisting rock into smaller particles (Fig. 3.2). This breakdown of rock into constituent particles is accelerated by the rock's reaction to water or ice. For example, rain falling on rock over time loosens the rock until particles break off and fall to the ground surface. At high elevations, in winter, the expansion

Fig. 3.2. Example of physical weathering of a granite to produce sedimentary particles of quartz, feldspar, and mica. (A) Granitic hillside in southern Arizona. (B) Roadside outcrop showing the cross-sectional characteristics of the granite. (C) Upon close inspection, the rock is very soft and crumbles easily into individual grains. (D) Pile of loose grains that have weathered out of the rock and been deposited at its base.

of water as it freezes in cracks in the rock further widens those cracks and breaks the rock into particles. Glaciers also can grind rock into smaller particles. In desert areas, a wide fluctuation between daytime and nighttime temperatures generates expansion–contraction stresses on rock, weakening it and breaking it into particles.

Once rock particles form, they are transported away from their site of origin by water (e.g., rivers), wind, and/or ice (e.g., glaciers) (Fig. 3.1). Eventually, the transported rock particles reach a final resting site within a "depositional environment" (Fig. 3.3). In that environment, new layers of particles are added repeatedly, one after another, and the underlying deposits are buried deeper and deeper. When they have been buried over geologic time, these deposits of siliciclastic material ultimately may become oil and gas reservoirs, as well as source rocks and seal rocks.

3.2.1.1 Texture

The primary basis for classification of siliciclastic sediments and sedimentary rocks is the size of the constituent grains or particles (Table 3.1). If one considers a grain as having long-, short-, and intermediate-size axes (diameters), the term particle size refers to the intermediate diameter. Sand-size particles can be further subdivided (Table 3.2). Sometimes an even further subdivision is used.

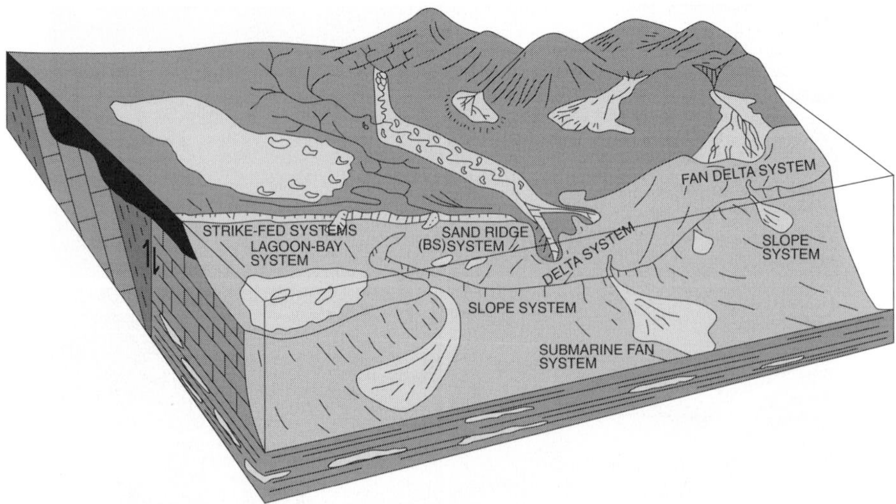

Fig. 3.3. Major clastic sedimentary environments. Details are provided in individual chapters on reservoir types. Modified from Fisher and Brown (1984). (Reprinted with permission of the Texas Bureau of Economic Geology.)

Table 3.1 Grain-size classification of sedimentary particles

Name of sediment	(Sedimentary rock type)	Particle size (mm)
Gravel	(Conglomerate)	>2
Sand	(Sandstone)	0.063–2
Silt	(Siltstone)	0.004–0.063
Clay	(Claystone)	<0.004
Mud	(Mudstone)	<0.063
Shale	Although it is a commonly used term, shale normally refers to a mudstone that has a specific fabric called fissility	<0.063

Table 3.2 Subdivisions of sand particles, by size

Name	Size (mm)
Very coarse sand	1–2
Coarse sand	0.5–1
Medium sand	0.25–0.5
Fine sand	0.125–0.25
Very fine sand	0.063–0.125

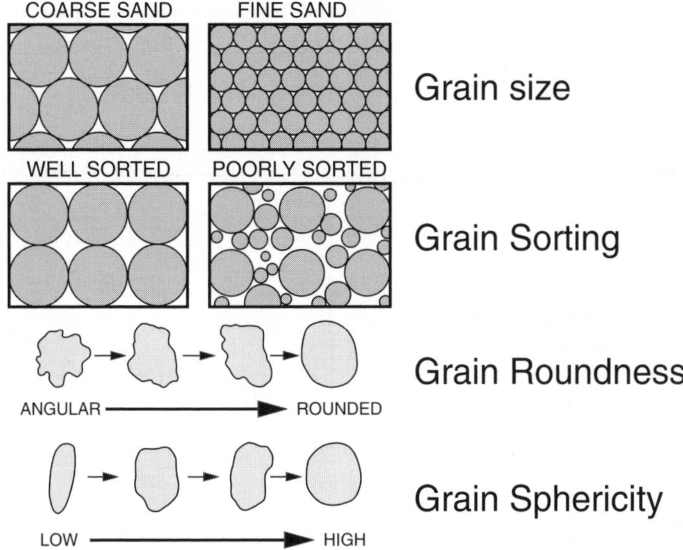

Fig. 3.4. Various textural properties of clastic sediments and sedimentary rocks. (Figure provided by T. Cross.)

The term 'upper' refers to the upper half of a sand-size range, such as upper fine sand (0.187–0.250 mm), and the term 'lower' refers to the lower half of a sand-size range, such as lower fine sand (0.125–0.187 mm).

Only rarely is a sediment composed entirely of grains of a single, uniform size – almost all sediments occur in a range of particle sizes. Sorting is an important property that describes the range of particle sizes comprising a particular sediment (Fig. 3.4). A sediment composed of grains of a wide range of sizes is termed "poorly sorted", and one with a narrow range of grain sizes is termed "well sorted". Two other measurable properties used to characterize sediments are "roundness" and "sphericity" (Fig. 3.4). Together, these properties comprise the main "textural attributes" of siliciclastic sedimentary rocks.

Figures 3.5–3.7 show some examples of the various properties of sediments and sedimentary rocks. All of these properties are important in controlling the storage and flow of fluids through the sedimentary rocks in a reservoir. Each property will be discussed in more detail in subsequent chapters.

3.2.1.2 Composition

The secondary basis for classification of siliciclastic sediments and sedimentary rocks is their mineral composition. The mineral components that comprise a siliciclastic sediment or sedimentary rock are related to the source rock, or "provenance", from which the sedimentary particles were derived by weathering (Fig. 3.2). The primary minerals that comprise siliciclastic sediments and sedimentary rocks are listed in Table 3.3.

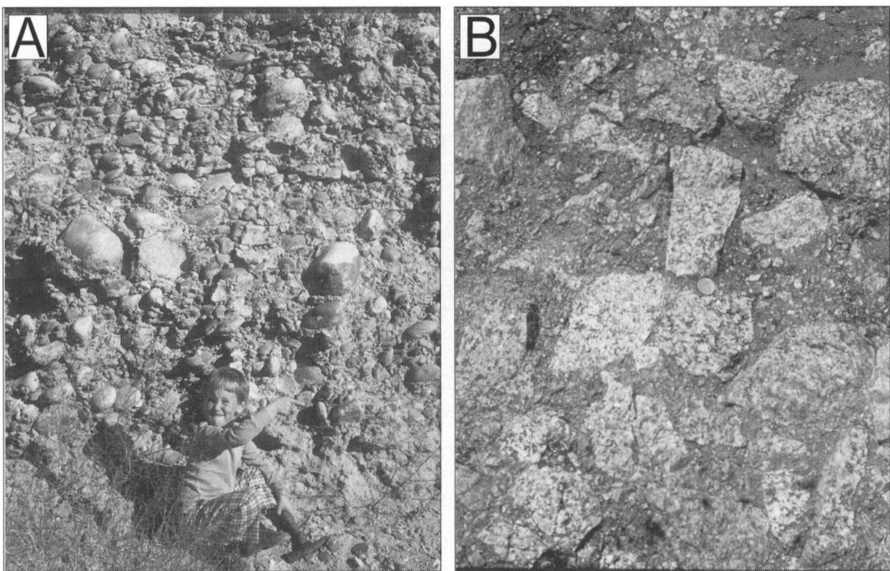

Fig. 3.5. (A) Poorly sorted conglomerate, with a boy for scale. Note the large size of the clasts that make up the conglomerate. Most of these are larger than 2 mm in size. (B) A special kind of conglomerate called a breccia. The size of the clasts all exceed 2 mm, but this rock differs from the conglomerate because individual clasts are very angular. The high degree of angularity indicates that the clasts did not travel very far from their source, for if they had, the edges would have been rounded off. There is a coin in the center of the photo for scale.

Fig. 3.6. (A) Sand on a beach. In this photo, the sand has partially buried an automobile. (B) Core photograph of a well-sorted, medium-grained sandstone. The circular hole is of a core plug that was extruded from the core for porosity and permeability measurement.

Fig. 3.7. (A) Mud flat during low tide. The geologists in this photo are looking at the lumps of mud (B) that have been produced by organisms that burrow into the mud. (C) Core from 4516–4529 ft within a Cretaceous interval. The interval above the red line at 4520 ft is light colored sandstone (SS). The interval below 4520 ft is darker colored mudstone (SH). Light colored blobs within the mudstone are silt-filled burrows.

Table 3.3 Primary mineral components of siliciclastic sediments and sedimentary rocks

Major components
Quartz
Feldspar (orthoclase, microcline, plagioclase)
Lithic clasts (multimineral fragments of the source rock)
Micas
Minor components
Heavy minerals (amphiboles, pyroxenes, garnet, etc.)
Clay minerals (smectite, illite, chlorite, kaolinite)
Cements and matrix (sedimentary rocks)
Calcite and dolomite
Quartz
Hematite
Clay minerals (same as above)

With the exception of cements, all of these minerals contain the framework element silica (Si). The word "clastic" means particles or fragments, thus the adjective "siliciclastic" means rock particles that contain silica.

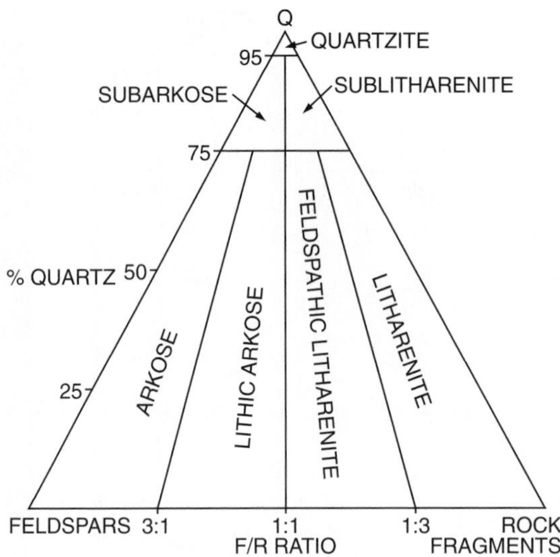

Fig. 3.8. Ternary classification of sandstones on the basis of their mineral composition. The three end members (100%) of the triangle are quartz, feldspar, and rock (lithic) fragments. To name a sandstone (arkose, litharenite, quartzite, etc.), the proportion of each member is determined and the resulting composition is plotted as a point on the triangle. Where the point plots on the triangle defines the name of the sandstone. Classification after Folk (1968).

Many classifications of siliciclastic sediments and sedimentary rocks are based on their texture and mineral composition (Fig. 3.8). For example, a quartz sandstone is a sandstone that is composed mainly of grains of the mineral quartz. A feldspathic sandstone is one that contains a large number of feldspar grains. A lithic sandstone is one that contains many lithic clasts.

3.2.1.3 Porosity and permeability

The two main microscopic-scale rock properties that control fluid storage and flow in a reservoir are porosity and permeability (Fig. 3.9). Collectively, these two properties are often referred to as "reservoir quality". The reservoir quality of a sedimentary rock is the product of the texture and composition of the original sediment. The original texture and composition are then modified by burial, compaction, diagenesis, and deformation. The geologic controls on reservoir quality are discussed in more detail in a subsequent chapter and are only summarized briefly here.

Unless it is eroded away, a sedimentary layer eventually becomes buried beneath another sedimentary layer (Fig. 3.10). This process of new deposition and subsequent burial is repeated throughout long periods of geologic time. Because temperature and pressure increase with depth beneath the Earth's surface, the sedimentary particles become susceptible to modification and lithification (the

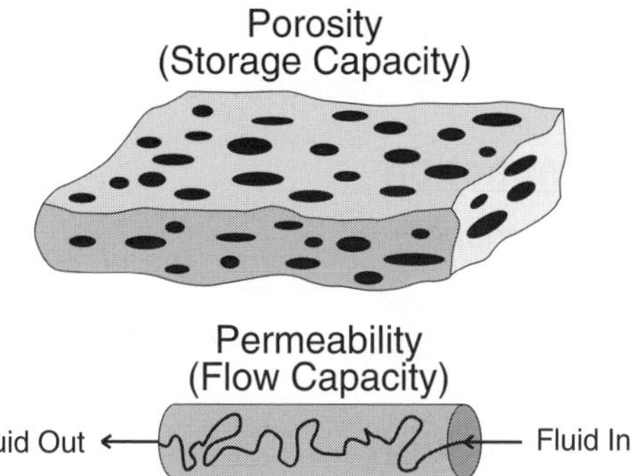

Fig. 3.9. (A) Porosity, or the percentage of the total rock volume that is pore (void) space. The pore spaces of the rock are where the fluids are contained. In this picture, the black ovals are pores and the yellow*/gray is rock matrix. (B) Permeability is a measure of the ease with which a fluid can flow through the pore spaces of a rock. The arrow shows a hypothetical flow path for reservoir fluids as they travel through pores. The rock in (A) has good porosity (high percentage of pore spaces), which can contain a large amount of fluid. However, none of the pores are interconnected, so the fluids will not be able to move from one pore to another (the rock has poor permeability). Better permeabilities result when more pores are interconnected. (Figure provided by T. Cross.)

Fig. 3.10. The principle of sediment burial is illustrated. Starting at Time 1 (below), a layer of sediment is deposited. At Time 2, a second layer has been deposited on top of Layer 1, and Layer 1 becomes partially compacted. At Time 3, a third layer is deposited, and the first layer begins to become lithified (turned into rock) by the increased pressure and temperature associated with the overlying layers. At Time 4, a fourth layer is deposited, and the lowermost layer of sediment has become lithified into sedimentary rock. This process continues in the same manner. The processes of burial and lithification may take millions of years for a sediment to be completely lithified.

---

*The indicated color is for a CD which contains all of the figures in color.

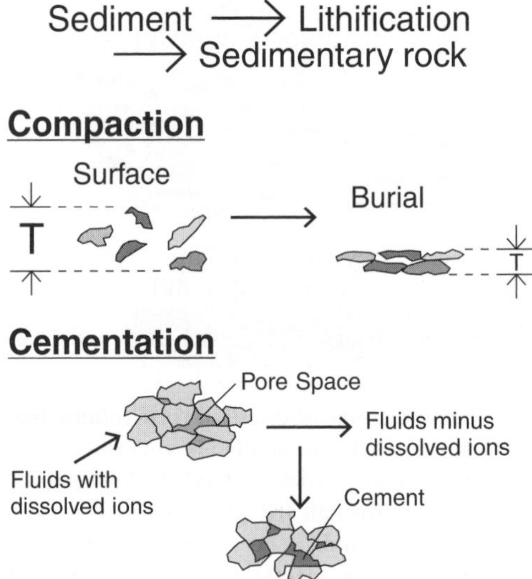

Fig. 3.11. The burial processes of compaction and cementation are illustrated in this figure. Initial compaction upon very early burial reduces the thickness of a deposit from *T* to *t*. But, not all of the pore spaces are eliminated. There is still pore space available for fluids to move through the sediment in the subsurface. These fluids contain dissolved ions that can precipitate in the pore spaces to form cement. Both of these processes lithify sediments.

processes by which a sediment becomes a sedimentary rock) as they are buried deeper and deeper.

Generally, the first effect of burial on a layer of sediment is simple compaction (Fig. 3.11). However, different minerals compact differently. For example, ductile minerals such as lithic fragments of shale will compact much more than will rigid grains, like quartz and feldspar. For this reason, plus the greater abundance of platy micas and clays, muds generally compact to a much greater extent than do sands. Compacted mud forms mudstone or shale when it is buried to sufficient depths (Fig. 3.7).

Because quartz and feldspar grains are more rigid than lithic fragments are, layers of sand retain more pore space during their burial compaction. At relatively shallow depths, subsurface fluids can flow through the pore spaces. If the physico-chemical conditions within the shallowly buried sediment are suitable, mineral cements, such as quartz, can form (Figs. 3.11 and 3.12). The quartz cement usually precipitates onto grains of quartz, forming "quartz overgrowths" (Fig. 3.12). With increased burial, additional cements (Figs. 3.13 and 3.14) or clay minerals (Fig. 3.15) can precipitate in the pore spaces, until the sediment is lithified.

Also, certain minerals are susceptible to chemical breakdown or dissolution in subsurface conditions of high temperature and pressure. For example,

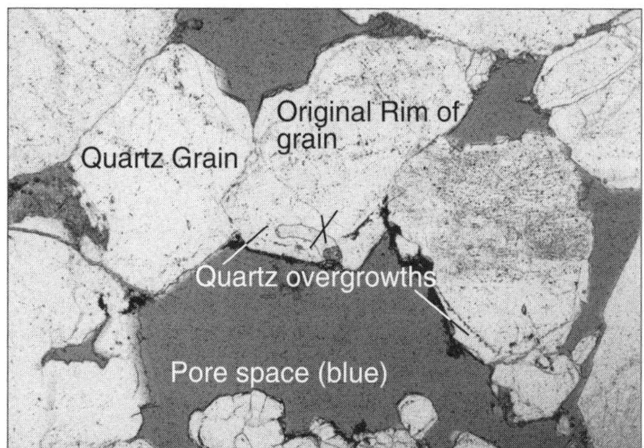

Fig. 3.12. Thin-section photomicrograph under a polarizing microscope of a quartz sandstone, showing the grains (white) and pore space (blue/dark-gray epoxy that was injected into the pore space prior to cutting the thin section). Note that individual grains are cemented together by quartz overgrowths (quartz cement that has precipitated around the original grain). If it were not for these overgrowths, fluids could move freely between pores. Thus the quartz overgrowth cement has reduced the permeability, as well as the porosity, of the sandstone.

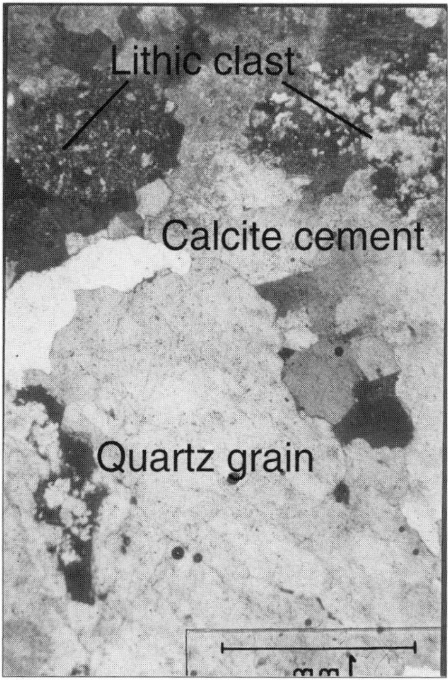

Fig. 3.13. Thin-section photomicrograph under a polarizing microscope of a sandstone cemented by calcite (yellowish/bright material). This sandstone contains grains of both detrital quartz and lithic or rock fragments (clasts).

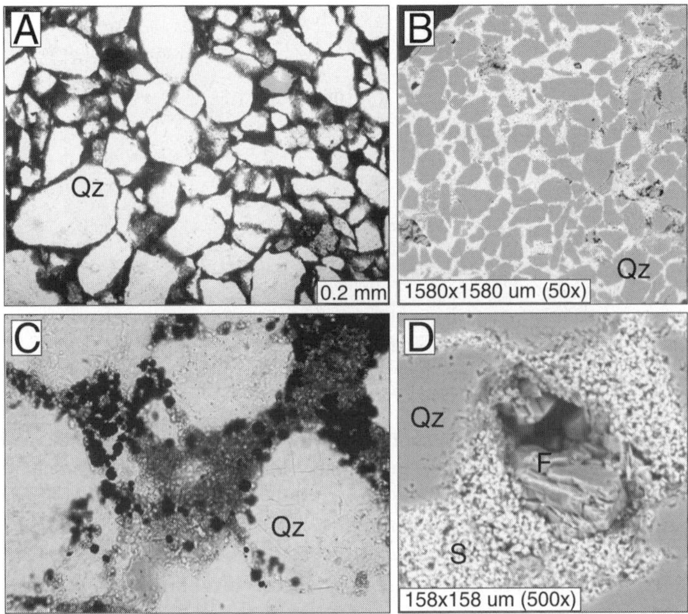

Fig. 3.14. (A) Thin-section photomicrograph of siderite-cemented (dark mineral), quartz (Qz) sandstone. (B) Cement and grains viewed through an electron microprobe. (C) Close-up of the thin section and the individual siderite crystals (orange/dark-gray). (D) Siderite crystals (S) that fill the pore space between a quartz grain and a feldspar grain, viewed through an electron microprobe. (Photographs provided by G. Romero.)

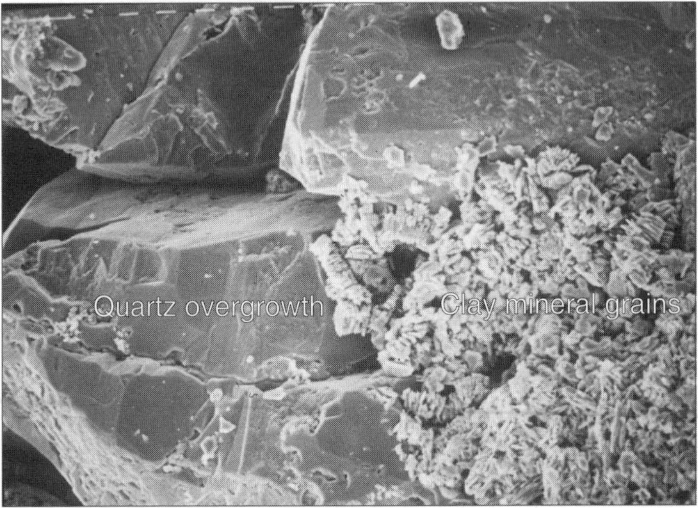

Fig. 3.15. Scanning electron microscope view of a quartz sandstone at very high magnification. Large crystal faces are quartz overgrowths (cement). Small crystals are clay minerals that have precipitated in the pore spaces.

Fig. 3.16. Chemically leached feldspar grain. The outline of the original grain is apparent, but almost all of the interior of the grain has been dissolved, as shown by the blue/dark-gray pore space (epoxy) within the grain outline.

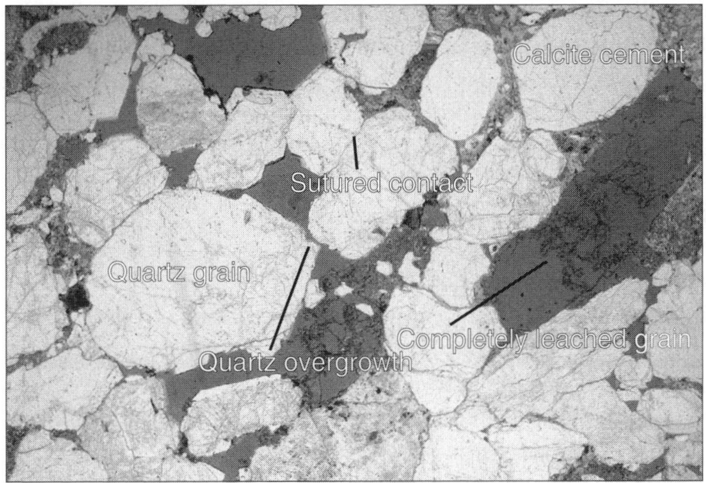

Fig. 3.17. Thin-section photomicrograph showing quartz grains with sutured contacts and quartz overgrowths. Brightly colored calcite cement and the outline of a completely leached grain are also visible. The shape of the pore suggests that it was a feldspar grain that was totally dissolved.

feldspar grains and calcite cement both are relatively unstable under such conditions and, given enough depth and time, their grains dissolve and create "secondary" porosity and permeability (Figs. 3.16 and 3.17). The dissolved elements are then free to react with other minerals in the sediment to form new ("authigenic") minerals, such as clays. One such important reaction is the dissolution

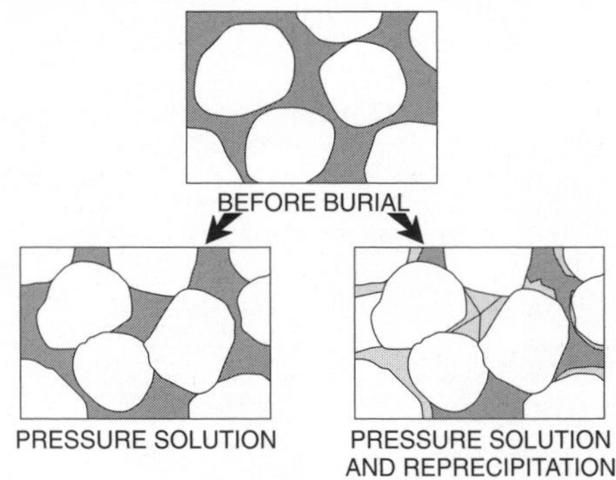

Fig. 3.18. Illustration of the cementing process called pressure solution. Pressure solution reduces the porosity and permeability of sandstones. (Figure provided by T. Cross.)

of feldspar, which releases potassium ions (K^+) into the subsurface waters. If the clay smectite (which swells when it absorbs water) is present in the sediment, the K^+ reacts to convert the smectite into the clay mineral illite (which does not swell). This reaction, in turn, frees some silica ions (Si^{+4}) into solution, and they can precipitate later to form quartz cement. This process generally occurs at significant burial depths, on the order of a few thousand meters.

Another common and important cementation process is pressure solution (Fig. 3.18). As burial of quartz-rich sediments proceeds, pressure and temperature can be sufficiently high at localized point contacts of grains, so the quartz-rich sediments dissolve at those contacts. The free Si^{4+} can then move into the adjacent pore space, where it reprecipitates as a quartz cement or "overgrowth" around another quartz grain (Fig. 3.18).

A trained petrographer can recognize combinations of these features in thin sections of rock (slices of the rock, generally 30 μm in thickness, cemented to a glass microscope slide) and can determine the order in which each of the processes occurred. For example, Fig. 3.19 shows the combined effects of compaction and cementation on a poorly sorted sandstone. The clay matrix was probably deposited at the same time as the mineral grains, and together they became compacted soon after initial burial. Two quartz grains in mutual contact became cemented by pressure solution at the sutured grain boundary. Figure 3.17 shows quartz grains with both quartz and calcite cement, sutured quartz grains, and secondary pore spaces that resulted from the dissolution of feldspar grains.

With continued burial and/or the application of tectonic stresses, brittle rocks can fracture, thereby creating fracture porosity and permeability (Fig. 3.20). Minerals in fluids circulating through the fractures can precipitate and seal the fractures from further fluid migration, or the fractures can remain open and act as conduits for hydrocarbon migration.

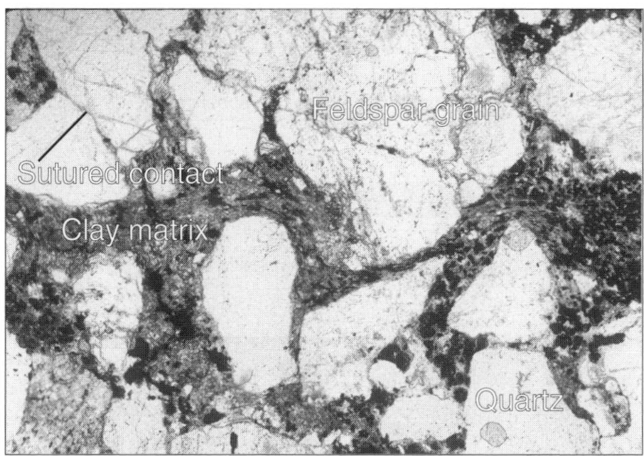

Fig. 3.19. Thin-section photomicrograph showing different quartz and feldspar grains and a matrix of darker-colored clay minerals. The "sutured" grain contacts in the upper left are a result of pressure solution. Because of the clay matrix in the rest of the rock, none of the other grains were in mutual contact during burial, so there are no other sutured grain–grain contacts. Also, the curvature of the matrix around some of the grains is a result of compaction of the clay matrix.

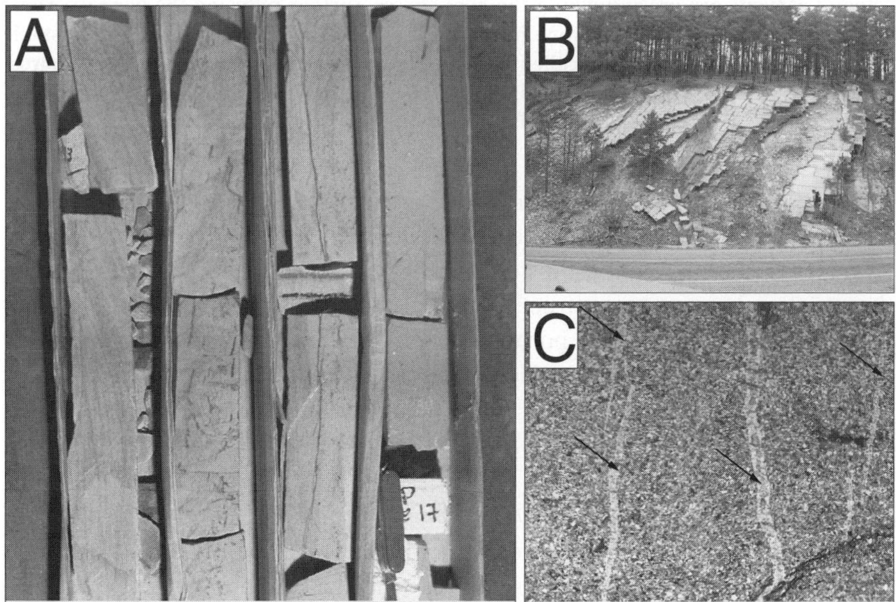

Fig. 3.20. (A) Core of quartz-cemented sandstone showing long vertical, and smaller random fractures that are filled with hematite cement (red/dark). (B) Fracture pattern on a bedding plane of brittle, quartz-cemented, quartz sandstone. (C) Thin-section photomicrograph of a fracture that is almost completely filled with mineral cement (white). Small, remaining open-fracture porosity appears as blue/dark (epoxy) areas within the fracture fill.

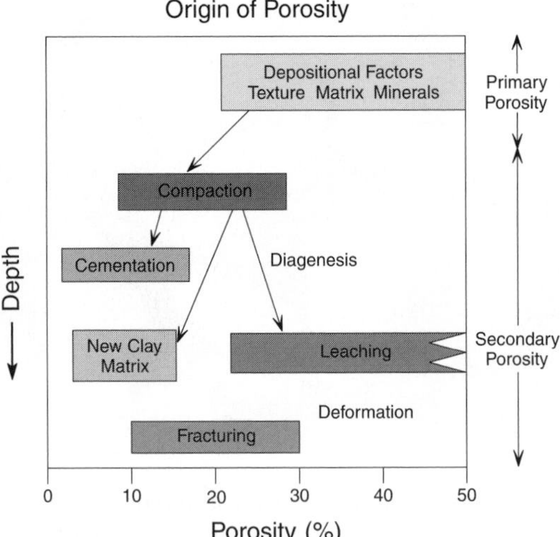

Fig. 3.21. Plot showing the several stages that a rock goes through as it is buried. The vertical axis is depth. The original porosity and permeability of sediment are those at the ground surface. With early burial, the sediment will compact, so that the porosity decreases substantially. With further burial, processes of chemical diagenesis replace physical compaction. Mineral cements are precipitated in the pore spaces, further reducing the porosity. However, some fluids can actually work in reverse by dissolving certain chemically unstable mineral grains. This process is called "leaching", and it actually forms secondary porosity (see Fig. 3.16). With further burial, the rock becomes brittle under the heavy weight of overlying rock, at which point it may fracture to produce fracture porosity.

All of the processes described above are collectively termed "diagenetic" processes, and they are associated with sediment burial and lithification. By noting crosscutting relationships of different diagenetic features in rocks, often the petrographer can reconstruct the burial history of the sediment, and from this, predict at which depths porosity and permeability might still be preserved to form a reservoir (Fig. 3.21). He or she may also be able to examine cuttings samples from a well to determine the vertical distribution of rocks with different diagenetic properties (Fig. 3.22).

3.2.1.4 Significance to reservoirs

The features described above have a major impact on reservoir performance. As will be discussed in more detail in a subsequent chapter, porosity and permeability are controlled by the geologic processes and features described here. Porosity provides the storage space for reservoir fluids, and permeability provides the fluid-flow properties of the reservoir. Some of the processes, such as development of secondary porosity, can be beneficial to a reservoir. Other processes, like cementation, are detrimental. Thus, it is important that these properties be included in a reservoir characterization.

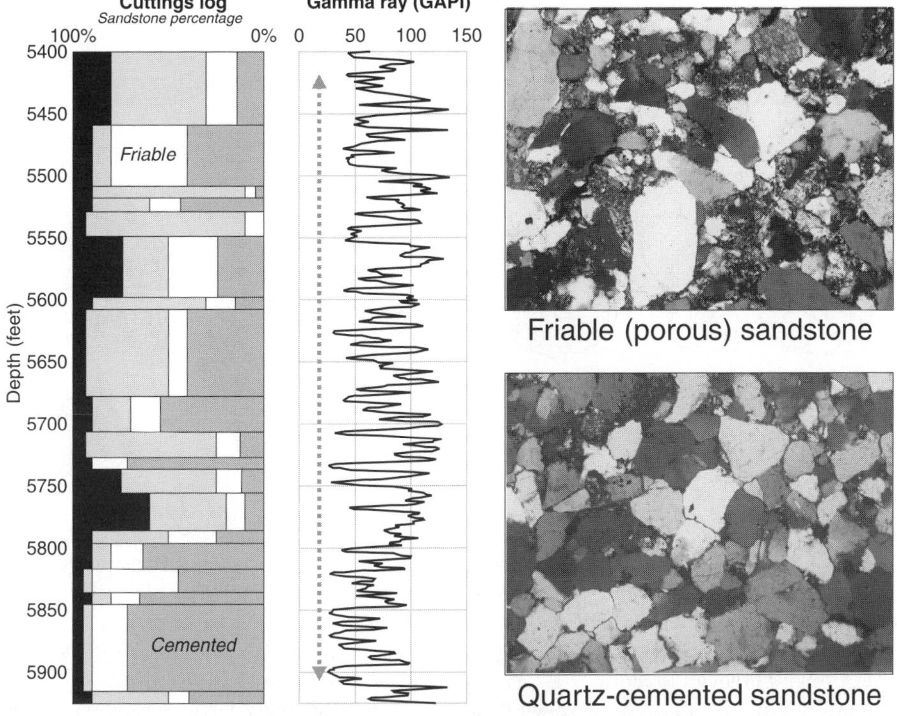

Fig. 3.22. In this example, different varieties and proportions of sandstones and shale in cuttings from 3.3-m (10-ft) intervals in a well were determined. The cuttings log shows the proportions of cemented sandstone, friable sandstone, mudstone (gray/bright-gray) and shale (black) in a 167-m (500-ft) interval in this well. Thin-section photomicrographs of the two types of sandstones are shown. After Romero (2004).

3.2.2 Chemical and biogenic sedimentary rocks

Chemical sedimentary rocks (e.g., halite) are those that form by direct precipitation of minerals from sea water or, under certain conditions, from freshwater. Biogenic sedimentary rocks are those that consist of the tests secreted (i.e., precipitated) by organisms such as foraminifera and diatoms. Neither of these types of sedimentary rocks is the main focus of this book, and they are only mentioned here.

The primary minerals that form by direct precipitation from sea water are calcite and aragonite. Both belong to the mineral group called "carbonates", because they contain the complex-anion molecule (CO_3). Microorganisms are capable of using these minerals to form their shells (Fig. 3.23). Larger organisms, such as corals, also use calcium carbonate ($CaCO_3$) in their skeleton and can build extensive coral reefs (Figs. 3.24–3.26) that, upon burial, can become

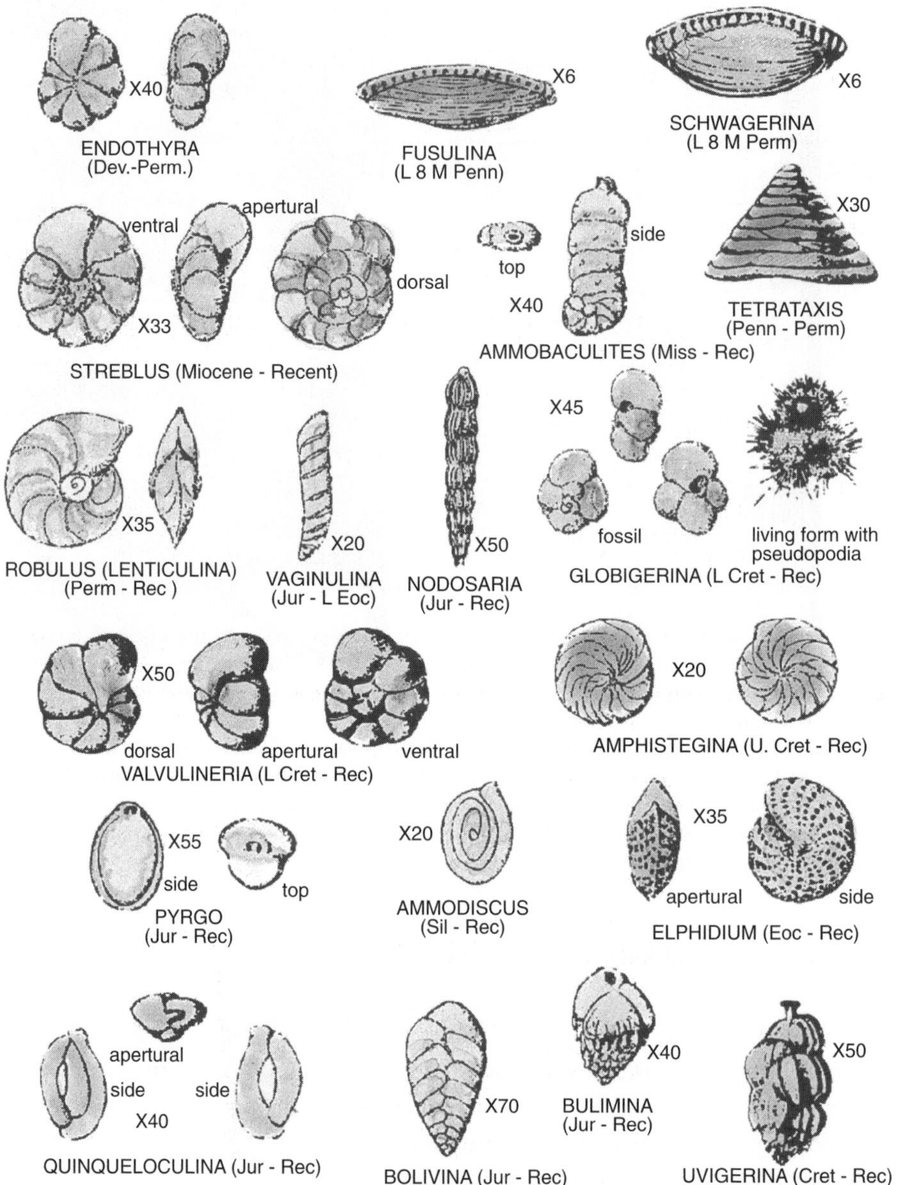

Fig. 3.23. Some of the many different types of plankton that live within the ocean water column. These are different species of foraminifera. Although they exhibit many sizes and shapes, their outer shells are composed of calcium carbonate (as the minerals calcite and aragonite). The soft body parts within the shells (stomach, etc.) contain organic molecules. When the foraminifera die and fall through the water column to the seafloor and become buried, the shells will be preserved, but the soft parts (guts) will decompose and can form hydrocarbon molecules. (Source of figure is unknown.)

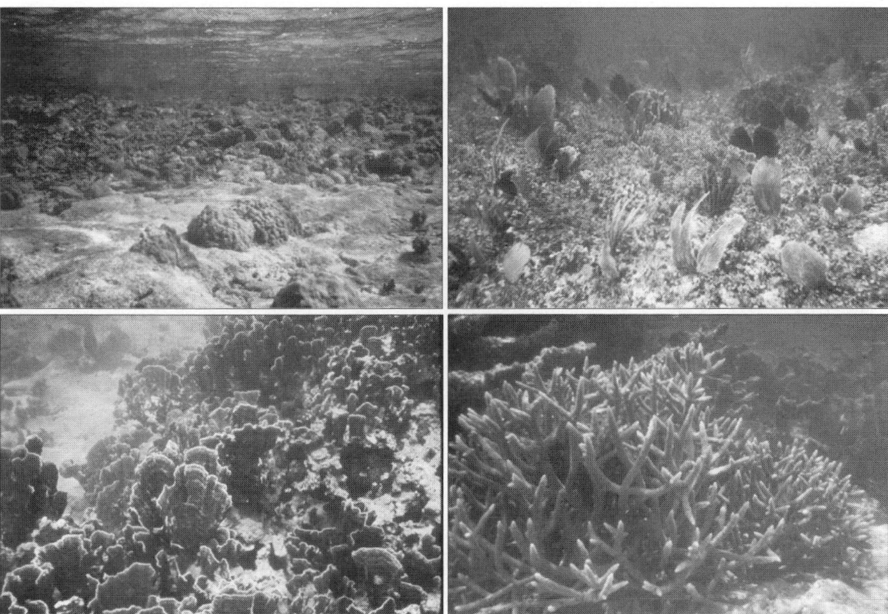

Fig. 3.24. The upper surface of a coral reef, as viewed from beneath the sea surface. Various types of living corals are shown in growth position. The fine dustlike material between patches of coral is sediment composed of pieces of coral and algae that have been broken off the living organisms by wave action and by feeding fish.

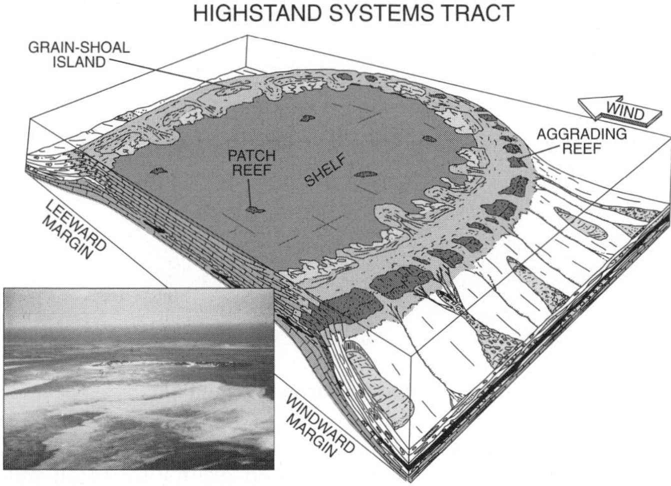

Fig. 3.25. Schematic diagram of a coral reef (red/darkest-gray) forming an atoll. Patch reefs form within the shallow waters encircled and protected by the reef. Inset shows a modern pinnacle reef. After Handford and Loucks (1993). (Reprinted with permission of AAPG, whose permission is required for further use.)

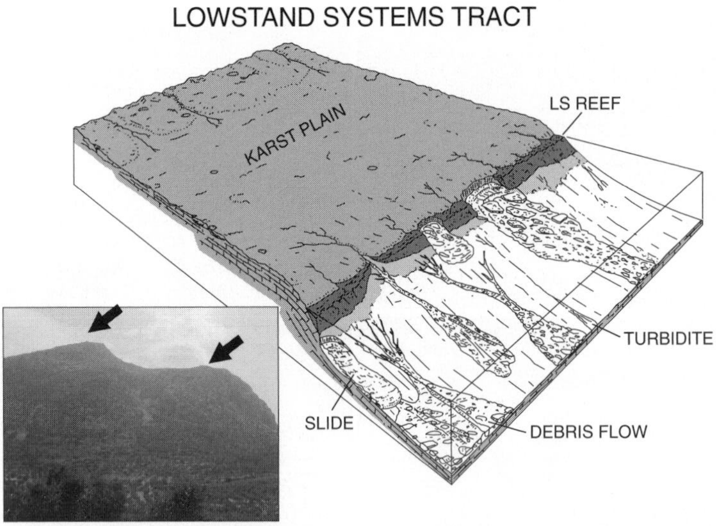

Fig. 3.26. Schematic diagram of a fringing coral reef (red/darkest-gray) that forms at the edge of the continental shelf. Inset shows the silhouette of two reef edges. After Handford and Loucks (1993). (Reprinted with permission of AAPG, whose permission is required for further use.)

Fig. 3.27. Modern carbonate mud flat.

oil and gas reservoirs. Chemically precipitated grains are often mud size and can form carbonate mudflats (Fig. 3.27).

A comparison of some of the important features of carbonates and siliciclastics is shown in Fig. 3.28. One important difference is the geographic locations in which the mineral groups form. Siliciclastics can form anywhere that

Carbonates	***Siliciclastics***
majority occur in shallow, tropical environments	occur woldwide and at all depths and in every climate
majority are marine	terrestrial and marine
grain size reflects size of skeleton and its calcified hard parts	grain size reflects hydraulic energy
presence of lime mud often indicates organic origin	presence of mud indicates settling out of suspension
shallow-water sand bodies result from localized physicochemical or biological fixation of carbonate	shallow-water sand bodies result from current and wave action
localized buildups without change in hydraulic energy	changes in sedimentary environment occur with changes in hydraulic environment
sediments are commonly cemented on the sea floor and subject to subaerial diagenesis	sediments are rarely cemented on sea floor and relatively unaltered by subaerial exposure
variable pore types at deposition	depositional porosity is intergranular

Fig. 3.28. Comparison of features of carbonate and clastic processes, sediments, and environments of deposition. After Sarg (1988). (Reprinted with permission of SEPM, Society for Sedimentary Geology.)

has source rock to be weathered (Fig. 3.1). Because chemical and biochemical reactions normally require warm water conditions, most carbonates form in shallow tropical waters (Fig. 3.29). However, in shallow marine waters where microorganisms with carbonate tests thrive, thick accumulations of the tests of dead microorganisms can form carbonate mudstones downslope, on the deeper seafloor. Upon diagenesis, these carbonate mudstones become chalks.

Carbonate rocks are also classified on the basis of their textural and compositional characteristics (Fig. 3.30). As the ratio of mud- to sand-size grains decreases, the carbonate rocks are referred to as mudstones, wackestones, packstones, and finally, grainstones (Table 3.1). These various classes of rocks occur in different depositional subenvironments of reefs, which allows geoscientists to reconstruct reef environments in ancient rock sequences (Fig. 3.31).

Carbonate minerals are relatively unstable when buried and can react with subsurface fluids to dissolve and/or to form new minerals. For example, at very shallow burial depths, the primary carbonate mineral aragonite converts to the more stable crystal structure of calcite, although both have the same mineral composition, $CaCO_3$. With deeper burial, dissolution of carbonate minerals produces secondary porosity (Fig. 3.32). Also, with burial and the reaction of $CaCO_3$ with seawater, dolomite ($Ca,MgCO_3$) forms. This subsurface mineral transformation results in a decrease in rock volume and an increase in porosity and permeability. Many hydrocarbon reservoirs are composed of dolomite.

Evaporites comprise another class of chemically precipitated minerals that are quite important to hydrocarbon reservoirs. Salt (NaCl), gypsum

Fig. 3.29. The warm-water, equatorial regions of the Earth, bounded by the two red/gray lines in the map, are where carbonate-bearing organisms thrive. The organisms build individual shells as personal homes (Fig. 3.23) by secreting aragonite or calcite. Other organisms, like algae, have small, needlelike skeletons made of the same minerals. These organisms live in colonies and sometimes grow into reefs, such as coral reefs (Figs. 3.24 and 3.25). After National Geographic magazine.

($CaSO_4 \cdot nH_2O$), and anhydrite ($CaSO_4$) are the three common evaporite minerals that form in environments that are conducive to evaporation of saline water. When the water becomes supersaturated with dissolved minerals, precipitation occurs. These precipitated minerals are important because they form impermeable rock masses such as salt pillows and domes, or extensive blanket deposits. In the former case, salt domes make excellent structural traps for hydrocarbons

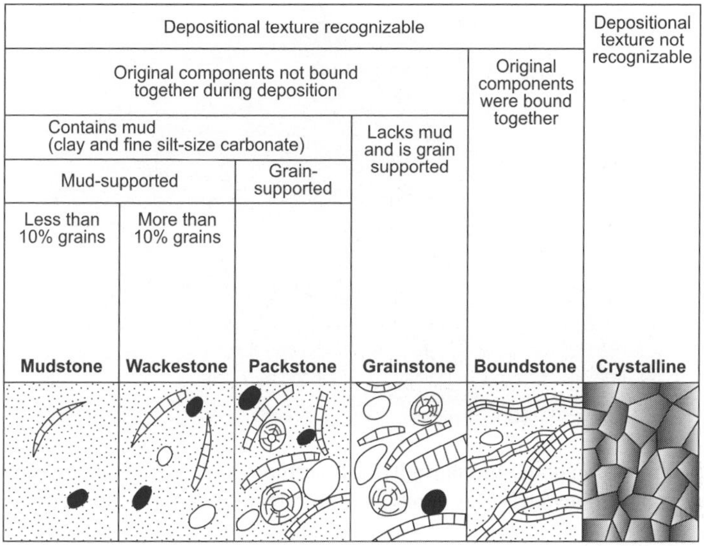

Depositional texture recognizable					Depositional texture not recognizable
Original components not bound together during deposition				Original components were bound together	
Contains mud (clay and fine silt-size carbonate)			Lacks mud and is grain supported		
Mud-supported		Grain-supported			
Less than 10% grains	More than 10% grains				
Mudstone	**Wackestone**	**Packstone**	**Grainstone**	**Boundstone**	**Crystalline**

Fig. 3.30. Carbonate rock classification of Dunham (in Morre, 2001). The classification is based primarily on depositional texture, with the five major classes being boundstone, grainstone, packstone, wackestone, and mudstone. Secondary basis of classification is on the proportion and type of grains.

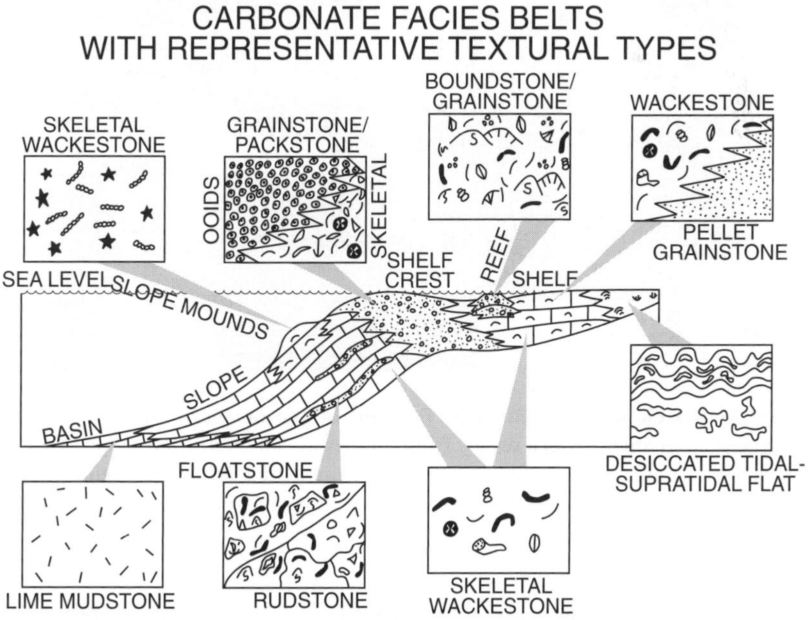

Fig. 3.31. Carbonate rock types and the locations within a carbonate shelf-to-basin environment in which they are deposited. After Sarg (1988). (Reprinted with permission of SEPM, Society for Sedimentary Geology.)

Fig. 3.32. Thin-section photomicrograph of a limestone showing moldic (discontinuous, dissolved) pore space (blue/dark-gray epoxy) within calcite (greenish-yellow/gray) grains.

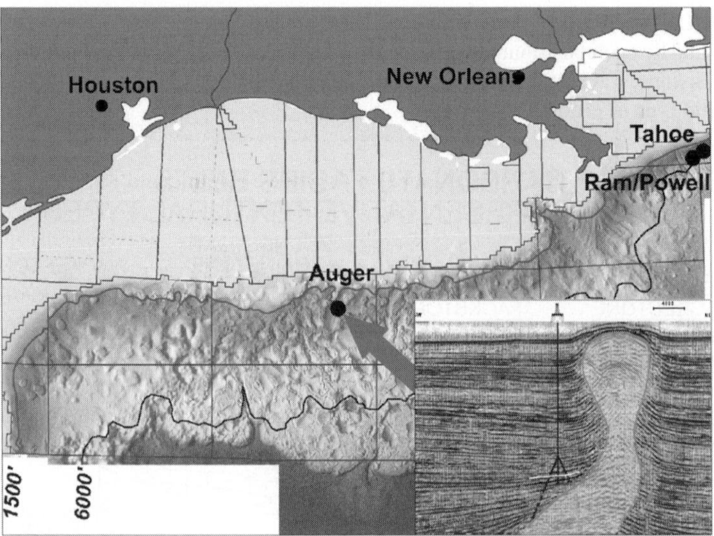

Fig. 3.33. Map view of the continental shelf off of the northern Gulf of Mexico. The pock-marked outer shelf surface results from numerous salt domes and pillows that underlie the sea floor. Inset is a seismic line showing a salt dome (highlighted in green/bright-gray). Map view is from Diegel et al. (1996). (Reprinted with permission of AAPG, whose permission is required for further use.)

in such places as the deepwater Gulf of Mexico and parts of offshore West Africa (Fig. 3.33). In the latter case, blanket deposits can become topseals to reservoirs. One example is the Permian Zechstein Salt, which forms an areally extensive topseal to the Permian Rotliegendes sandstone reservoir in parts of

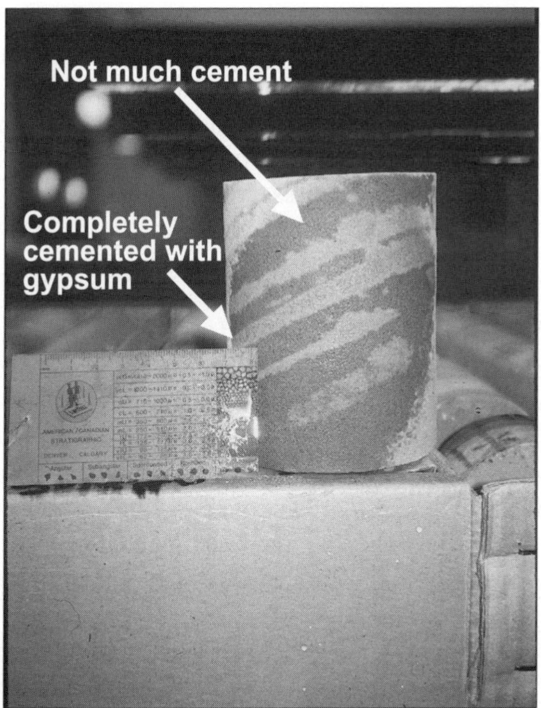

Fig. 3.34. Core slab from a Permian-age (Rotliegendes) eolian (sand dune) reservoir in the North Sea. The sandstone is finely laminated. The white streaks are thin individual laminae that have been tightly cemented with the evaporite gypsum. The orange (darker gray) laminae are porous and permeable, because they have not been tightly cemented by gypsum. In this case, the actual reservoir rock (which is porous and permeable) is the thin orange/darker gray laminae. The white laminae act as micropermeability barriers to the free flow of reservoir fluids.

the North Sea. Not only is the Zechstein Salt a topseal, but, in the geologic past evaporite-bearing fluids have permeated the sandstones to form gypsum cement within them (Fig. 3.34).

3.3 Sedimentary structures and their significance

Sedimentary structures are treated separately in this section, because most of the same types of structures are found in siliciclastic, chemical, and biogenic sedimentary rocks. There are three categories of sedimentary structures: (1) those that are produced by physical interaction of sedimentary particles with their environment, (2) those that are produced by organisms interacting with the sediment, and (3) those that result from chemical interactions of fluids and sediments.

Many books and papers have been published on the subject of sedimentary structures, and only a brief review is presented here. The primary goal is to explain the importance of sedimentary structures in reservoir characterization, rather than to present a thorough understanding of their formative processes. Sedimentary structures are extremely important in reservoir characterization, because they provide information on the nature of the depositional environment, which in turn controls such reservoir properties as size, areal extent, geometry, thickness, internal architecture, and reservoir quality.

3.3.1 Physical sedimentary structures

3.3.1.1 Structures formed by currents and waves

Many different physical sedimentary structures form as a result of sedimentary particles interacting with their surrounding environment. One simple example is a modern lake that has dried up, so that the mud on the bottom of the lake has lost its water by evaporation and has desiccated and developed a pattern of polygonal cracks (Fig. 3.35A). Dead fish or other organisms can litter the dry mud bottom (Fig. 3.35B). When new sediment is deposited on top of this horizon, such as when the lake is replenished with muddy water, the polygonal

Fig. 3.35. (A) Modern mud cracks that have formed on a dry lake bed. When the lake went dry, the mud on the lake bottom lost water from within its pore spaces and mineral structures and then shrank, creating mud cracks. (B) Modern mud cracks with dead fish atop the mud. At some future time, if new mud is deposited on top of the fish and mud cracks, imprints of these fish skeletons will remain. (C) Surface of a Paleozoic mudstone bed, showing lithified mud cracks.

desiccation cracks are filled with deposited mud, and eventually, the cracks are preserved by lithification (Fig. 3.35C). The skeletal remains of the fish, or at least a cast of the shape of the remains, can also be preserved as a body or trace fossil (described below). Polygonal mud cracks can also form on tidal flats that are deprived of water seasonally, so that in dry seasons the mud desiccates.

Wind is an important agent of transport of both sand-size and mud-size sedimentary particles (Fig. 3.36A). In any given area there is a single dominant wind direction, and often that direction can easily be interpreted by the telltale direction of cross-bedding in sedimentary rocks (Fig. 3.36B,C). Knowing the direction of transport of wind-blown (eolian) sand can be important for predicting the areal extent and trend of eolian reservoirs, which are discussed in a later chapter. The process of wind transport tends to segregate rock grains into a limited range of particle sizes, thus providing excellent sorting, which in turn can provide excellent reservoir quality of an eolian reservoir (unless it is heavily cemented).

Water is the most widespread and important agent of transport of sedimentary particles. As a result, many types of sedimentary structures are formed by

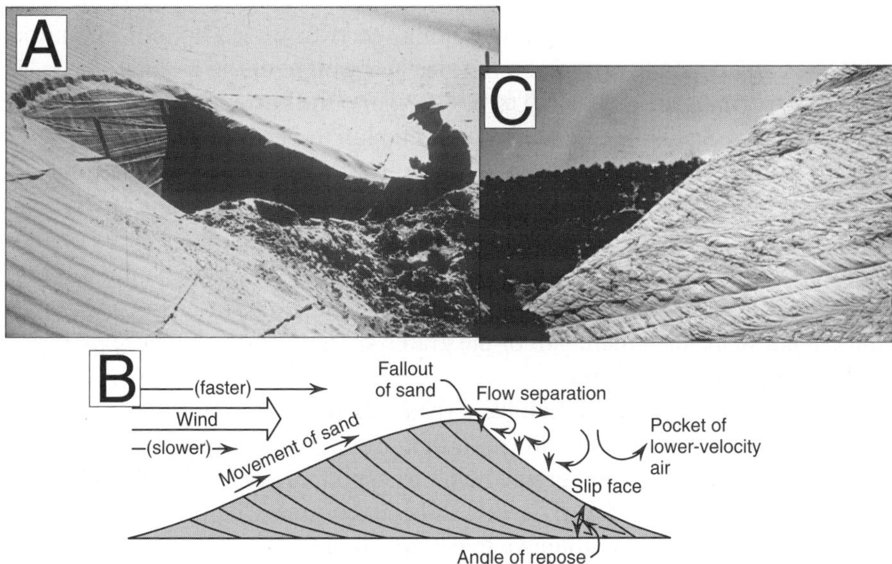

Fig. 3.36. (A) The geologist in this picture has excavated the face of a sand dune. Note the different angles of the cross-beds. This indicates that the sand dune formed in an area of alternating wind directions. (B) Schematic cross-section that illustrates the movement of sand grains to form sand dunes. Grains move across the top of the windward side of the dune and fall over the steeper lee side. Over time, the continual downwind movement of grains causes the entire dune to migrate. (C) Windblown (eolian) sedimentary rock. Direction of transport of the sand by the wind was from left to right, as shown by the orientation of the dune's cross-stratification.

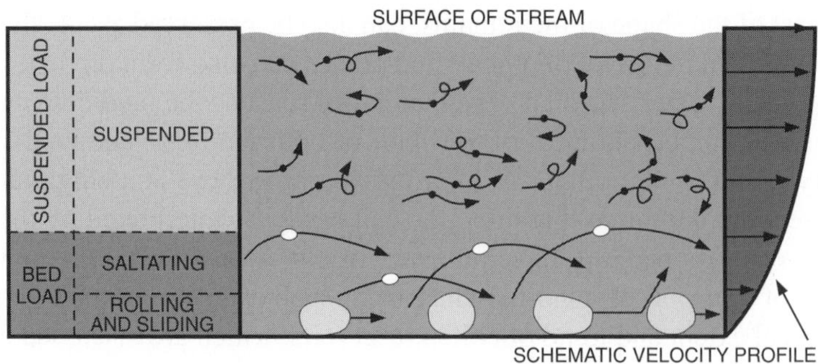

Fig. 3.37. Movement of sedimentary particles in a stream by rolling/sliding, saltation, and suspension. Different-size particles are transported within the current by these different processes, depending upon the weight and shape of the particles. (Source of figure unknown.)

sediment that has been transported by and deposited in water. Only a few types are discussed here, and others are discussed in later chapters on specific types of reservoirs. Grains are transported along the river or sea bottom by various processes that depend partly upon the size and weight of the grains (Fig. 3.37). Gravel-size grains are too heavy to be lifted into the current, so they roll along on the bottom. Sand-size grains are light enough to be projected into the water column but heavy enough to follow a downward trajectory under the influence of gravity; when the grains hit the bed again, they are projected upward and the process is repeated. This process of forward movement by a series of short intermittent jumps is called "saltation" (from the Latin word meaning to jump) (Fig. 3.37). Silt- and clay-size grains are light enough to be lifted into the current and remain suspended there. All of the grains are transported by these processes in the downcurrent direction.

When currents move sand along a river bed or the sea bed, the sand tends to form structures, called "bedforms", that are dependent upon the velocity of the flow and the size of the grains being transported. As a collection of grains moves downcurrent on the river or sea bed, the grains bunch into bedforms. Flow velocities are classified into a lower flow regime and an upper flow regime, according to their strength (Fig. 3.38A). Generally, the faster the flow, the larger the bedform that develops in sandy deposits (increasing in size from ripples, to sand waves, to dunes; Fig. 3.38B). This is because the faster the flow velocity, the harder the grains hit the bed, the farther up into the flow they are projected, and the longer their downward trajectory is to where they hit the bed again. The bedforms take the shape of asymmetric ripples and waves, with a shallow-dipping side and a steeper downcurrent side (Fig. 3.38B). Examples of bedforms are shown in Fig. 3.39.

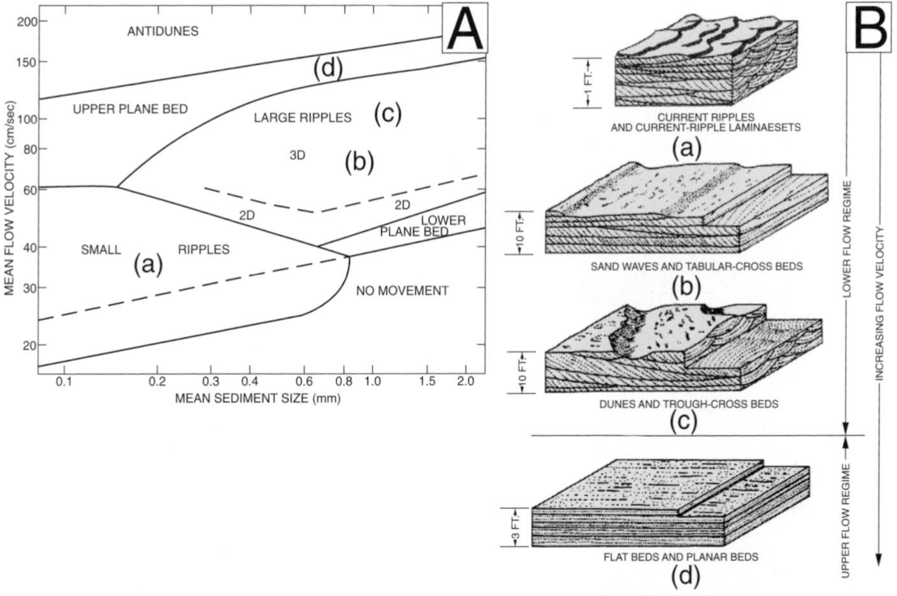

Fig. 3.38. (A) Graph showing the different types of water-deposited bedforms that form as a function of current flow velocity and sediment grain size. (B) Block diagrams showing the types of bedforms. Different bedforms are labeled (a)–(d). (Source of figures is unknown.)

A cross-bed is the configuration of grains, within a bedform, that results from the above processes. Cross-beds are inclined at an angle to the main planes of stratification and are a common product of flow along a river or sea bed. The process of grain movement is similar to that described above for the movement of wind-blown sand. Like the beds in eolian dunes, the cross-beds of water-deposited sands dip downward in the direction that the current is moving when viewed internally, parallel to the current direction (Figs. 3.36–3.38). When they are viewed internally, but perpendicular to the flow direction, the cross-beds can take several forms, depending upon how much the strength and direction of the current varied at the time of deposition (Figs. 3.40 and 3.41).

Although river currents are unidirectional (they flow in one general direction), marine currents can be unidirectional or bidirectional. Bidirectional currents result from tidal action, where the current moves in one direction during the incoming tide and in the opposite direction during the outgoing tide. The resulting bedding surface take the shape of symmetrical ripples (with equal dip on both sides of the ripple crest) (Figs. 3.42A,B,C), and internally, the cross-beds dip in opposing directions (Fig. 3.42D).

Bedforms and internal sedimentary structures both provide important clues to the environment of sediment deposition. Vertical variations in sedimentary

Fig. 3.39. (A) Ripples moving in a stream bed. (B) Asymmetric ripples exposed on a tidal flat. Direction of the current is from right to left (steep sides of ripples are in the down-current direction). (C) Sand waves forming at the mouth of a river channel as it enters the marine environment. Direction of the current was from upper right to lower left. (D) Asymmetric ripples exposed on a sand flat; direction of the current was from right to left.

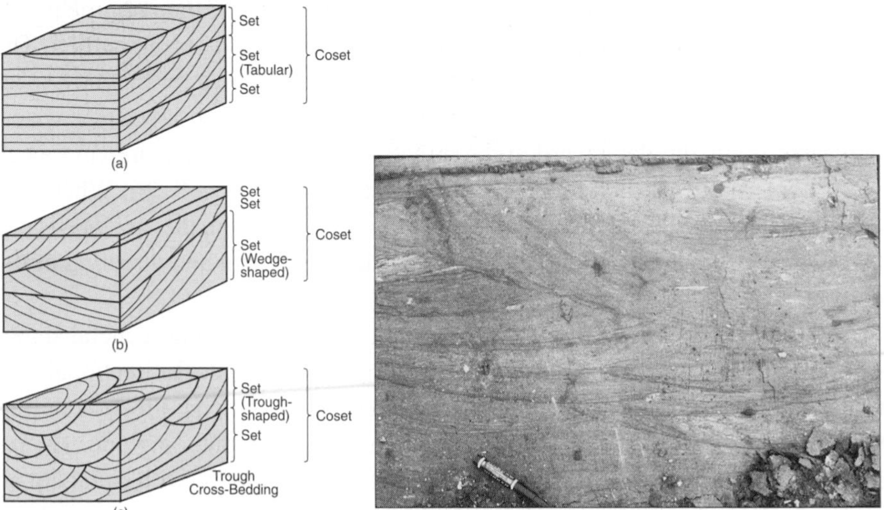

Fig. 3.40. Three-dimensional views of different types of cross-beds and an outcrop of trough cross beds (c). After Blatt et al. (1972). (Reprinted with permission of Prentice-Hall, New Jersey.)

Fig. 3.41. Cross-bedded sand exposed in a trench on a beach. The dominant current direction was from left to right, as indicated by the direction of dip of the cross-beds.

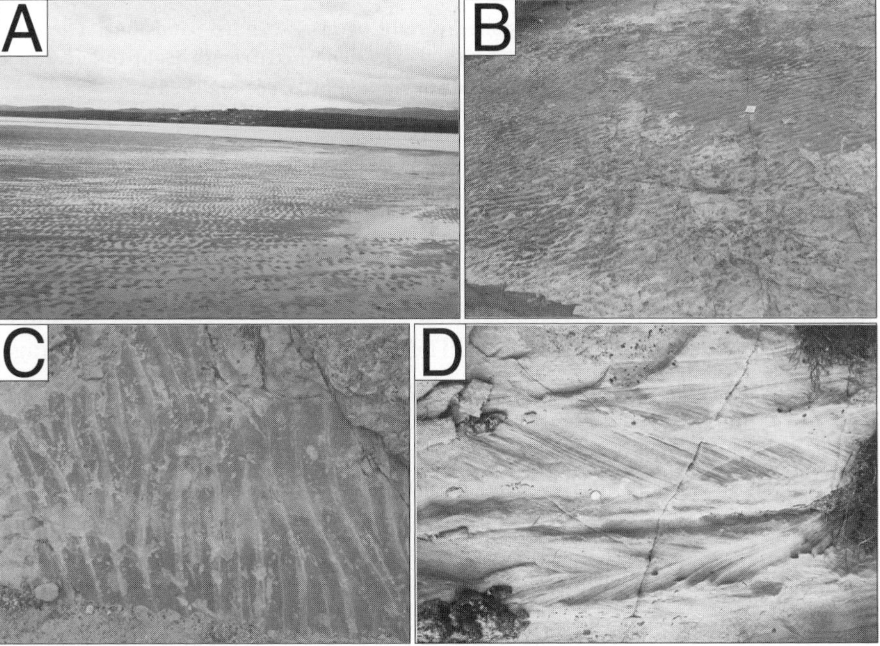

Fig. 3.42. (A) Ripples exposed on a tidal sand flat during low tide. Note the washboard-like appearance. (B) Symmetrical ripples on the surface of a Cretaceous-age sandstone. (C) Close-up view of the symmetrical ripples on the top of the sandstone bed shown in B. (D) Herringbone cross-bedding in the vertical dimension of a sand bed, indicating alternating directions of currents, as occurs on a tidal sand flat.

Fig. 3.43. Horizontally-bedded sandstone overlain by cross-bedded sandstone. These two sedimentary structures indicate that the sand was deposited from a right-to-left flow of progressively decreasing velocity (plane beds to rippled beds, see Fig. 3.38).

texture and bedding style are also important indicators of processes and environments of deposition. Figure 3.43 illustrates vertical variations in sedimentary structures; these variations are a result of the decrease in velocity of the current that, in this case, deposited sandstone at that one spot on the sea floor or river bed. Two sedimentary structures occur: a lower plane-bedded sandstone and an upper ripple cross-bedded sandstone. The grain size of the lower unit averages 0.2 mm and that of the upper unit averages 0.15 mm. Comparing these sizes with the graph in Fig. 3.38 shows that the upper flow regime's plane-bedded sedimentary structure, which formed the lower unit, must have been deposited from a flow of velocity on the order of 80 cm/s. The flow velocity then diminished to less than 60 cm/s, giving rise to the lower flow regime's small ripples being deposited atop the plane-bedded sandstone.

When a wide range of particle sizes is carried within a flow, progressive reduction in flow velocity at one location on the river or sea bed produces an upward decrease in the size of grains that are deposited over time (Fig. 3.44). The well-known "Bouma Sequence" (Bouma, 1962) is an example of progressively upward-fining sediment deposition at one location on the deep seafloor, as a result of combined turbulent suspension and traction as flow velocity diminishes progressively from early high-energy flow to subsequent calmer flow

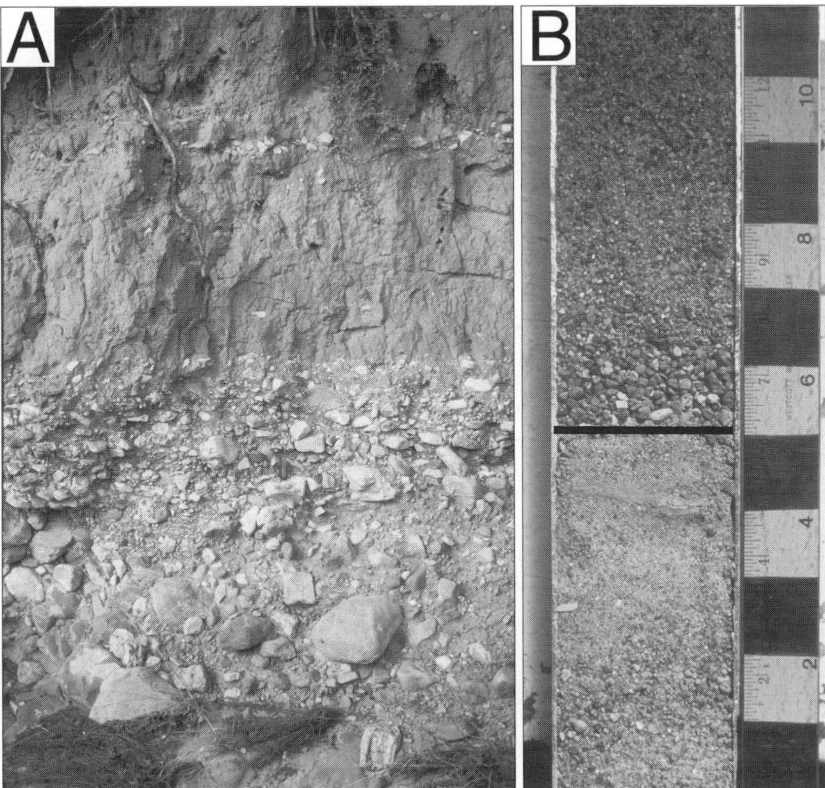

Fig. 3.44. (A) Fluvial (river) channel deposit. The size of grains comprising the sediment decreases (fines) upward, indicating a progressive loss of river power or energy, through time, within the river channel. In this way, finer and finer-grained, lighter-weight grains could be deposited as the flow velocity decreased progressively. (B) A cored interval composed of two individual gravelly-sand beds, each one becoming finer-grained upward. The boundary between the two beds (red/dark line) is a "bedding plane" or surface between two individual deposits. Each of the two beds was deposited from flows of progressively waning velocity, so that progressively finer-size grains were deposited over coarser grains.

(Fig. 3.45). Bouma defined five rock divisions of the Bouma sequence, which he called T_a to T_e. T stands for turbidite, and "a" through "e" represent the various divisions. T_a is the coarsest-grained part of the bed and is deposited under fairly high current velocities, so that only the coarsest of grains could be deposited. Bouma T_a beds are characterized by "size grading", wherein grains become finer from the base of T_a to the top of T_a. T_b is slightly finer grained and characterized by parallel bedding. T_c and T_d are even finer grained. T_c is characterized by cross-bedding. T_d is characterized by parallel laminations, as is T_e, which represents deposition of clay from suspension during a quiescent period of time between deposition from sandy turbidity currents.

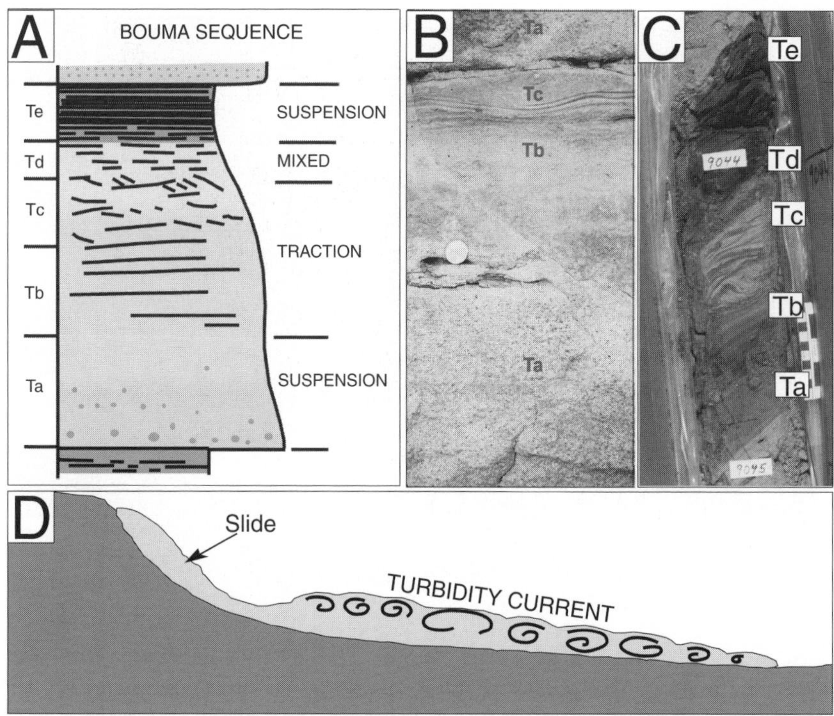

Fig. 3.45. (A) The typical type of deposit that is associated with deposition of sediment from a turbidity current in deep ocean waters. The sediment records a progressive decrease in flow velocity over time, so that progressively finer-grained sediment is deposited over coarser-grained sediment. This sequence of sediments or rocks is called a Bouma Sequence, after Arnold Bouma, who first described the rocks (Bouma, 1962). (B) Rock outcrop showing the Bouma divisions. (C) Core showing the Bouma divisions. In both B and C, the grain size of the sand decreases upward, as it does in A. Also note the different types of sedimentary structures in both the cartoon and the rocks. (D) Process of generation of a turbidity current (turbulent flow) down a slope in the marine environment; part or all of the Bouma Sequence commonly is the product of deposition from turbidity currents. Figure A is from Jordan et al. (1991). Figure D is modified from Morris (1971). (A – Reprinted with permission of the Dallas Geological Society. D – Reprinted with permission of AAPG, whose permission is required for further use.)

3.3.1.2 Structures formed by sediment loading

Another group of physical sedimentary structures comprises "load structures". Load structures form when a layer of sediment is deposited on top of another, softer sediment layer. If the overlying layer is denser than the underlying layer (such as sand overlying mud), the sand can sink into the underlying mud, forming a characteristic pattern of downward protrusions of sand called load structures or load casts. The mud preserves this multiple-protrusion load structure, so that it becomes lithified along with the flat, horizontal sandstone and mudstone layers (Fig. 3.46). In rock formations that have been tilted or overturned by tec-

Fig. 3.46. Outcrop showing a "loaded" sandstone. The various bulbous features were originally a sand bed that was deposited on top of a wet, soupy mud. The sand was heavier than the mud and began to sink into the mud. The resulting detached, bulbous-shaped sandstone features now are encased in the reddish mudstone. Such features indicate very rapid deposition of sand onto a mud flat.

tonic activity, load structures indicate whether a particular sedimentary bed has been overturned or is in the same position that it was in when it was deposited (right-side up).

3.3.1.3 Erosional sedimentary structures

Not all physical sedimentary structures are depositional in nature; some can be erosional. In fact, strong currents and waves can erode sediment that was previously deposited on river or sea beds. Figure 3.47 shows a large erosional surface that downcuts an underlying sandstone from a different depositional environment. Figure 3.48 shows a smaller erosional scour surface on a sandstone bedding plane. Ripples occur on part of the surface, indicating that currents that eroded the sediment surface also were capable of moving the eroded sediment.

Tool and groove marks are another set of features produced by erosion of sediments. Figure 3.49A shows a geologist looking underneath a sedimentary bed, because this is where tool and groove marks tend to be located. Figure 3.49B shows how an erosional groove mark might form. A shell on the beach acts as a barrier to incoming surf. When the surf reaches the shell, turbulent flow is generated on the downcurrent side of the shell, eroding the sand there. If another layer of sediment is deposited on top of the eroded sand, lithification will preserve that feature beneath the overlying sand bed. Figure 3.49C shows tool marks on the underside of a sandstone bed. These marks were produced by a pebble being moved along by a current on a muddy seafloor; the pebble scoured into the mud, and later the scour was filled in with a sand bed, which,

Fig. 3.47. Two sandstone bodies separated by an erosional surface (red/black dashed line) produced when distributary channel sands (above the erosional surface) eroded into underlying shoreface sands (beneath the erosional surface).

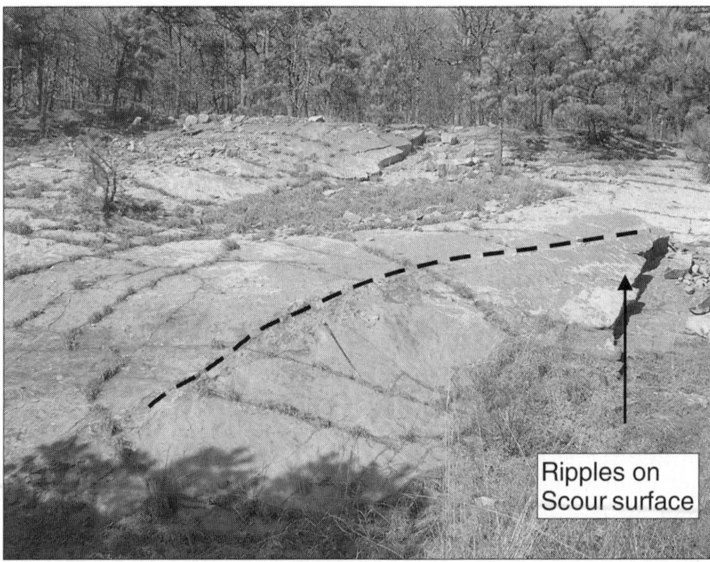

Fig. 3.48. Erosional scour carved into the top of a sand (now sandstone) bed. Red/black dashed line shows the surface of the scour. Within the scoured surface are ripples, indicating that moving currents transported the sand within the scour, once the scour was formed. Black walking stick for scale.

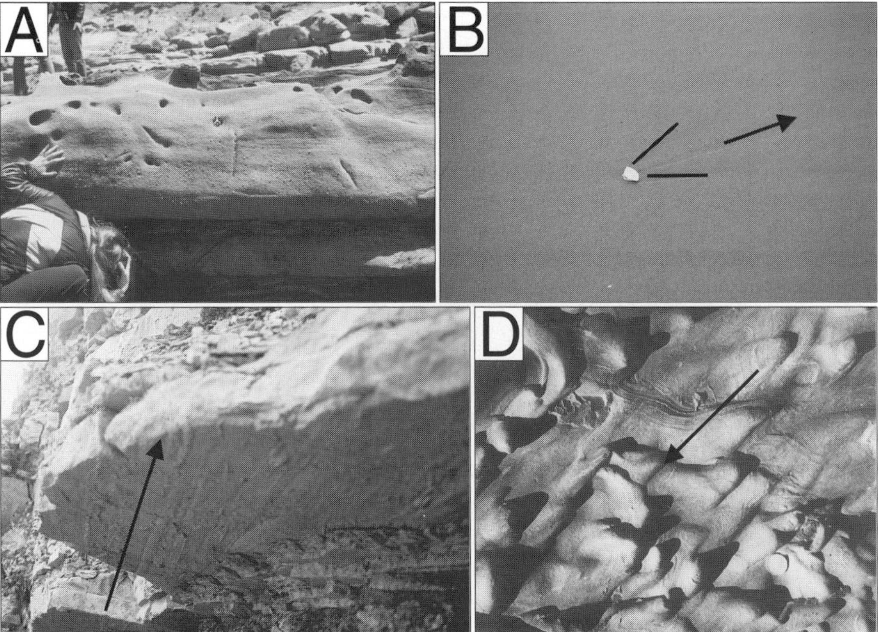

Fig. 3.49. (A) Geologist looking at the underside of a pebbly sandstone bed for tool and groove marks. (B) Shell on a modern beach creates turbulent flow on its downcurrent side (arrow), resulting in a small scour surface that splays outward in the downcurrent direction. (C) Tool marks on the underside of a sandstone bed. Arrow points to the direction of the current. (D) Flute marks on the underside of a sandstone bed. The direction of splaying of the structure indicates the downcurrent direction (arrow).

when lithified, preserved the trace of the tool's movement. Figure 3.49D shows a commonly found groove mark on the underside of sandstone beds originally deposited in deep marine environments. The feature is called a "flute mark". Turbulent flow on the bed eroded out a scour surface that splays outward in the downcurrent direction. Subsequent infilling by sand, followed by lithification, preserved this feature. Tool and groove marks are excellent indicators of the direction of sand transport and can provide important clues to the direction in which sand might have accumulated to form a reservoir.

3.3.1.4 Sandstone injectites

Sandstone injection dikes (injectites) are an unusual but very important physical sedimentary structure (Fig. 3.50) that can increase reservoir performance above expectations in fields where they are abundant. Sandstone dikes form in rapidly deposited sediment, when the sediment load on a wet sand body is sufficient to force the sand into an overlying (or sometimes underlying) wet mud. In several North Sea fields, examination of cores has revealed the presence of sandstone injection dikes (Fig. 3.51) that apparently crosscut shales and connect

Fig. 3.50. Sandstone injection dike that connects the light-colored sandstone bed below to the light-colored, thicker sandstone bed above. Thin shale beds (darker color) have been intruded by the sandstone dike.

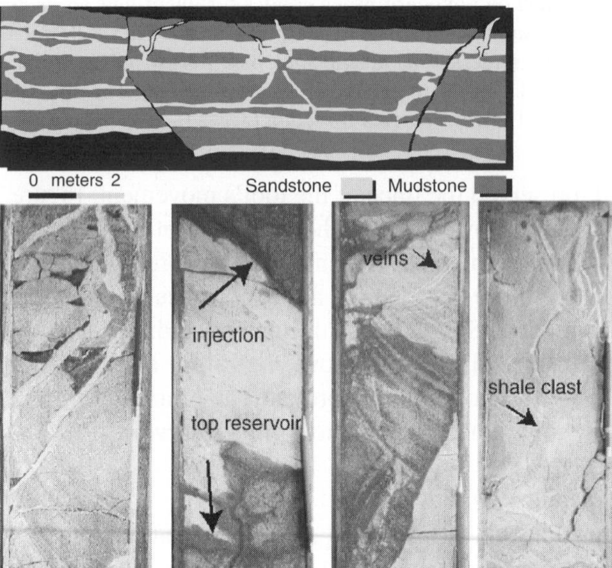

Fig. 3.51. (A) Schematic illustration of how sandstone injection dikes can connect sandstone beds (yellow/bright) that would otherwise be isolated by shale beds (brown/dark). Modified from Cossey (1994). (B) Core photographs showing sandstone injection dikes and related features. After Lonergan et al. (2000). (Reprinted with permission of SEPM, Society for Sedimentary Geology.)

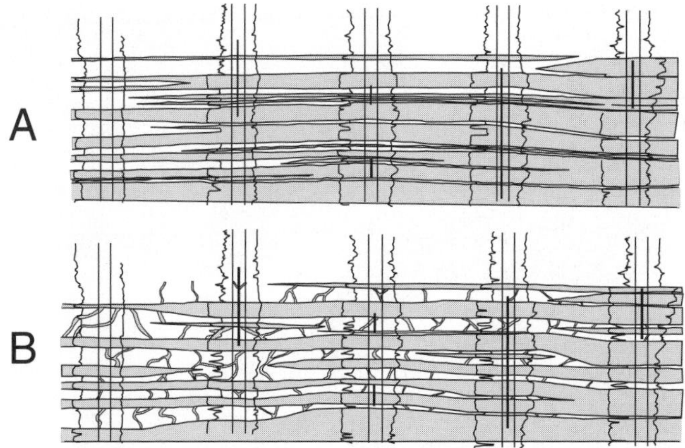

Fig. 3.52. (A) Interpreted sandstone distribution in a reservoir prior to recognizing the presence of sandstone injection dikes. (B) Reinterpreted sandstone distribution with the dikes connecting individual sandstone beds. Modified from Cossey (1994). (Reprinted with permission of SEPM, Society for Sedimentary Geology.)

sandstone beds that otherwise would have been separated by the shale interbeds (Fig. 3.52) (Cossey, 1994; Lonergan et al., 2000). Thus, reservoir connectivity and production are enhanced by their presence.

3.3.2 Biogenic sedimentary structures

3.3.2.1 Body fossils

Biogenic sedimentary structures are those produced by organisms when they were living in sediment or on the sediment surface. Body fossils, such as the Cretaceous-age oyster shells shown in Fig. 3.53, are direct indicators of the presence of organisms. There are many body fossils contained within rocks, such as dinosaur bones, mammal teeth and tusks, fish scales, and shells.

3.3.2.2 Trace fossils

There also are indirect indicators of the presence of organisms living in and on sediment that has since lithified. Such indirect indicators are called "trace fossils". The footprints of dinosaurs in Figs. 3.54 and 3.55 are one example; bird footprints on a beach are another example (Fig. 3.56A). A leaf impression is a third example (Fig. 3.56B).

Unlike these examples, however, most trace fossils are small and were produced by small organisms such as worms and crabs (or their ancient equivalents). These small trace fossils generally fall into one of two groups: those that are now found on bedding planes of rocks (Fig. 3.57) and those found perpendicular to the bedding surface (Fig. 3.58). There are environmental reasons for

Fig. 3.53. Two oyster shells within a Cretaceous sandstone. The organisms died while living in the sand, and the shells were preserved when the sand was lithified. Coin for scale.

Fig. 3.54. The "bulge" in this Jurassic age rock is the infilled footprint of a large dinosaur. The dinosaur was walking on the sand bed (red/black X) and left a footprint that was later filled in with younger sand. Dinosaur Ridge, Colorado.

Fig. 3.55. Footprints of two types of dinosaurs, on the surface of a Cretaceous muddy sandstone bed. Dinosaur Ridge, Colorado.

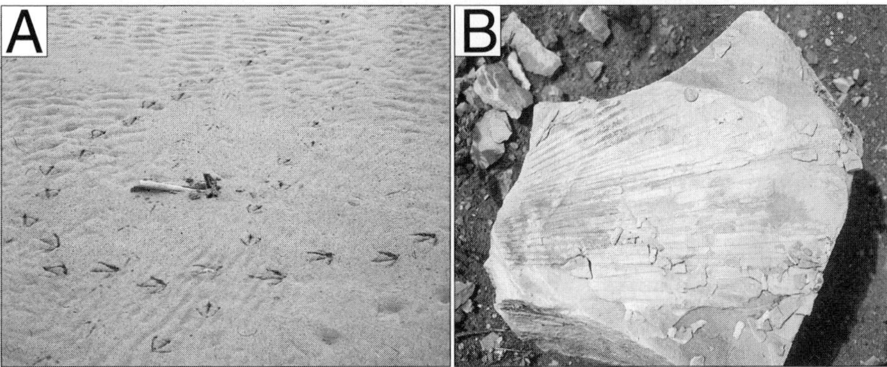

Fig. 3.56. (A) Footprints of a bird walking on a sandy beach. When this sand layer is covered by another layer, the footprints will be preserved as a trace fossil. (B) Palm leaf imprint in a rock. The imprint was preserved when the leaf was buried by a layer of sediment and the organic remains decayed, leaving a mold that was later filled with sediment, thus preserving the cast of the leaf.

these two groups. Organisms such as worms, living in a relatively sheltered, quiet water environment, such as might occur on a mud flat, are free to roam that surface looking for food. They will leave the trace of their movement on

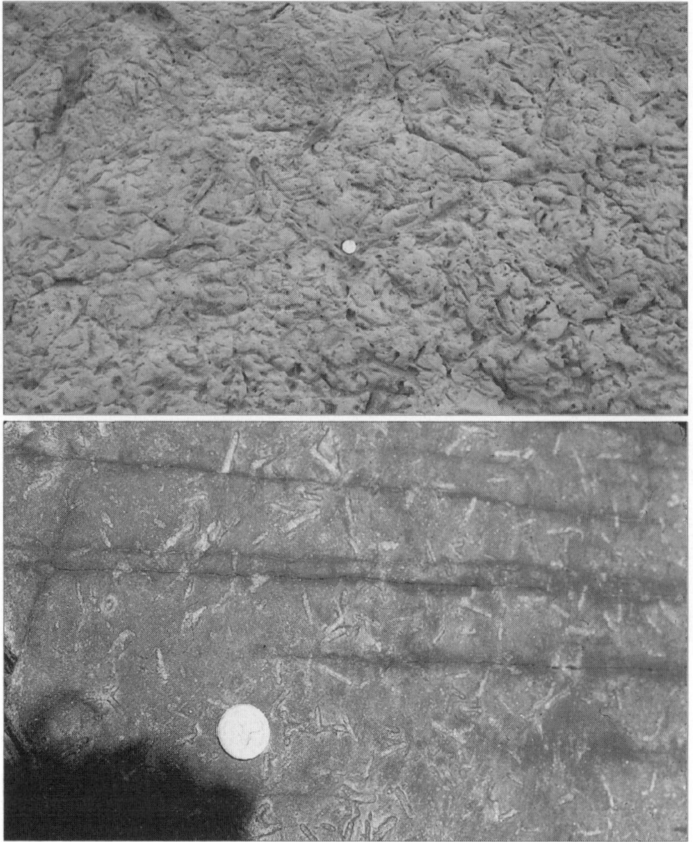

Fig. 3.57. Horizontal burrows on the surfaces of muddy sandstone beds. Coins for scale.

the sediment surface. But, in a higher-energy environment, such as a tidal flat that is subjected to wet (high tide) and dry (low tide) fluctuations, the organism is not free to roam the surface and instead burrows beneath the surface and waits for food to come to it on the incoming tide. Such an organism will leave as its trace a more or less vertical burrow. Thus, trace fossils are very good indicators of the environment of deposition of the sedimentary rock in which they are found.

There are many different types of trace fossils, with different sizes and shapes. Scientists who have studied trace fossils have been able to categorize them into groups called "ichnofacies", which are defined as assemblages of trace fossils that are diagnostic of a particular set of conditions in the depositional environment, such as water depth, current energy, and tidal influence (Fig. 3.59).

When one is examining cores through a reservoir, all trace fossils should be identified, because they provide knowledge of the depositional environment

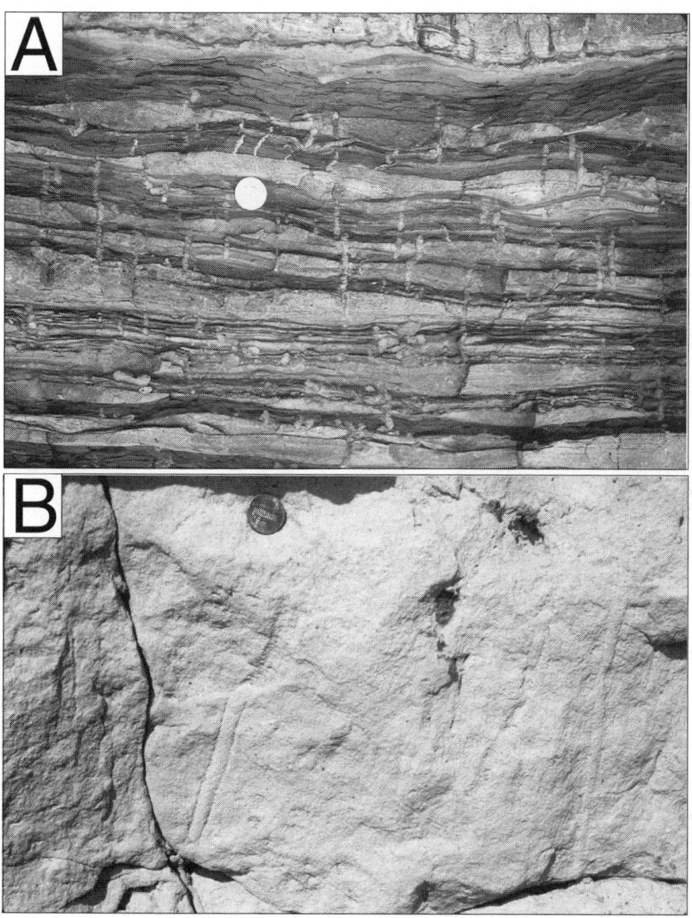

Fig. 3.58. (A) Vertical section of alternating thin sandstones and shales deposited in a tidal-flat environment. The abundance of vertical burrows indicates that many organisms occupied this environment, and they had to burrow to prevent themselves from being washed away by strong tidal currents. The burrows are now filled with sand. (B) This vertical section of rock has vertical trace fossils called *Ophiomorpha*. It is thought that the organism that produced this burrow was a wormlike creature that burrowed into the sand. It would stick its head out of the sand to catch food that washed by in the marine currents. When the organism left its burrow to find a new home, the burrow became filled with sediment. Vertical burrows usually occur in areas of strong marine wave or current action. The organism must burrow into the sediment to prevent itself from being washed away. Coins for scale.

(Fig. 3.59). For example, *Cruziana* and *Skolithos* ichnofacies (Fig. 3.60) are very good indicators of conditions within a shallow marine-shoreface depositional environment, and they also provide a means of differentiating subenvi-

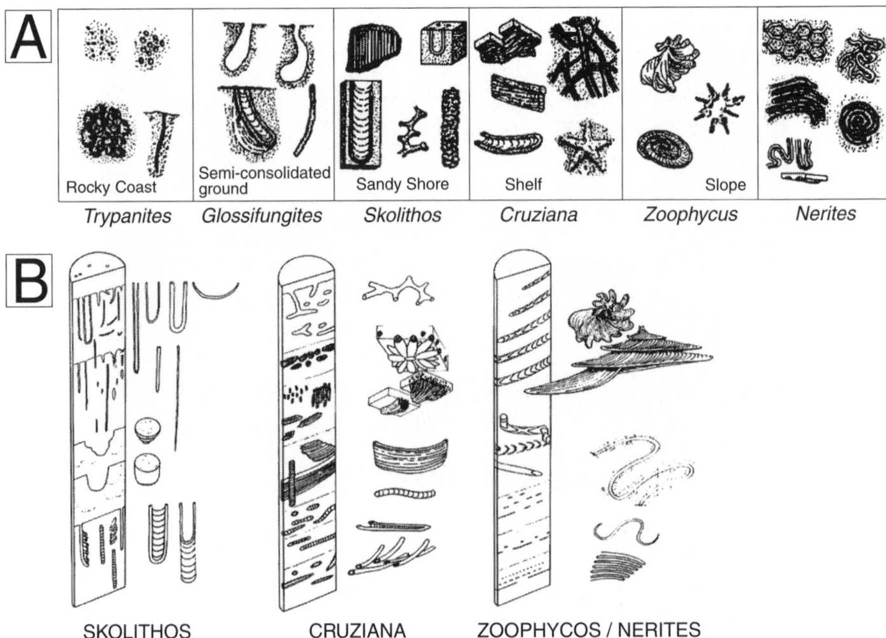

Fig. 3.59. (A) Various groups of trace fossils form ichnofacies. The ichnofacies are diagnostic of different sedimentary environments that usually are based on wave and current energy. In this figure, the current energy and marine water depth generally decrease from left to right (B) Ichnofacies as observed in core. After Pemberton (1992). (Reprinted with permission of SEPM, Society for Sedimentary Geology.)

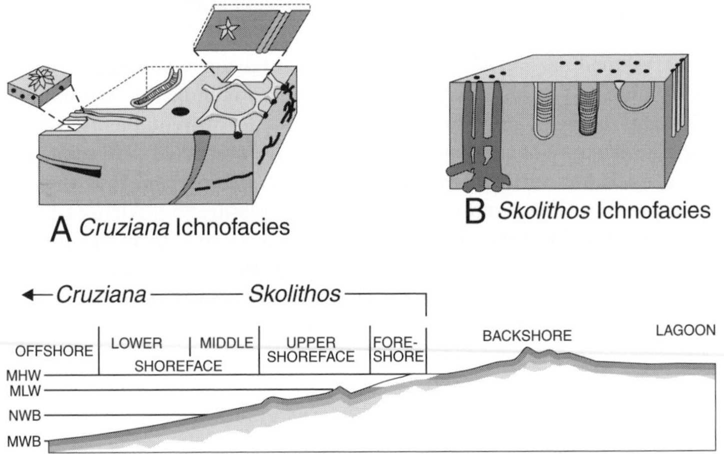

Fig. 3.60. Cruziana and Skolithos ichnofacies and the shallow marine environments in which these trace fossils occurred. After Pemberton (1992). (Reprinted with permission of SEPM, Society for Sedimentary Geology.)

ronments (Fig. 3.61). This is very important in reservoir characterization. The *Skolithos* ichnofacies is indicative of a sandier depositional environment than is a *Cruziana* ichnofacies (Figs. 3.60 and 3.61). When Skolithos is found in a core of a reservoir sandstone, predictions can be made as to the direction in which more sand might be present.

Trace fossils also can affect the reservoir quality and the thickness and continuity of reservoir sandstones. For example, if sand infills burrows significantly on an underlying mud bed, the effective thickness of the bed is increased (Fig. 3.62A). If a sandstone bed contains an abundance of mud-filled burrows,

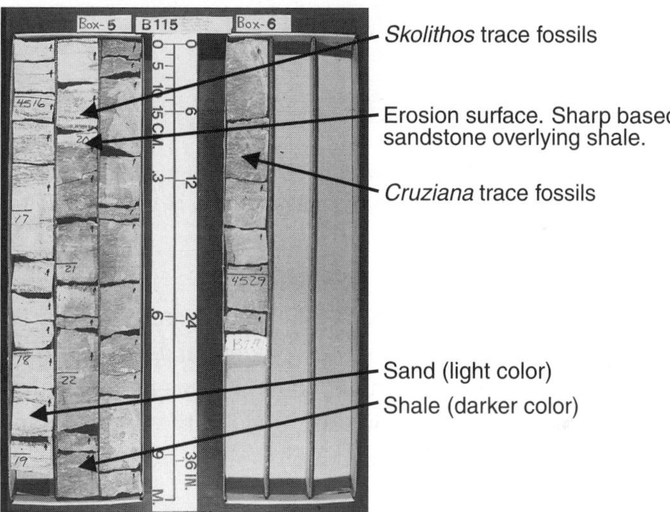

Fig. 3.61. Core of Cretaceous rock showing different physical and biogenic sedimentary features, including darker (shale) and lighter (sandstone) rocks, a sharp erosional surface just beneath the number 20, and numerous examples of trace fossils (burrows) in both the sandstone and shale (light dots and lines in the darker matrix rock).

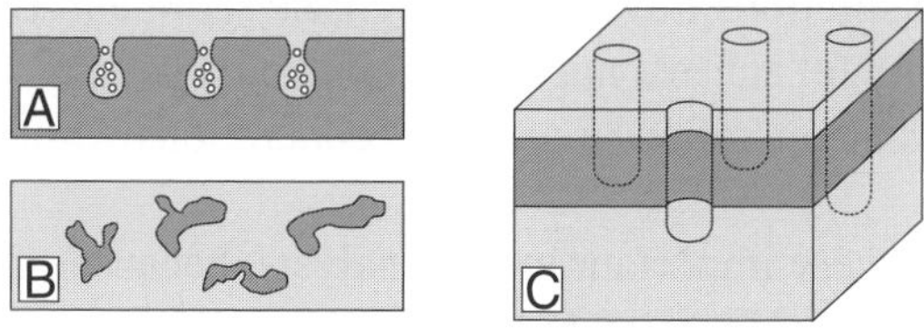

Fig. 3.62. Schematic illustrations of the effect of thickness and biogenic sedimentary structures on permeability, porosity, and connectivity of sedimentary beds. (A) Vertical burrows on the upper part of a shale bed have been infilled with sand, effectively increasing the thickness of the sandstone bed. (B) Burrows within a sandstone bed have been infilled with shale, effectively reducing the net sandstone. (C) Vertical sedimentary structures can either connect or disconnect sandstone beds, depending upon whether the burrows are filled with sandstone or shale.

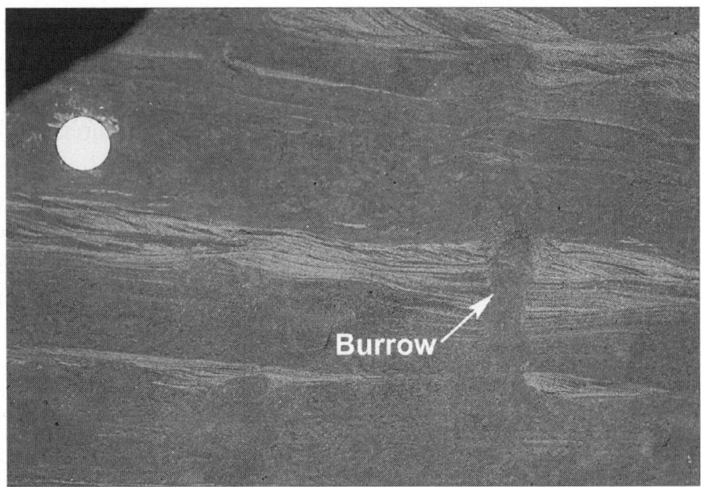

Fig. 3.63. Vertical burrow that is filled with dark colored mud, thus dividing the lighter colored sand bed into two disconnected parts on either side of the burrow in this Miocene rock.

the effective permeability and porosity of the bed are reduced (Fig. 3.62B). If thin interbeds of sandstone and shale contain sand-filled burrows that extend across shale beds, then the vertical connectivity of beds can be increased (Fig. 3.62C). Alternatively, if burrows are filled with mud, the sand beds can become discontinuous (Figs. 3.62C and 3.63).

3.3.3 Chemical sedimentary structures

Chemical sedimentary structures are the product of chemical precipitation of minerals that occurs within a sediment before it lithifies. Concretions are the most common of these structures (Fig. 3.64). Concretions form by precipitation of a mineral from solutions traveling through the sediment. Often, a sand grain or a shell fragment will act as a nucleus onto which such precipitation will proceed. The result is a concretion or series of concretions within the sedimentary bed (Fig. 3.65).

Concretions are very important in log analysis, because often they are composed of the mineral siderite, which is an iron-rich carbonate. Because the density of this mineral is greater than that of the surrounding sandstone, the concretion can provide erroneous or misleading well-log responses. Also, concretions can form a layer within a sedimentary rock sequence (Fig. 3.65); if the layer contains enough concretions, it can act as a baffle or permeability barrier to fluid flow (Fig. 3.66). Concretions are easily seen in core or on borehole-image logs (Fig. 3.67).

Fig. 3.64. The bulbous features shown here are concretions. They form by chemical precipitation of calcite (usually) within a sand soon after the sand is deposited. For a concretion to form, fluids containing dissolved calcite (ions of Ca, CO_3) must be moving through sand that contains a nucleus, such as a fossil shell, onto which the calcite cement initially precipitates. Once precipitation begins on a nucleus, the process will continue and the concretion will grow within the sediment.

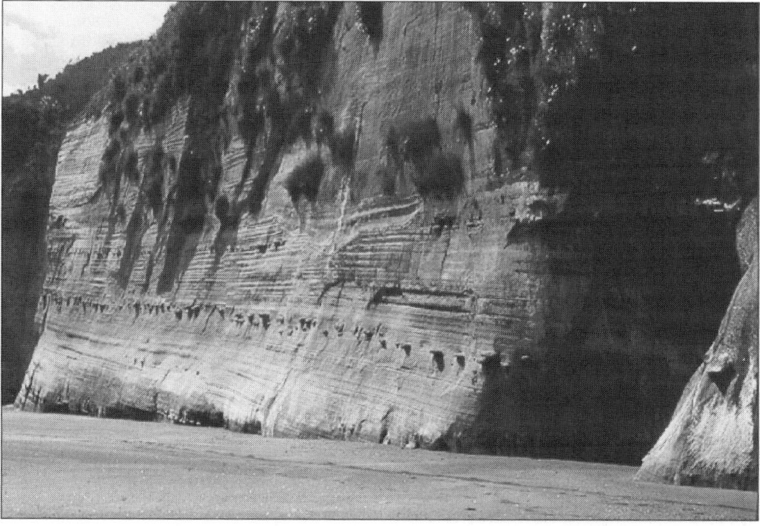

Fig. 3.65. Row of concretions (globular features extruding from the cliff face) along a single bedding plane. Miocene Mt. Messenger Formation, New Zealand. After Browne and Slatt (2002).

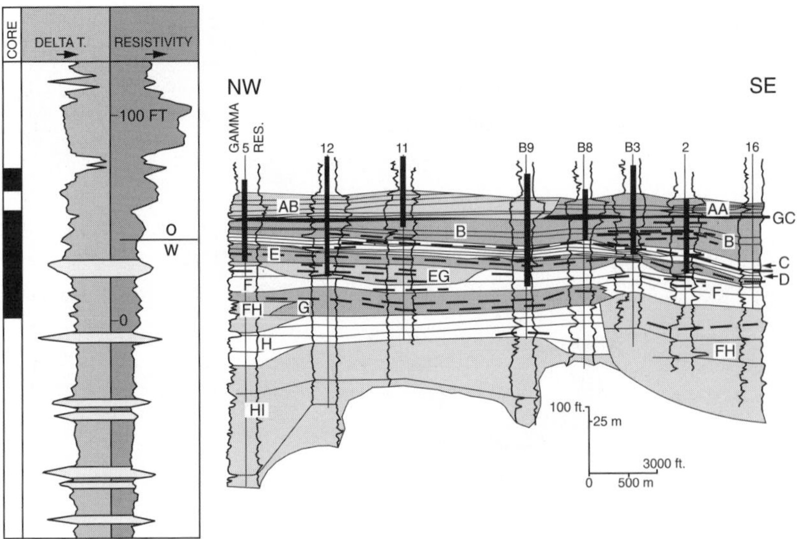

Fig. 3.66. Concretion zones in the Balmoral field, North Sea (Slatt and Hopkins, 1991). The concretion zones are characterized by anomalous spikes on sonic and resistivity logs (blue/bright-gray on the logs). These spikes can be correlated across several wells (dashed lines), indicating that the concretions form as continuous layers such as that shown in Fig. 3.65.

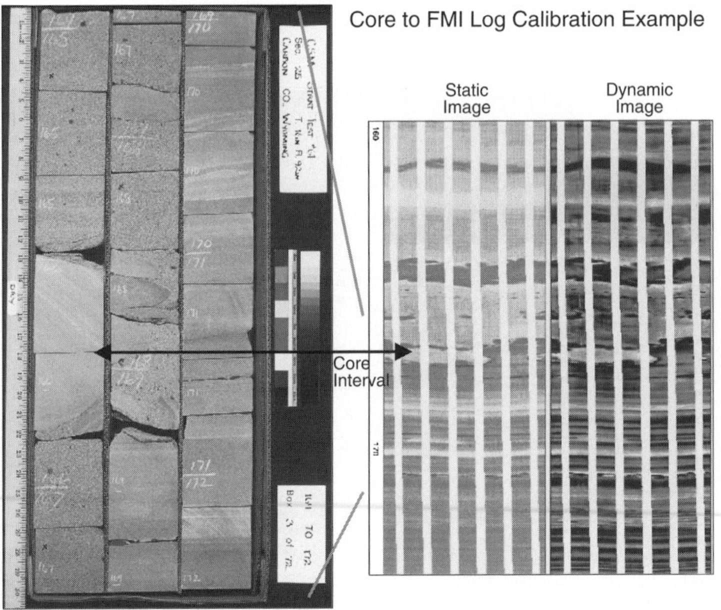

Fig. 3.67. Core and a borehole-image log, through the cored interval in the wellbore, both showing a spherical, calcareous concretion. Dad Sandstone Member, Lewis Shale, Wyoming. (Picture provided by S. Goolsby.)

3.4 Summary

In summary, physical, biogenic, and chemical sedimentary structures are important to many aspects of reservoir characterization and should be included in every characterization, whether the analyst is using cores, borehole-image logs, or an analog outcrop. Sedimentary structures provide important information about the depositional environment of the reservoir rock, and from that information one can determine the extent and geometry of the reservoir, its trend, and any likely impediments to hydrocarbon production. Porosity and permeability and, in particular, fluid-flow paths, are also affected and guided by how the sediment grains are arranged into specific structures. Finally, one should bear in mind that some sedimentary structures can produce misleading or erroneous well-log results.

Chapter 4

Geologic time and stratigraphy

4.1 Introduction

The concept of geologic time is central to an understanding of all geologic phenomena and features. Many nongeologists find it difficult to grasp the idea that the earth is a dynamic environment that changes over long time intervals, because daily, and even lifetime, changes on and beneath the earth's surface are imperceptibly small. However, there is a large body of evidence that geologic processes do operate over vast periods of time to continually shape and reshape the earth. This chapter presents an overview of some fundamentals of geologic time, so that the reader can better comprehend topics in later chapters concerning the origins of sedimentary environments and their deposits (including reservoirs) and postdepositional processes that modify these deposits. Also, the broad field of biostratigraphy is touched upon in this chapter, because it is so indelibly linked with geologic time.

Table 4.1 lists some of the key events in the origin of the earth and its inhabiting organisms, including humans. The earth is thought to have formed about 4.5 billion years ago, although this date is subject to change as new evidence is uncovered. The oldest age-dated rocks (again, subject to change as older rocks are discovered and verified) are thought to have formed on the order of 3.8 billion years ago, suggesting a long time interval between the earth's formation in a molten state and its cooling to form solid rock. The atmosphere formed a few hundreds of thousands of years afterward.

Life on earth is thought to have begun about 3.3 billion years ago, with the birth of single-cell plant organisms. Between that time and approximately 540 million years ago (Ma), simple life forms evolved. Significant evidence of more-advanced life forms does not appear in the rock record until about 540 million years ago. Rocks deposited since that time contain a record of progressively developing life forms, up to and including modern plant and animal species. Global environmental events such as mountain-building and glacial cycles, and extraterrestrial events such as meteor impacts, periodically have led

Table 4.1 Data from Davidson et al. (2002). (Reprinted with permission of Prentice-Hall, New Jersey)

Millions of years ago (Ma)	Major event
4,550	Formation of planet Earth
3,800	Oldest dated rocks
3,500	Oceans and atmosphere begin to form
3,300	Oldest dated fossils; single-cell plants
540	Abundant fossil record; first multi-cell and shelled organisms
440	First land plants
400	First amphibians
265	First reptiles
245	Pangea supercontinent begins to form
200	Pangea supercontinent formed
65	Extinction of dinosaurs
50	Ancestors of horses appear in North America
40	Himalayas begin to form by collision of India and Asia
2+	Great ice age; humanoids appear

to the decline and sometimes even the destruction of one group of life forms (i.e., to "mass extinctions") and to the development of other groups that fill the new or recently vacated environmental niches. One example is the extinction of dinosaurs about 70 million years ago and the subsequently rapid and rich expansion of mammalian species.

4.2 North American geologic time scale

The North American geologic time scale (Table 4.2) provides a classification of geologic time on the basis of important events in the evolution of earth and its inhabitants. The time scale is divided into three "post-Precambrian" Eras: the Paleozoic, Mesozoic, and Cenozoic. These Eras are subdivided into a number of Periods. The Tertiary and Quaternary Periods are further subdivided into Epochs. These names are used to denote the times in which various geologic events occurred. For example, a rock deposited during Cretaceous time would be a rock originally deposited as sediment somewhere during the time interval 144–65 Ma (Table 4.2). A mountain-building event that occurred about 180 Ma would have occurred during Jurassic time (Table 4.2).

4.3 Determining the time frame in which a rock formed

The age range during which a sedimentary rock was originally deposited can be determined either by direct measurement or by relative dating techniques that establish the order in which geologic events occurred.

Table 4.2 Data from Davidson et al. (2002). (Reprinted with permission of Prentice-Hall, New Jersey)

Start of interval: millions of years ago (Ma)	Era	Period	Epoch
0.01	Cenozoic	Quaternary	Holocene
1.8			Pleistocene
5.3		Tertiary	Pliocene
23.8			Miocene
33.7			Oligocene
54.8			Eocene
65			Paleocene
144	Mesozoic	Cretaceous	
206		Jurassic	
251		Triassic	
286	Paleozoic	Permian	
325		Pennsylvanian	
360		Mississippian	
410		Devonian	
440		Silurian	
505		Ordovician	
544		Cambrian	

4.3.1 Radiometric age dating ("the clocks in rocks")

Many of the Earth's natural chemical elements are stable once they are formed. Other elements have more than one species. Chemical species that contain the same number of protons in the nucleus that the stable elemental form has, but that contain a different number of neutrons are called isotopes (Jackson, 1997). Because of slight differences in physiochemical properties of isotopes relative to their stable elemental relatives, radioactive isotopes tend to be unstable and over time will "decay" and form more stable "daughter" elements or products. Radiometric dating is a process that determines the age in years of geologic materials by using various techniques that are based on nuclear decay of naturally occurring radioactive isotopes.

Some examples of unstable isotopes and their stable daughter products are given in Fig. 4.1. For example, an isotope of uranium (U) is U^{235}. Its daughter product is lead (Pb^{207}). The conversion from U^{235} to Pb^{207} also releases helium (He) atoms. It has been determined that it takes 704 m.y. for half of an original amount of U^{235} isotope to convert to Pb^{207}. Thus, if a rock contains equal amounts of U^{235} and Pb^{207}, the rock must have formed 704 Ma. The amount of time necessary for a radioactive isotope to lose half of its radioactivity is termed the "half-life" of the isotope. Carrying the U^{235} example further, if a rock con-

AGE DATING OF ROCKS
(THE CLOCKS IN ROCKS)

PRINCIPLE:
CERTAIN UNSTABLE ELEMENTS (ISOTOPES) BREAK DOWN (DECAY) TO PRODUCE STABLE ELEMENTS

SOME EXAMPLES:

$Rb^{87} \rightarrow \beta \rightarrow Sr^{87}$
$U^{235} \rightarrow Pb^{207} + 7He$
$U^{238} \rightarrow Pb^{206} + 8He$
$Th^{232} \rightarrow PB^{208}$
$K^{40} + e \rightarrow Ar^{40}$
$K^{40} - \beta \rightarrow Ca^{40}$

Fig. 4.1. Primary rocks (igneous) contain a certain amount of unstable elements (radioactive isotopes) that break down (decay) with time to form new, more-stable elements (and in the process, the isotopes emit radioactive particles). Experimental studies have allowed scientists to determine the rate at which the radioactive elements break down to form stable elements. Thus, by measuring the amount of a radioactive element and the amount of the stable "daughter" product in a rock, scientists can determine when (how long ago) the rock formed. Some examples are provided in the figure and are discussed further in the text.

tains 3 times the amount of Pb^{207} that it does of U^{235}, the rock formed 1,408 Ma (i.e., two half-lives of U^{235} have elapsed).

Different isotopes have different half-lives. Of those listed in Fig. 4.1, Rb^{87}–Sr^{87} has a half-life of 48,800 m.y., K^{40}–Ar^{40} has a half-life of 1,300 m.y., U^{238}–Pb^{206} has a half-life of 4,470 m.y., and Th^{232}–Pb^{208} has a half-life of 14,000 m.y.

In practice, minerals containing these elements are extracted from a rock and the ratios are measured from the mineral. The numerical age that is determined from radioisotope analysis is the time of formation of the radioactive mineral in the rock; that is, the time at which the mineral (and rock) solidified. Because siliciclastic sedimentary rocks are formed by breakdown of preexisting rocks into mineral components, absolute age-dating of the minerals in a sedimentary rock does not give the time that the sedimentary particles were deposited. Rather, such determinations are of the age of the source rock from which the sedimentary particles were derived. The exception to this rule is for chemically precipitated minerals, such as those precipitated directly from sea water, those forming the shells of organisms, and those comprising cements that precipitated out of subsurface waters moving through sediments.

Some isotopes have short half-lives. One important example is carbon (C^{14}), which has a half-life of 5,730 years. This short half-life, coupled with the fact that carbon accumulates in plant and shell material that then becomes incorporated into a sedimentary deposit, allows the dating of deposition of very young Holocene (Table 4.2) sediments.

4.3.2 Relative age-dating

Because of the limitations of using radiometric age determination to identify the time during which sedimentary rocks formed, relative age dating techniques commonly are used. Several geologic "laws" have been developed to date the relative age of formation of a sedimentary rock.

The Law of Original Horizontality states that because stratified rocks are deposited as sediments from water, ice, or air, they must be deposited with an originally horizontal attitude on an originally horizontal surface (Fig. 4.2).

The Law of Superposition states that in an undisturbed sequence of rocks, the bottom layer is oldest, and the top layer is youngest (Fig. 4.3). When one is examining a suite of stratified rocks in order to date their relative ages of formation, it is important first to ascertain that the rocks are right-side-up and have not been overturned as a result of tectonic activity. This can be accomplished by looking for various sedimentary features that occur only on the tops or bottoms of individual beds. Some of these features were discussed in Chapter 3 – for example, mud cracks (Fig. 3.35), ripples (Figs. 3.39–3.43), size-graded beds such as Bouma sequences (Figs. 3.44 and 3.45), load structures (Fig. 3.46), erosional

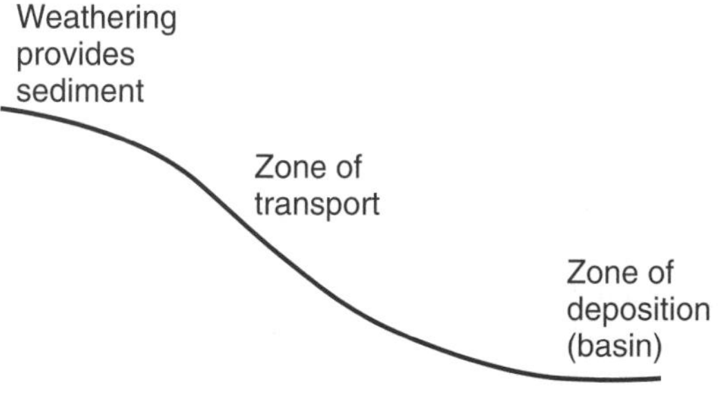
Because stratified rocks are depositied as sediments from water or air, they must be deposited with an originally horizontal attitude
Weathering provides sediment
Zone of transport
Zone of deposition (basin)
LAW OF ORIGINAL HORIZONTALITY

Fig. 4.2. The Law of Original Horizontality states that sediments must be deposited with an originally horizontal attitude. The sedimentary particles originate in mountainous areas as a result of weathering processes, are transported to lower areas by wind, water, and/or ice and, under the influence of gravity, are deposited onto a horizontal surface. The upper inset shows steeply dipping beds that have been tilted by tectonic activity, rather than having been deposited on a steep slope. The lower inset shows a sequence of horizontally bedded strata; these strata have not been tilted significantly following their deposition.

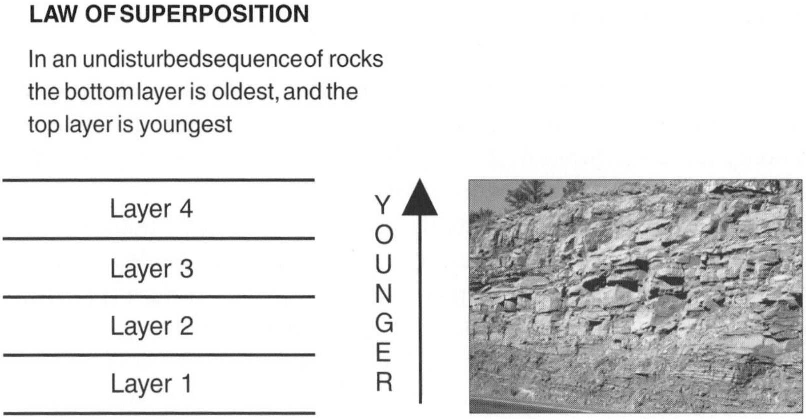

LAW OF SUPERPOSITION

In an undisturbed sequence of rocks the bottom layer is oldest, and the top layer is youngest

Layer 4

Layer 3

Layer 2

Layer 1

YOUNGER

Fig. 4.3. The Law of Superposition states that in a sequence of sedimentary rocks, the lowest layer was deposited before overlying layers and therefore must be older. In the figure, the oldest or earliest deposited layer is Layer 1, and the youngest is Layer 4.

LAW OF FAUNAL SUCCESSION

Characteristic fossils have succeeded one another over geological time

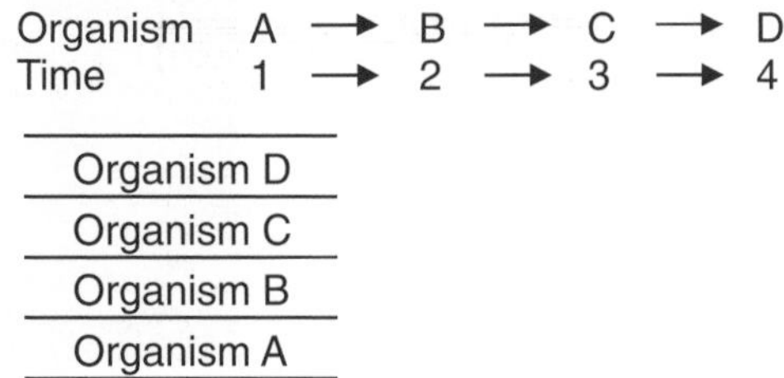

Fig. 4.4. The Law of Faunal Succession relates to the relative ages of fossils. It is well known from study of ancient fossils that particular organisms, such as clams, oysters, and the like, have changed over time. According to the Law of Superposition, the layer that contains Organism A is the oldest, and the layer that contains Organism D is the youngest. By examining the fossil forms from the oldest to the youngest layer, any temporal changes or progressions in the life form can be noted. The importance of this is that particular organisms (A, B, C, or D in this case) live during a particular time interval. By knowing that time interval, it is possible to determine the age of the rock in which the organism occurs.

scour surfaces (Figs. 3.47 and 3.48), flute marks (Fig. 3.49), and some trace fossils (Figs. 3.54–3.60).

The Law of Faunal Succession states that characteristic fossils have succeeded one another throughout geologic time (Fig. 4.4). By knowing the order of succession of fossils in a suite of sedimentary rocks, it is possible to determine the relative ages of the strata that contain the fossils.

An "unconformity" is an important rock surface that is defined as a substantial break, or time gap, in the geologic record. An unconformity occurs either when sediment was not deposited or when it was deposited and then eroded away. Thus, an unconformity is the surface where one rock unit is overlain by another rock unit that is not next in stratigraphic or age succession. There are three types of unconformity: disconformity, angular unconformity, and nonconformity (Figs. 4.5 and 4.6). It is critical to recognize unconformities in the stratigraphic record, because they represent an interval of geologic time for which there is no rock record at a particular location. During such a time, geologic processes like tectonic uplift or sea-level fluctuation may have occurred that, in fact, may have been responsible for the formation of the unconformity surface.

Another useful law for dating rock sequences is the Law of Crosscutting Relations. This law sometimes can be used in conjunction with radiometric age dating to apply an absolute age (or age range) to the formation of a sedimentary rock or rock succession. This law states that if a rock crosscuts another rock, the crosscutting rock must be younger than the rock being cut (i.e., a rock must

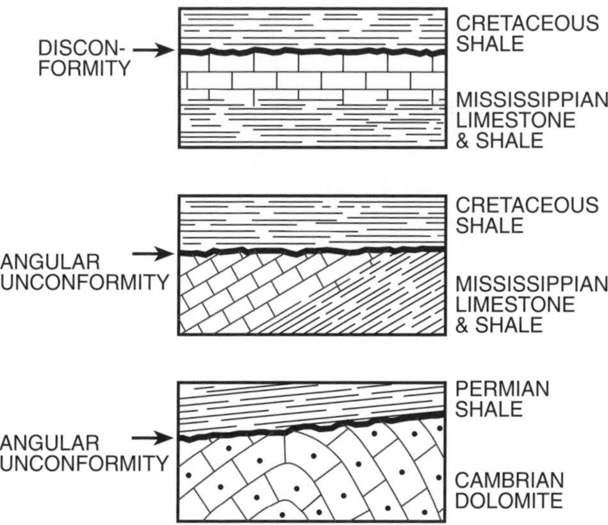

Fig. 4.5. Types of unconformities. A disconformity records a surface of erosion or nondeposition during a particular interval of geologic time. A time gap exists between the underlying and overlying strata, even though the beds dip at the same angle. An angular unconformity is a surface representing a gap in geologic time, during which older rocks initially were deposited and then deformed or tilted, next eroded (to form the unconformity), and finally covered by deposition of horizontal strata above. Thus, it is a situation in which younger sediments rest upon the eroded surface of tilted or folded older rocks (Jackson, 1997). Angular unconformities normally imply a long time interval between deposition and tilting of underlying strata, and deposition of the horizontal overlying strata. (Reprinted with permission of American Geological Institute.)

Fig. 4.6. (A) The famous Hutton (angular) Unconformity at Siccar Point, Scotland. (B) Angular unconformity of underlying Pennsylvanian Jackfork Group deepwater sandstones, which dip at 55° and are overlain by Cretaceous fluvial strata (that are horizontally bedded). The white color of the Jackfork, which normally is highly cemented and brittle, is due to weathering, erosion, and soil formation during the time interval between tectonic tilting of the beds and deposition of the Cretaceous strata. Such a porous zone directly beneath an angular unconformity capped by a nonporous rock can form an excellent trap for hydrocarbons. (C) Nonconformity between Precambrian-age granites and overlying Pennsylvanian Fountain Formation fluvial sandstones, Colorado.

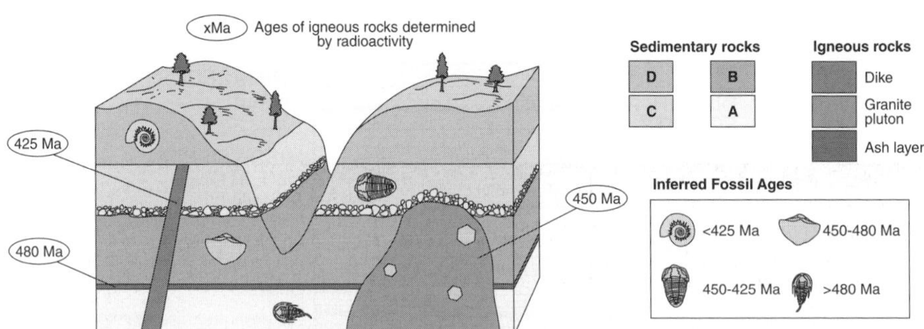

Fig. 4.7. Illustration of the Law of crosscutting relationships. Three igneous rocks have been radiometrically dated at 480, 450, and 425 Ma. Because rock A underlies a 480-Ma ash layer, it must have been deposited more than 480 m.y. ago. Rock B has been intruded by the 450-Ma granite pluton and crosscuts the 480-Ma ash layer, so it must have been deposited between 480 and 450 m.y. ago. Rock C is intruded by the 425-Ma dike, so it must have been deposited more than 425 m.y. ago. But, because Rock C is not intruded by the 450-Ma granite pluton, it must have been deposited after the pluton was intruded. Thus rock C was deposited sometime in the interval 450–425 Ma. Rock D has not been intruded by the dike, so it must have been deposited less than 425 Ma. In this case, a combination of absolute and relative age-dating techniques, coupled with analysis of how the rocks formed, have provided a geologic history of the area. After Davidson et al. (2002). (Reprinted with permission of Prentice-Hall, New Jersey.)

have been present for it to have been cut by another rock). This law is particularly useful when the crosscutting rock is of igneous origin, so that it can be radiometrically dated. The example shown in Fig. 4.7 illustrates this principle.

4.4 Micropaleontology and biostratigraphy in reservoir characterization

The relative and absolute age dating of marine microorganisms is particularly important to reservoir characterization, as well as to oil and gas exploration. Both plant (microflora) and animal (microfauna) microorganisms live in the oceans (Fig. 4.8). Important groups of organisms are foraminifera (Fig. 3.23), radiolaria, palynomorphs, and nannoplankton. Microorganisms that live within the ocean water column are called planktonic microorganisms, and those that live on the seafloor are called benthonic microorganisms. In the following sections, some applications of biostratigraphy are presented.

4.4.1 High-resolution biostratigraphic zonation (biozones)

In some parts of the world, the stratigraphic distribution of ocean-dwelling microorganisms has been documented in great detail. The northern Gulf of Mexico is one such location, and a detailed Cenozoic biostratigraphic zonation has been established for foraminifera, nannoplankton, radiolaria, and palynomorphs, by analyzing numerous oil and gas wells (Lawless et al., 1997). In addition to bio-

Fig. 4.8. Planktonic microorganisms living in the ocean water. Beneath the diver's fins is the side of an iceberg. The blurry material in the water is the microorganisms, some of which were captured in a jar (inset).

stratigraphic zonation, measurements of oxygen isotopes from the tests (hard external structures or 'homes') of these microorganisms have provided an oxygen isotope age scale that reflects paleoclimatic variations in ocean water temperatures. This oxygen isotope age scale can be related to glacial–interglacial cycles. Also, a magnetostratigraphic absolute age scale has been developed on the basis of radiometric age dating and magnetic polarity of volcanic strata interbedded with sedimentary rocks. These combined scales of relative and absolute ages provide a very detailed record of the age range in which specific microorganisms were present in Gulf of Mexico waters during the Cenozoic.

Biostratigraphers who are familiar with the Cenozoic chronostratigraphic (time) and biostratigraphic scales can examine the microorganisms in sediments from well cuttings (Fig. 2.35) or cores and, on the basis of which organisms are present, determine the age of deposition of the sediment in which the microorganisms are found. In practice, a biostratigrapher will make particular note of the first and last depths at which certain diagnostic microorganisms occur in well cuttings, because these depths document the time span during which the organism lived. The stratigraphic interval in which a specific suite of microorganisms occurs is called a biozone. By comparing it with the biostratigraphic and chronostratigraphic time scales, the age range of a biozone can be determined. This method of determining first and last occurrences of diagnostic microorganisms is a particularly useful means for correlating strata in different wells in places like the northern Gulf of Mexico. Common nicknames or abbreviations for specific diagnostic microorganisms used for such correlations include "Cib Carst", "Big Hum", and Upper "Cris I". In cases where only the first occurrence of the microorganism is the correlation horizon, the term "Top Cib Carst" (for example) is used.

4.4.2 High-resolution well log and seismic correlation from biostratigraphy

Once a particular biozone is identified on a series of well logs, either the entire biozone or its top represents a time-stratigraphic correlation marker. If the absolute age of the biozone horizon is known from the microorganisms it contains, that horizon's geologic age of deposition also becomes known (Fig. 4.9). If the well crosses a seismic reflection profile, a 1D synthetic seismogram of the well can be generated and the dated horizon can be located on the seismic profile. Applying ages to seismic reflections is important for mapping geologic time horizons over longer distances than can be done from wells alone. In tectonically complex areas, determining the geologic ages represented by seismic reflections is the only way in which horizons can be correlated across structures (Fig. 4.10).

High-resolution or "high-impact" biostratigraphy generally is used at the exploration scale, but it also can be used to correlate stratigraphic intervals

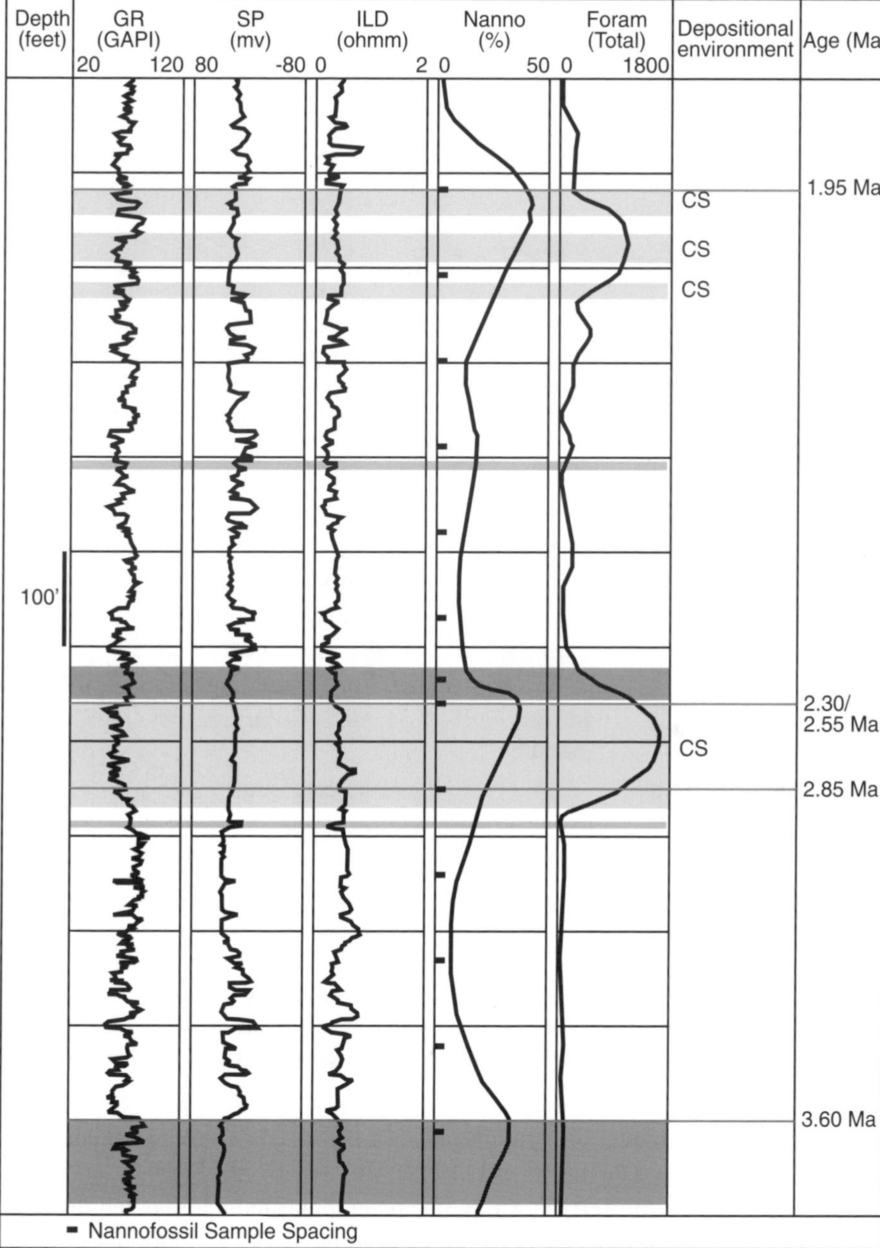

Fig. 4.9. Conventional well logs, and nannoplankton and foraminiferal abundances, in a northern Gulf of Mexico well. Age determinations were made by using biostratigraphy. Four ages are shown at different horizons. Two strong abundance peaks and one inconclusive peak (with only nannoplankton abundance) are shown. CS – condensed section.

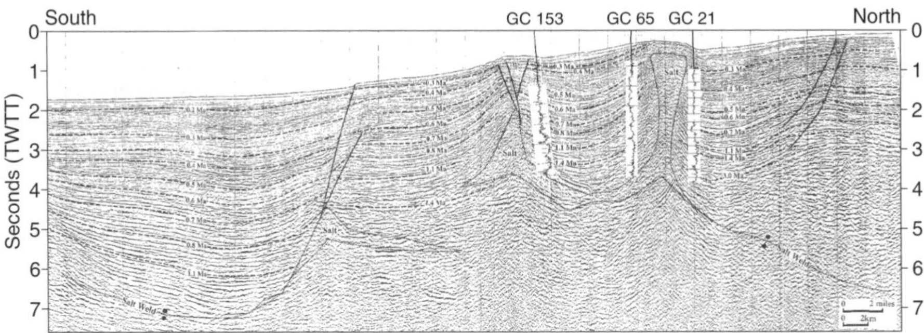

Fig. 4.10. Seismic reflection profile across salt minibasins in the northern Gulf of Mexico. Dashed lines are biostratigraphically age-dated horizons from well logs which have been correlated to seismic amplitude horizons. This method allows correlation of seismic horizons and intervals on both sides of structures. After Weimer et al. (1998). (Reprinted with permission of AAPG, whose permission is required for further use.)

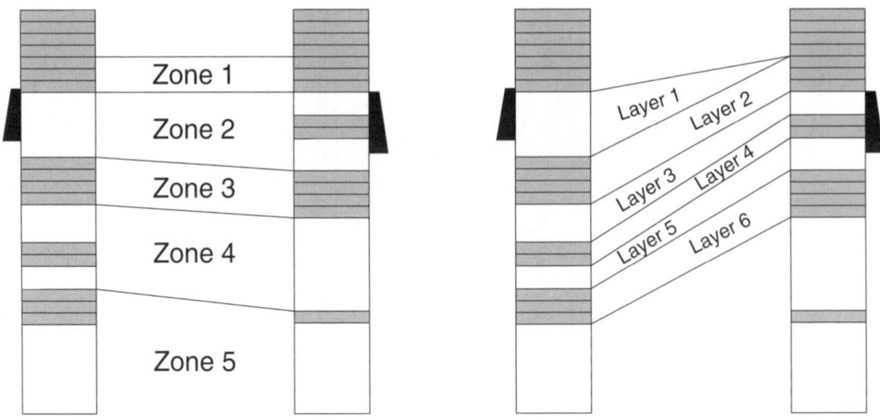

Fig. 4.11. (A) Lithostratigraphic correlation across two wells. (B) The same two wells, but with a more refined and accurate biozonation based upon micropaleontologic analysis. After Payne et al. (1999). (Reprinted with permission of Geological Society of London.)

in reservoirs, if a good biostratigraphic zonation has been established for the particular strata (Payne et al., 1999). In the North Sea, for example, detailed biozonation of Tertiary strata has been identified and has led to more-refined correlations among wells within reservoirs. This has important implications for determining lateral continuity of sandstones (reservoirs) and shales (seals and barriers) (Fig. 4.11). A particularly good example of using high-resolution biostratigraphy to characterize a Paleocene reservoir is Grane field, Norway (Mangerud et al., 1999). On the basis of the log response to lithofacies in four wells spaced 3–7 km apart, it appeared to be easy to correlate the

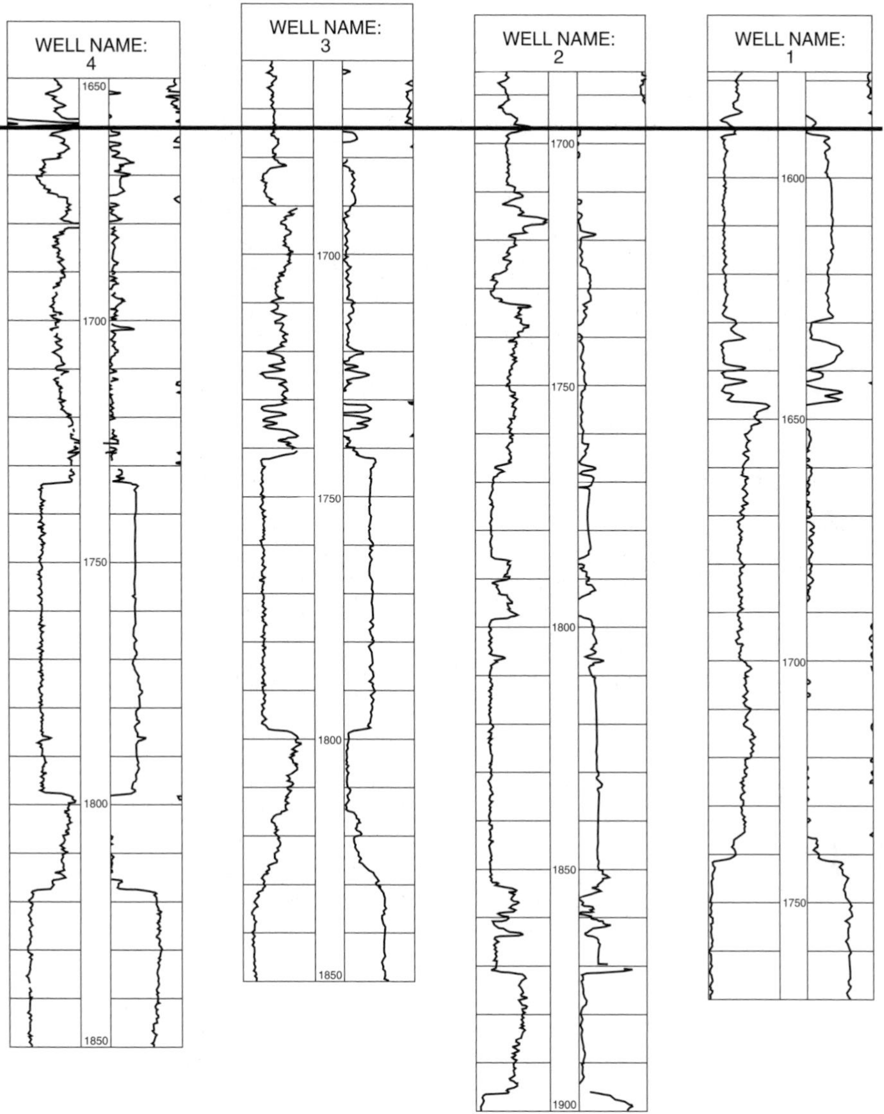

Fig. 4.12. Four-well, north–south cross-section in Grane field, Norway. The cross-section is approximately 15 km across. Lithostratigraphic datum is shown by heavy black line. After Mangerud et al. (1999). (Reprinted with permission of Geological Society of London.)

thickest sandstone interval between wells 4 and 3 (Fig. 4.12). One of several possible lithostratigraphic correlations is shown in Fig. 4.13. However, biostratigraphic analysis revealed a significantly different correlation pattern, with the thickest sandstone interval in wells 4 and 3 being within different biozones

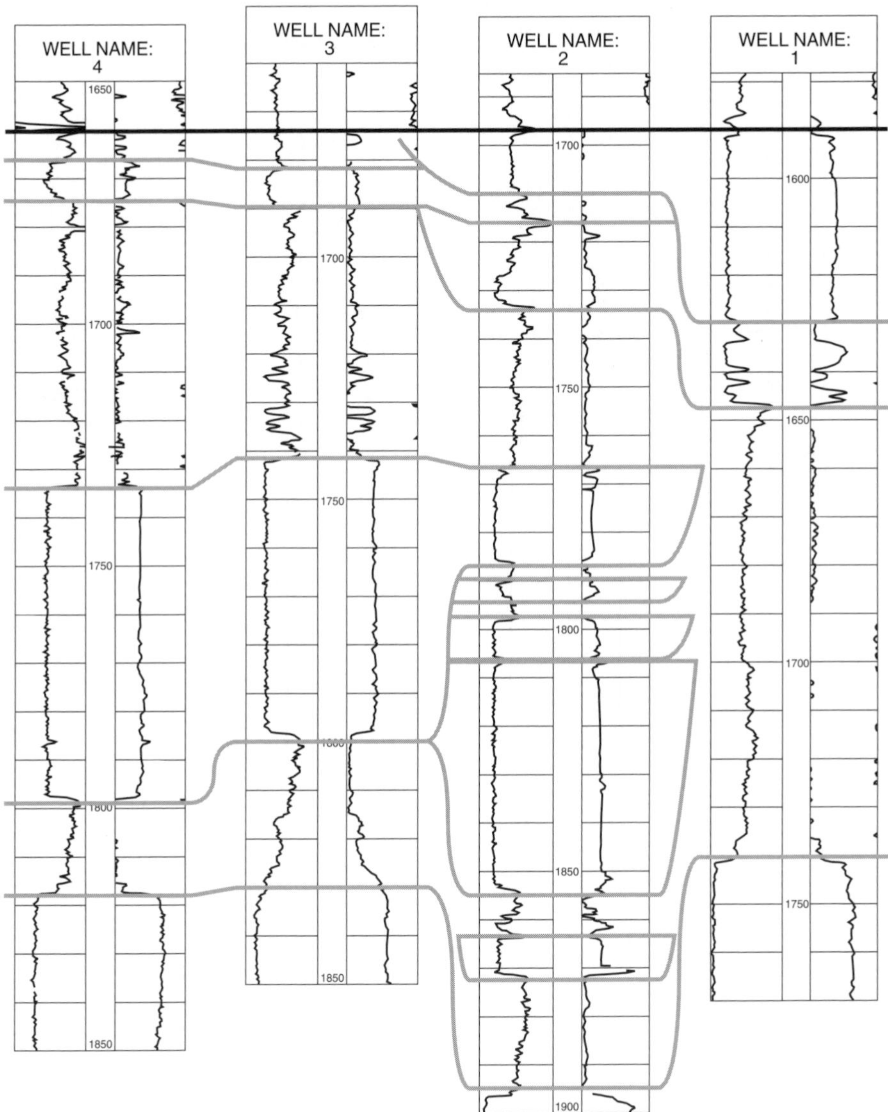

Fig. 4.13. One of several possible lithostratigraphic correlations of the strata in Grane field, shown in Fig. 4.12.

and separated by a continuous shale (Fig. 4.14). This correlation pattern is particularly significant to secondary or tertiary hydrocarbon recovery processes in similar fields, because the sandstones comprising the two biozones are compartmentalized by the shale. Also, the correlation pattern indicates progradation or younging of sediment toward the right.

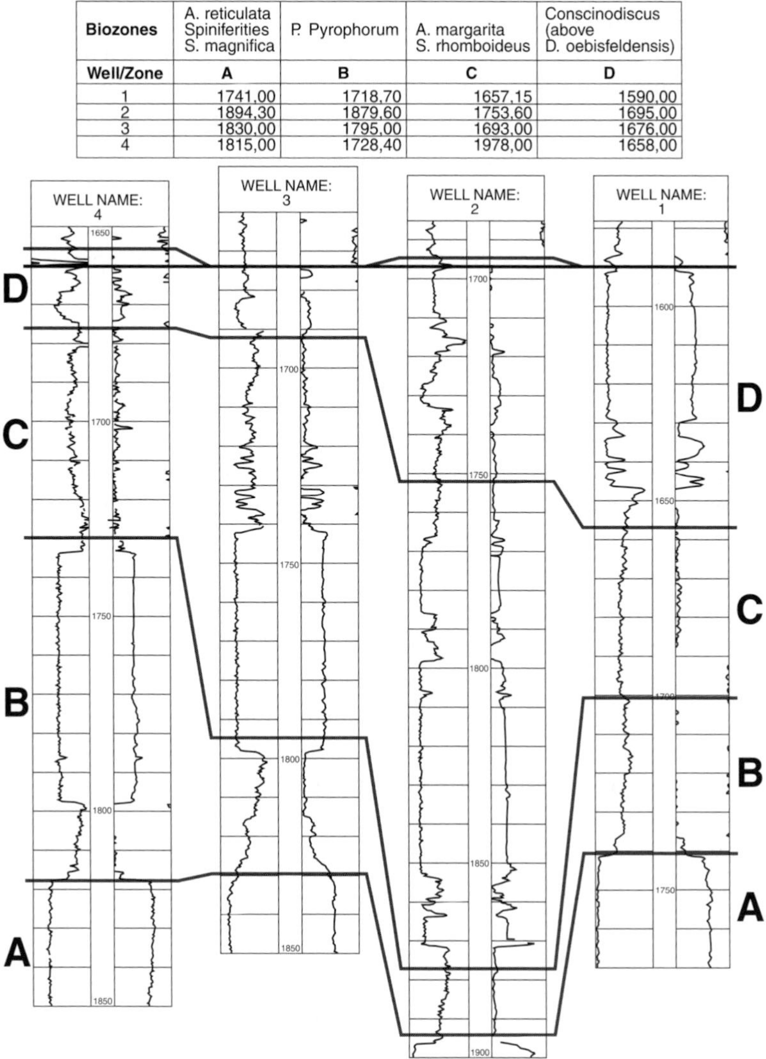

Biozones	A. reticulata Spiniferities S. magnifica	P. Pyrophorum	A. margarita S. rhomboideus	Conscinodiscus (above D. oebisfeldensis)
Well/Zone	**A**	**B**	**C**	**D**
1	1741,00	1718,70	1657,15	1590,00
2	1894,30	1879,60	1753,60	1695,00
3	1830,00	1795,00	1693,00	1676,00
4	1815,00	1728,40	1978,00	1658,00

Fig. 4.14. Biostratigraphic zonation of the four wells shown in Fig. 4.12. Four biozones are defined: A–D. This more precise biozonation crosscuts the lithostratigraphic correlation of Fig. 4.13 and shows that individual, thick reservoir sandstones are compartmentalized by shale between wells 4 and 3. This correlation also demonstrates a north-to-south progradational pattern of sedimentation, with younging to the south.

4.4.3 Determining sedimentation rates from biostratigraphy

It is possible to estimate the rate of sediment deposition, if the geologic ages of deposition of two stratigraphic horizons within a well or outcrop are known. For example, depositional ages of 2.85 and 3.60 Ma for strata in the well shown in

Fig. 4.9 are separated by 100 m (333 ft). Thus, the average sedimentation rate for that 0.75-m.y. time interval is $100/750{,}000 = 0.13$ m$/1000$ yr. This sample calculation does not take into account the amount of compaction that the sediment column has undergone since deposition, so the calculated sedimentation rate is a minimum rate. In this example, if the sediment had been compacted by 30%, the decompacted sedimentation rate would be 130 m$/750{,}000 = 0.17$ m$/1000$ yr. Determination of sedimentation rates is important when one is establishing a depositional sequence stratigraphic framework for an area; this topic is discussed in more detail in a later chapter.

4.4.4 Biostratigraphy and condensed sections

Condensed sections are stratigraphic intervals that record a long period of relative sediment starvation in a particular marine area, where only the finest-grained, lightest-weight siliciclastic, organic and biogenic particles are deposited. Condensed sections form on the seafloor during periods of marine transgression, when most siliciclastic sediments are deposited on a landward-stepping shoreline. Because the rate of siliciclastic sedimentation is very low, and only the finest particles reach the seafloor, condensed sections are very shaly or clay-rich and contain an abundance of microorganisms (as a result of the relative enrichment of the microorganisms during times of low siliciclastic sedimentation). In some instances, the microorganisms can comprise the majority of the sedimentary particles, giving rise to a calcareous or siliceous stratigraphic interval.

Because of the unique characteristics and the depositional timing of condensed sections, they are very important not only for correlation purposes, but also for age-dating. Condensed sections are recognized by biostratigraphers on the basis of both the greater abundance and the greater diversity of microorganisms in the stratigraphic interval. These two characteristics, along with the fine grain size, define a condensed section. Figure 4.15 gives an example of the abundances of planktonic foraminifera and nannoplankton in a northern Gulf of Mexico well. Three strong and three weaker abundance peaks occur within the stratigraphic interval.

The well in Fig. 4.9 shows strong microorganism abundance peaks at about 1.95 Ma and at 2.30–2.85 Ma, and a less conclusive peak (a nannoplankton peak only) at 3.60 Ma. Thus, these intervals record geologic time periods of marine transgression and condensed-section formation. Such determinations are important when one is establishing a depositional sequence stratigraphic framework for an area; this topic is discussed in more detail in a later chapter.

Because of the slow sedimentation rates that are characteristic of their deposition, condensed sections also may be enriched in chemically precipitated minerals such as phosphorites, glauconite, and siderite. Bentonites, which are

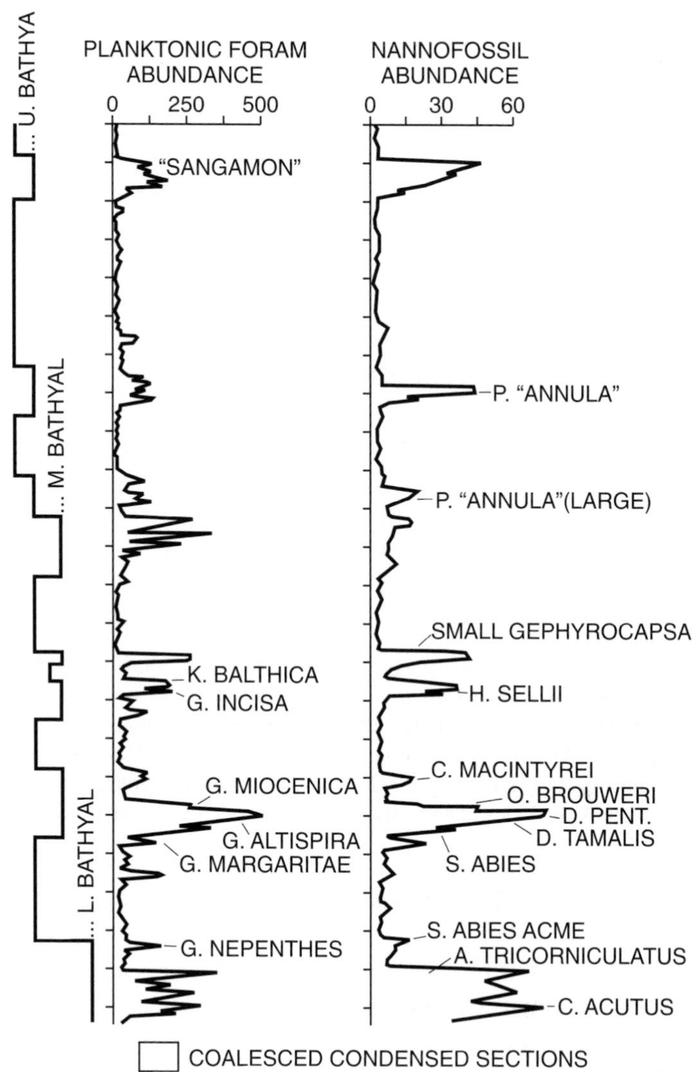

Fig. 4.15. Planktonic and nannofossil abundance plots in a northern Gulf of Mexico well. Three strong abundance peaks and three weaker peaks define condensed sections within this stratigraphic interval. After Shaffer (1990). (Reprinted with permission of the Gulf Coast Association of Geological Societies.)

volcanic ash layers, may enrich condensed sections during times of volcanic activity. Because organic material is lightweight, it tends to be deposited with the finest-grained, detrital clay fraction of siliciclastic sediments. Organic matter in condensed sections can chelate and concentrate radioactive elements (as can some clay minerals), and this often gives rise to an anomalously high gamma-ray count on well logs (Fig. 4.16).

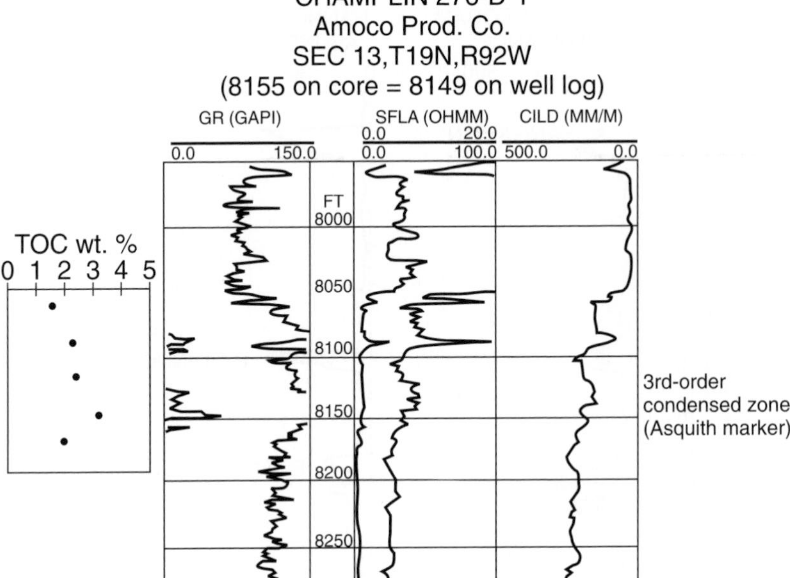

Fig. 4.16. Gamma-ray and resistivity logs of part of the Lewis Shale in the Greater Green River Basin of Wyoming, showing an anomalously high gamma-ray count in a basin-scale shale called the "Asquith Marker" (Pyles, 2000). Total organic carbon (TOC) measurements on cores through the interval reveal TOC concentrations of almost 4%, which is high enough to provide the high level of radioactivity relative to overlying and underlying strata that are not as organic-rich. After Pyles (2000). (Reprinted with permission of David R. Pyles.)

4.4.5 Biostratigraphy and depositional environments

Micropaleontologists and biostratigraphers have developed a classification of water depths based primarily on diagnostic benthic or bottom-dwelling microorganisms. In the broadest terms, the three main divisions or zones of water depth are neritic (from high tide to the edge of continental shelf, at 200 m or 600 ft), bathyl (from the edge of the continental shelf to the base of the continental slope, at 2000 m or 6000 ft), and abyssal (the basin plain, deeper than 2000 m or 6000 ft) (Fig. 4.17). Further subdivisions of these three divisions are also used (Fig. 4.17). By identifying microorganisms in cores, cuttings, or outcrops that have been determined to be diagnostic of different depth zones, it is possible to identify the approximate water depth of a sediment's deposition.

In a rock sequence, variations in water depth over geologic time intervals can be identified, so that temporal sea-level fluctuations can be documented. An example is provided in Fig. 4.18, which shows two wells and a correlation of water-depth zones between them for a specific time interval A. During this time interval, water depth at Well #1 was deeper than that at Well #2. During this

CLASSIFICATION OF MARINE ENVIRONMENTS

HIGH TIDE
LOW TIDE
SEA LEVEL
150' ± (50M.)
300' ± (100M.)
600' ± (200M.)
1500' ± (500M.)
3000' ± (1000M.)
6000' ± (2000M.)
SUPRALITTORAL
TERRESTRIAL
PARALIC
LITTORAL
SHALLOW INNER NERITIC (SIN)
DEEP INNER NERITIC (DIN)
SHALLOW MIDDLE NERITIC (SMN)
DEEP MIDDLE NERITIC (DMN)
SHALLOW OUTER NERITIC (SON)
DEEP OUTER NERITIC (DON)
UPPER BATHYAL
MID BATHYAL
LOWER BATHYAL
ABYSSAL
IN UNDIFF.
MN UNDIFF.
ON UNDIFFERENTIATED
BATHYAL UNDIFFERENTIATED

Fig. 4.17. Classification of marine environments on the basis of water depths, from sea surface to seafloor. After Armentrout (1991). (Reprinted with permission of Springer Science and Business Media.)

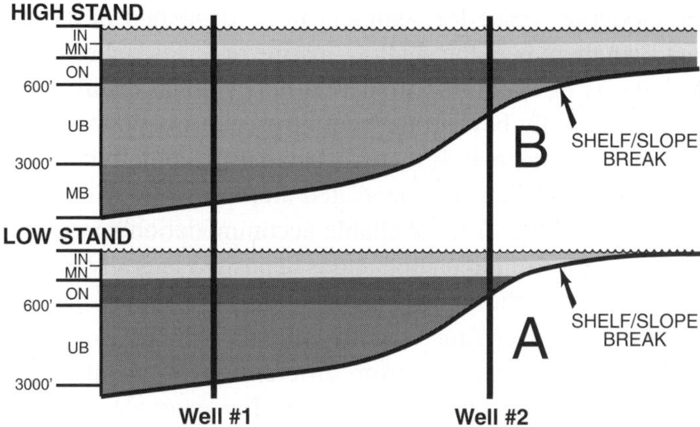

Fig. 4.18. Change in sea level is recorded by a change in water depths in strata from two wells. In the lower figure (A), sea level is near the shelf-slope break, and neritic strata are deposited at the location of Well #2 in the upper slope environment. With a rise in sea level (upper figure (B)), there is a thicker succession of bathyal strata in Well #1 than was the case within the older strata, and there now are upper bathyal strata in Well #2. This upward-deepening pattern indicates a landward shift in the depocenter between the lower and upper intervals, or a marine transgression. After Armentrout (1991). (Reprinted with permission of Springer Science and Business Media.)

time, upper bathyl strata were deposited at Well #1, and primarily outer neritic strata were deposited at Well #2. This information alone defines the spatial position of the shoreline relative to the offshore environment during time interval A. The upper part of the figure (time interval B) shows changes in water depth at the two well sites as a result of a relative rise in sea level, or a landward migration of the sediment's depocenter. The resulting sediment deposited at the site of Well #1 would be predominantly middle bathyl strata, and at Well #2, some upper bathyl sediment would be deposited. Thus, from the bottom to the top of the strata in both wells, a deepening of water, or marine transgression, is recorded.

4.5 Walther's law and the succession of sedimentary facies

A sedimentary facies is defined as "The aspect, appearance, and characteristics of a rock unit, usually reflecting the conditions of its origin; especially as differentiating the unit from adjacent or associated units.... A mappable, areally restricted part of a lithostratigraphic body, differing in lithology or fossil content from other beds deposited at the same time and in lithologic continuity.... A distinctive rock type, broadly corresponding to a certain environment or mode of origin." (Jackson, 1997). During intervals of geologic time, sedimentary facies (i.e., lithofacies) are deposited within depositional environments. The physical space in which facies are deposited is called "accommodation space". There is more accommodation space for sediment to be deposited in the deep-marine environment than there is in the shallow-marine or terrestrial environments.

Sediment that is newly transported to a depositional environment must be deposited either laterally beyond the spot of previous deposition, or vertically, on top of the previous deposit. If there is abundant accommodation space at the depositional site, the sediment is deposited atop previous sediment and stacks vertically. If there is little or no available accommodation space, the sediment bypasses the prior zone of deposition and is deposited laterally, resulting in basinward migration of the shoreline (Fig. 4.19). In this manner, the environments also migrate laterally over time. Lateral migration of environments, coupled with vertical stacking of deposits from different environments, gives rise to a vertical stratification sequence or succession. For example, the succession in Fig. 4.20 shows an upward increase in sandstone content and bed thickness, indicating an environmental change over time from one in which shaly sediment was deposited (under low-energy water conditions) to one in which thick sands were deposited (under high-energy conditions). Such a vertical change probably represents a shallowing of marine water over time.

The general law that describes the origin of depositional sequences is Walther's Law of Succession of Facies, which is illustrated in Fig. 4.21. According to the original statement by Walther, "The various deposits of the same

Fig. 4.19. An example of deposition over time. A river is shown entering the sea. Waves are shown on the right. The various linear features highlighted with the arrows point to old shorelines that have built outward over time. In this case, the sediment is transported by the river to the river mouth, where waves then move the sand to the south along the shoreline. In this way, the shoreline builds in the seaward direction over time. The oldest shoreline is the one outlined by the red*/gray arrow, and the youngest is the shoreline outlined by the blue/black arrow.

Fig. 4.20. Rock succession composed of fine-grained, shaly beds deposited in relatively deep, quiet marine water. With progradation, additional sand beds (thicker, lighter colored beds) were deposited over the older beds. This gives rise to what is called a "thickening- and coarsening-upward progradational sequence". The succession is capped by additional shaly beds, indicating a return to deeper water (transgressive) conditions.

facies area and, similarly, the sum of the rocks of different facies areas were formed beside each other in space, but in a crustal profile we see them lying on top of each other . . . it is a basic statement of far-reaching significance that only

* The indicated color is for a CD which contains all of the figures in color.

J. WALTHER'S LAW OF SUCCESSION OF FACIES (1894)

"Facies sequences observed vertially are also found laterally". We see rocks in vertical sequence that were deposited beside each other at the same time.

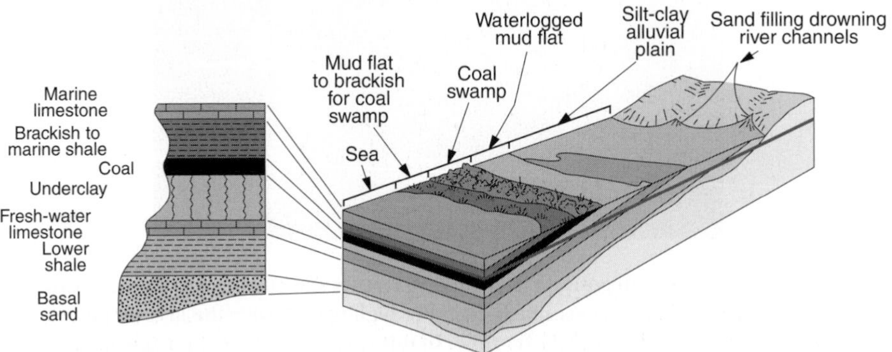

Fig. 4.21. Walther's Law is an important principle upon which the origin of vertical rock successions is explained. Sediments are deposited in environments that change over time as a result of relative sea-level fluctuations. As the environments change, so does the nature of the sediments deposited at any one location. The vertical succession thus records the lateral changes in environments over time.

those facies and facies areas can be superimposed, primarily, that can be observed beside each other at the present time" (Blatt et al., 1972, p. 187). Stated more simply, facies sequences that are observed vertically are also found laterally. The vertical sequence shown in Fig. 4.21 records an upward change in sedimentary environments over one particular location over a period of time (i.e., it is important to emphasize here that any vertical succession of rocks or sediments seen in outcrop or in a well represents deposition over a time interval at one particular location, be it somewhere on the seafloor, shoreline, or on land – or a combination thereof). In Fig. 4.21, the upward change is from nonmarine facies (basal fluvial sandstone), to muddy alluvial plain, to shoreline environments (mudflats, coal swamps), to marine facies (shale and limestone). This vertical stacking pattern records a period of marine transgression or landward migration of the shoreline and its related environments of deposition.

A second example, this time of delta progradation, is illustrated in Fig. 4.22. In this example, the shoreline and offshore environments migrate progressively seaward over time, as sediment is deposited (from Time 1 to Time 3), so progressively shallower-water sediments are deposited atop deeper-water sediments. As the environments change over time at one location, they migrate laterally. Strata deposited in each environment are deposited in a horizontal fashion (Law of Original Horizontality). In this manner, geologic time lines

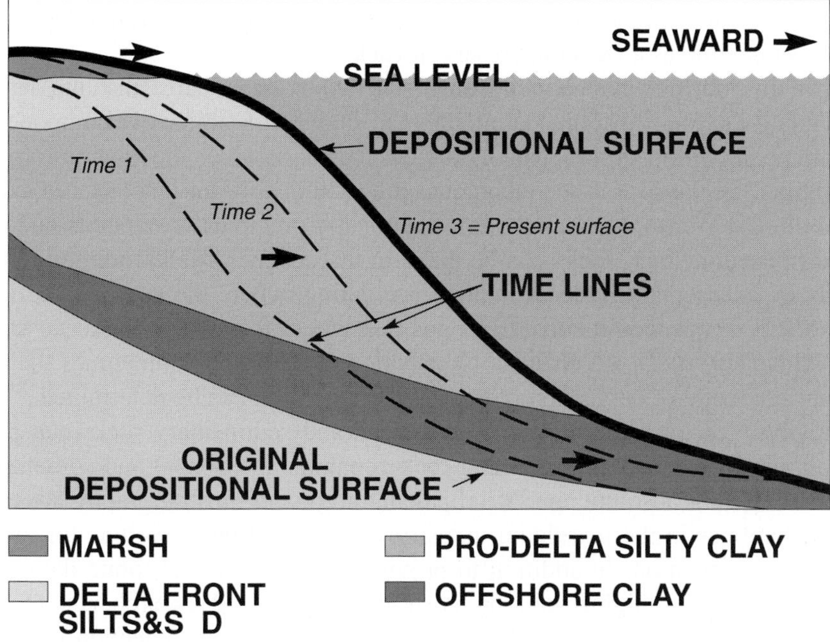

Fig. 4.22. Illustration of the principle of progradation of sediments over time. At any one instant in time, the seafloor becomes the depositional site for sediments. Normally, with an increase in water depth, progressively finer-grained sediment is deposited. In this picture, pink is sand, brown is silt, and red is mud. Time 3 in the picture shows the present-day ocean floor in front of a delta. Because the delta builds outward as well as upward with time as a result of progradation, the positions of some older seafloor surfaces are shown (Times 1–3). As the sediments prograde with time, each of the sedimentary deposits is laid down in a flat or horizontal manner, even though the time lines (or old seafloor surfaces) are inclined. The seaward buildup of the delta deposit is the result of the process of "progradation".

crosscut lithostratigraphic boundaries and facies. This important point is elaborated upon in a later chapter

4.6 Summary

The concept of long periods of time being required for reservoirs to assume their present form is difficult to grasp, particularly for those individuals who track daily oil and gas production from reservoirs. However, the lengthy formative processes for hydrocarbon reservoirs can be understood, and this understanding is important for proper knowledge of why a reservoir is built the way it is.

The geologic time scale is divided into a series of time intervals that are based on significant events in the geologic record. Various temporal names applied to rock units commonly are used and must be recognized by people studying reservoirs. For a simple example, a Cretaceous reservoir rock was not deposited at the same time as a Devonian reservoir rock.

The time during which a rock formed is dated by two means: absolute dating and relative dating. Absolute dating refers to analysis of radioactive components in a mineral (within a rock), which provides the age at which the mineral formed (solidified) in the rock. Such techniques are used mainly for igneous rocks that cool directly from magma, but some chemically precipitated minerals and cements in sedimentary rocks can be dated in this manner. More common to the study of sedimentary rocks is relative age dating, where the age of a particular rock is determined relative to its position within a stratigraphic succession. If sedimentary rocks are crosscut by datable igneous rocks, sometimes the absolute age range of deposition of the sedimentary rock can be determined.

Analysis of microorganisms in sediments and sedimentary rocks can provide a useful means of establishing rock zonations (biozones) and sometimes for determining absolute age. Micropaleontology and biostratigraphy are critical disciplines in the petroleum industry, for exploration as well as for reservoir characterization. In addition to providing a means for absolute dating of sedimentary rocks, high-resolution biostratigraphy can aid researchers in (1) interpreting stratigraphic intervals and their ages on seismic reflection profiles, (2) correlating between-well stratigraphic and temporal relationships, (3) determining sedimentation rates, and (4) determining depositional environments and changes in environments over time.

Walther's Law of Succession of Sedimentary Facies is key to understanding the origin of sedimentary deposits and reservoirs. It is a fundamental principle that is the backbone of stratigraphy. Stratigraphic sequences, such as those that comprise reservoirs, exhibit systematic and somewhat predictable vertical stacking patterns that are explained by Walther's Law. By understanding the vertical stratigraphy of a reservoir, such as observed in core or well logs, one can make improved interpretations of the lateral (dis)continuity of reservoir intervals.

Chapter 5

Geologic controls on reservoir quality

5.1 Definitions

Porosity is defined as the interstitial void space in a rock. Permeability is a property of rock (usually sedimentary rock) that characterizes the ease with which fluid can flow through it in response to an applied pressure gradient. Both of these properties control the ability of a rock to store (via porosity) and transmit (via permeability) reservoir and nonreservoir (injected) fluids. Capillarity is defined as "the attraction of the surface of a liquid to the surface of the solid with which it is in contact. Capillarity affects the recovery of oil from a reservoir as it hinders the oil from flowing through the pores of the rock" (Hyne, 1991).

This chapter emphasizes the important geologic factors that control reservoir quality (porosity and permeability) in oil and gas reservoirs. These factors include primary depositional texture (grain size and grain sorting) and postdepositional processes (physical–chemical) of burial diagenesis. Diagenesis controls the final geometry of the pore structure, grain orientation and packing, and the degree of cementation and clay filling of pore spaces.

5.2 Examination and measurement of porosity and permeability

5.2.1 Direct observation

Pore spaces can be examined directly in sedimentary rocks, either in a hand sample or by cutting a "thin section" (0.030 mm thick) of the rock. To prepare a thin section, a small slab of rock is cut from the larger sample and is placed in a chamber containing colored epoxy (normally, either blue or red). Then, either pressure or a vacuum is applied to the sealed chamber until the epoxy fills the pores and pore throats. This procedure is done at elevated temperature to lower the viscosity of the epoxy. The rock is allowed to cool, and the thin section is cut. Figure 5.1 shows a thin section of a sandstone with pale gray-pink quartz grains.

Fig. 5.1. Thin-section photomicrograph of a quartz-rich sandstone. Quartz sand grains are pale colored and the blue*/dark-gray is dyed epoxy in pore space. Brown/black material is fine cement crystals and clay minerals that partially fill pore spaces and pore throats (narrow spaces between grains). Some of the quartz grains exhibit quartz cement rims. Sand grains are about 0.150 mm in diameter.

The quartz grains exhibit quartz cement rims. Blue/dark-gray areas are pore spaces filled with epoxy. This sandstone has good porosity and permeability. In two dimensions, it does not appear as if many pores are in communication across pore throats, but in three dimensions, more connected pores are visible. Another thin section is shown in Fig. 5.2. This sandstone contains virtually no porosity or permeability, because the spaces between sand grains are filled with muddy matrix. Rock examples of porosity and permeability are shown in the following four figures.

Pores and pore throats, as well as grain boundaries, also can be observed with a scanning electron microscope. Figure 5.3 is a high-magnification, 3D scanning electron photomicrograph of quartz grains comprising a sandstone. Clay mineral crystals have accumulated in the pore throats of the sandstone and on the faces of the quartz grains. Clay crystals can grow by precipitation, sometimes lining a pore throat between two sand grains and almost bridging the grains (Fig. 5.4). By occurring within and across pores, clay minerals can significantly reduce permeability. In addition, the clay mineral chlorite is sensitive to hydrochloric acid and can dissolve in a reservoir when the reservoir is acidized. The dissolution process generates free Fe^{2+}, which then reprecipitates in the pore throats as iron hydroxide or iron hydrous oxide.

At a somewhat larger scale, cements within sandstones can be observed either by the naked eye or with a hand lens. Figure 5.5 shows a slabbed sandstone

*The indicated color is for a CD which contains all of the figures in color.

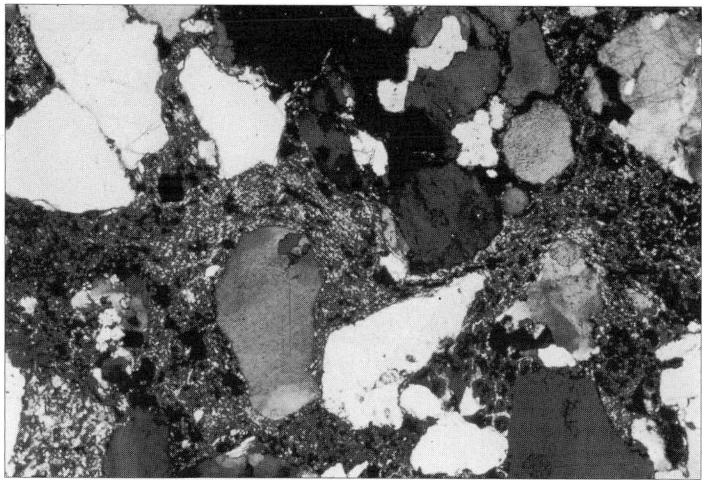

Fig. 5.2. Thin-section photomicrograph of a sandstone with abundant mud matrix (brown-yellow/bright-gray material) encasing quartz grains. Sand grains are about 0.150 mm in diameter. Sand grains are angular and do not exhibit cement overgrowths.

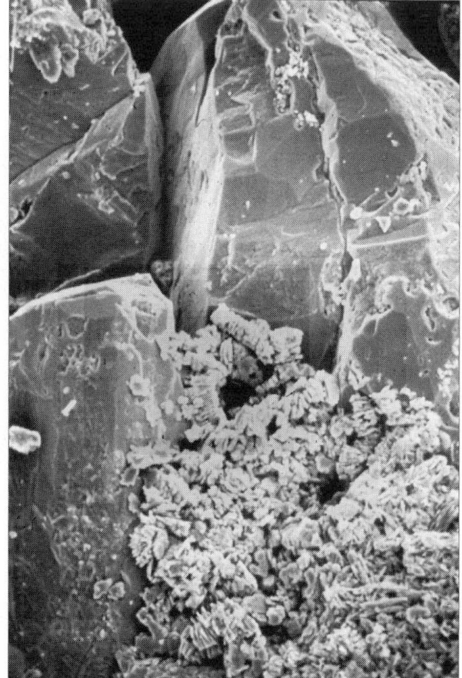

Fig. 5.3. High-magnification scanning electron photomicrograph of quartz grains with crystalline overgrowths upon which lies a cluster of small clay minerals within the pore spaces. The quartz grain in the upper right corner is about 0.200 mm in diameter.

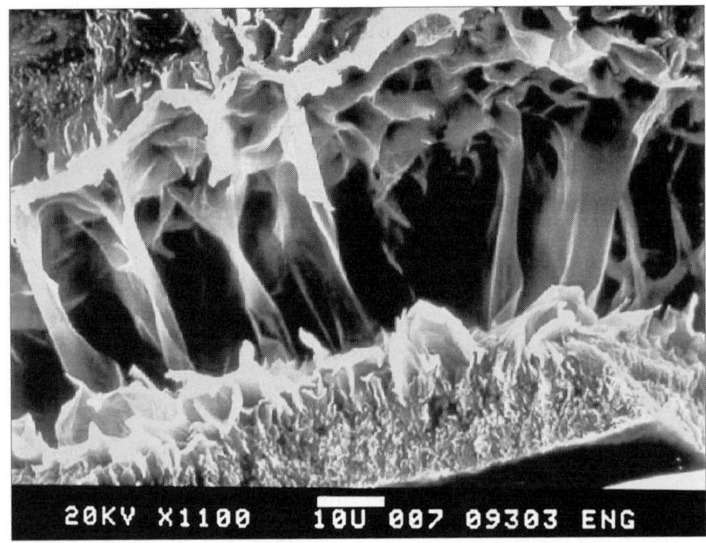

Fig. 5.4. Scanning electron photomicrograph of clay mineral particles lining a pore throat between two sand grains and bridging the grains. Such bridging reduces permeability and porosity of sandstones. White scale bar is 10 microns in length.

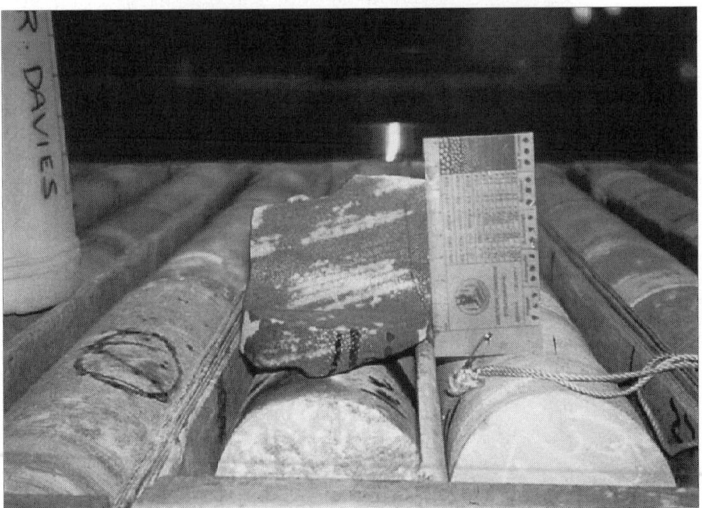

Fig. 5.5. Photograph of a slabbed sandstone core from an aeolian (sand dune) reservoir in the North Sea. One piece of slab has been tilted vertically to illustrate the discontinuous nature of cement in this sandstone. Some, but not all, individual laminae have been cemented by gypsum (white), giving rise to a complex distribution of permeability and porosity. Plastic ruler for scale.

core. Individual laminae have been differentially cemented by gypsum (white); laminae that originally were more permeable now contain more gypsum cement than do the originally less-permeable laminae, resulting in a locally complex stratification of porosity and permeability.

5.2.2 Direct measurement

Using a thin section, it is possible to estimate the area of pore space relative to the area of rock by "point counting" pore spaces using a polarizing microscope. In practice, 200 or 300 equally spaced points on a thin section are counted as "grain", "cement", or "pore", and the areal percentage of pore space is calculated.

The more traditional manner in which sample values for porosity and permeability are obtained is through analysis of "plugs" obtained from rock cores (Fig. 5.6). Two and a half cm (1 in) diameter, 3.75 cm (1.5 in) long core plugs are the standard size. These plugs are drilled after a core has been obtained at the well site and has been cleaned and transported to a commercial core-analysis facility. Core plugs normally are obtained at equally spaced intervals along the core, typically every 0.3 m (1 ft). Various laboratory methods determine porosity (after removal of soluble hydrocarbons). Rock permeability to both fluids and gas can be measured in the laboratory, but the flow rate of air into the rock

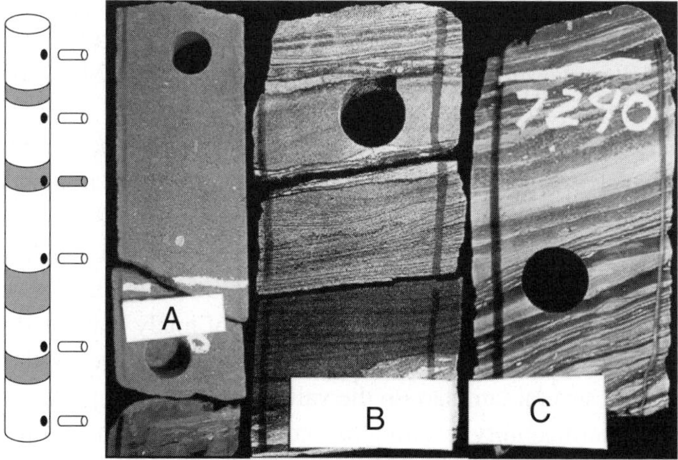

Fig. 5.6. Cartoon illustrating 3.75 cm (1.5 in) long, horizontal core plugs extracted from a full-diameter core for routine core analysis. The photograph of three pieces of core show where core plugs were extracted. The plug from core A is representative of the entire piece of homogeneous sandstone. The plug from core B is representative only of the thin sandstone from which it was extracted. The plug from core C crosscuts different sandstone (light-colored) and shale (dark-colored) laminae, so it is representative only of the average reservoir quality of the combined laminae.

is the most commonly used method because it is easy to determine. The numerical values of porosity and permeability measurements are partly a function of the bedding of the rock relative to the positioning of the plugged core. Figure 5.6 provides three examples to illustrate this point. The plug in core A was obtained from a sandstone with uniform properties, thus, the measurements of porosity and permeability will be representative of that length of uniform core. The plug in core B is representative of the 2.5 cm (1 in) thick sandstone bed through which the core plug was obtained, but the numerical values of porosity and permeability will not be representative of the greater length of the core. The plug in core C was obtained through both sandstone and shale layers, and the numerical values of porosity and permeability will be some average of those two lithologies, but also will not represent the greater length of the core. Sometimes, core plugs are obtained in which the long axis of the plug is perpendicular to bedding; such measurements are called "vertical permeability" measurements. Normally, vertical permeability values are lower than horizontal permeability measurements from the same bed, because the vertical measurements crosscut, rather than parallel, stratification.

When one is using porosity and permeability data from a cored well, it is important to have photographs of the core available that show where the plugs were obtained (as in Fig. 5.6) so that the rock representation/limitations of the numerical values are known. A "whole-core analysis" of the permeability of a large piece of unslabbed core also provides a more accurate representation of the porosity and permeability of a section of rock. More information on porosity and permeability measurement techniques can be found in Morton-Thompson and Woods (1992).

In recent years, a rapid, reliable method has been developed to measure permeability over a very small sample of rock using a laboratory minipermeameter. This instrument measures the rate of flow of air from a small-diameter tube (approximately 1-mm aperture) into and through the rock. This air-flow rate can then be related to rock permeability through calibration. In Fig. 5.7, a core-plug permeability value has been obtained on the slab of finely laminated sandstone; a single core-plug permeability is 19 md (millidarcys). However, the core plug was cut through several laminae, so the value is not really representative of the heterogeneous permeability resulting from this fine-scale lamination. Individual spot values of permeability shown on the diagram were taken with a minipermeameter. There are at least two orders of magnitude of variation in permeability within the laminae (from <0.5 to 38.5 md). The spot measurements provide a more accurate representation of the vertical permeability's heterogeneity than does the single core-plug measurement.

Unconsolidated sands in cores present a special problem, because insertion of the plugging tool into the sands will destroy the in situ arrangement and packing

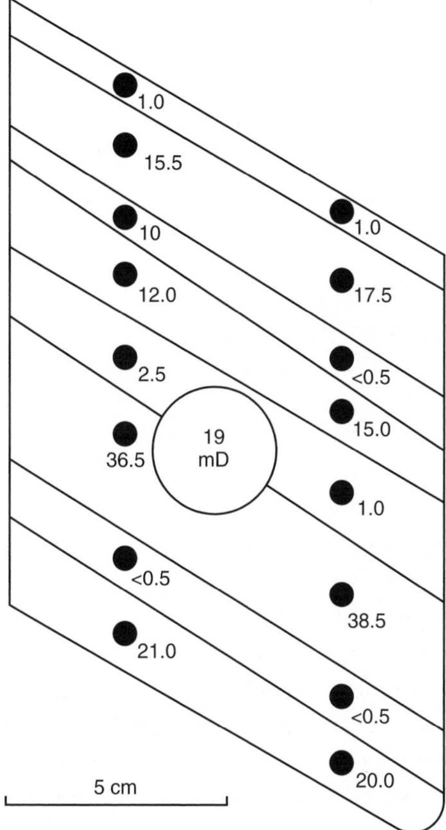

Fig. 5.7. Diagram of a slab of core through which a core plug permeability value of 19 md has been obtained. Individual spot values of permeability shown on the diagram were taken with a minipermeameter. There are at least two orders of magnitude of variation in permeability within the laminae (from <0.5 to 38.5 md). Diagram is after Weber (1987). (Reprinted with permission of the Society of Sedimentary Geology (SEPM).)

of the grains and thus the porosity and permeability. An example of plugs taken from unconsolidated cores is shown in Fig. 5.8. The extent to which the original reservoir-quality values are modified by the plugging process is not known, but undoubtedly this modification is quite variable and not systematic.

A similar problem is encountered when one is attempting to obtain reservoir-quality measurements on sidewall cores. Forceful insertion of the plug into the borehole wall can severely modify the in situ rock fabric. Any measurements of porosity and permeability from sidewall cores should be considered suspect.

When core samples are submitted to a commercial laboratory for measurement of porosity and permeability, unless otherwise stated, measurements will be made at "benchtop" or atmospheric pressures rather than at pressures that resemble reservoir conditions of elevated pressure. Under increased pressure,

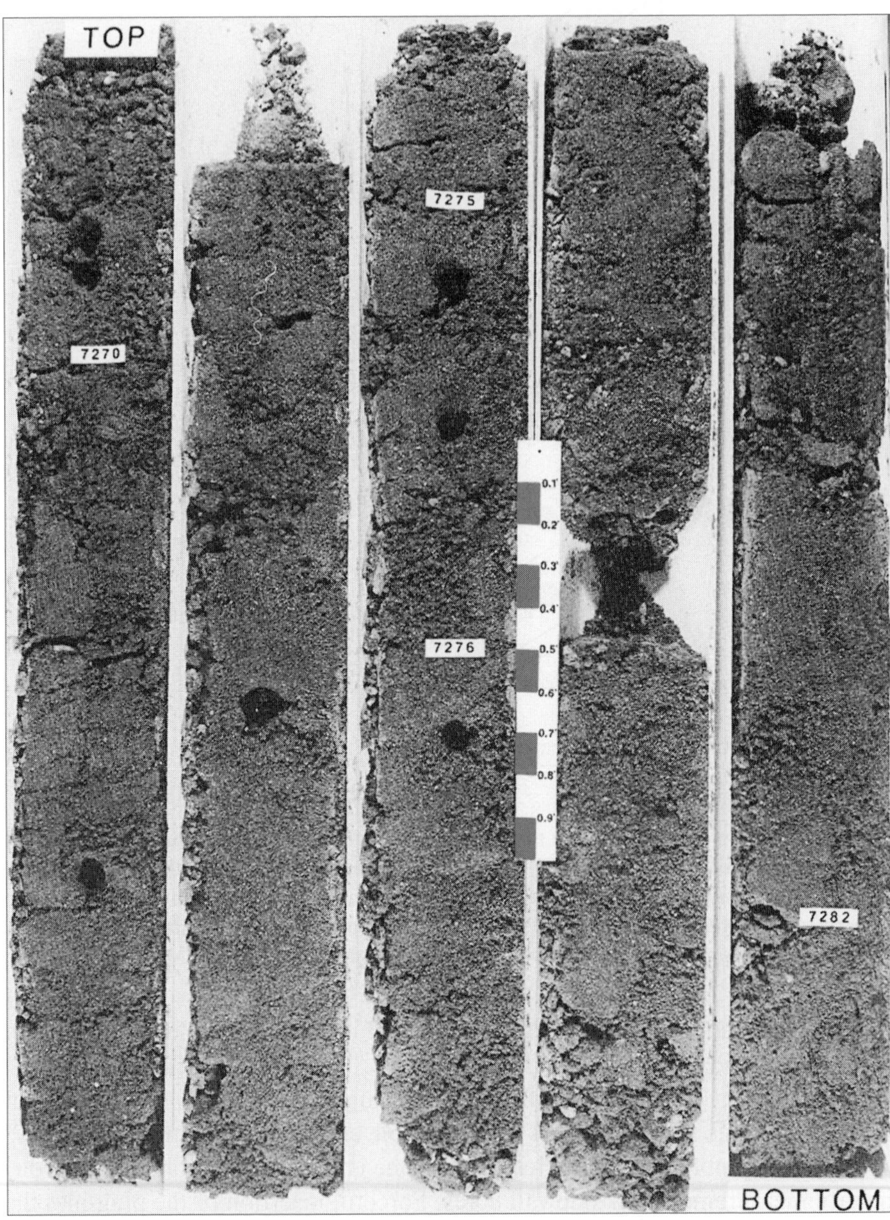

Fig. 5.8. Five core slabs, each slab 1 m (3 ft) long. The cores are all of unconsolidated sand. The holes are where core plugs were obtained. The plugging process caused sand grain movement, thus leading to a new packing of grains and a change in the original porosity and permeability.

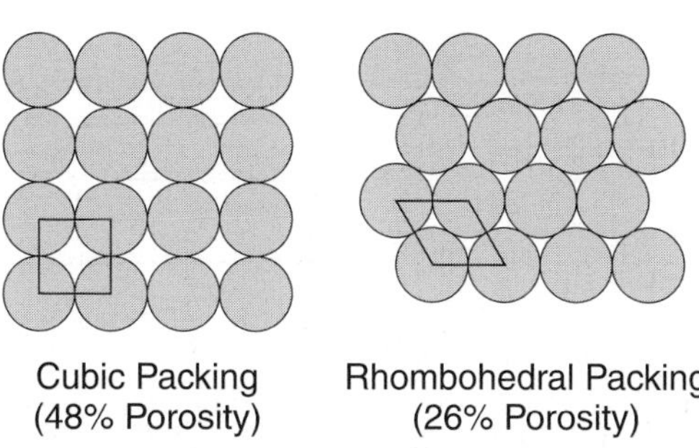

Fig. 5.9. Variations in porosity as a function of cubic and rhombohedral packing of sand grains. With cubic packing, each pore is outlined by four grain-to-grain contacts (in 2D view). In the tighter rhombohedral packing, each pore is defined by three grain-to-grain contacts, so within a given area (or volume, in 3D space), the porosity and permeability will be reduced. (Figure provided by T. Cross.)

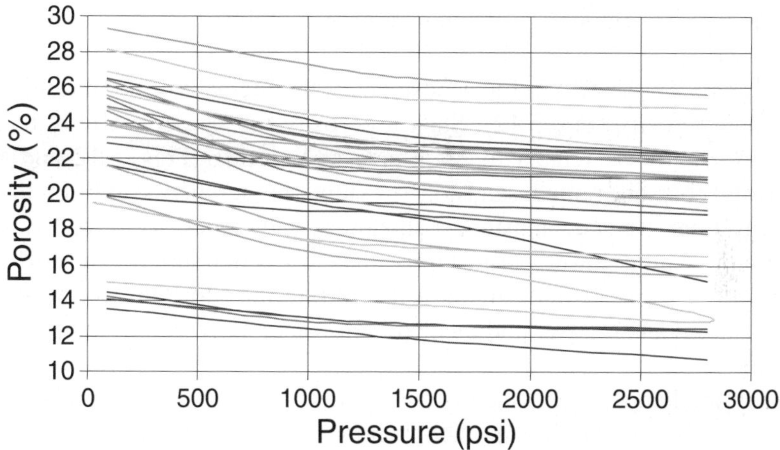

Fig. 5.10. Series of curves showing the systematic decrease in measured porosity from a suite of core plugs when the plugs were subjected to increased pressure during measurement. Porosity values generally decrease by more than 2% over the pressure range measured (0–3000 psi). (Figure provided by C. Jenkins.)

sands become compacted, so grain-to-grain contacts become more frequent and grains become more closely packed, thus reducing the pore volume (Fig. 5.9).

A more expensive "special core analysis" is required to obtain porosity and permeability measurements under reservoir conditions. Figures 5.10 and 5.11 illustrate the reduction in porosity and permeability that accompanies measurement at increases in pressure on the plug. The reservoir-quality value is thereby reduced.

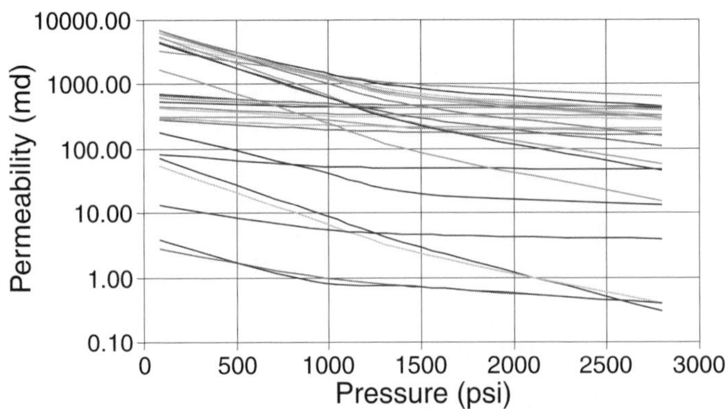

Fig. 5.11. Series of curves showing the systematic decrease in measured permeability from a suite of core plugs when the plugs were subjected to increased pressure. Permeability values may decrease by an order of magnitude over the pressure range measured (0–3000 psi). (Figure provided by C. Jenkins.)

5.3 Primary grain-size control on reservoir quality

A commonly observed trend in sandstones is the increase of permeability with increasing porosity (Fig. 5.12). Such a porosity versus permeability relationship can usually be related to grain size and sorting (Figs. 5.12 and 5.13). Relatively coarser-grained sandstones exhibit higher permeabilities than do relatively finer-grained sandstones, siltstones, and shales. More poorly-sorted sandstones exhibit lower permeability than do better-sorted sandstones, because, in the former, small grains can infiltrate into pore throats of adjacent grains. Some quantitative studies of oilfield sands have confirmed these relationships.

Reedy and Pepper (1996) clearly demonstrated the sensitive control that grain size exerts on permeability in unconsolidated turbidite reservoir sands in the Gulf of Mexico. A plot of porosity versus permeability reveals no apparent positive or negative correlation (Fig. 5.14). However, a plot of permeability versus grain size (determined by laser grain-size analysis) clearly shows a log-linear correlation of increasing permeability with increasing grain size (Fig. 5.15). The grain-size frequency distributions of individual samples from which permeability measurements were made are unimodal with similar degrees of grain sorting (Fig. 5.15). Figure 5.16 shows the relationship between modal grain size and permeability for three samples. The purple grain-size curve is for the finest-grained sand (the mode is at 0.08-mm diameter); this sand has a permeability of 102 md. The blue grain-size curve is for the coarsest-grained sand (the mode is 0.2-mm diameter); this sand has a permeability of 5,078 md, which is extremely good reservoir quality. The orange curve is for intermediate grain size (with a mode of 0.1-mm diameter) and the permeability is 1,957 md.

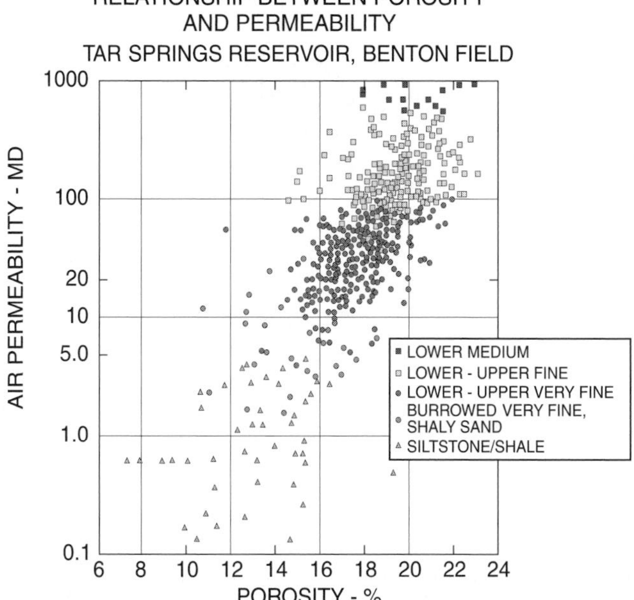

Fig. 5.12. Crossplot showing the typical porosity versus permeability trend for clastic strata. Permeability increases with increasing porosity, usually as a result of an increase in grain size and improvement in sorting. After Sneider (1987). (Reprinted with permission of AAPG, whose permission is required for further use.)

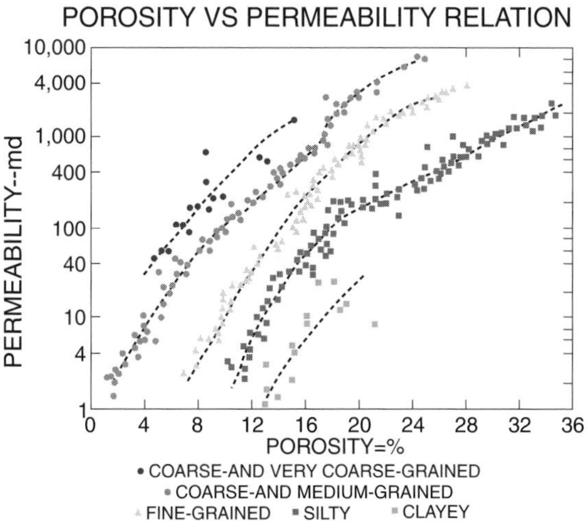

Fig. 5.13. Figure illustrating the relationship of porosity and permeability for sets of sands and muds of differing grain size. A constant sorting is assumed for individual core plugs measured. For any given grain size, there is a clearly positive correlation between porosity and permeability. After Sneider (1987). (Reprinted with permission of AAPG, whose permission is required for further use.)

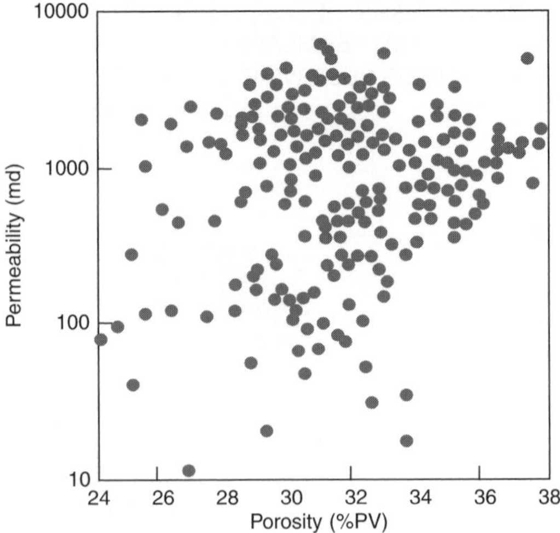

Fig. 5.14. Crossplot of porosity and permeability measurements from core plugs obtained from a Gulf of Mexico reservoir (Reedy and Pepper, 1996). The permeability scale is logarithmic and the porosity scale is arithmetic. There is no clear relationship between porosity and permeability for these core plug samples. After Reedy and Pepper (1996). (Reprinted with permission of the Society of Petroleum Engineers (SPE).)

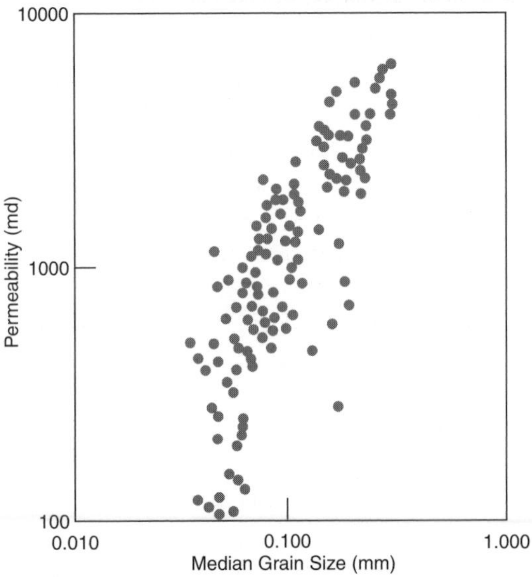

Fig. 5.15. Median grain size versus permeability (log–log) crossplot for the same samples shown in Fig. 5.13. Grain-size analysis was conducted on the plugs after the porosity and permeability measurements had been made. There is a clear trend of increasing permeability with an increase in grain size of the plug sample. After Reedy and Pepper (1996). (Reprinted with permission of the Society of Petroleum Engineers (SPE).)

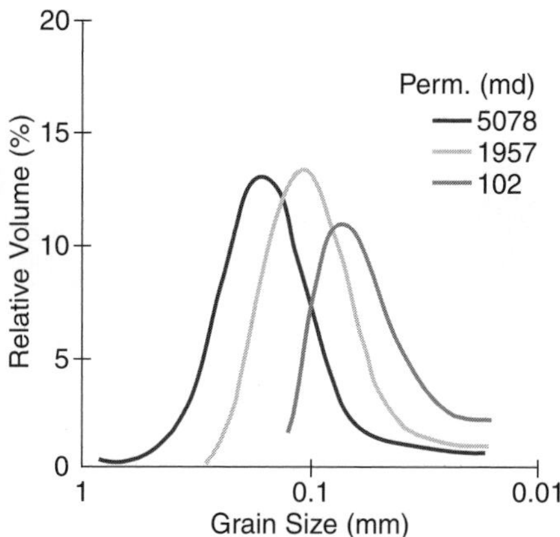

Fig. 5.16. Figure showing three of the grain-size analyses conducted on the suite of samples shown in Fig. 5.15. Each analysis is a frequency curve showing the distribution of particle sizes within the sample. The median grain size is the most frequently occurring grain size in the sample and is shown by the peak in the frequency curve. For the three samples shown here, one has a median grain size of 0.2 mm, another is 0.1 mm and the third is 0.08 mm. Permeability measurements made on these three plug samples are also provided. The coarsest grained sand (0.2-mm median grain size) has a permeability of 5,078 md, the middle sand (0.01-mm median grain size) has a permeability of 1,957 md, and the finest grained sand (0.08-mm median grain size) has a permeability of 102 md. After Reedy and Pepper (1996). (Reprinted with permission of the Society of Petroleum Engineers (SPE).)

Slatt et al. (1993) showed the causal relationship between grain size, sorting, porosity, and permeability. Figure 5.17 displays four histograms of porosity and permeability for two sets of genetically related, unconsolidated sands – termed "thick-bedded" (more than 2 ft thick) and "thin-bedded" (less than 2 ft thick) sands – from the Wilmington oilfield of southern California. The porosities of both sets of sands are similar, but permeabilities are significantly different. Grain-size analysis of a subset of samples, using sieve methods, shows that the average grain size of the thick-bedded sands is slightly coarser (0.160 mm) than the average grain size of the thin-bedded sands (0.148 mm), yet the average numerical values of sorting are equivalent. Thus, there is a positive correlation among bed thickness (thick- versus thin-bedded), grain size, and permeability, but not porosity.

A visual explanation of these relationships is presented in Fig. 5.18. Two equal-area squares are shown (equal-volume cubes could also be used for this demonstration). Yellow/bright-gray circles represent sand grains and purple areas are pore spaces. The thick-bedded sands contain coarser grains (larger yellow/bright-gray circles) than do the thin-bedded sands, yet the sorting is equal for both (perfect sorting, with all grains in each set being of equal size). If one were to calculate the ratio of grains to pores (i.e., porosity) for both the thick-bedded and thin-bedded areas, the values would be the same. However,

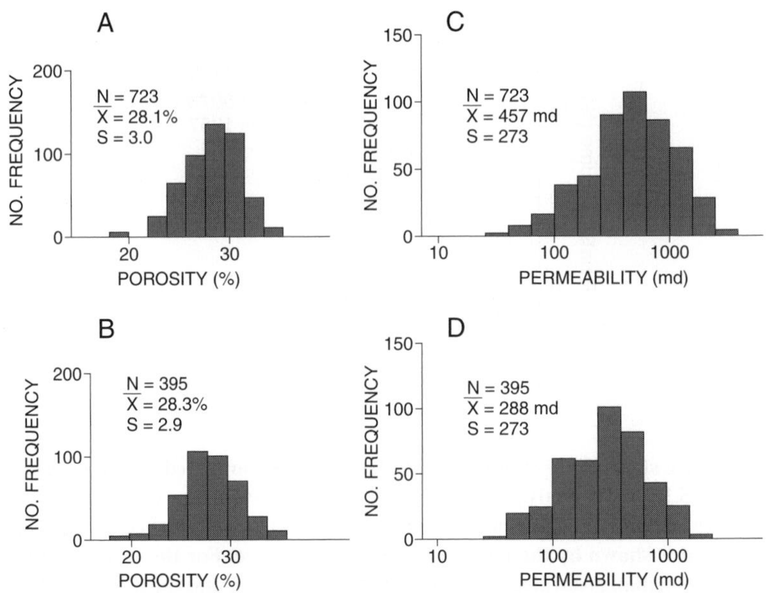

Fig. 5.17. Four histograms showing the (A) distribution of unstressed and (B) stressed core porosity and (C) unstressed and (D) stressed core permeability measurements. Number of core plugs analyzed (*N*), average values (*X*), and standard deviations (*S*) are also provided for each histogram. Plug samples are from core of the Long Beach Unit, Wilmington Field, California. Modified from Slatt et al. (1993). (Reprinted with permission of Springer-Verlag Publishing Co.)

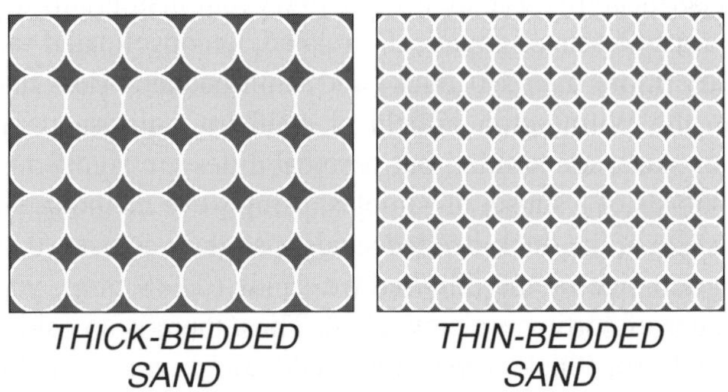

Fig. 5.18. Figure showing two equal areas with sand grains (yellow/bright-gray) and pore spaces (blue/black). The thick-bedded sand has coarser grains than the thin-bedded sand, so more grains can fit into an area of the thin-bedded sand than in an equal area of the thick-bedded sand. The same would be true if we were considering 3D volume rather than 2D area. The area of grains and pores in both squares is equal, so the porosity is the same for both areas. However, because there are fewer grain-to-grain contacts per unit area in the thicker-bedded, coarser-grained sand, the permeability (flow of fluids from pore to pore) will be higher in the coarser-grained sand than in the same area of the finer-grained sand, which has more grain-to-grain contacts.

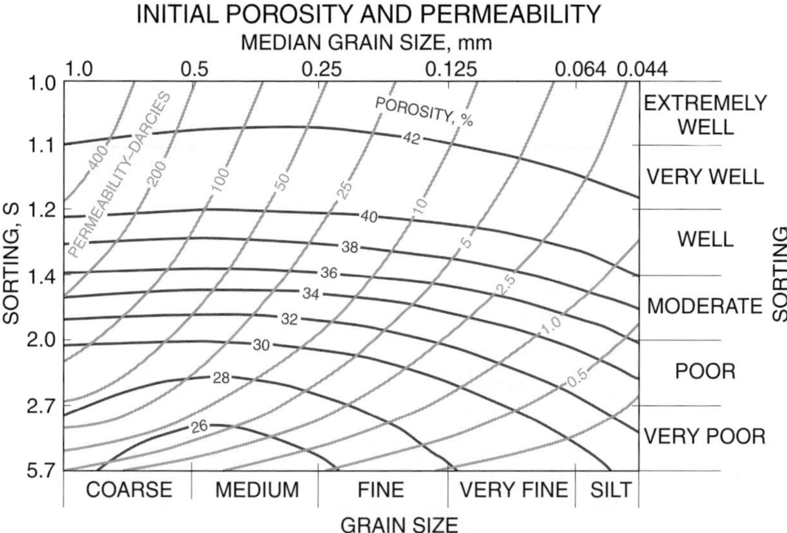

Fig. 5.19. Complex figure illustrating general relations among grain size, sorting, porosity, and permeability. See text for explanation. After Sneider (1987). (Reprinted with permission of AAPG, whose permission is required for further use.)

there are fewer grain-to-grain contacts in the coarser-grained sand than in the finer-grained sand, therefore permeability will be higher in the coarser-grained sand than in the finer-grained sand.

Grain size has also been shown to correlate with bed thickness of turbidite deposits, such as those discussed here (Potter and Scheidegger, 1966). This is because coarser-grained sands deposited from relatively higher-energy flows often will form thicker beds than those formed by finer-grained sands deposited from lower-energy flows.

Figure 5.19 illustrates the complex relationships among grain size, sorting, porosity, and permeability. For example, for any given grain size, permeability and porosity will increase with an improvement in sorting. Thus, for medium-grained sand, the permeabilities may be 10 md if the sand is poorly sorted and 50 md if it is very well sorted. For the same sands, porosities are 29% and 41% for the poorly- and well-sorted sands, respectively. The causal relationship is one of clogging pore spaces with finer-than-average-size particles. If the sand is very well sorted, there are no finer-size particles to fill pore spaces, but if the sand is poorly sorted (i.e., it contains an abundance of fines), the fines will filter through the sand and clog the pore spaces (throats) and reduce porosity and permeability. Beard and Weyl (1973) wrote a standard reference paper relating texture to reservoir quality.

The reader should note that one measurement of porosity or permeability within a single sandstone bed does not necessarily capture all of the variability

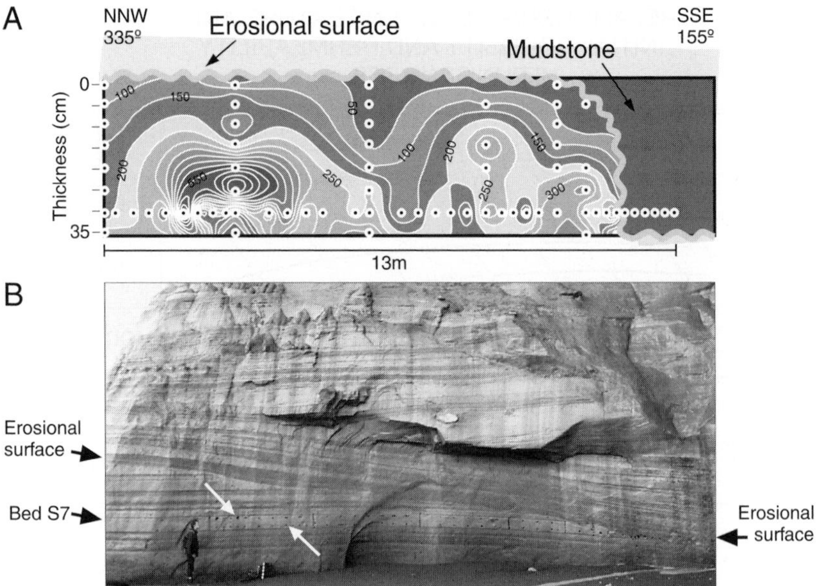

Fig. 5.20. (A) Figure showing the distribution of permeability within a single 35-cm-thick sand bed, Bed S7, shown in (B). The permeability distribution is quite complex and would be difficult to define adequately with a single measurement, such as from a core. The variability in permeability is due to small-scale variability in grain size of the sand in the lateral dimension. Miocene Mt. Messenger Formation, New Zealand. After Browne and Slatt (2002). (Reprinted with permission of AAPG, whose permission is required for further use.)

within that bed. For example, internal variations in grain size resulting from progressive sand deposition from a waning flow can result in an order of magnitude of vertical variation within the bed (Fig. 5.20).

5.4 Diagenesis and reservoir quality

Diagenesis is defined in the Glossary of Geology (Jackson, 1997) as "All the chemical, physical, and biologic changes undergone by a sediment after its initial deposition, and during and after its lithification, exclusive of surficial alteration (weathering) and metamorphism.... It embraces those processes (such as compaction, cementation, reworking, authigenesis, replacement, crystallization, leaching, hydration, bacterial action, and formation of concretions) that occur under conditions of pressure (up to 1 kb) and temperature (maximum range of 100–300°C) that are normal to the surficial or outer part of the Earth's crust".

Diagenetic processes and their products are discussed in Chapter 3. These processes are particularly important in oil and gas reservoirs, because they affect the development of porosity and permeability. Although the ultimate porosity and permeability of a sandstone depends largely upon primary depositional processes and sediment texture, as was discussed above, diagenetic processes certainly exert a strong secondary influence on the final porosity and permeability of a reservoir rock. Bloch (1991) presents an excellent discussion on the controls on porosity and permeability in sandstones. In carbonate rocks, diagenesis has a far greater effect on ultimate reservoir quality because of the greater potential for chemical reactions during burial. A good discussion of the effects of diagenesis on reservoir quality is provided by Grier and Marschall (1992).

An example of the influence of diagenesis in reservoir development is the Pennsylvanian Jackfork Group turbidites of western Oklahoma. In outcrop, two distinct sedimentary facies are present: deepwater sheet sandstones and channel-fill sandstones (Fig. 5.21) (Omatsola, 2003). Each of these two types exhibits different porosity and permeability characteristics. The channel-fill sandstones

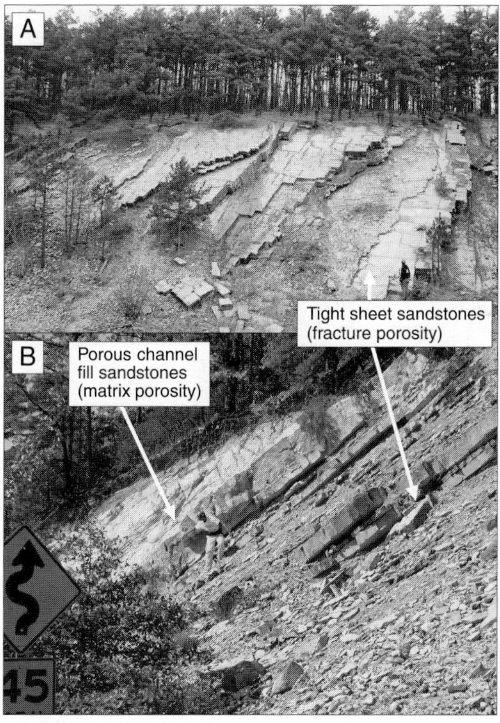

Fig. 5.21. (A) Deepwater sheet sandstones of the Pennsylvanian-age Jackfork Group in eastern Oklahoma, showing orthogonal fracture sets. These quartz-rich sandstones are tightly cemented with quartz, resulting in a brittle rock that is susceptible to fracturing. Fractures are open, providing fracture porosity and permeability. (B) An outcrop near the outcrop shown in (A). The lower strata are well-cemented and fractured sheet sandstones. Stratigraphically above the person (for scale) are channel-fill sandstones that exhibit significant matrix porosity and permeability. These sandstones are not as highly quartz-cemented as those with fracture porosity and permeability, but are in stratigraphic proximity. After Omatsola (2003).

are porous and permeable, whereas stratigraphically adjacent sheet sandstones are tightly quartz-cemented and exhibit distinct, porous and permeable, open fracture sets. Thin-section and microscopic analysis of cuttings samples in subsurface wells through the Jackfork reveal three main sandstone types: (1) porous and permeable sandstones, which tend to be poorly sorted and fine to medium grained, with abundant matrix, (2) moderately well-sorted, fine-grained quartz sandstones that are quartz-cemented, and (3) siderite-cemented sandstones. Thin sections of the porous and permeable sandstones and the quartz-cemented sandstones are shown in Fig. 5.22, along with a cuttings analysis of a subsurface well that contains these two sandstone types. It is interpreted that the quartz-cemented sandstones underwent normal burial cementation from silica-rich fluids, whereas the greater matrix content of the poorly sorted sandstones

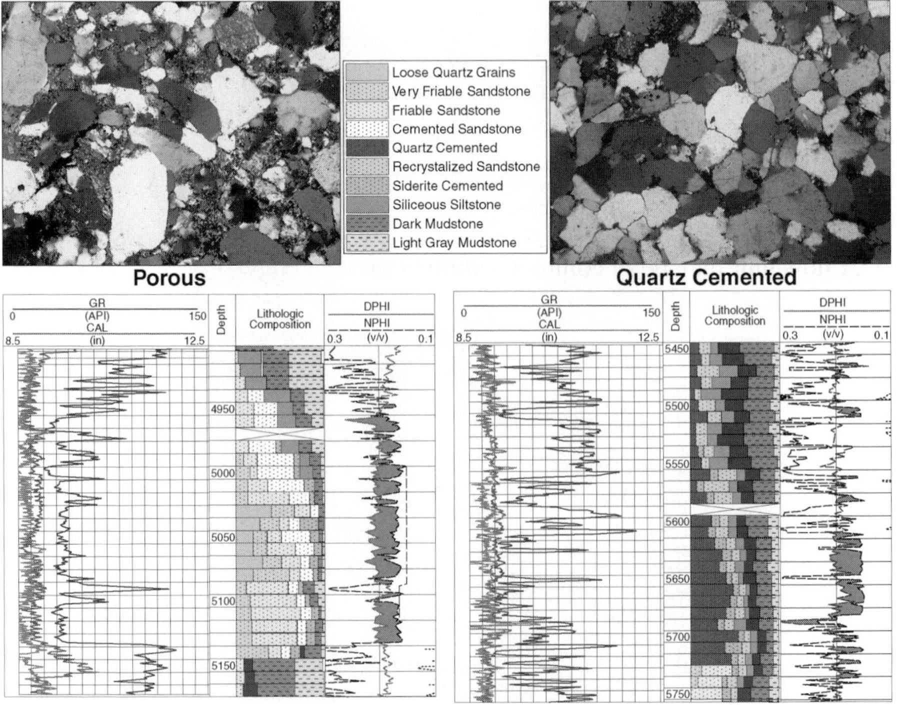

Fig. 5.22. The two upper figures are thin-section photomicrographs of cuttings of the porous and quartz cemented sandstones shown in Fig. 5.21. Note the poor sorting and somewhat coarser grain size of the porous sandstone and the good sorting of the somewhat finer-grained, quartz-cemented sandstone. The lower figures are sections of well logs for the two sandstone types. The corresponding lithologic logs were determined by cuttings analysis at 3.3-m (10-ft) intervals (Romero, 2004). Neutron-density crossover (gas indicator) is shown by red/dark-gray. Note the different gamma-ray log (GR) shape of the two different sandstone types. The blocky log pattern on the left is a typical channel-sandstone pattern, whereas the more serrated, interbedded character on the right is of sheet sandstones.

prevented significant cements from forming. Thus, there was a primary depositional facies control on the secondary diagenetic processes, and this control resulted in sandstones with matrix porosity and permeability and sandstones with fracture porosity and permeability within the same stratigraphic interval (Garich, 2004; Romero, 2004).

5.5 Flow-unit characterization for correlation and upscaling

Flow units have become a popular means of characterizing or zoning a reservoir. A flow unit is defined as "a mappable portion of the total reservoir, within which geological and petrophysical properties that affect the flow of fluids are consistent and predictably different from the properties of other reservoir rock volumes" (Fig. 5.23). (Ebanks et al., 1992.)

Flow units have the following characteristics in common:

- A flow unit is a specific volume of a reservoir; it is composed of one or more reservoir-quality lithologies and any nonreservoir-quality rock types within that same volume, as well as the fluids they contain.
- A flow unit is correlative and mappable at the interwell scale.
- A flow unit zonation is recognizable on wireline logs.
- A flow unit may be in communication with other flow units. (However, flow units based on lithostratigraphic characteristics are not always in pressure communication (Fig. 5.24)).

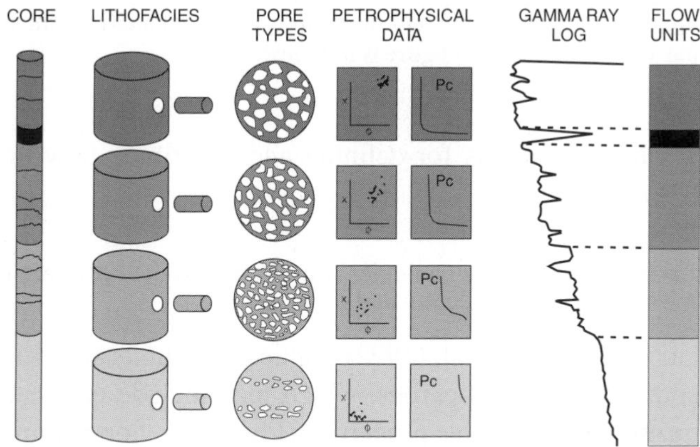

Fig. 5.23. Cartoon showing some of the various parameters that are used in defining geologic flow units (Ebanks et al., 1992). In this cartoon, four flow units are defined on the basis of lithofacies, pore types, porosity, and permeability crossplots, capillary pressure measurements, and gamma-ray log response. After Ebanks et al. (1992). (Reprinted with permission of AAPG, whose permission is required for further use.)

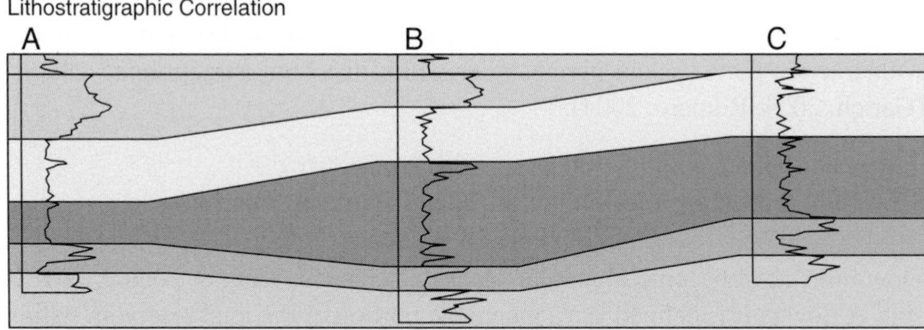

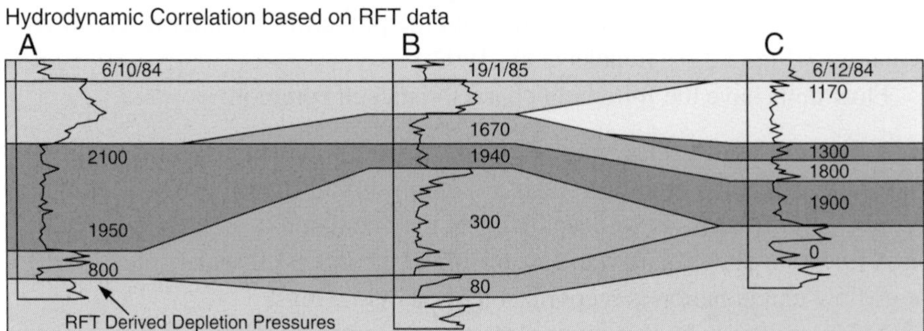

Fig. 5.24. Upper diagram shows a lithostratigraphic correlation of sandstones across three wells (A, B, and C). The lower diagram shows the same three wells, but with correlations based upon repeat formation tester (RFT) pressure measurements. Note that the pressure-derived correlations crosscut the lithostratigraphic boundaries and define a greater degree of compartmentalization than had been interpreted originally on the basis of lithostratigraphic parameters. (Source of figure is unknown.)

There are various methods for defining and describing flow units. Two of those methods are presented.

5.5.1 Flow units that combine geological and petrophysical properties

As the definition of Ebanks et al. (1992) implies, flow units are identified on the basis of a combination of properties, which may include qualitative geologic facies and quantitative reservoir properties (Fig. 5.23). For example, Balmoral field in the North Sea (Fig. 5.25) was subdivided into deepwater channel and lobe sandstones, on the basis of excellent core control (Slatt and Hopkins, 1991). The field could also be subdivided into five distinct flow units, on the basis of a combination of geological and petrophysical properties (Table 5.1; Fig. 5.26). The flow units cross-cut sedimentary facies boundaries (compare

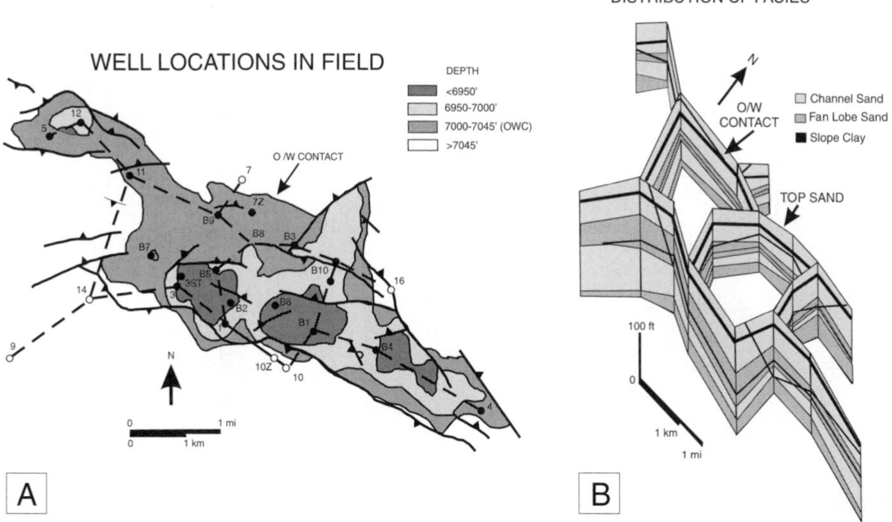

Fig. 5.25. (A) Balmoral field, North Sea, and well locations (Slatt and Hopkins, 1991). The different colors refer to different structural elevations, as shown by the inset. The oil–water contact is at a depth of 7,045 ft. The outline of the fence diagram shown in (B) is illustrated as a series of dashed line segments. (B) A fence diagram of the 3D distribution of the main facies in the field, channel sand, fan-lobe sand, and slope clay.

Table 5.1 (Slatt and Hopkins, 1991)

Flow unit	Abbre- viation	Perme- ability (md)	Porosity (%)	Medium grain size (mm)	Medium pore throat size (mm)	Sb@ 200 psi (%)	Facies
Excellent	E	>1000	23–34	0.18–0.30	0.01	6–12	Massive, sandy channel facies
Good	G	100–1000	20–34	0.08–0.24	0.007	11–24	Massive, sandy channel and lobe facies
Poor I	Pi	0.01–1000	7–32	0.10–0.23	0.002	31	Interbedded sand- stone and shale of channel and lobe facies
Poor C	Pc	0.01–1000	4–28	0.11–0.25	0.002	30–37	Massive sand- stone with calcite cemented zones; channel and lobe facies
Poor M	Pm	Imperm.	Nonpor.	–	–	–	Clay/Shale

Abbreviations: i – interbeds; c – cemented; m – shale; Sb – Brine saturation.

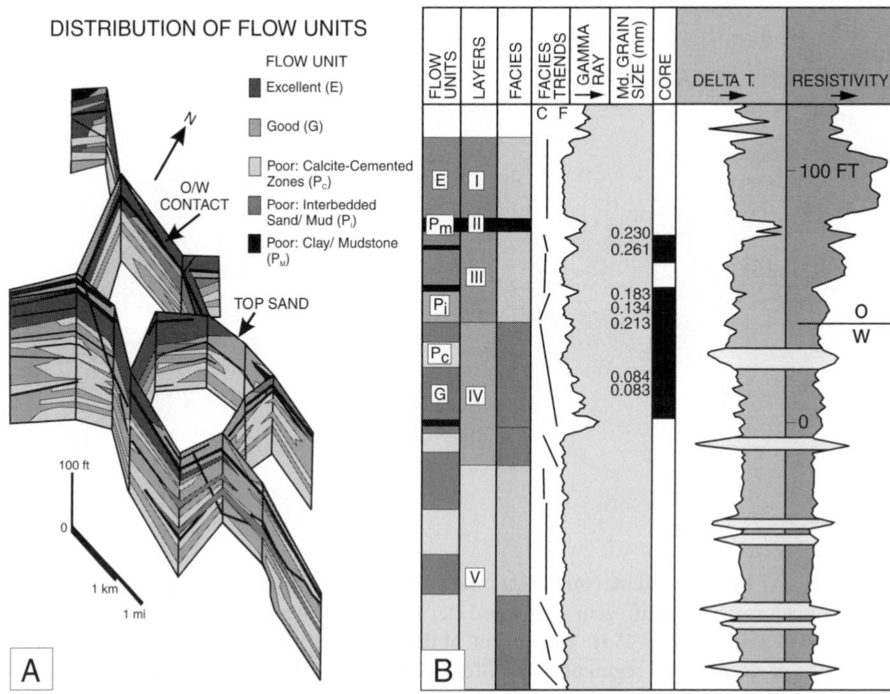

Fig. 5.26. (A) Fence diagram shown in Fig. 5.25, but showing the 3D distribution of flow units based on a combination of geological and petrophysical properties (Slatt and Hopkins, 1991). Note the greater complexity of the flow-unit stratification over the facies stratification shown in Fig. 5.25. (B) Various properties from a single well, including gamma-ray, sonic, and resistivity logs, some gram-size measurements (in millimeters), and the vertical distribution of facies, layers (a zonation based upon average porosity and permeability values), and flow units. Calcite concretionary intervals are shown in blue/bright-gray by low sonic transit time and high resistivity.

Figs. 5.25 and 5.26), but this classification provides a better means of subdividing the reservoir into intervals based upon similar fluid flow properties.

5.5.2 Gunter et al.'s (1997) method of flow-unit characterization

Gunter et al. (1997) described a technique for combining porosity, permeability, and bed thickness data for flow-unit identification. Gunter et al. (1997) utilize the Stratigraphic Modified Lorenz (SML) plot for characterization. This method of flow-unit determination is quite useful, because it requires only routine porosity and permeability data (from logs and/or cores), is independent of facies identification, and uses simple crossplotting techniques. The SML plot is a crossplot of "cumulative flow capacity" – defined as the product of average permeability and thickness of an interval (kh) – versus "cumulative storage capacity" – defined as the product of average porosity and thickness of the same interval (Φh). The equation for obtaining one cumulative-flow-capacity value is as follows (Maglio-Johnson, 2000),

$$(kh)cum = k_1(h_1 - h_0) + k_2(h_2 - h_1) + \cdots + k_i(h_i - h_{i-1}) / \sum k_i(h_1 - h_{i-1}),$$

where k – permeability (md), h – thickness of the sample interval.

A similar equation is used to determine a single cumulative-storage-capacity value,

$$(\Phi h)cum = \Phi_1(h_1 - h_0) + \Phi_2(h_2 - h_1) + \cdots + \Phi k_i(h_i - h_{i-1}) / \sum \Phi k_i(h_i - h_{i-1}),$$

where Φ – fractional porosity.

An example calculation of these parameters is provided in Table 5.2, using hypothetical porosity and permeability measurements made on core plugs sampled every 0.3 m (1 ft) over a 1.5-m stratigraphic interval. Table 5.2 lists the products of porosity (Φ) and thickness (h) and of permeability (k) and thickness (h), calculated for each 0.3-m interval. Next, individual products are summed to give total porosity-thickness and total permeability-thickness values. A fractional porosity-thickness value is calculated for each interval by dividing the product of that interval by the total porosity-thickness. The same procedure is used to calculate a fractional permeability-thickness value for each interval.

Values are added to each successive increment, for both porosity-thickness and permeability-thickness, to obtain cumulative porosity-thickness (cumulative storage capacity) and cumulative permeability-thickness (cumulative flow-capacity) values. These values are crossplotted, beginning with the deepest stratigraphic interval at the base (Fig. 5.27). Straight-line segments on the plot define individual flow units.

Application of the Gunter et al. (1997) technique is described here for a research well drilled in Wyoming called the CSM Strat Test Well #61. This well was drilled from the ground surface to a depth of 567 m (1,700 ft) through the Upper Cretaceous Dad Sandstone member of the Lewis Shale. Conventional well logs down to a depth of 400 m (1,200 ft) are shown in Fig. 5.28. Samples of continuous core were obtained from the depth intervals of 50–200 m (150–600 ft) and 290–305 m (860–915 ft), from which core plug and minipermeameter permeabilities were measured at closely spaced intervals (Fig. 5.28).

Because the entire interval was not cored, it was not possible to develop a flow-unit zonation directly from this data set. Porosity logs were available for the entire interval, but permeability values were determined only for the cored intervals. A neural-network approach was used to develop a synthetic, continuous permeability log from the conventional well logs and then calibrated to the core permeability measurements (Maglio-Johnson, 2000). A nuclear magnetic resonance (NMR) clay-corrected porosity log (determined from an NMR

Table 5.2 Calculation of cumulative storage and flow capacities

	Step 1			
Depth (ft)	Porosity	Porosity* (h)	Permeability	$k(h)$
1805	0.15	(0.15)(1) = 0.15	10	(10)(1) = 10
1804	0.20	(0.20)(1) = 0.20	20	(20)(1) = 20
1803	0.15	(0.15)(1) = 0.15	10	(10)(1) = 10
1802	0.10	(0.10)(1) = 0.10	5	(5)(1) = 5
1801	0.05	(0.05)(1) = 0.05	2	(2)(1) = 2
Summation		0.65		47

	Step 2	
	Fractional (Porosity) $(h) = (X)/0.65$	Fractional $(k)(h) = (Y)/47$
1805	(0.15)/0.65 = 0.23	(10)/47 = 0.21
1804	(0.20)/0.65 = 0.31	(20)/47 = 0.43
1803	(0.15)/0.65 = 0.23	(10)/47 = 0.21
1802	(0.10)/0.65 = 0.15	(5)/47 = 0.11
1801	(0.05)/0.65 = 0.08	(2)/47 = 0.04
Summation	1.00	1.00

	Step 3	
	Cumulative (Porosity) (h) Cumulative storage capacity	Cumulative $(k)(h)$ Cumulative flow capacity
1805	0.23	0.21
1804	0.54	0.64
1803	0.77	0.85
1802	0.92	0.96
1801	1.00	1.00

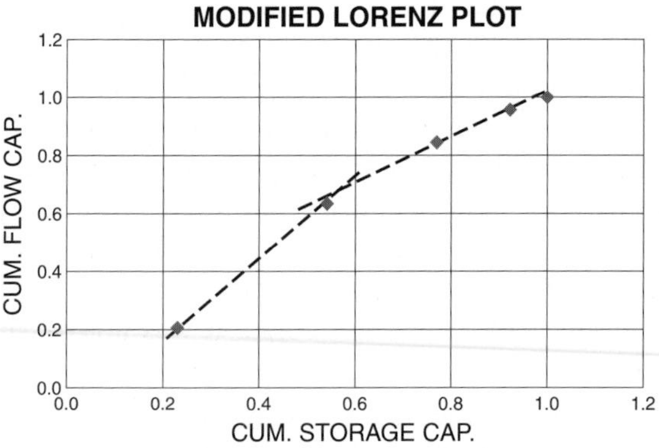

Fig. 5.27. SML plot of cumulative storage capacity versus cumulative flow capacity for the data presented in Table 5.2 and described in the text. Two flow units are defined on the basis of the five data points.

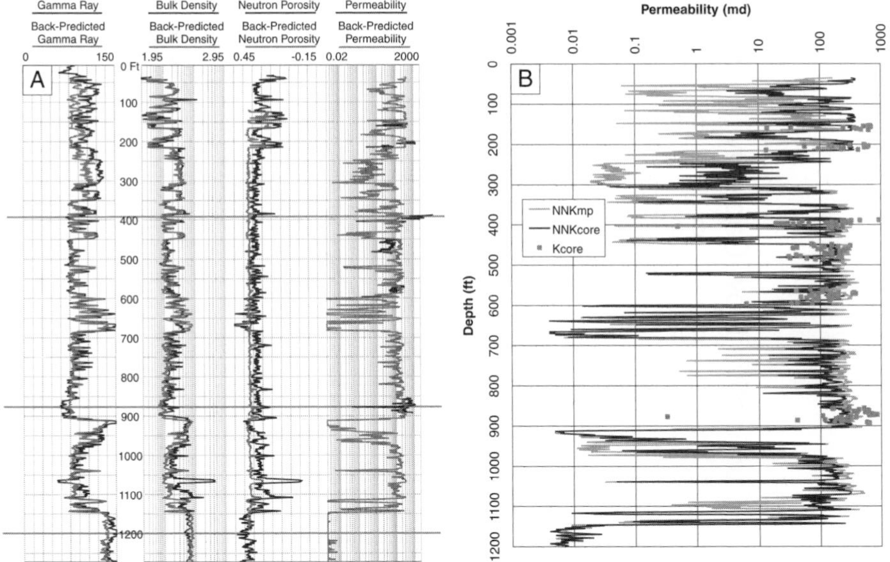

Fig. 5.28. (A) Shown by blue/black curves are gamma-ray, bulk density, and neutron porosity logs from the logged interval in the CSM Strat. Test #61 well from the Cretaceous Lewis Shale. (Pyles and Slatt, 2000). Blue/black intervals on the permeability log are from core plug data. (B) shows core plug permeability measurements in green/dark-gray. A neural-network-derived permeability curve was developed for the entire cored interval based on minipermeameter measurements from the core (blue/black curve) and from minipermeameter measurements of permeability (purple/gray curve). The resulting permeability log is shown in red/gray (A), along with neural-network-derived gamma-ray, bulk density, and neutron porosity curves (also shown in red/gray). Training points are at the green/dark-gray lines in (A). Modified from Maglio-Johnson (2000).

log run in the well) was found to approximate true porosity more closely than a density-log-derived porosity did. On this basis, 10 flow units were identified (Fig. 5.29). Table 5.3 lists average porosity and permeability of the flow units as well as the percentage contribution of each flow unit to the total cumulative storage capacity and cumulative flow capacity. The stratigraphic (depth) distribution of these flow units is shown in Fig. 5.30. On the SML plot, shaly intervals plot as low-angle to horizontal-trending flow units (Fig. 5.29, flow units 3, 5, and 8). By contrast, the flow units that contribute the most to the overall cumulative storage capacity and cumulative flow capacity are sandy intervals (Fig. 5.28) that exhibit a steeper gradient-slope on the SML plot (Fig. 5.29; particularly flow units 2, 4, 6, and 7).

It is also possible and desirable to develop a 3D flow-unit characterization of a reservoir. To do this, it is necessary to identify flow units from one cored well, and then use that information to correlate well logs from other wells. Figure 5.31 shows an example of a cored well and the corresponding core-gamma-ray log of the reservoir interval. Most of the interval was cored, so a flow-unit model

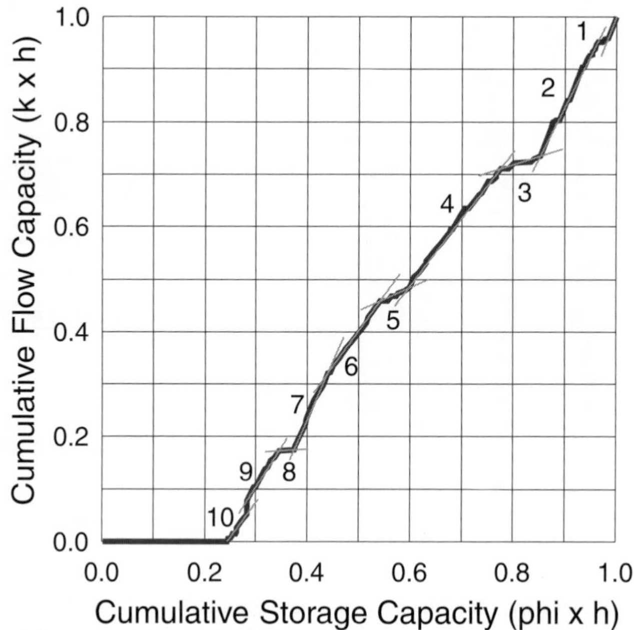

Fig. 5.29. SML plot of cumulative storage capacity versus cumulative flow capacity for the well shown in Fig. 5.28. Ten flow units are defined for this stratigraphic interval on the basis of a combination of well log, core-derived, and neural-network-derived porosity and permeability values.

Table 5.3 Cumulative storage and flow capacity for CSM #61 well

Flow unit	Depth (ft)	Ave. porosity (%)	Ave. permeability (md)	% Porosity × h	% Permeability × h
1	0–85	28.0	138	4	5
2	85–225	27.8	191	11	22
3	225–350	19.9	23	7	2
4	350–590	25.7	119	18	23
5	590–690	20.3	37	6	3
6	690–930	25.5	129	11	14
7	830–920	25.1	188	6	13
8	920–980	17.3	18	3	1
9	980–1,060	24.2	141	6	9
10	1,060–1,160	15.8	97	4	<1
	1,160–base	15.6	0.03	0	0

could be developed from the porosity and permeability data (a neural network approach was not used here to develop a permeability log in the uncored portion of the well). Twelve flow units (labeled A–L in Fig. 5.31) were identified from

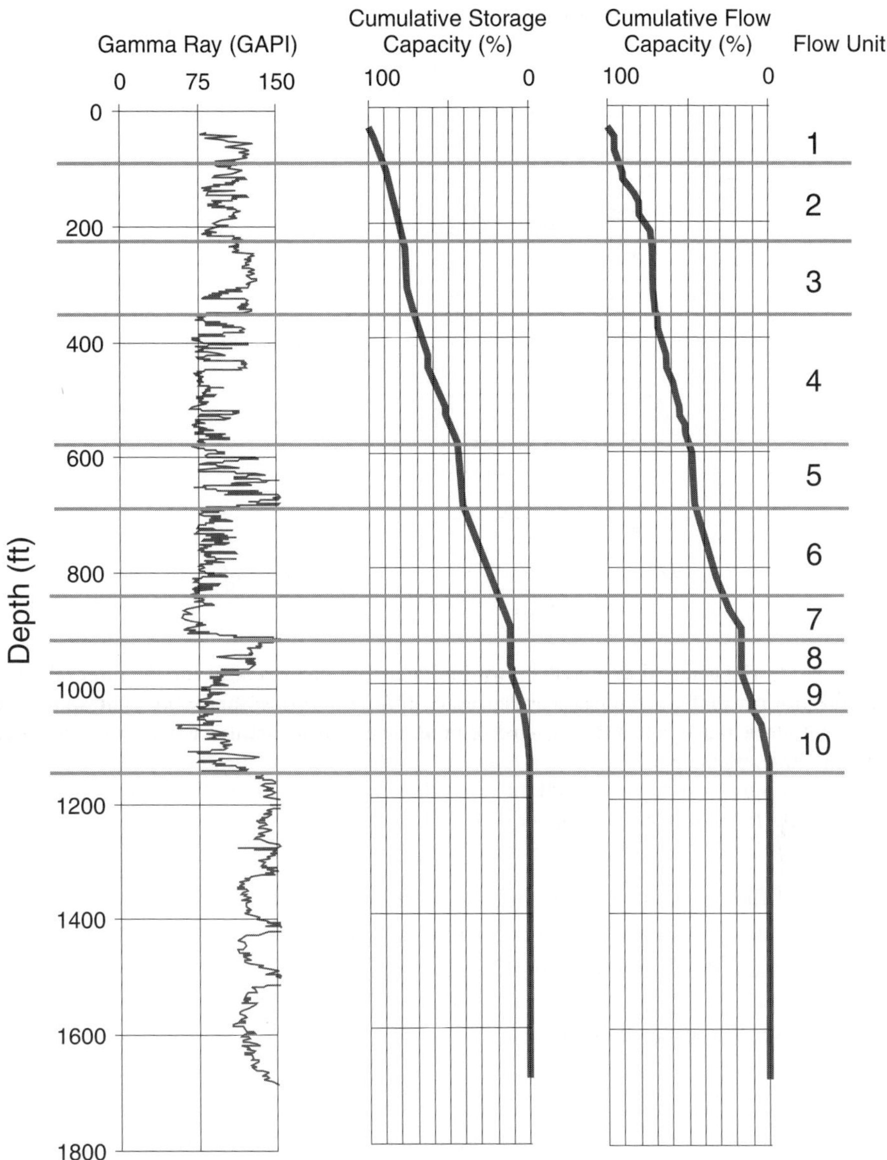

Fig. 5.30. Depth plot of gamma-ray log, cumulative storage capacity, and cumulative flow capacity for the CSM Strat. Test #61 well. The stratigraphic distributions of the 10 flow units defined in Fig. 5.29 are shown. Average properties of the flow units are listed in Table 5.3.

the flow-unit characterization. Logs from two other wells in the field, positioned 267 m (800 ft) and 533 m (1,600 ft) from the cored well, were then correlated to the gamma-ray log in the cored well, and that allowed correlation and identi-

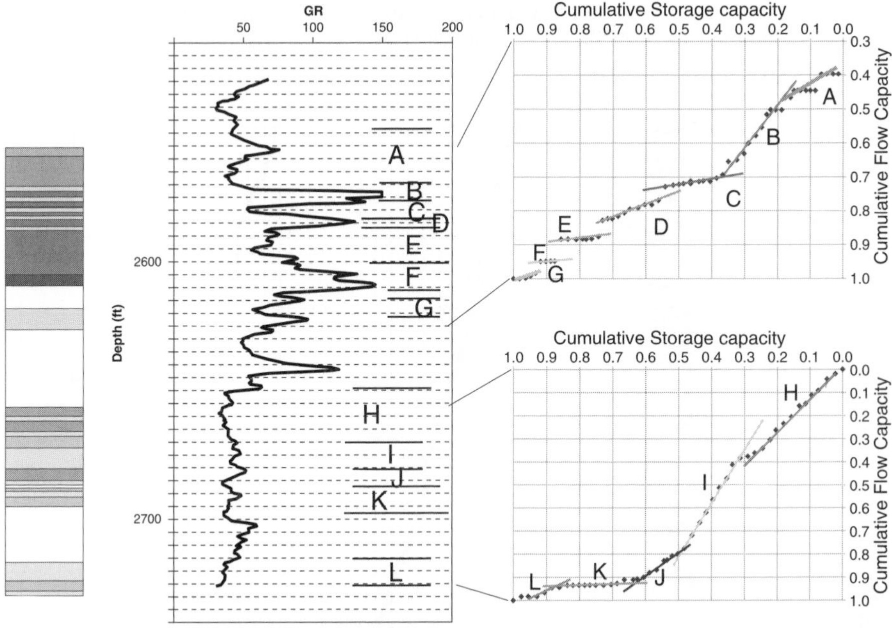

Fig. 5.31. Cored interval in a well shown beside a core gamma scan for the interval. Yellow/bright-gray and green/dark-gray represent different types of sandstones, and shades of gray and black represent different types of mudstones. White areas are those without core. The two SML plots on the right show a flow-unit subdivision for the upper and lower cored intervals. Upper muddy zones contain seven flow units (A–G), and the lower sandy zone contains five flow units (H–L). (Figure provided by D. Restrepo.)

fication of flow units in the uncored wells (Fig. 5.32). In this instance, the three wells formed a triangular area in the field, so the correlated flow units and their thicknesses could be mapped in 3D space (Fig. 5.33).

5.5.3 Upscaling using flow units

For reservoir characterization, it would be ideal to incorporate all geological and petrophysical data at the scale at which the data are available. However, computing time, costs, and capabilities all limit our ability to build a data-rich characterization that can be used for reservoir fluid-flow simulation. Therefore, it is necessary to group data into a smaller set of attributes that are intended to portray the most significant aspects of the reservoir. This process, called "upscaling", is defined by Stephen et al. (2001) as "grid coarsening, enabled by the calculation of effective flow properties using analytical (e.g., arithmetic, geometric or harmonic averages, streamlines, etc.) and numerical (single- and two-phase flow) simulation". An upscaled reservoir characterization is then used in reservoir simulation modeling.

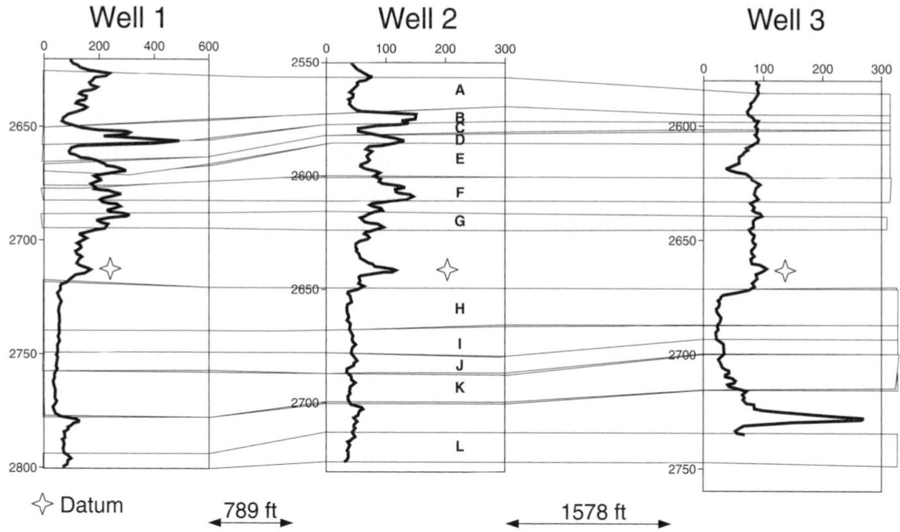

Fig. 5.32. Gamma-ray logs from two wells (1 and 3) correlated with the core gamma-scan of well 2. The twelve flow units A–L (Fig. 5.31) are shown. Distance between the three wells is given, as is a shale correlation datum. (Figure provided by D. Restrepo.)

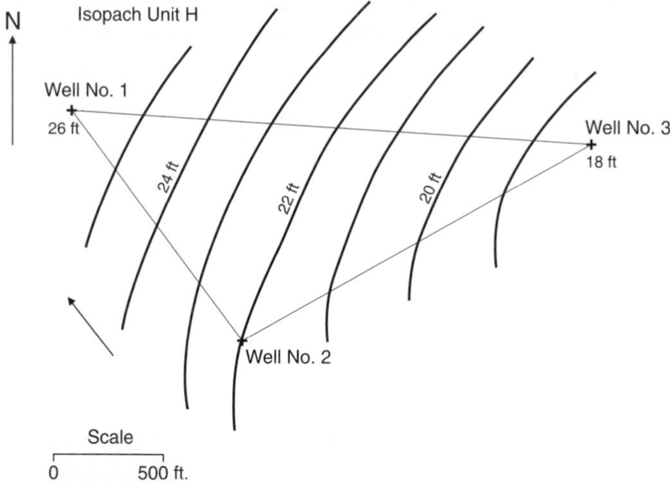

Fig. 5.33. The map view locations of the three wells in Fig. 5.32 are shown. Because the three wells do not form a straight line, it is possible to map the distribution of individual flow units between the wells, as is shown in this figure for Flow Unit H. Contour interval is 0.3 m (1 ft). (Figure provided by D. Restrepo.)

The SML plot provides a relatively easy technique for upscaling a reservoir on the basis of only three parameters: porosity, permeability, and thickness. For the CSM Strat. Test #61 example, the 10 flow units represent an upscaling of more than 365 m (1,100 ft) of strata, on the basis of diverse reservoir parameters. If fewer flow units were required, flow units could be combined in the following manner: Flow Unit I = Flow Units 10–5 with Flow Unit 8 a shale break; Flow Unit II = Flow Units 4 and 3, where Unit 3 is a shale break; Flow Unit III = Flow Units 1 and 2 (Figs. 5.29 and 5.30). For the three-well example (Fig. 5.32), flow units can be grouped in the following manner: Flow Unit I = Flow Unit A with Flow Unit B being a shale break, Flow Unit II = Flow Units C, E, and G with Flow Units D and F being shale breaks, Flow Unit III = Flow Units H and I, Flow Unit IV = Flow Units J and K, and Flow Unit V = Flow Unit L.

5.6 Capillary pressure and its applications to reservoir characterization

5.6.1 Principles of capillary pressure

An excellent paper on the principles and applications of capillary pressure from a geologic perspective is provided by Vavra et al. (1992). Key parts of that paper are summarized here, with particular emphasis on the set of calculations required to evaluate reservoir properties for characterization purposes.

Two competing pressures (forces) control hydrocarbon entrapment (Fig. 5.34A). Buoyancy acts to displace less-dense fluids (such as oil) upward, whereas capillary pressure acts to displace denser fluids (such as water) downward. In a reservoir, if P_b (buoyancy pressure) is less than P_c (capillary pressure), oil cannot migrate upward to displace water in the pore spaces and the

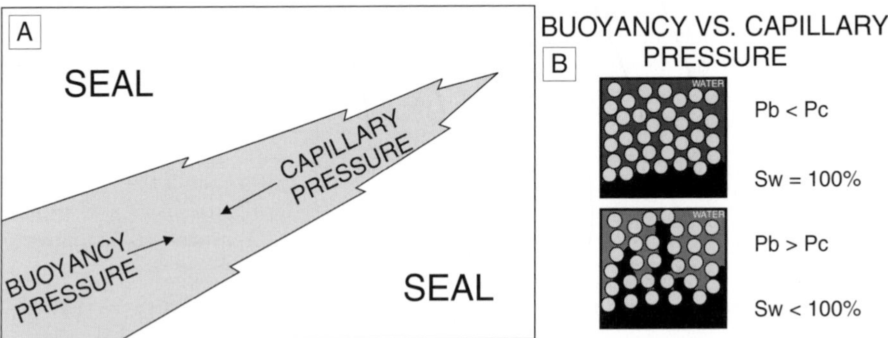

Fig. 5.34. (A) Illustration of the principle of capillary and buoyancy pressures. The upward migration of hydrocarbons is driven by buoyancy (density difference between water and hydrocarbons). Buoyancy pressure is opposed by capillary pressure (displacement pressure of largest pore throat; Vavra et al., 1992). (B) shows variations in reservoir saturations (S_w) depending upon whether buoyancy (P_b) or capillary pressure (P_c) is greater. When $P_b < P_c$, then hydrocarbons cannot migrate into reservoir pore spaces. When $P_b > P_c$, upward buoyancy drives hydrocarbons into the reservoir by replacing water. After Vavra et al. (1992). (Reprinted with permission of AAPG, whose permission is required for further use.)

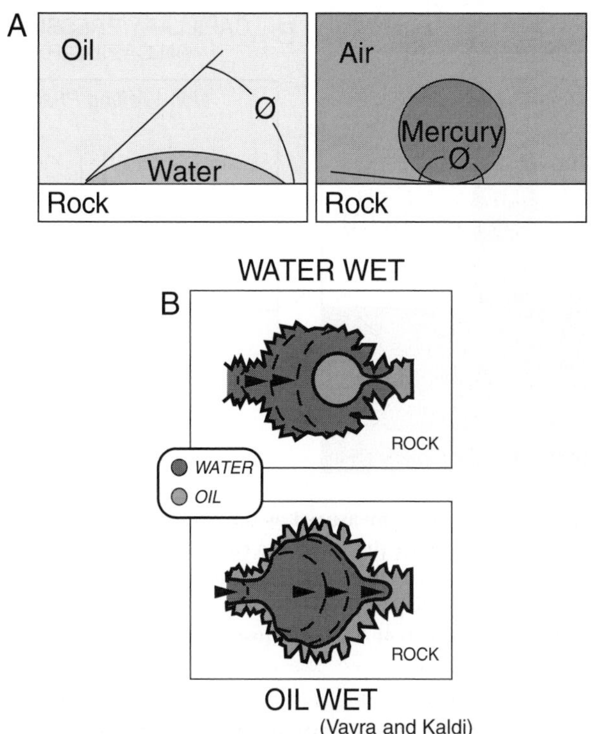

Fig. 5.35. (A) The interaction of adhesive and cohesive forces on wettability. If adhesive forces are greater than cohesive forces, the fluid spreads out on the surface and is referred to as being "wetting" (left figure). If cohesive forces are greater than adhesive forces, the liquid beads up and is termed "nonwetting" (right figure). The contact angle is the measure of wettability. (B) When water lines the pore space, the rock is said to be water-wet. When oil lines the pore space, then the rock is said to be oil-wet. After Vavra et al. (1992). (Reprinted with permission of AAPG, whose permission is required for further use.)

spaces are entirely filled with water ($S_w = 100\%$) (Fig. 5.34B). When P_b is greater than P_c, the pore spaces will contain hydrocarbons that have moved upward and displaced water in the pore spaces ($S_w < 100\%$). Wettability of a fluid–rock system is dependent upon interactions of adhesive and cohesive forces (Fig. 5.35). Adhesive forces (AF) are those between the fluid and solid (pore walls composed of grains). Cohesive forces (CF) are those between fluids. If cohesive forces exceed adhesive forces, the liquid beads up and is said to be "nonwetting". When adhesive forces exceed cohesive forces, the fluid spreads out on the pore walls (grain surface) and is said to be "wetting" (Fig. 5.35). A rock can be approximated by a bundle of capillaries (pores), with formation water being the wetting phase and oil being the nonwetting phase.

Capillary pressure (P_c) (i.e., buoyancy or displacement pressure) is the difference in pressure measured across the interface of a capillary, or the amount

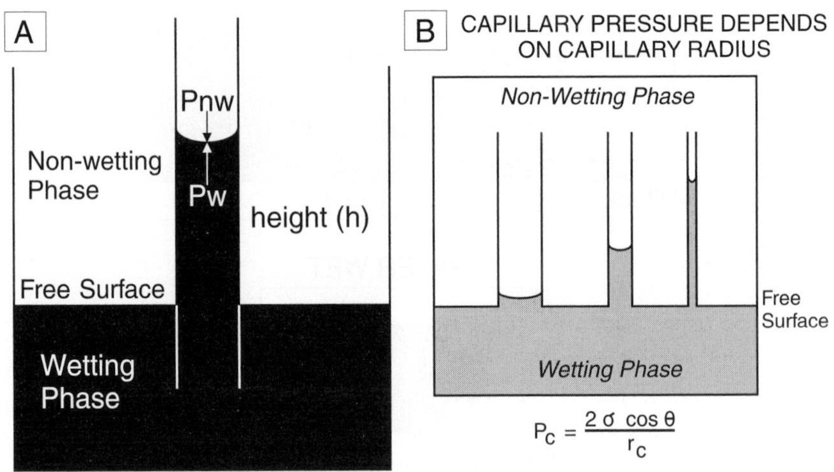

Fig. 5.36. (A) Capillary pressure is measured as the pressure required for a wetting-phase fluid to displace a nonwetting-phase fluid within a capillary tube. (B) Variation in the rise of wetting phase fluids within capillary tubes is a function of tube radius. Larger radii require less capillary pressure for fluid displacement than do smaller-radii tubes. After Vavra et al. (1992). (Reprinted with permission of AAPG, whose permission is required for further use.)

of pressure required to force the nonwetting phase (oil) to displace the wetting phase (water) in a capillary (Fig. 5.36A). Mathematically, it is defined as:

$$P_c = \frac{2\sigma(\cos\theta)}{r_c},$$

where σ – interfacial tension (dynes/cm), θ – contact angle, r_c – pore (capillary) radius (cm) (Fig. 5.35).

Thus, P_c is inversely proportional to pore size. In smaller pores, greater P_c is required to displace wetting-phase (water) with nonwetting-phase (oil) fluids; thus, wetting phase fluid more readily remains in or is incorporated into the pore spaces (that is, it rises higher in a capillary tube (Fig. 5.36B)). In larger pores, nonwetting-phase fluid is more readily incorporated into the pore spaces. Because movement of fluids within a rock depends more upon the size of pore throats than on the size of pores, there is a direct relationship between permeability (which is controlled by pore-throat size) and P_c, as discussed below.

Capillary pressure (a) controls the original static distribution of reservoir fluids and (b) provides a mechanism for hydrocarbon movement through the reservoir (Vavra et al., 1992). Its measurement can be used to calculate or estimate:

- reservoir rock quality
- net pay or nonpay, and classification of pay types

- seal capacity of rocks and faults
- expected maximum hydrocarbon column
- thickness and location of transition zone
- fluid saturations at different levels in the reservoir
- recovery efficiency of rocks of varied pore types
- residual oil saturation after primary or secondary recovery.

5.6.2 Routine laboratory measurement of capillary pressure

The principle of measuring capillary pressure is illustrated in Fig. 5.37A. A rapid means of determining P_c is by mercury (Hg)-injection capillary-pressure measurement. The procedure is to inject mercury into an evacuated, cleaned, and extracted core plug. Mercury-injection pressure is increased in a stepwise fashion, and the percentage of the rock's pore volume that becomes saturated with mercury at each step is recorded after equilibration (Fig. 5.37B). Pressure is then plotted against mercury saturation. The capillary pressure (P_c) applied in core test measurements to displace water with hydrocarbons is equivalent to the buoyancy pressure in a reservoir when hydrocarbons migrate into and charge the reservoir (Heymans, 1998). In the example provided in Fig. 5.38, each of the four different rocks (A–D) has a different entry-level pressure (P_c).

5.6.3 Relationship of P_c to pore-throat size and size distribution

The different entry-level pressures are inversely related to the sizes (radius, or r_c) of pore throats. The larger the pore-throat size, the smaller the P_c required for nonwetting phase fluid (oil) to displace wetting phase fluid (water). In air–Hg systems, $2\sigma(\cos\theta)$ approximates 107.6 (after conversion of units to psi and microns), so the capillary pressure equation becomes

$$P_c = 107.6/r_c,$$

where P_c – is in psi (pounds per square inch), r_c – is in microns.

Thus, the entry pressure (P_c) required for nonwetting-phase fluid to enter a pore of size r_c is:

r_c (microns)	P_c (psi)
10,000	0.011
1,000	0.108
100	1.076
10	10.760
1.0	107.600
0.1	1,076.000
0.05	2,000.000

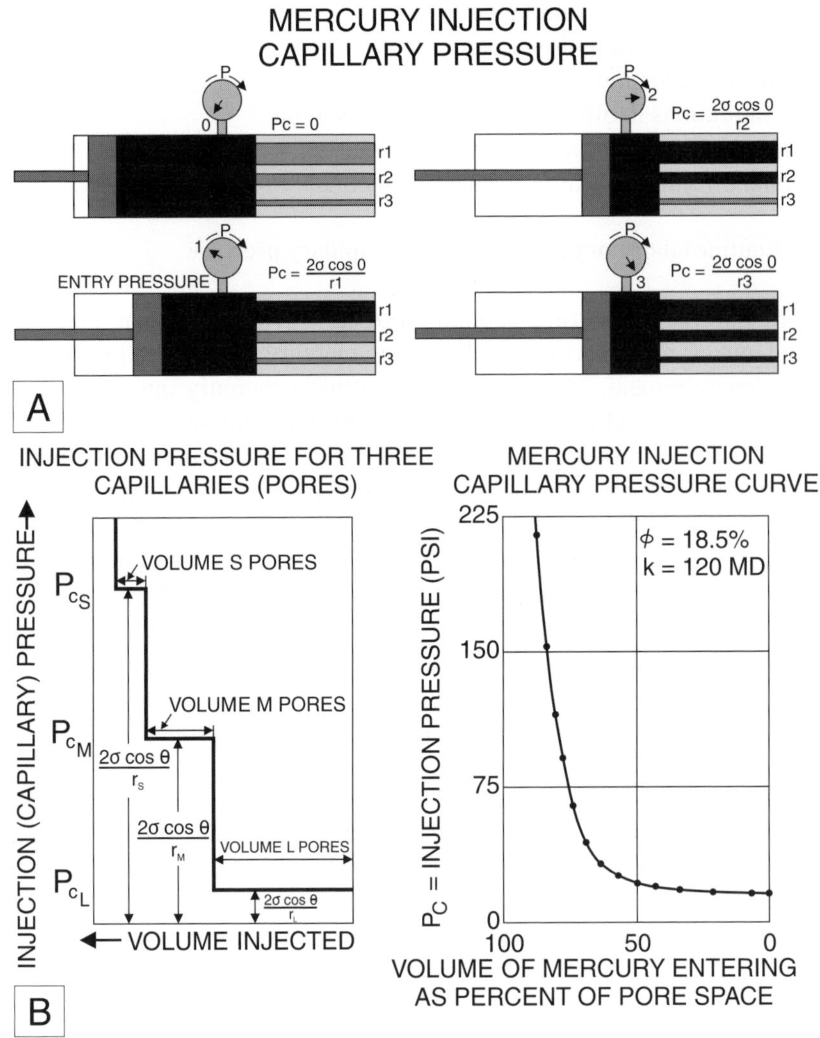

Fig. 5.37. (A) Illustration of the concept of capillary pressure measurement. A cylinder contains slots (pores) of three different diameters: large (r_1), intermediate (r_2), and small (r_3). The cylinder is filled with oil. At $P_c = 0$ (upper left figure), the oil remains in the cylinder and does not enter the pores. As pressure ($P_c = 2\sigma(\cos\theta)/r_1$) is applied with the piston, oil fills the largest pore space (r_1) because only a low pressure is required (lower left diagram). With continued increase in pressure to $P_c = 2\sigma(\cos\theta)/r_2$, the intermediate pore (r_2) becomes filled with oil. With continued increase in pressure to $P_c = 2\sigma(\cos\theta)/r_3$, the smallest ($r_3$) pore becomes filled with oil (Vavra et al., 1992). (B) When results are plotted with the volume of mercury injected (nonwetting phase) as the horizontal axis and pressure as the vertical axis, stepwise pressure increases are shown. In reality, the pressure increase is progressive, giving rise to a smoother, typical capillary pressure curve, as is shown on the right curve. After Sneider (1987). (Reprinted with permission of AAPG, whose permission is required for further use.)

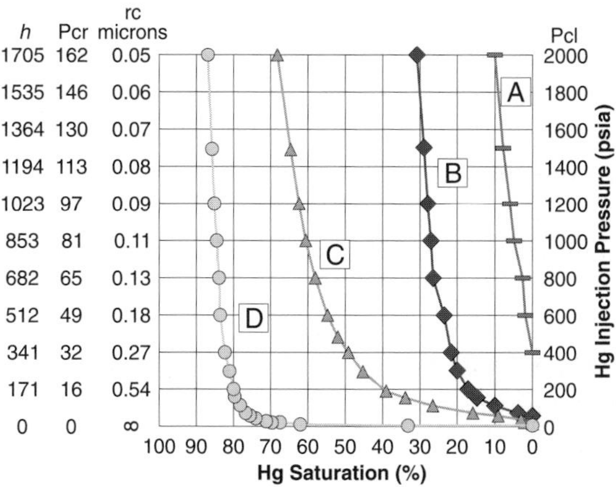

Fig. 5.38. Capillary pressure curves of four samples, A–D. The horizontal axis is nonwetting phase (Hg) saturation. The vertical axis shows the primary measurement, Hg-injection pressure (Fig. 5.37) and derived vertical scales of pore-throat size in microns, P_{cr} (oil-brine capillary pressure), and *h* (height above free water level). See text for derivations of these parameters.

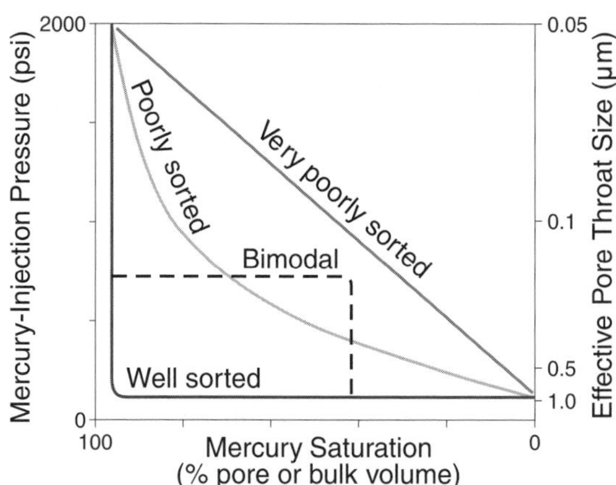

Fig. 5.39. Three idealized Hg-injection capillary-pressure curves for rocks with different pore-size distributions. All of the curves have identical displacement pressures and minimum unsaturated pore volumes, but the saturation profiles differ dramatically as a result of differences in pore-throat size distribution. After Vavra et al. (1992). (Reprinted with permission of AAPG, whose permission is required for further use.)

This scale of pore throat sizes can be placed on the vertical axis of a capillary-pressure curve (Fig. 5.38).

The distribution of pore-throat sizes within a rock also affects the capillary properties of the rock. In the hypothetical capillary-pressure curves shown in Fig. 5.39, the capillary pressure curves differ as a result of different pore-throat size distributions, even though the entry pressure is the same for all three samples. The better sorted the size distribution (i.e., the greater the uniformity of pore-throat sizes), the easier it is for the nonwetting-phase fluid (oil) to displace the wetting phase fluid (water) (i.e., a lower P_c is required for nonwetting-phase fluid to enter the pore spaces).

5.6.4 Relationships among porosity, permeability, pore-throat size, and P_c

Reservoir quality and entry-pressure P_c for rock types A–D (Fig. 5.38) are given in Table 5.4. Rocks with smaller porosity and permeability values require higher P_c (entry or displacement pressures) and are associated with smaller pore-throat sizes.

5.6.5 Relations among capillary pressure, grain-size distribution, and water saturation (S_w)

As discussed earlier in this chapter, permeability is directly related to grain-size frequency-distribution parameters. Thus, it is not surprising that there is a direct relationship between P_c and grain size. In Fig. 5.40, capillary pressure curves are shown for three different rocks: a very coarse-grained sandstone, a medium-grained sandstone, and a very fine-grained sandstone. Entry pressures are lower for the coarser-grained rocks, because the permeability will be higher in them than it is in finer-grained rocks. Thus, wetting-phase fluid saturations (S_w) vary according to grain size and permeability. Figure 5.41 illustrates the variation in S_w with permeability for the Gulf of Mexico sands discussed earlier (Figs. 5.14–5.16). In this reservoir, S_w is related directly to permeability, but it is really the pore-throat-size distribution (which is controlled by grain size) that has provided the varying S_w values.

The variation in S_w as a function of grain size and permeability is an important point, particularly when one is calculating reserves in a stratigraphic interval composed of thin sandstones of variable grain size. Fluid saturation (S_w) will vary within individual beds, according to the grain-size distribution. Even at the same structural elevation in a reservoir, rocks will exhibit different fluid

Table 5.4 Reservoir properties of four rocks

Rock type	A	B	C	D
Porosity (%)	3.1	12	21	27.5
Permeability (md)	0.009	0.25	13	714
P_c (entry) Pressure (psi)	400	75	30	10
r_c (microns)	0.27	1.43	3.59	10.8
P_{cr}	32.4	6.08	2.43	0.81

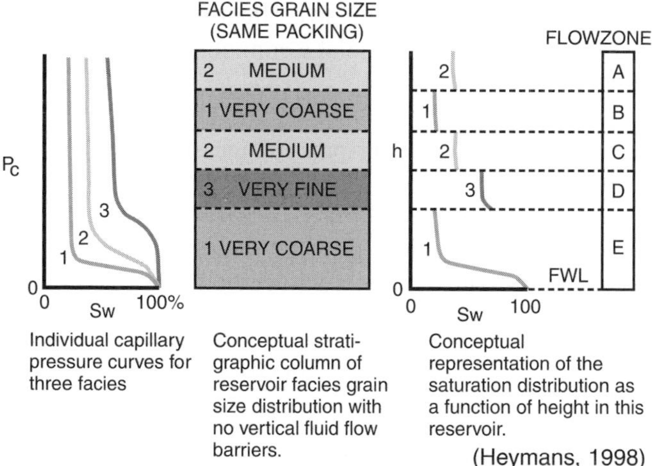

Fig. 5.40. Variations in S_W of sandstones of different grain size, which is controlled by capillary pressure (Heymans, 1998). The left figure shows individual capillary-pressure curves for three facies. The middle figure shows a conceptual stratigraphic column of the three facies and their grain size. There are no vertical fluid-flow barriers (shales) between these facies. The right figure shows the conceptual representation of the S_W distribution as a function of height in this reservoir. After Heymans (1998).

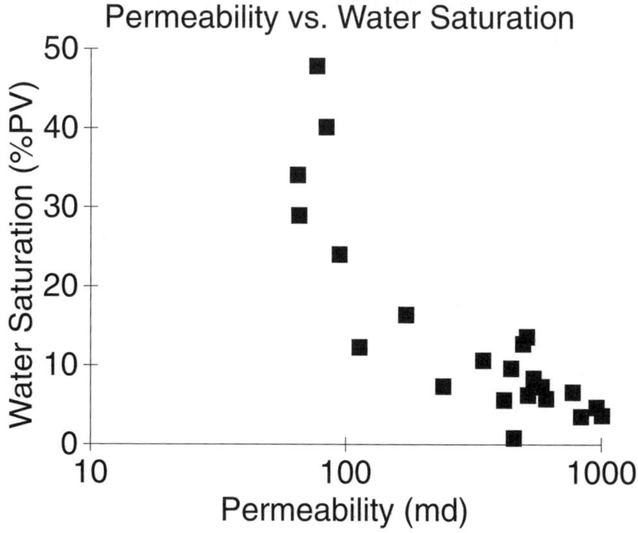

Fig. 5.41. Crossplot of S_W versus permeability for the Gulf of Mexico turbidite reservoir discussed in Figs. 5.14–5.16. More-permeable rocks exhibit a lower S_W, because hydrocarbons can replace water more readily in the permeable rocks. The capillary pressure measurements were adjusted to a height of 67 m (200 ft) above the oil–water contact. After Reedy and Pepper (1996). (Reprinted with permission of the Society of Petroleum Engineers (SPE).)

saturations (S_w) because of the grain-size effect (Fig. 5.40). When one is upscaling stratigraphic intervals, the use of a common S_w can result in misleading reserve calculations.

5.6.6 Conversion of air–Hg capillary-pressure measurements to reservoir conditions

In order to relate air–Hg capillary pressure data to reservoir conditions, it is necessary to convert the data to brine–hydrocarbon values, according to the following equation:

$$P_{cr} = P_{cl} \times \frac{\sigma_r \cos\theta_r}{\sigma_l \cos\theta_l},$$

where P_{cr} – capillary pressure of brine–oil reservoir system, P_{cl} – capillary pressure of air–Hg system, σ_r – interfacial tension of reservoir system, σ_l – interfacial tension of air–Hg system, θ_r – contact angle of reservoir system, θ_l – contact angle of air–Hg system.

Interfacial tension (σ) and contact angle (θ) are dependent upon many factors, including API gravity of the oil, temperature, viscosity, and pressure. The following are useful approximations:

$$\sigma_r \cos\theta_r = 30,$$

$$\sigma_l \cos\theta_l = 370.$$

Thus, $P_{cr} = P_{cl}(30)/(370) = P_{cl}(0.081)$.

Once this conversion is made, a P_{cr} value scale can be placed on the vertical axis of the capillary-pressure plot (Fig. 5.38 and Table 5.4).

5.6.7 Free water level and fluid saturations in a reservoir

Figure 5.42 shows the relationship between oil and water in a reservoir. The free water level or surface is that depth at which $P_c = 0$ and there is 100% water production. This structural elevation can be determined from routine analysis of conventional well logs. The height above free water level is the height of the hydrocarbon column (i.e., the height above the free water level, where $P_c = 0$) required to attain a particular pressure P_c. It is possible to determine the height above the free water level (h) from P_c information.

The relationship between h and P_c is defined by

$$h\ (\text{ft}) = \frac{P_{cr}}{0.433(\rho_b - \rho_{hc})},$$

where 0.433 – in psi/ft is the pressure gradient of pure water at ambient conditions, ρ_b – density of brine (normal range is 1.0–1.2 g/cm^3; a commonly

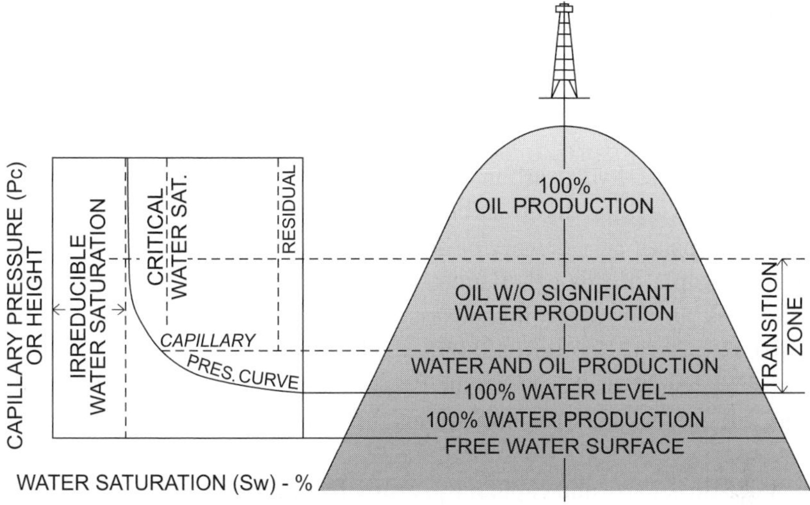

Fig. 5.42. Relationships among capillary pressure, permeability, and oil accumulation. The free water surface is defined as that surface beneath which there is 100% water production. It corresponds to $P_c = 0$ on a capillary-pressure curve. The entry pressure is the pressure at which a hydrocarbon first enters the reservoir, so that both water and oil are produced. The irreducible water saturation is that saturation (height above free water surface) above which no more water – only oil – is produced. The zone between 100% oil production and 100% water production is the transition zone. After Sneider (1987). (Reprinted with permission of AAPG, whose permission is required for further use.)

used value is 1.069), ρ_{hc} – density of hydrocarbon (the normal range for oil is 0.51–1.00; a commonly used value is 0.850).

Because height above free water level (h) is related to P_c, after the appropriate conversions, it can be scaled onto the vertical axis of a capillary-pressure plot. This was done in Fig. 5.38 with the equation $h = P_{cr}/0.095$. So, for example, it is possible to determine the nonwetting phase (S_O) saturation 227 m (682 ft) above the free water level for the four rocks as: $A = 3\%$, $B = 26\%$, $C = 58\%$, and $D = 83\%$. As mentioned above, in a highly interstratified, thin-bedded interval in which grain size varies, the fluid saturations can be quite variable.

5.6.8 Capillarity and seal capacity

Seal capacity is the capacity of a rock to hold a hydrocarbon column of a given height without leaking. The maximum seal capacity (H_{max}) is the height of column that can be held before the seal leaks hydrocarbons. H_{max}, which varies with rock type, is calculated by

$$H_{max} = \frac{P_{ds} - P_{dr}}{0.433(\rho_b - \rho_{hc})},$$

where P_{ds} – brine–hydrocarbon (P_{cr}) displacement or entry pressure of the seal (psia), P_{dr} – brine–hydrocarbon (P_{cr}) displacement or entry pressure of the reservoir (psia), ρ_b – density of brine, ρ_{hc} – density of the hydrocarbon.

As an example of this calculation, if rock A (Fig. 5.38) is assumed to be a seal and rock D is a reservoir sandstone, then $H_{max} = P_{cr}D - P_{cr}A/(1.069 - 0.850)(0.433) = (32.4 - 0.81)/0.095 = 333$ ft. In other words, as long as the hydrocarbon height in this reservoir is less than 333 ft (100 m), the seal will prevent vertical leakage of hydrocarbons through it.

5.6.9 Pore-throat size and capillary pressure from conventional core-analysis data

Kolodzie (1980), applying earlier unpublished research by Winland, demonstrated that pore-throat size could be accurately predicted from conventional core-analysis data. His empirically derived equation is

$$R_{35} = 5.395\frac{K^{0.588}}{\Phi^{0.824}},$$

where R_{35} – pore throat size in microns at 35% nonwetting phase saturation capillary pressure tests, K – air permeability (md), Φ – porosity (%).

Several workers have generated similar equations (Coalson et al., 1994) or have refined the original work (Pittman, 1992). The method has been found to work well with intergranular or intercrystalline rock types with conventional "Archie" pore systems. It is less reliable in rocks with moldic, vuggy, and fracture porosity.

Figure 5.43 compares R_{35} values of a suite of sandstones based on actual measurements and calculation using the Winland, Coalson, and Pittman equations. There is generally good correlation between measured and calculated data for each equation, with R^2 always greater than 0.8. However, the calculated values are consistently lower than the measured values. Probably a major reason for this is the low permeability of these particular sandstones when measured (<10 md), relative to the larger database (with a wider range of permeability values) used in developing the empirical equations. Nevertheless, this test shows the feasibility of developing a pore-throat size, and thus a capillary-pressure relationship from conventional core-analysis data, if a set of capillary-pressure measurements are made to determine whether an appropriate correction factor is required and if so, what that factor should be.

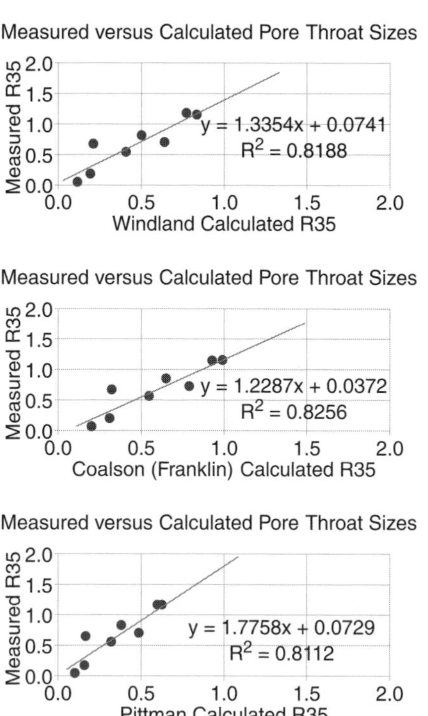

Fig. 5.43. Measured versus calculated R_{35} values for three different methods (Winland, Coalson, and Pillman). Correlation coefficients are significant in each of the three cases. See text for explanation of calculations. (Figures provided by S. Goolsby.)

5.7 Seismic porosity measurement

With improvements in seismic acquisition and processing, it has become possible for workers to estimate and map porosities from seismic data. An early example of this is provided by Dorn et al. (1996) for Pickerill field in the North Sea (Fig. 5.44). In this example, acoustic-impedance logs were calculated from sonic and density logs (Fig. 5.45). A plot of log-derived porosity versus log-derived acoustic impedance for one well shows a clear relationship of increasing porosity with a decreasing numerical value of impedance (Fig. 5.45). This relationship suggests that a linear relationship also should exist between reflection amplitude extracted directly from the seismic data, and reservoir porosity, because other factors affecting seismic amplitude could be factored out in this instance (Dorn et al., 1996). The phase of the seismic data was corrected and a linear relationship was established between the reflection amplitude at the top of the reservoir reflector and average log-derived porosity for each well in the field (red/gray dots in Fig. 5.46). Blue/black dots (Fig. 5.46) are average porosity values from wells drilled after the analysis, and they indicate the validity

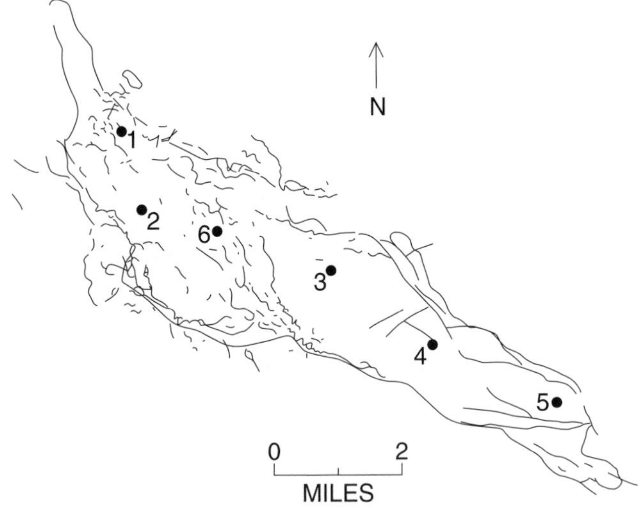

Fig. 5.44. Pickerill field in the North Sea, showing the locations of the first six wells drilled. The black lines are faults mapped from an early 2D seismic survey. The first five wells delineated the extent of the reservoir. The sixth well was a production well.

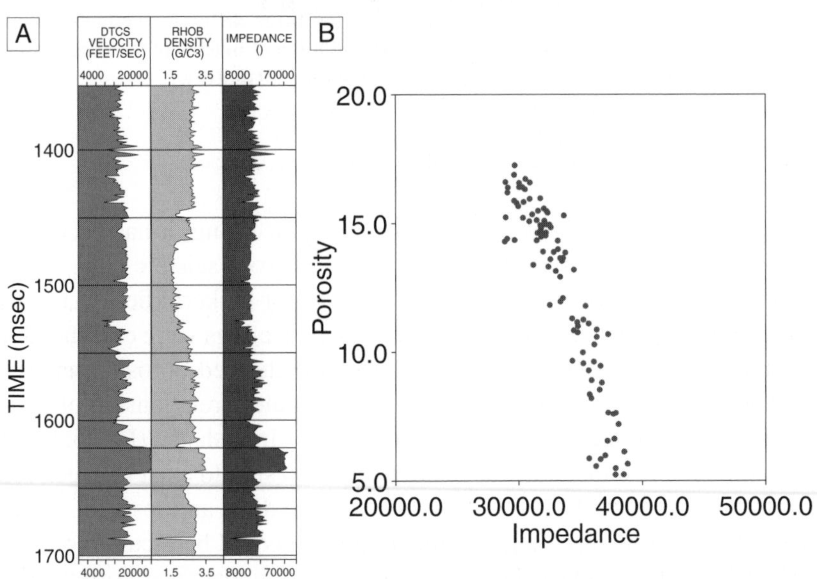

Fig. 5.45. (A) A typical sonic velocity and density log from one of the wells in Pickerill field, and a derivative acoustic impedance log. (B) QA crossplot of acoustic impedance and log-derived porosity for the well. After Dorn et al. (1996). (Reprinted with permission of AAPG, whose permission is required for further use.)

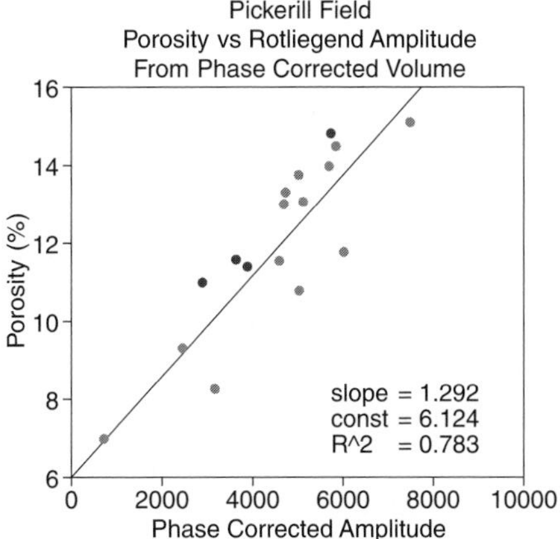

Fig. 5.46. Crossplot of phase-corrected amplitude derived from seismic records versus porosity. The red/gray dots are data points used to generate the trend line. The blue/black dots are measured values determined from wells drilled after the trend line was established. After Dorn et al. (1996). (Reprinted with permission of AAPG, whose permission is required for further use.)

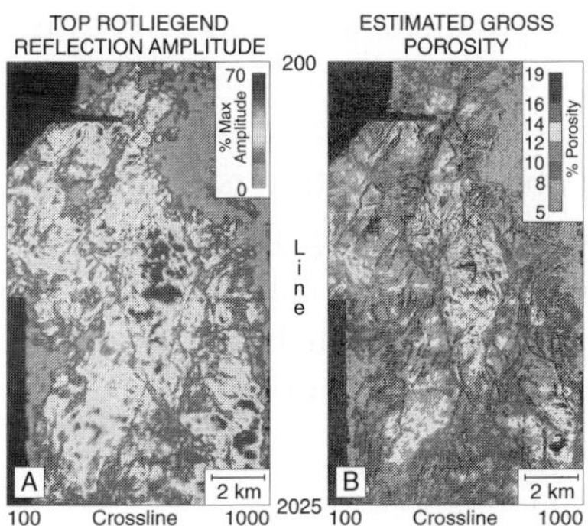

Fig. 5.47. (A) Seismic reflection amplitude map of the top of the reservoir interval in Pickerill field. (B) Estimated gross porosity map based on the seismic amplitude versus porosity trend. After Dorn et al. (1996). (Reprinted with permission of AAPG, whose permission is required for further use.)

of the linear relationship established prior to drilling these wells. On this basis, a seismic amplitude map of the top reservoir (Fig. 5.47A) was converted to a gross porosity map of the reservoir (Fig. 5.47B). Relatively high porosity zones are shown by the hot (red/black, orange/bright-gray) colors. Bounding faults are also clearly highlighted on the figure. In this example, generation of the porosity map proved to be very valuable in locating wells more accurately for penetrating porous reservoir intervals.

5.8 Summary

Reservoir quality controls the storage, distribution, and flow of fluids within a reservoir. Porosity and permeability are key parameters that are readily measured on rock samples and from well logs. If core material is obtained from a well, and porosity and permeability measurements are made on the core, the values can be compared with porosity logs and a permeability log can be developed. Although flow units can be determined using a suite of geological and petrophysical parameters, Gunter et al.'s (1997) method uses only the three easily obtained parameters of porosity, permeability, and thickness to calculate flow units in terms of their capacity to store and transmit fluids within the reservoir. Three-dimensional flow-unit models of a reservoir can be used for reservoir fluid-flow and performance simulation. Flow units can be upscaled, as needed, to meet the requirements of computing time and capability. Also, gross porosity estimates can be made for a reservoir from 3D seismic reflection surveys.

Capillary properties of a rock also affect the storage and flow of fluids through the rock. Capillary properties are routinely measured and used to determine fluid saturations, height of the oil column above the free water level, and maximum height of the column that can be retained by a reservoir topseal. These are very important parameters for characterizing a reservoir.

Values of porosity, permeability, and capillarity will vary not only according to the nature of rocks comprising a reservoir, but also according to the way in which the values were obtained. Caution is the key to interpreting laboratory-derived data, and it is worth knowing just how and where on a rock sample the measurements were made prior to using them for reservoir characterization. Also, upscaling or averaging values such as S_w can provide misleading results, particularly in thin-bedded stratigraphic intervals.

Chapter 6

Fluvial deposits and reservoirs

6.1 Introduction

Fluvial deposits are sediments that are transported and deposited by rivers in a continental environment (Fig. 6.1). There are several types of fluvially derived deposits, including (1) alluvial fans, which are fan-shaped sediment bodies that form at the bases of mountain slopes at the mouths of rivers; (2) fan deltas, which also form at the bases of mountain slopes, but which are deposited very near a marine shoreline and in marine waters; (3) braided-river deposits, which form at and beyond the bases of mountains, where the gradient of the ground surface is relatively steeply inclined, (4) meandering-river deposits, which form

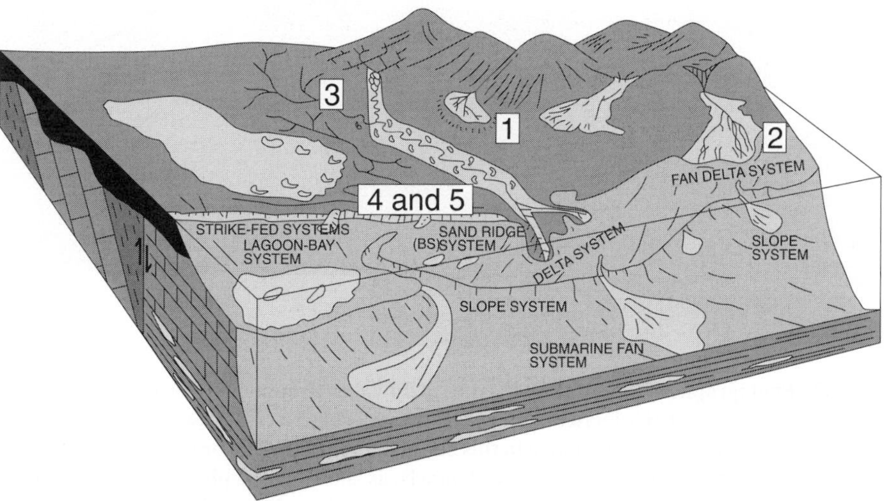

Fig. 6.1. Block diagram showing the distribution of different types of nonmarine (continental) depositional environments, including (1) alluvial fan, (2) fan delta, (3) braided river, and (4 and 5) incised or nonincised meandering river. (Reprinted with permission of F. Brown.)

on more gently inclined floodplains, and (5) incised-valley-fill deposits which fill preexisting continental valleys (Fig. 6.1). Each of these types of deposits exhibits a unique set of properties that distinguish it from the others, including grain size, sand-body geometry, orientations, flow barriers, and the like (Fig. 6.2). An understanding of these differences is important to the evaluation of a subsurface reservoir, because these properties affect fluid flow and ultimately, reservoir performance. It is not sufficient to know that your reservoir is a "fluvial" reservoir; you must also know the type of fluvial reservoir and its defining characteristics.

This chapter contains basic information on the processes by which the different types of deposits form, with emphasis on braided-river, meandering-river, and incised-valley-fill deposits (Fig. 6.2). Examples of modern and outcrop fluvial deposits are mentioned and compared with analog reservoirs.

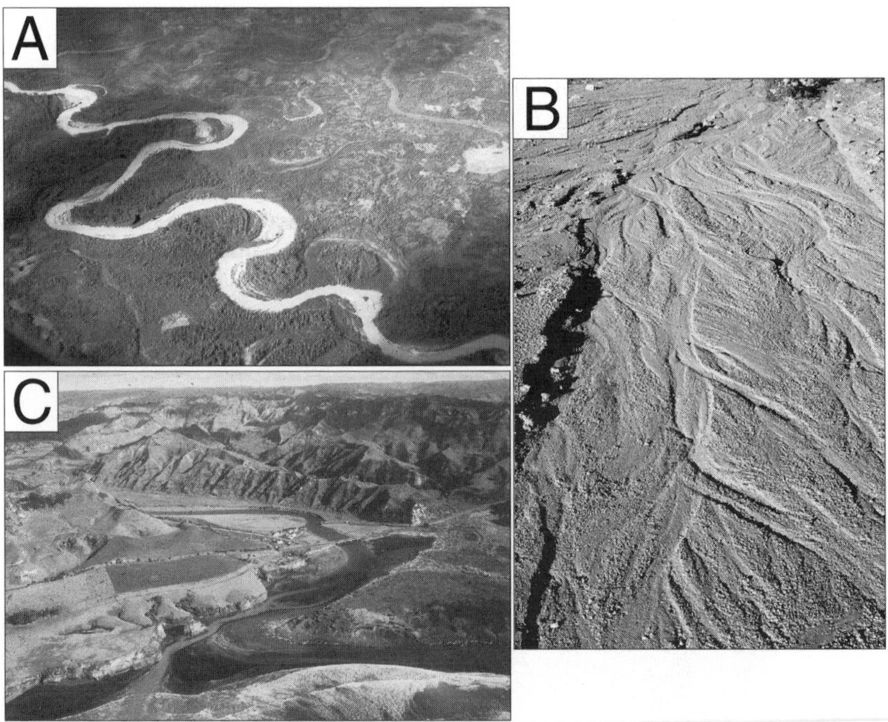

Fig. 6.2. Photographs of the three main types of fluvial environments: (A) meandering river, (B) braided river and (C) incised valley. The meandering river derives its name from the sinuous nature of the river, as shown in this aerial photograph. The braided river derives its name by the braided nature of the river channels, as shown in the photograph. The incised valley derives its name from the fact that the valley has been incised into preexisting strata that now form valley walls, as shown in the photograph. The valley is shown at low tide, when the meandering river is flowing toward the ocean (surf zone in foreground). At high tide, this reach of the valley is inundated with marine water and forms an estuary.

6.2 Braided fluvial (river) deposits and reservoirs

6.2.1 Processes and deposits

Braided fluvial deposits form where sediment-laden rivers debouch from mountains and the slope gradient decreases over a relatively short distance, but is still steep enough that the rivers flow at relatively high velocities (Fig. 6.3). Under these conditions, coarse-grained sedimentary particles (gravel) move along the river floor by rolling and sliding, coarse sand may move by saltation, and finer-grained sand and mud remain in suspension and are carried downstream beyond the confines of the braided-river system (Fig. 6.4). Thus, the typical deposit of

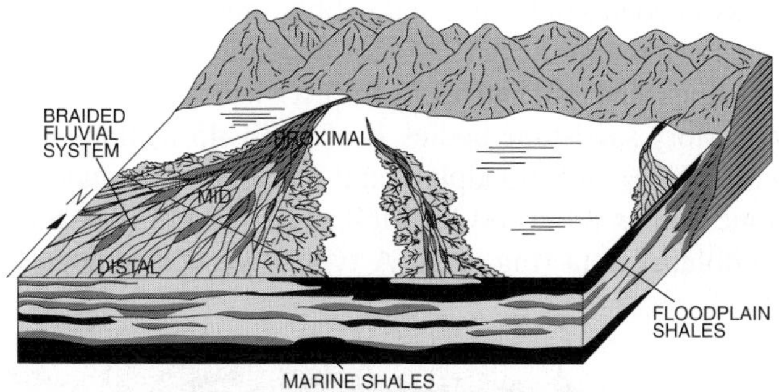

Fig. 6.3. Block diagram of a braided fluvial river system showing the proximal, middle, and distal portions of the system. Grain size of the deposits generally decreases from proximal to distal subenvironments. After Atkinson et al. (1990). (Reprinted with permission of Springer-Verlag.)

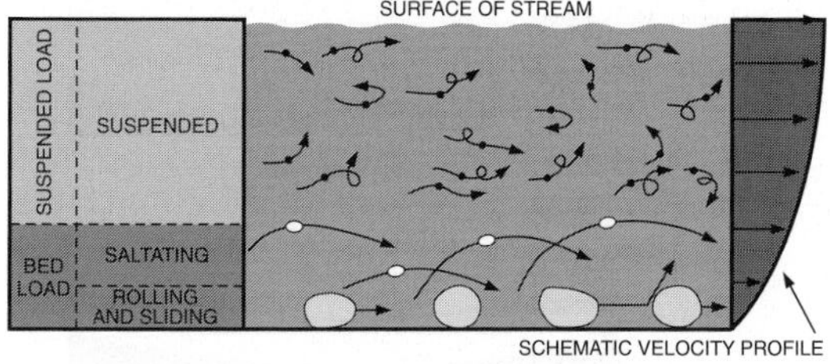

Fig. 6.4. Vertical profile through a river or stream, showing the three main processes of sediment transport. The largest, heaviest grains (gravel) move along the river by rolling and sliding. Intermediate size and weight grains (sand) move by saltation (jumping). The sediments that are transported by saltation and rolling/sliding are collectively termed bedload. Suspended load consists of the finer-grained particles (mud) that remain in suspension during downstream transport. Friction between the river column and the river bed results in reduction of velocity near that interface. River-flow velocities are faster in the upper parts of the column. (Source of figure is unknown.)

a braided river is coarse-grained, with very little mud. Also, gravel and sand deposits typically are laterally continuous and vertically connected. Braided rivers often flow intermittently because of variations in rainfall in the adjacent mountains. With new flows, a river may change its course repeatedly, thereby giving rise to the braided pattern that is diagnostic of this type of deposit (Fig. 6.3). The Salt Wash Sandstone Member of the Upper Jurassic Morrison Formation provides an excellent outcrop example of a braided-river deposit (Robinson and McCabe, 1997). The Salt Wash Member is composed of low-sinuosity, sandy, braided-stream deposits and associated finer-grained, abandoned-channel fill, along with overbank/floodplain strata that were deposited on a broad alluvial plain. The Salt Wash Member is divided into lower and upper intervals. Its hierarchy of sandstone and shale bodies consists of-from the base, upward – (1) trough cross-bedded sandstones and pebbly sandstones as thick as 1 m (3 ft) and as wide as 1–15 m (3–45 ft); (2) fining-upward, single-story sandstone bodies 3 m (9 ft) thick and 10–50 m (30–150 ft) wide, with some abandonment shale; (3) multistory sandstone bodies 1–15 m (3–45 ft) thick and tens to hundreds of meters wide, with floodplain and abandonment shale; and (4) multistory sandstone bodies 20 m (60 ft) thick and 1–10 km (0.6–6 mi) wide, with overbank/floodplain strata (Fig. 6.5). A typical vertical sequence of a single-

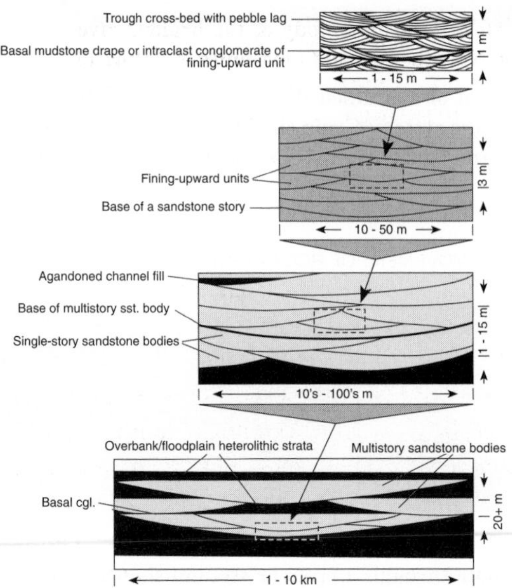

Fig. 6.5. Hierarchy of sandstone and related rock bodies in the braided river deposit of the Salt Wash Member of the Morrison Formation. At all four hierarchical levels, lenticular sandstone is the dominant geometrical form. Thickness and width of the lenticular bodies increase from the individual bed scale to the single- and multistory scales. After Robinson and McCabe (1997). (Reprinted with permission of AAPG, whose permission is required for further use.)

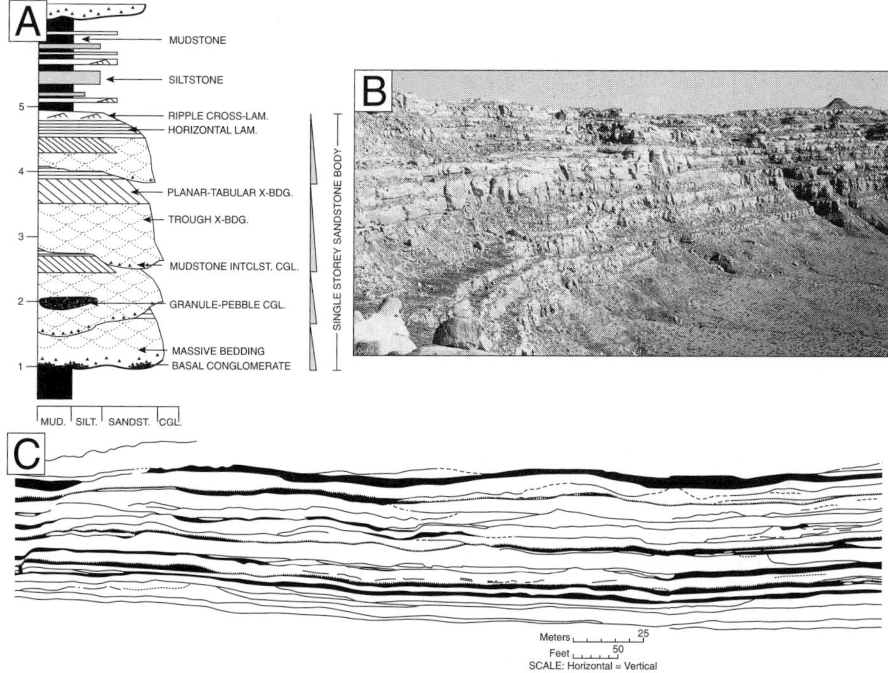

Fig. 6.6. (A) Single-story sandstone body of the braided river Salt Wash Member of the Morrison Formation. Note the overall fining-upward nature of the body, and the mudstone cap. (B) Outcrop photograph of the Salt Wash Member. Lateral continuity of sandstone bodies appears to be good, but internal discontinuities, such as those shown by the outcrop photomosaic in (C), can result in some compartmentalization of bodies. After Robinson and McCabe (1997). (Reprinted with permission of AAPG, whose permission is required for further use.)

story sandstone body becomes somewhat finer-grained upward and is capped by overbank/floodplain shale (Fig. 6.6A).

Figure 6.6B shows an outcrop photomosaic of a typical 300-m (900-ft)-long, 65-m (195-ft)-thick set of multistory sandstone bodies. Although many of the overbank/floodplain shales are not continuous laterally across the outcrop, some shales are continuous and could form permeability barriers or baffles in an analog reservoir. Width-to-thickness values for single-story sandstones, overbank/floodplain mudstones, and abandoned-channel mudstones average 59:1, 68:1, and 28:1, respectively (Fig. 6.7). Such calculations are important for estimates of how far braided-river deposits extend laterally away from a wellbore, when only sandstone thicknesses measured from the well data are available. For example, a single-story, 10-m (30-ft)-thick sandstone might be expected to be 590 m (1,770 ft) wide.

As was discussed in Chapter 5, trends in permeability often parallel trends in grain size. Such is the case for the Salt Wash sandstone, as shown by miniper-

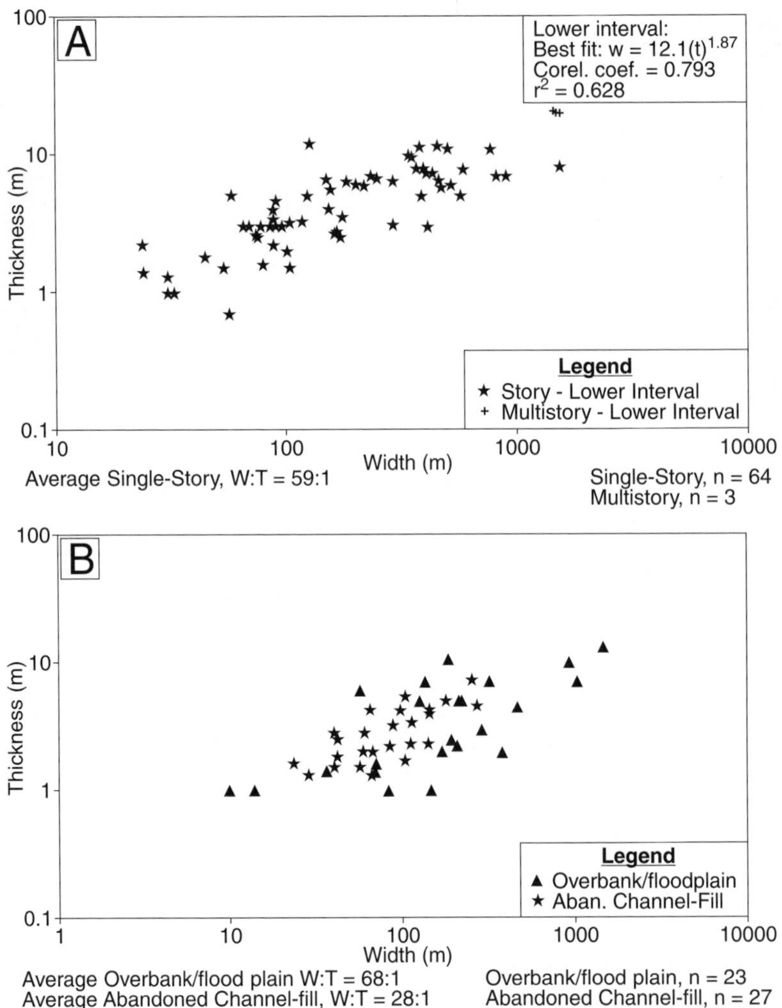

Fig. 6.7. (A) Width (*W*) versus thickness (*T*) plot for single and multistory sandstone bodies in the lower interval of the Salt Wash Member. Average *W*:*T* values are given. (B) Width versus thickness plot for overbank/floodplain and abandoned channel fill in the Salt Wash Member. Average *W*:*T* values are given. After Robinson and McCabe (1997). (Reprinted with permission of AAPG, whose permission is required for further use.)

meameter measurements at 5-cm (2-in) intervals in a core obtained from behind an outcrop (Fig. 6.8). Permeabilities are generally greater than 1,000 md for medium-grained and coarser sandstones, and less than 1,000 md for finer-grained sandstones.

On the basis of the data set from this outcrop deposit, Robinson and McCabe (1997) were able to develop a 3D geologic model of the Salt Wash Member in this area and then conduct reservoir performance simulations on the model.

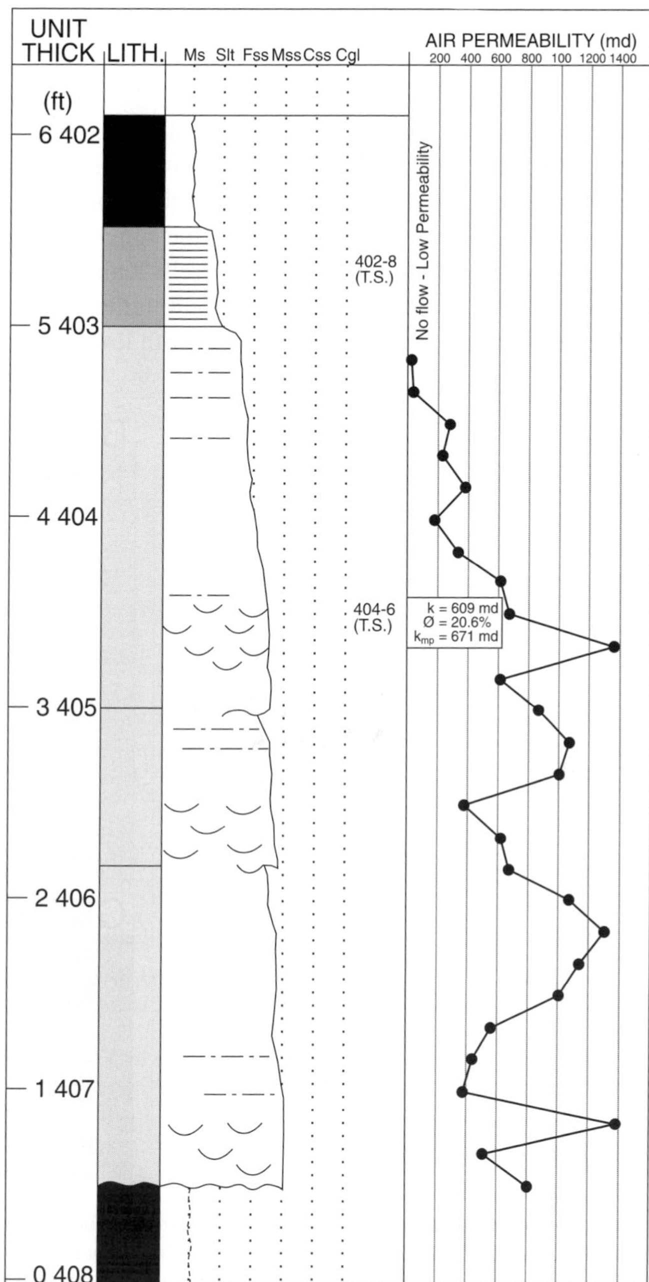

Fig. 6.8. Upward vertical decrease in permeability of a sandstone body. Permeability varies by an order of magnitude between the top and bottom of the body. Note that the upward decrease parallels the upward decrease in grain size of the sandstone, from medium-grained sand (Mss) at the base to mudstone (Ms) at the top. After Robinson and McCabe (1997). (Reprinted with permission of AAPG, whose permission is required for further use.)

6.2.2 Reservoir examples

6.2.2.1 Braided river field, North Sea

A field in the North Sea provides one example of a Paleozoic braided fluvial gas reservoir (Green and Slatt, 1992). The field is composed mainly of conglomerates and sandstones (Fig. 6.9). Permeability of the sandstones and some of the conglomerates is generally low, whereas others of the conglomerates have permeabilities in the hundreds of millidarcys. Those wells with good permeability also have excellent gas-flow rates, on the order of 28 million cubic feet of gas per day (MMCFGPD) (Fig. 6.9). In contrast, wells with low-permeability con-

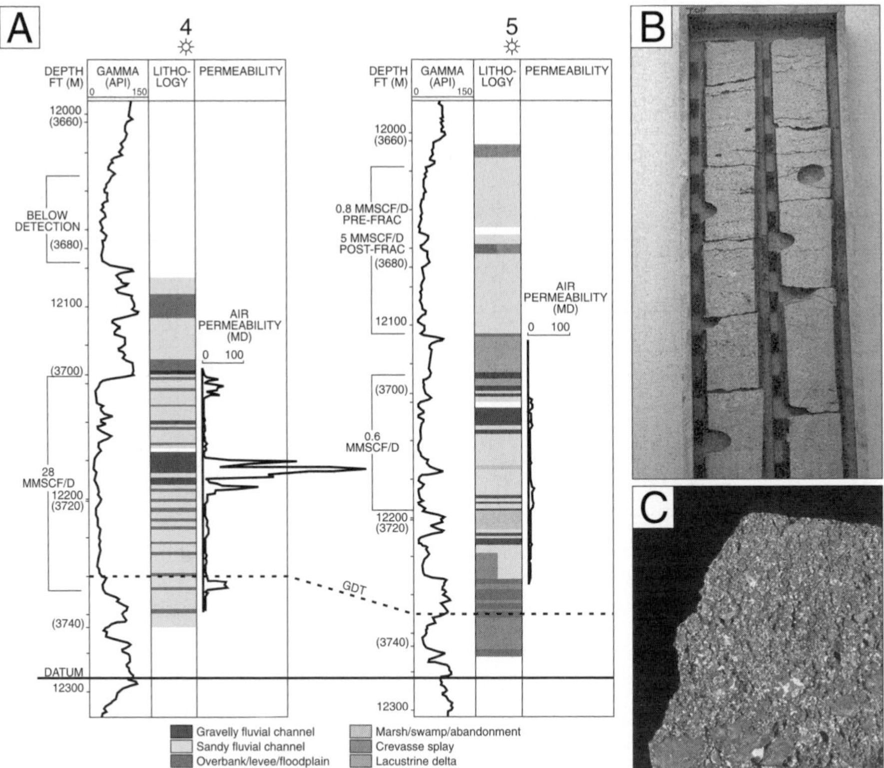

Fig. 6.9. (A) Gamma-ray, core-lithology, and permeability profiles for two wells (4 and 5) in a field, North Sea. Also shown are initial gas-flow rates. Conglomerates in well #4 have high permeabilities and good gas-flow rates, whereas conglomerates in well #5 have poor permeability and low flow rates. (B) Core of conglomerate with high permeability. (C) Core piece of conglomerate with poor permeability. The reddish*/black color of the conglomerate in (C) is due to the presence of reddish/black shale clasts, which are not present in the conglomerate in (B).

* The indicated color is for a CD which contains all of the figures in color.

glomerates have much lower flow rates, approximately less than 2 MMCFGPD. This difference occurs because the high-permeability conglomerates are composed of quartz, feldspar, and crystalline lithic fragments, whereas the low-permeability conglomerates are enriched in red-shale clasts. These differences indicate an additional provenance for the shale-clast-rich conglomerates, even though the wells in which they occur are located only a few kilometers from wells with high-permeability conglomerates. The ductile shale clasts were deformed during burial, and ultimately they plugged most of the pore throats. Thus, in this case, diagenesis and provenance have played key roles in the performance of individual wells. Mapping the distribution of the low- and high-permeability conglomerates within the field would reduce the risk of drilling infill wells in the low-permeability conglomerates.

6.2.2.2 *Braided river field, north Africa*

A second example of a braided fluvial reservoir that has been influenced by a different type of diagenesis occurs in north Africa (Mitra and Leslie, 2003). It is a multi-billion-barrel oil field. In the late 1990s, the field was producing approximately 20,000 bbl/day of oil from several wells. The field is a northeast–southwest-trending, elongate, fault-bounded anticline.

A 290-m (870-ft) continuous core, from which routine core-plug analyses were obtained at 0.3-m (1-ft) intervals, provides us with an opportunity to develop a flow-unit zonation for this 4-BBOIP reservoir. Four lithofacies comprise the 260-m interval (Fig. 6.10). Beginning at the top, Facies 1 is composed of very-fine-grained, well-sorted, cross-bedded, quartz sandstone with *Tigillites* (marine) burrows and fractures. This facies is interpreted to be a shoreface-shallow-marine deposit. Facies 2 lies below Facies 1 and is similar to it, but Facies 2 is a slightly coarser-grained, perhaps more proximal-marine sandstone. Facies 3 is a series of cycles of 1- to 6-m-thick, fining-upward, cross-bedded sandstones; each cycle is medium-grained at the base and fine-grained at the top. This facies is considered to be a transitional facies of interbedded marine and underlying nonmarine, fluvial deposits. At the bottom lies Facies 4, which is composed of coarse-grained braided fluvial sandstones and pebbly sandstones that are trough cross-bedded and have a muddy matrix. The sandstones are enriched in granitic lithic clasts and feldspar grains, many of which have been partially dissolved (via burial diagenesis – see Chapter 3) to give isolated, moldic porosity.

A crossplot of porosity versus permeability shows two clear trends: one with a steep slope (relatively high permeability for low porosity) and the other with a shallow slope (relatively high porosity for low permeability) (Fig. 6.11). Some intermediate porosity and permeability values appear between these two trends. The trend with the steep slope is related to Facies 2 and 3, and the shallow-sloping trend is related to Facies 4 (Fig. 6.12).

An SML (Stratigraphic Modified Lorenz) plot for flow-unit definition (Chapter 5) was constructed from the porosity and permeability data (Figs.

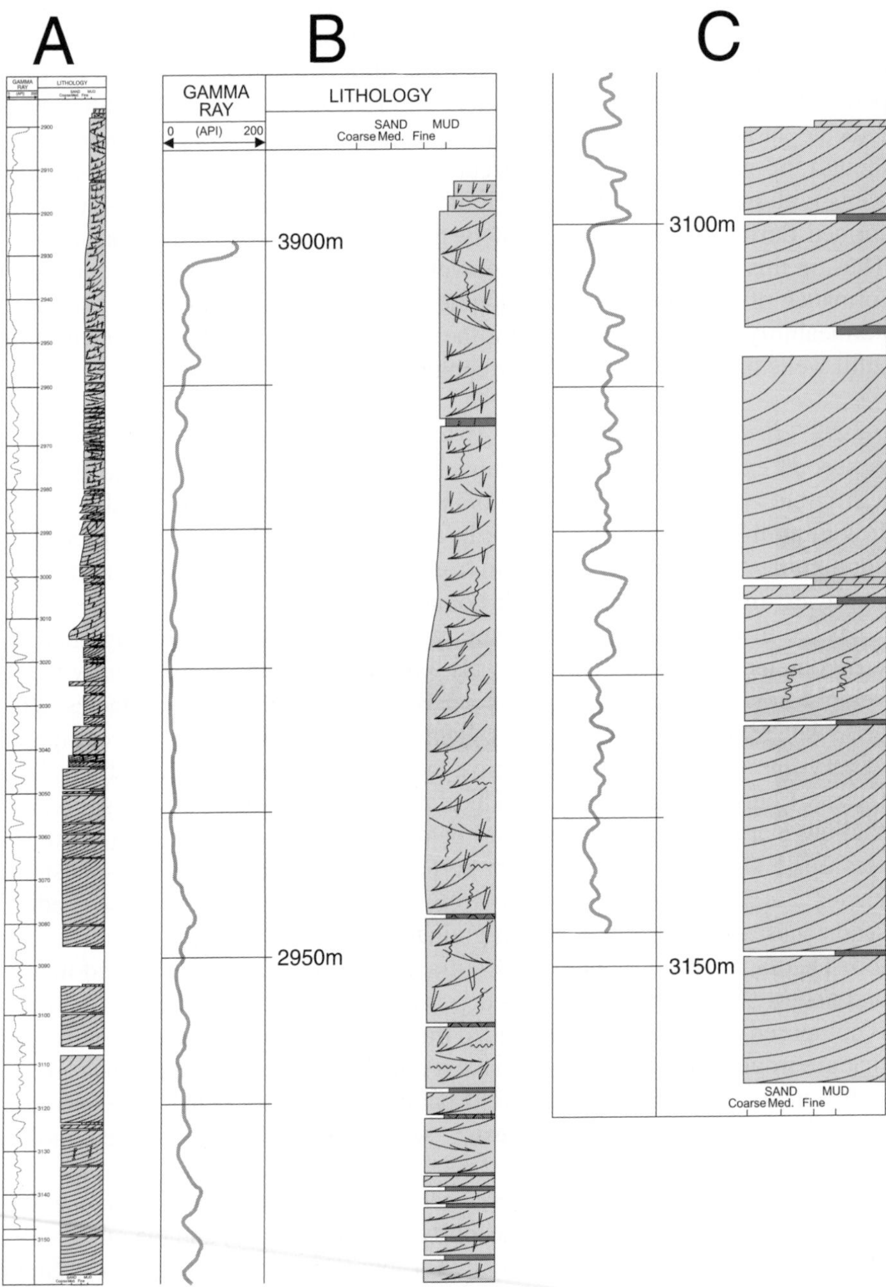

Fig. 6.10. (A) Gamma-ray log and description of a 250-m (750-ft) core in the north African field. (B) Gamma-ray log and 70-m cored interval showing the upper marine facies. Symbols refer to cross-beds and *Tigillites* vertical burrows. (C) Gamma-ray log and 60-m cored interval showing the lower fluvial facies. Symbols refer to large-scale trough cross-beds.

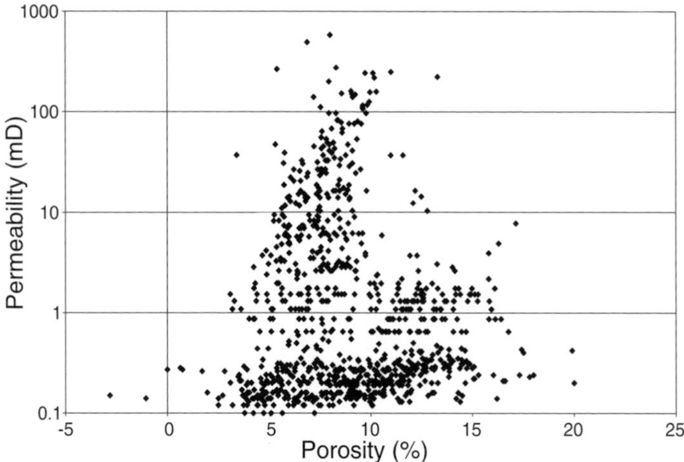

Fig. 6.11. Porosity versus permeability cross-plot for the cored interval shown in Fig. 6.10. Note the two distinctly different trends, with a more diffuse, intermediate trend between them. Further description is provided in the text.

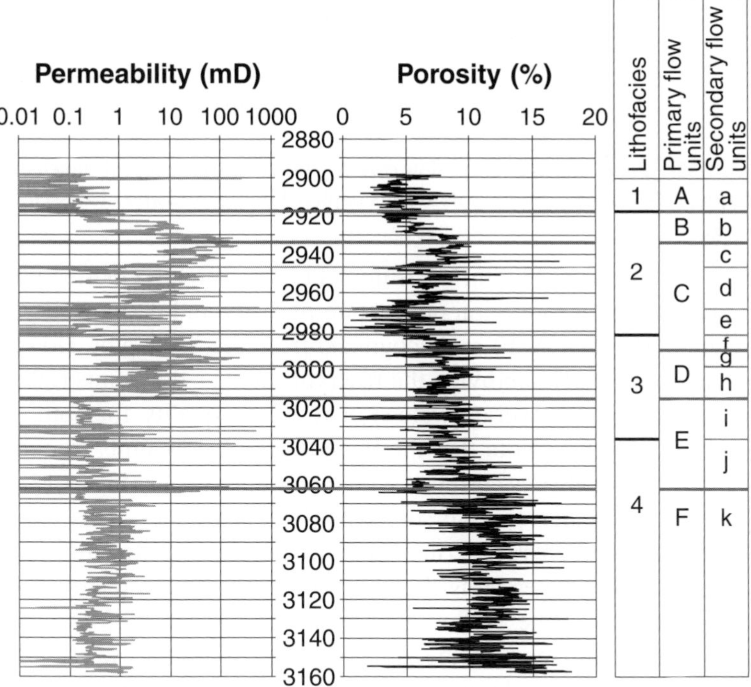

Fig. 6.12. Porosity and permeability versus depth plots for the cored well shown in Fig. 6.10. The vertical distributions of the four lithofacies described in the text are displayed. Also shown are the vertical zonations of primary and secondary flow units discussed in the text and appearing in Fig. 6.13.

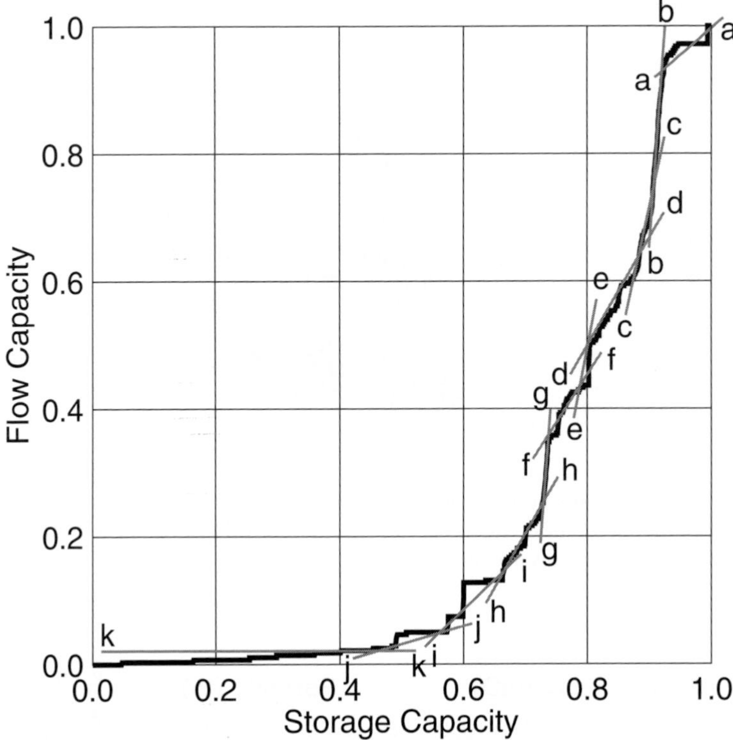

Fig. 6.13. SML plot of storage versus flow capacity for the cored interval shown in Fig. 6.10. Small letters refer to the secondary flow units depicted in Fig. 6.12 and discussed in the text.

6.12 and 6.13). The core-plug measurements are of matrix porosity only and do not provide any information on fracture porosity and permeability. Six major flow units (labeled A–F) are recognized from the SML plot. However, it is possible to more-finely subdivide the reservoir into 11 flow units (labeled a–k) (Figs. 6.12 and 6.13). Average values of reservoir quality for both zonations are listed in Table 6.1. Note that the boundaries of flow units in either set of subdivisions do not always correlate with facies boundaries (Fig. 6.12); instead, flow-unit characteristics overlap somewhat between facies. This is an advantage that comes with characterizing reservoirs in terms of flow units instead of facies. However, it is interesting that the boundary between relatively high and low permeabililties in the 11-flow-unit zonation is at the base of Flow Unit i, and in the 6-flow-unit zonation the same boundary is at the base of Flow Unit E. Thus, the location of this boundary differs by 22 m (Table 6.1). The relationship between the best flow capacity, which is in the stratigraphically higher, more marine-influenced flow units, and the best storage capacity, which is in the lower, braided fluvial facies, is a result of both stratigraphic and diagenetic factors. The upper marine facies is relatively fine-grained and quartzose. By

Table 6.1 Two flow unit zonations for the cored interval in Figs. 6.12 and 6.13, north Africa, Algeria

Depth (m)	Flow unit	Thickness (m)	X % Por.	X Perm. (md)
2898.00				
	a	20.00	4.81	3.53
2918.00				
	b	18.00	6.32	33.58
2936.00				
	c	8.00	8.20	32.77
2944.00				
	d	22.00	6.80	11.03
2966.00				
	e	16.00	4.77	12.98
2982.00				
	f	8.00	7.75	20.36
2990.00				
	g	6.00	8.03	46.37
2996.00				
	h	16.00	8.09	11.86
3012.00				
	i	27.00	7.93	9.90
3039.00				
	j	22.00	8.53	0.76
3061.00				
	k	97.00	11.25	1.11
3158.00				
2898.00				
	A	20.00	4.81	3.53
2918.00				
	B	18.00	6.32	33.58
2936.00				
	C	84.00	6.56	16.20
2990.00				
	D	22.00	8.07	21.38
3012.00				
	E	49.00	8.20	5.88
3061.00				
	F	97.00	11.25	1.11
3158.00				

contrast, the coarser-grained fluvial facies contains larger individual feldspar grains and granitic clasts that are also feldspathic. Many of the feldspar grains are highly or completely corroded, indicating that they were subjected to burial chemical diagenesis through dissolution. Some of the moldic pores that now occupy the spaces where feldspar grains once resided are lined with kaolinite, a common byproduct of chemical dissolution of feldspar. The result of this combination has been a lower unit that has significant storage capacity, but the storage is in isolated pores. This is in contrast to the upper units, which have intergranular porosity and better pore connectivity.

6.2.2.3 Prudhoe Bay field, Alaska

Prudhoe Bay field is the largest oil field in North America. Discovered in the 1960s by Arco Oil and Gas Co., it began production in 1977 at more than 70,000 BOPD (Fig. 6.14). Production declines in 1979 and 1981 prompted waterflood startup, which improved production until 1984, when another waterflood was initiated. Since that time, production has declined steadily, and several enhanced-recovery techniques have been applied to the field.

The field produces from the Permian-Triassic Ivishak Sandstone, which is part of the Sadlerochit Group (Fig. 6.15). The field sits atop the Barrow Arch, where a combined unconformity and fault form its northern limit (Figs. 6.15 and 6.16). The Prudhoe Bay Unit is divided into a western operating area and an eastern operating area (Fig. 6.16).

Gross thickness of the Ivishak Sandstone ranges to 200 m (600 ft) in the field area, and net sand is as high as 90% (Fig. 6.17). Most of the Ivishak Sandstone is a sandy and conglomeratic braided-river deposit, but a lower, finer-grained

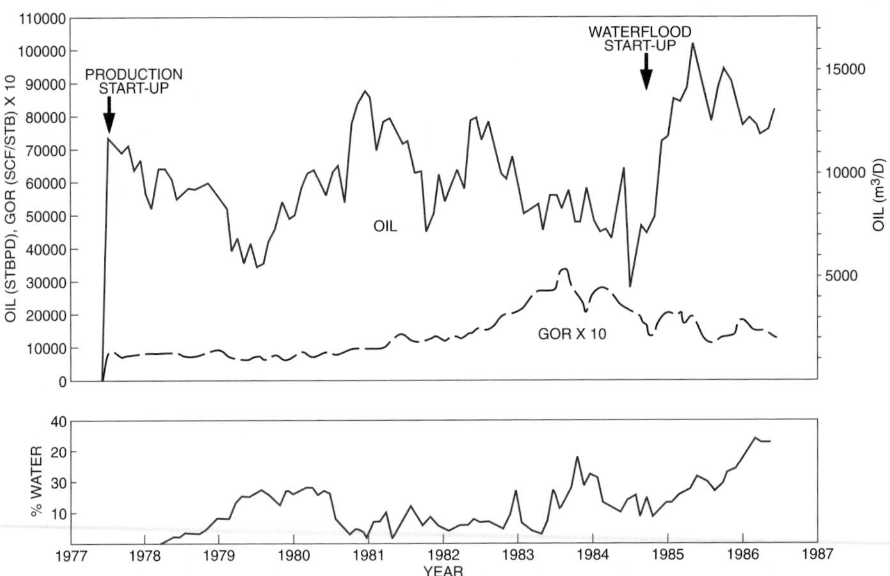

Fig. 6.14. Graph of the production history of Prudhoe Bay field, beginning at startup in 1977. Notice how the field began a slow decline in 1981 (after an earlier decline in 1979), which prompted a waterflood startup that increased production. The field is now on a substantial decline. After Atkinson et al. (1990). (Reprinted with permission of Springer-Verlag.)

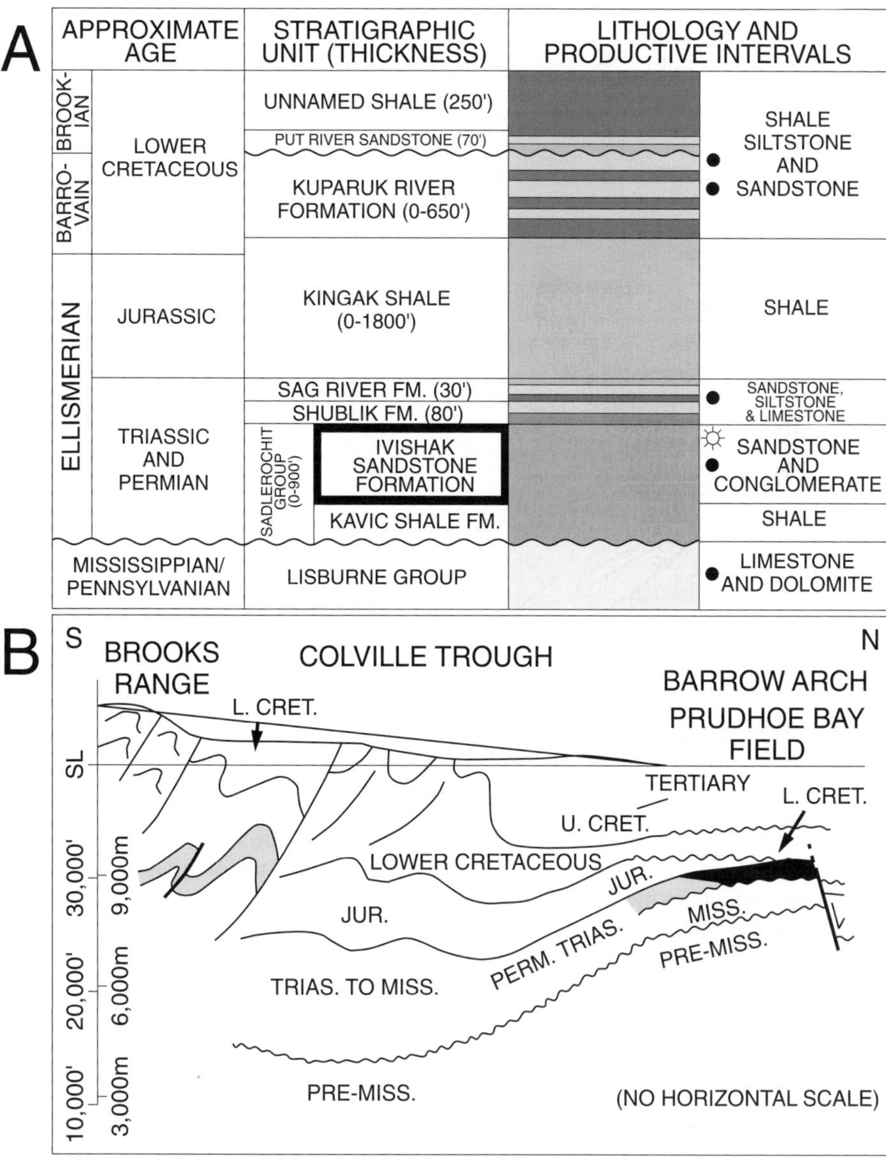

Fig. 6.15. (A) Geologic column with the ages of the different formations that comprise Prudhoe Bay field. (B) The main reservoir sand is the Ivishak Sandstone. The Ivishak Sandstone is composed of both sandstone and conglomerate. The overlying Jurassic Kingak Shale and Lower Cretaceous Kuparuk River Formation form shaly top seals. Sandstones within the Kuparuk River Formation are major reservoir rocks in the giant Kuparuk River field, which is also on the North Slope of Alaska. After Atkinson et al. (1990). (Reprinted with permission of Springer-Verlag.)

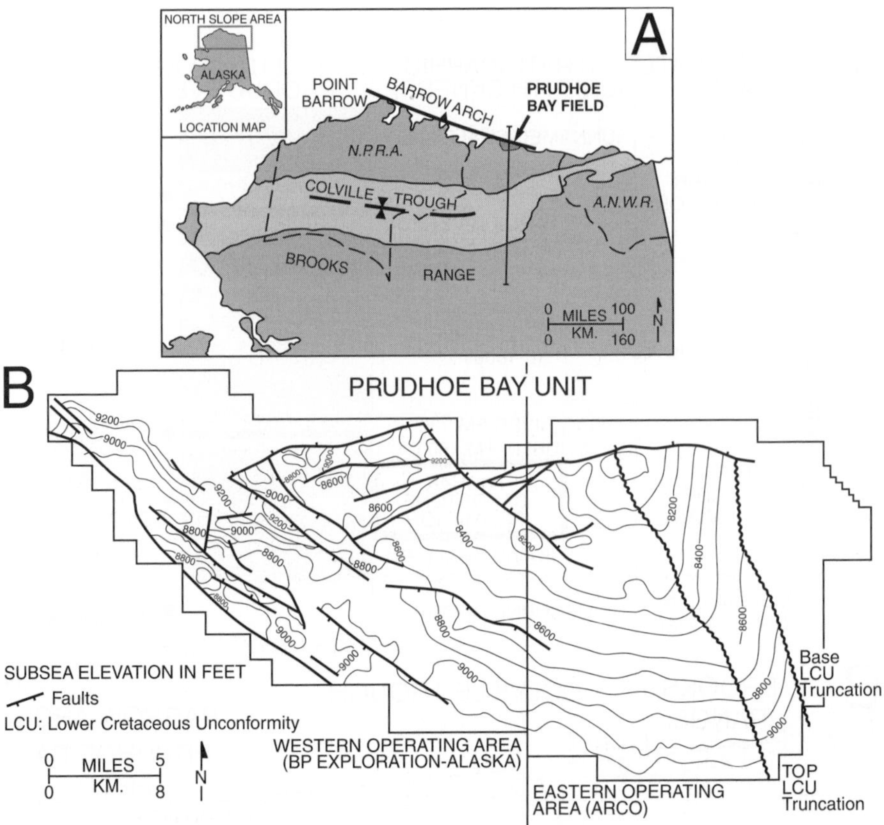

Fig. 6.16. (A) Location of Prudhoe Bay field in relation to the Barrow Arch, which is a major regional anticline that forms the structure of the field and that aided in the trapping of oil and gas in the geologic past. (B) Map of the structure of Prudhoe Bay field, including major faults (black lines). The field is divided into two parts, a Western Operating Area (run by BP) and an Eastern Operating Area (formerly run by ARCO, now run by Conoco-Phillips). The Lower Cretaceous Unconformity (LCU) on the east side of the Eastern Operating Area truncates the field. After Atkinson et al. (1990). (Reprinted with permission of Springer-Verlag.)

and shalier interval is composed of delta-front strata (Figs. 6.18 and 6.19) (Tye et al., 1999). Stratigraphically, the field is divided into a basal delta-front sandstone/shale sequence, followed upward by mainly distal braided-river deposits, then mid-braided-river deposits, which are capped by more-distal braided-river deposits (Figs. 6.18 and 6.19; also see Fig. 6.3 for subenvironments of a braided-river system). Thick floodplain shales isolate some of the sandstones. Porosity and permeability are facies-controlled; pebbly and coarse-grained sandstones of the mid-braided-river facies exhibit the highest porosity and permeability, followed by medium-grained sandstones of the distal-braided-river facies, and

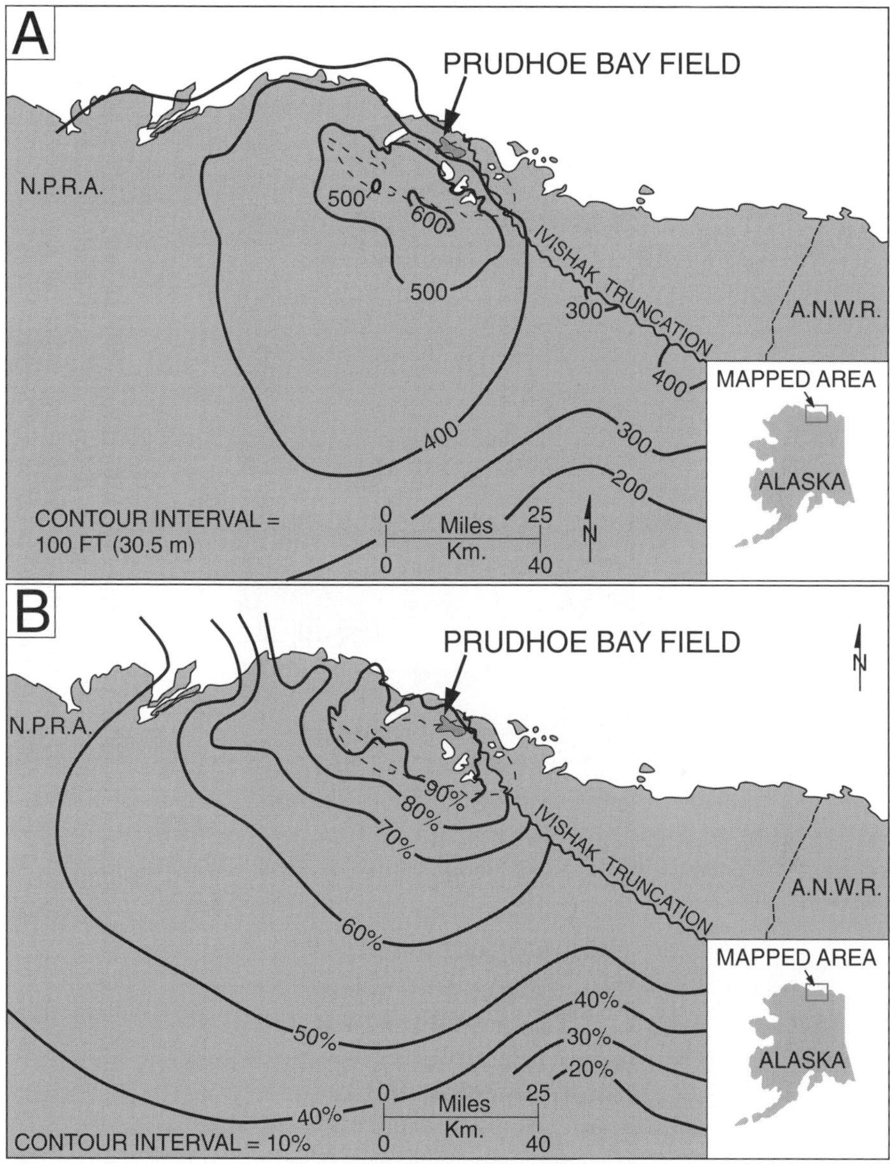

Fig. 6.17. (A) Gross-thickness isopach map of the Ivishak Sandstone. Note that at its thickest, the Ivishak is more than 200 m (600 ft) thick. (B) Net-sand isopach map of the Ivishak Sandstone. At its thickest, it consists of about 90% or 180 m (540 ft) of sandstone. After Atkinson et al. (1990). (Reprinted with permission of Springer-Verlag.)

the lowest porosity and permeability values are found in the fine- to very-fine-grained sandstone delta-front facies (Fig. 6.20).

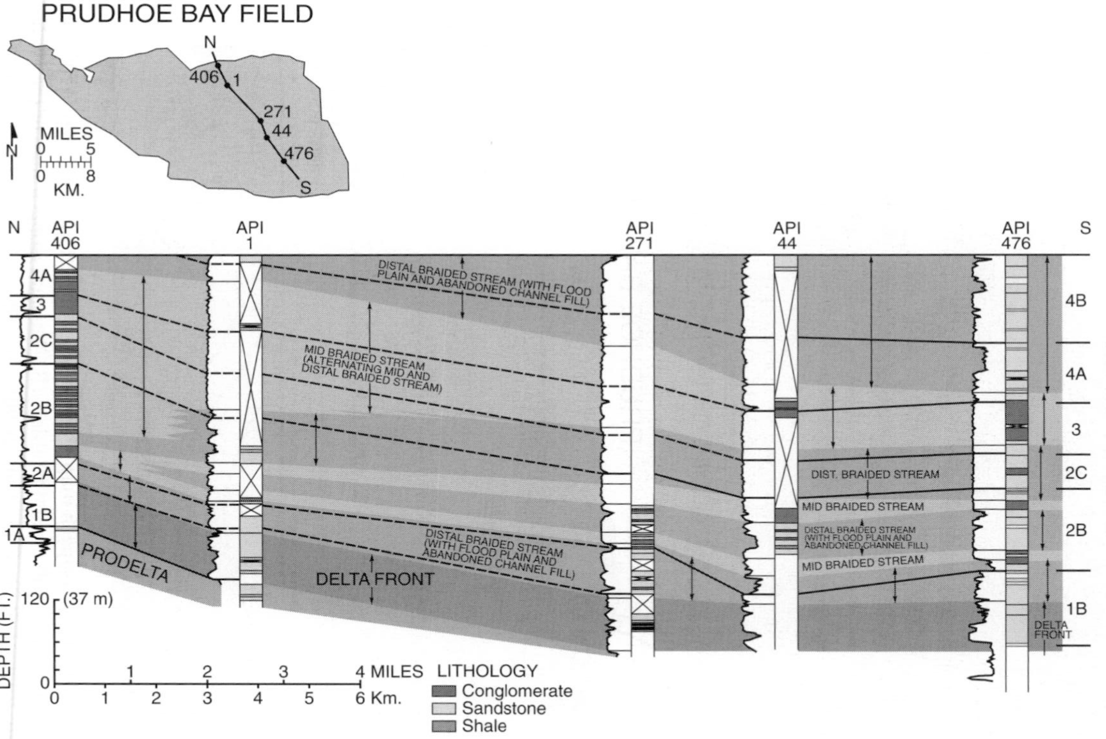

Fig. 6.18. **Northwest-to-southeast stratigraphic cross-section of the different rock types and sedimentary environments, as represented by the different colors. The number and letter symbols on the left and right (e.g., 4A, 2C, etc.) refer to the reservoir zonation scheme that the operators use. There is a tendency for a preponderance of conglomerate toward the north (blue-green/dark-gray color) and a preponderance of sand toward the south (yellow/bright-gray color). After Atkinson et al. (1990). (Reprinted with permission of Springer-Verlag.)**

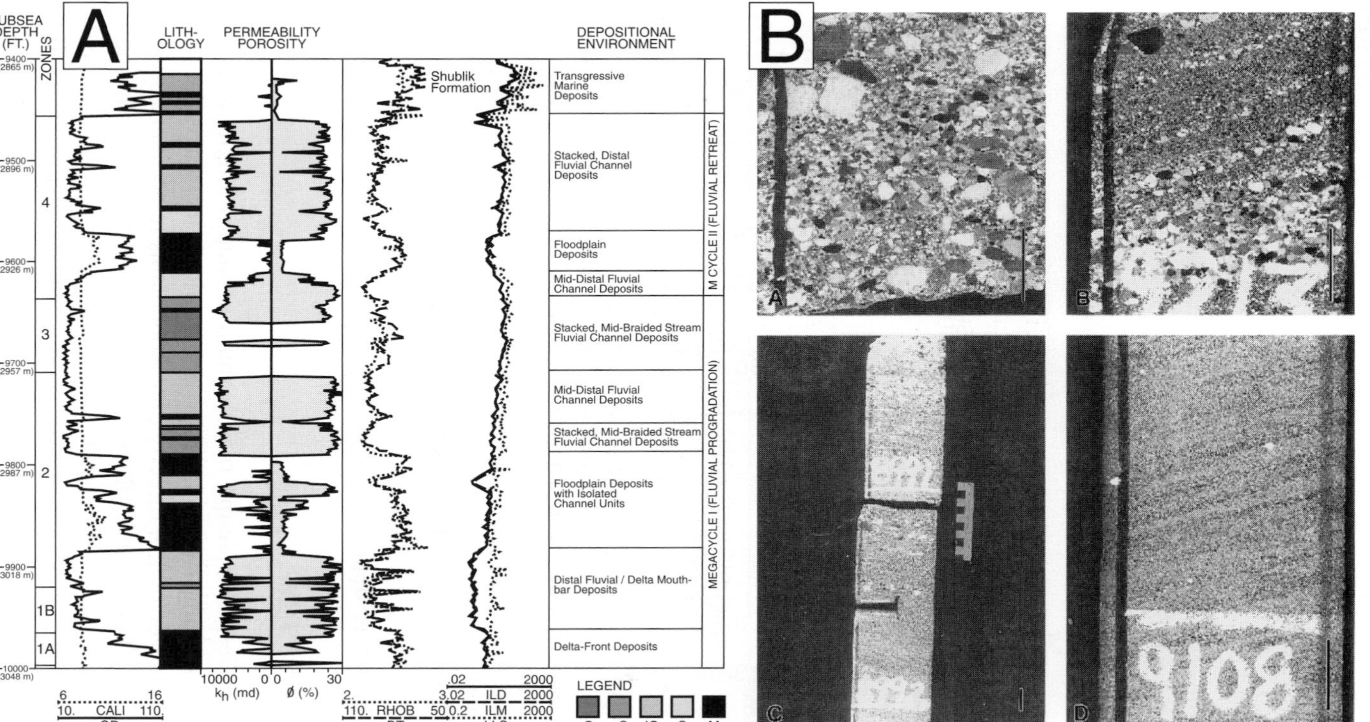

Fig. 6.19. (A) Well logs, core description, and porosity and permeability profiles of a reservoir interval. For the color codes, G – gravel, gS – gravelly sand, IG – intermediate gravel, S – sand, and M – shale. Note that different rock types are stacked vertically in the well, but in all but the shaly intervals, porosities and permeabilities are very high. Depositional environments are shown on the right column. (B) Conglomerate and coarse-grained sandstone cores of reservoir intervals. After Atkinson et al. (1990). (Reprinted with permission of Springer-Verlag.)

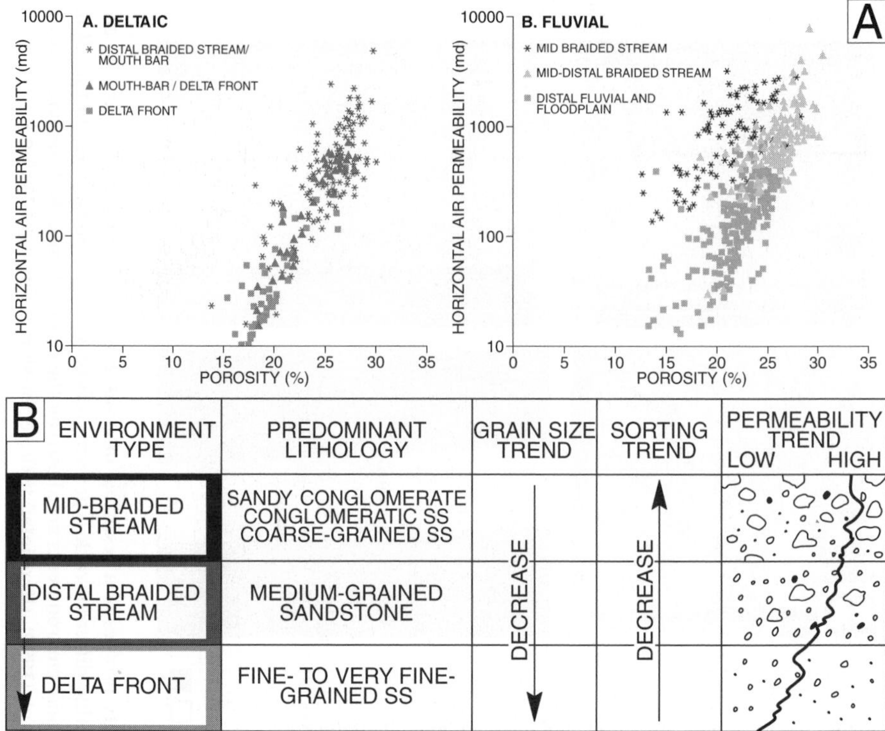

Fig. 6.20. (A) Crossplot of porosity versus permeability for different lithofacies of deltaic and fluvial deposits in the Ivishak Sandstone. Coarser-grained, middle-braided-river deposits are the most permeable, followed downward (in permeability) by finer-grained distal-braided-river deposits, and finally delta-front deposits. (B) Summary of grain size, sorting, and permeability trends for middle- and distal-braided-river and delta-front deposits of the Ivishak Sandstone. The lower figure shows the different types of deposits interpreted in another well. After Atkinson et al. (1990). (Reprinted with permission of Springer-Verlag.)

6.3 Meandering-river deposits and reservoirs

6.3.1 Processes and deposits

As their name implies, meandering rivers exhibit a sinuous longitudinal trend (Fig. 6.21A). Meandering rivers can form either within incised valleys (discussed in more detail below) or directly onto floodplains in nonincised valleys (Posamentier, 2001). For lowstand fluvial systems, the critical factor in whether incision occurs is the gradient of the alluvial plain and adjacent shelf. If the gradient is steeper than the associated alluvial profile, valley incision occurs. If the gradient is gentler than the alluvial profile, valley incision does not occur.

In this section, nonincised meandering-river deposits are discussed first, followed by incised-valley-fill deposits and reservoirs. Meandering rivers typically

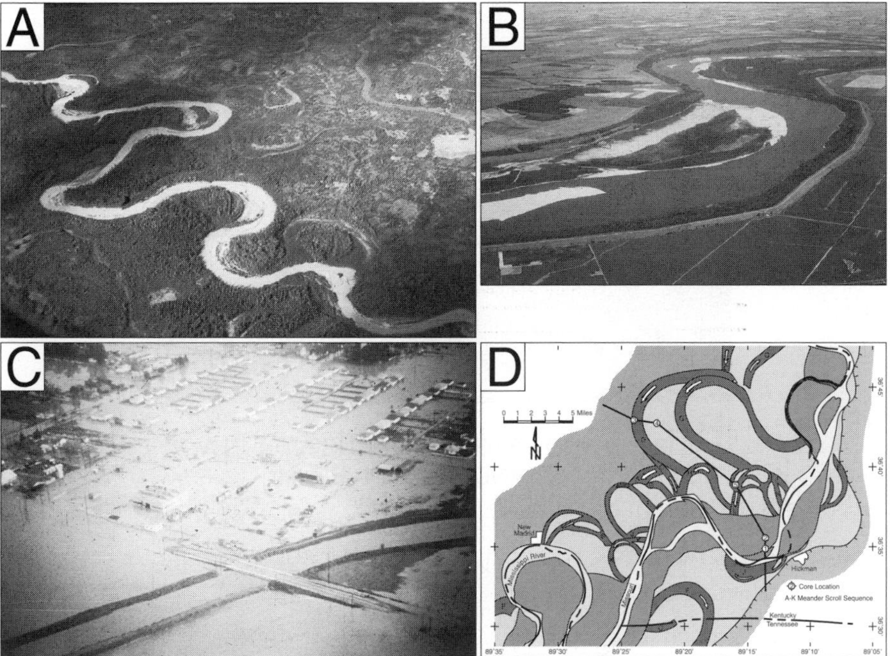

Fig. 6.21. (A) Meandering river and adjacent floodplain. Note the abandoned sinuous channels highlighted by vegetation on the floodplain. (B) Point-bar sand on a meander bend in the Mississippi River. (C) River at flood stage. Note that the main channel is outlined by the levees on both of its sides. (D) Map of part of the upstream Mississippi River fluvial system. Note the complexity of the point-bar deposits (yellow/bright-gray) and the fact that they are separated by muddy channel-fill plugs (green/dark-gray). The modern, active channel is shown in blue. (Figures (B) and (D) courtesy of D. Jordan.)

occur in the lower reaches of a river's drainage system, where the topographic gradient is low. Consequently, the transportational and depositional energy is less in a meandering-river system than in a braided-river system, resulting in finer-grained sediments being more prevalent in the meandering-river system. Although a meandering river might appear tranquil during most times of year (Fig. 6.21A and B and Fig. 6.22A), during flood stage the river is very dynamic (Fig. 6.21C). Gravel and sand-size particles are transported downcurrent as bedload (Fig. 6.4), whereas the finer-grained suspended load may overtop the channel margins and be deposited on the adjacent floodplain (Fig. 6.21C). The river bends (i.e., the meander loops) form by a combination of erosion and deposition during times of such high-energy flow.

The flow structure within a river channel is not uniform either horizontally or vertically (Fig. 6.22). Rather, once a bend begins to form, water flowing downcurrent on the outside of the bend will flow faster (will have a higher flow velocity) than the water has on the inside bend. Thus, the outside bend is subjected to

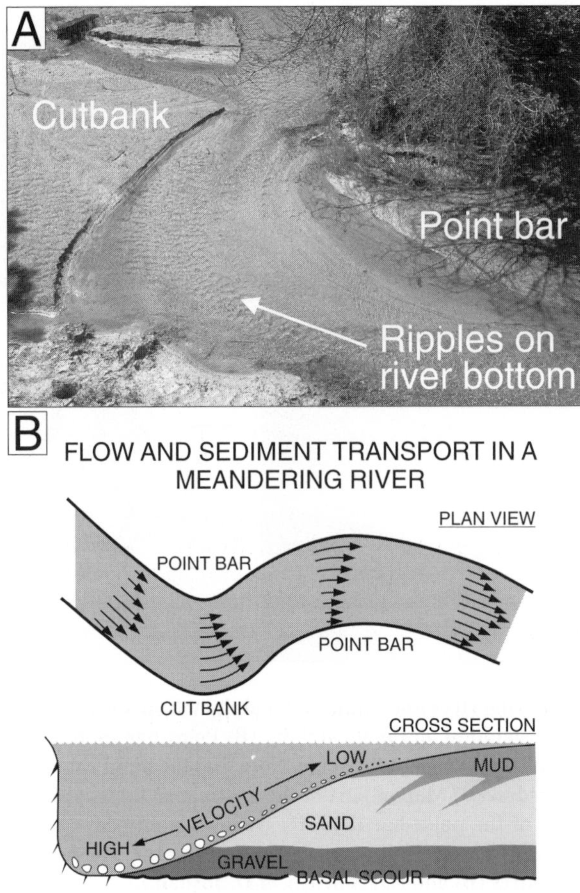

Fig. 6.22. (A) Flowing stream, showing the cutbank and point-bar sides of the meander bend. Note the rippled stream bed. Water and sediment transport is from top to bottom. (B) Plan view and cross-sectional view of flow in a meandering-river channel. Flow velocities are higher on the cutbank side of the river than they are on the point-bar side, leading to erosion on the cutbank side and deposition on the point-bar side. Except for frictional forces near the river bed, vertical flow velocities tend to decrease laterally across the channel, from the cutbank to the point-bar side. (Source of figure is unknown.)

erosion, giving rise to the name "cutbank side". The inside bend, or "point bar", is where deposition occurs (Fig. 6.21B).

Over time, the cutbank side loses sediment and the point-bar side gains sediment; the net result is the lateral migration of the sediment body across its floodplain (Fig. 6.23). During particularly energetic flows, the river can cut across a point bar and floodplain to form a new channel. The old channel becomes inactive and eventually fills with mud and organic matter. A meandering-river system can be quite complex, containing a series of sandy point-bar deposits truncated by muddy channel-fill deposits (Fig. 6.21D). Commonly, grain size

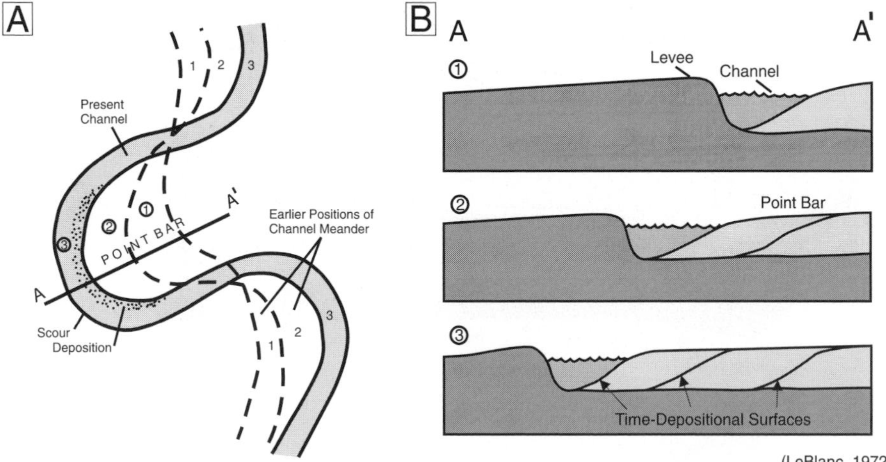

Fig. 6.23. (A) Plan view and (B) cross-sectional view illustrating the processes by which a meander bend migrates over time in a lateral direction. As sediment is eroded from the cutbank side of the river, it is deposited downcurrent on the point-bar side, resulting in lateral migration of the river bend over time. In these figures, migration is from right to left. (Source of figures is unknown.)

decreases upward from the base to the top of the point bar because of the combination of lower flow velocities in the upper parts of the river's profile (Fig. 6.22) and because, over time, the point bar migrates farther from the river channel (Figs. 6.22 and 6.23).

These processes have several important implications for reservoir characterization and performance. (1) The point-bar accumulations of sand form the main reservoir in a meandering-river system. (2) These accumulations can be highly compartmentalized as a result of crosscutting, mud-filled channels (Figs. 6.21D and 6.24). (3) The fining-upward trend of point-bar deposits (Fig. 6.24) means that permeability will also decrease upward, as was explained in Chapter 5. For this reason, a waterflood will tend to be more sweep-efficient in the basal, higher-permeability parts of a point-bar sandstone and will leave unswept oil in the upper, finer-grained parts (Fig. 6.25).

6.3.2 Reservoir examples

6.3.2.1 Rulison field, Colorado

Rulison field in the Piceance Basin of Colorado provides an example of the value of detailed reservoir characterization for extracting more gas from an existing meandering-river reservoir (Kuuskraa et al., 1997). Several improved technologies for reservoir characterization led to rejuvenation of this gas field and increased its estimated ultimate recovery (EUR) from less than

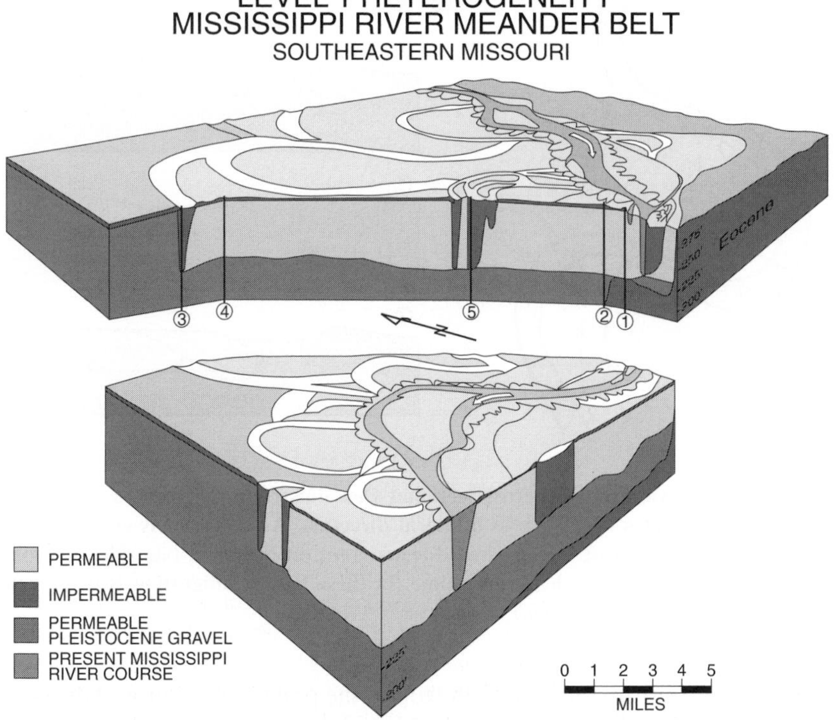

Fig. 6.24. Block diagrams showing the internal complexities characteristic of the portion of the Mississippi River illustrated in Fig. 6.21D. At this location, the meandering-river deposit sits atop Pleistocene gravel. The active channel is shown in light blue/black. Permeable point-bar sand is shown in yellow/bright-gray. When a particular channel becomes abandoned, it fills with mud and organic matter, which, when buried, will become impermeable shale. These shale or clay plugs, as they are often called, provide vertical barriers to fluid migration in an analog reservoir and horizontally compartmentalize the point-bar sands. (Illustration courtesy of D. Jordan.)

1 BCFG/well to approximately 1.9 BCFG/well, a 90% increase in recovery! A contour map of gas production per section (i.e., per square mile) in the southern Piceance Basin places Rulison field in the basin's area of highest gas production (Fig. 6.26).

Production in Rulison field is from Upper Cretaceous sandstones of the Williams Fork Formation. The Williams Fork is a prolific gas producer in this area. Production is from a series of stacked, lens-shaped sandstones that originally were deposited in a fluvial, point-bar setting. Each sandstone is 6.7–20 m (20–60 ft) thick and 500 m (1,500 ft) wide, but they have internal discontinuities. Clays and cements in the sandstones also influence production by giving rise to low matrix porosity and permeability (Fig. 6.27). The higher values of well-test permeabilities compared with core-plug permeabilities (1–2 orders of

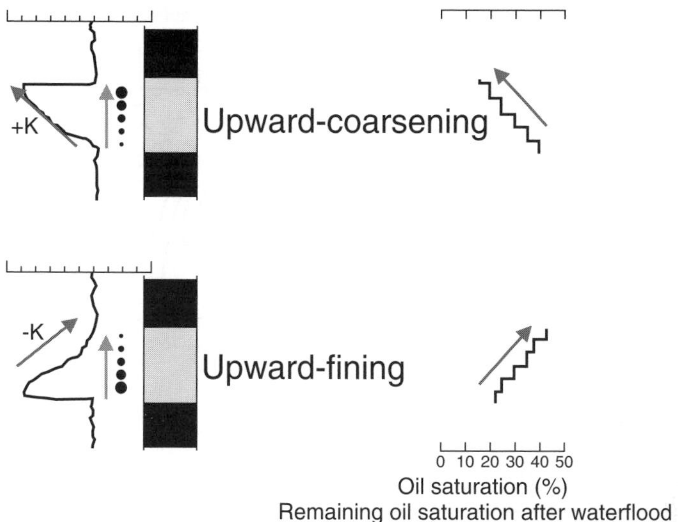

Fig. 6.25. Schematic illustration of gamma-ray logs of two sandstone beds or deposits. The black dots refer to the relative size of sand grains comprising the deposits. The upper sand becomes coarser grained from the bottom to the top, and the lower sand becomes finer grained from the base upward. Because of the grain-size effect, the permeability will increase upward in the upward-coarsening sand, and it will decrease upward in the upward-fining sand. The two stairstep plots on the right show oil saturations remaining after a waterflood. The principle illustrated here is that water was injected into the reservoir to sweep more oil to producing wells. In both cases, the distribution of remaining oil after the waterflood was not uniform but instead followed the permeability/grain-size trends. For the upward-coarsening sand, there was less remaining oil in the top of the reservoir after waterflood than there was at the bottom of the sand, because the injected water preferentially sought the more permeable zone at the top of the reservoir. Thus, upon completion of the waterflood, there was less remaining oil in the more-permeable top than there was in the less-permeable bottom of the sand. The same principle is shown for the lower, upward-fining sand. The least amount of oil remaining after waterflood is in the lower, coarser-grained, more-permeable part of the sand, which was better swept by the waterflood. (Source of figures is unknown.)

magnitude higher) indicates that this tight gas sandstone field has significant fracture permeability. The trend, or orientation, of open natural fractures was determined using aerial magnetic and seismic surveys and well-log analysis. A specific trend, called the "Productive Trend", was discovered in which open fractures were anomalously abundant. Estimated ultimate recoveries (EURs) from the Productive Trend were calculated to be about three times higher than EURs outside of this trend. Also, redetermination of R_W, including the effects of the clay minerals illite and smectite, increased calculated gas saturation from the original estimate of 40% to 64–85%.

Prior to 1993, wells were completed only in well-defined productive zones. In 1993, more zones were tested and opened to production, which led to iden-

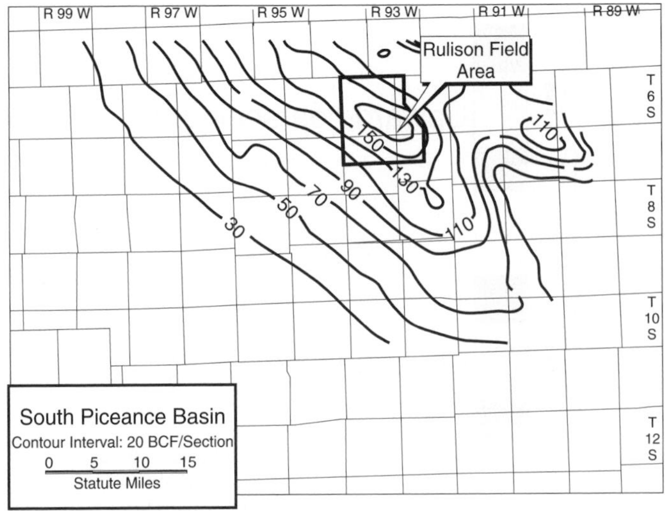

Fig. 6.26. Contour map of gas production per section (square mile) in the southern Piceance Basin. Rulison field lies in the area of highest gas production. After Kuuskraa et al. (1997). (Reprinted with permission of Rocky Mountain Association of Geologists.)

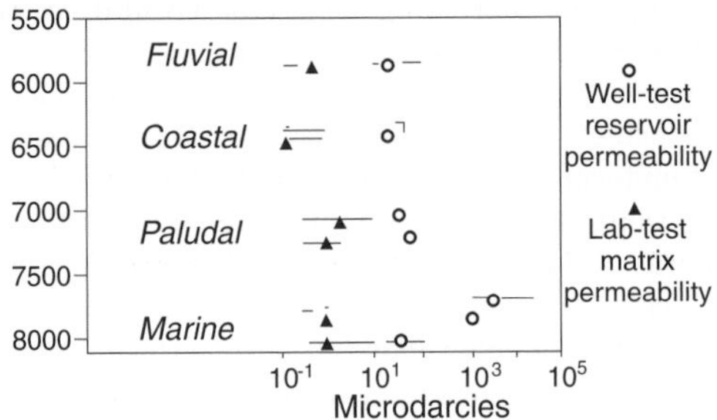

Fig. 6.27. Plot of laboratory-measured matrix permeability and well-test permeabilities for the four different sandstone types (which were deposited in different sedimentary environments). Well-test permeabilities are one or more orders of magnitude higher than matrix permeabilities, indicating that a significant component of permeability is to the result of open fractures in the field. After Kuuskraa et al. (1997). (Reprinted with permission of Rocky Mountain Association of Geologists.)

tification of a greater number of productive intervals that could be penetrated by a single vertical or slant well. However, to determine the optimal well spacing, it was necessary to estimate the spacing of compartmentalized point-bar sandstones. On the basis of point-bar-lens measurements in analog outcrops, coupled with study of orientations of sandstones in Rulison field, geoscientists estimated that each point-bar lens has an average of 250 m (750 ft) of connected width along a northwesterly to southeasterly trend (Fig. 6.28). Table 6.2 lists the

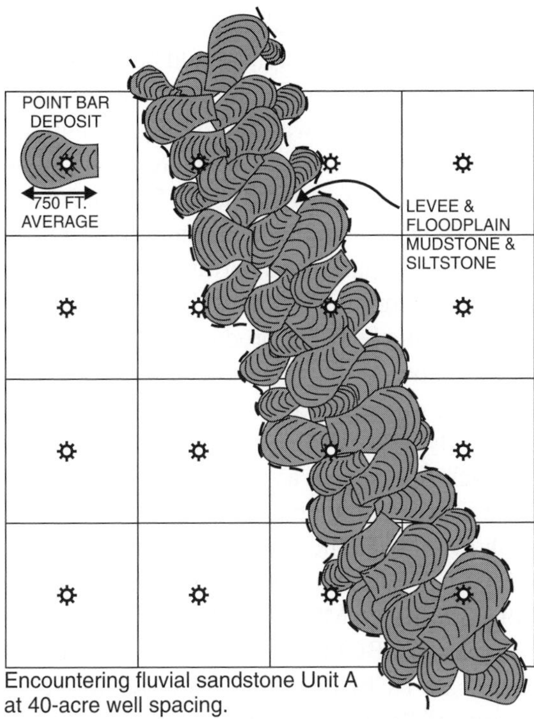

Fig. 6.28. Map showing the interpreted distribution of point-bar reservoir sandstones in Rulison field, based on dimensional measurements of outcrops. The entire square is equivalent to a one 1-mile-square section of land, so each small square is 1/8 of a section, or 40 acres. After Kuuskraa et al. (1997). (Reprinted with permission of Rocky Mountain Association of Geologists.)

Table 6.2 Recovery efficiency at different well spacings, Rulison Field, Colorado

Well placement		No. wells in same sandstone body	No. wells not in same sandstone body	Recovery efficiency in % GIP
Acres/well	#/Section			
160	4	0	4	7
80	8	2	16	13
40	16	4	12	26
20	32	8	24	

number of calculated point-bar sandstones that would be penetrated at different well spacings, using this width value for each point-bar lens. At 160 acres/well, each of the four wells in a section (640 acres) would be in a separate sandstone body. At 80 acres/well, of the eight wells per section, two wells would penetrate the same sandstone body and six would be in separate sandstone bodies. At a 40 acre/well spacing, four wells would be in the same sandstone and 12 wells would be in different sandstones. Finally, at a 20 acre/well spacing, 24 wells would be in different sandstones and eight would be in a single sandstone. Originally, wells were widely spaced (one well per 160 acres). Because of the high number of dry holes, well spacing was later reduced to 40 acres per well, and the area was tested later for 20-acre spacing. Interestingly, bottom-hole pressure measurements indicated that new wells at about 50 acres of equivalent distance from productive older wells encountered original reservoir pressures, which further documented the closely spaced compartments.

A trade-off exists in determining proper well spacing. The closer the spacing, the more likely one is to hit a "pay" or productive sandstone. But, the downside is that each well is expensive, so the minimum number of wells necessary should be drilled. In this field, wells drilled to drain an area of 160 acres each are calculated to recover only 7% of the gas resource, whereas wells drilled to drain an area of 40 acres each can recover as much as 26% of the gas (Table 6.2).

The net economic impact of the updated analysis is summarized in Fig. 6.29. The vertical axis shows total billion cubic feet of gas (BCFG) produced. At a price of $1.50/MCFG (thousand cubic feet of gas), and with finding or replacement costs of only $0.50/MCFG at that time, 42 economic wells were drilled for a recovery of 1.75 BCFG, 19 wells were drilled that were marginally economic, providing 1.20 BCFG, and only 18 uneconomic wells were drilled. Seventy-seven percent of the wells were deemed economic under those price constraints.

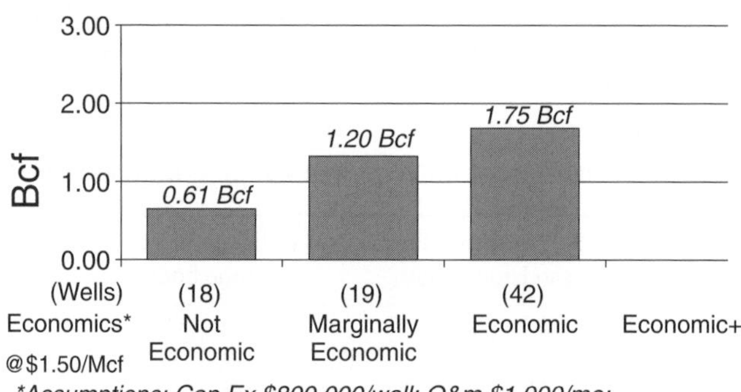

Fig. 6.29. Bar graph showing the net impact of using the new reservoir – characterization information for improved drilling. The vertical axis shows total billion cubic feet of gas (BCFG) produced. At a price of $1.50/MCFG (thousand cubic feet of gas), and with certain listed expenses, 42 economic wells were drilled in Rulison field for a total recovery of 1.75 BCFG; 19 wells were drilled that were marginally economic, providing 1.20 BCFG; and only 18 uneconomic wells were drilled. Seventy-seven percent of the wells were deemed economic under the price constraints listed. After Kuuskraa et al. (1997). (Reprinted with permission of Rocky Mountain Association of Geologists.)

6.3.2.2 Stratton field, Texas

Stratton field in south Texas was discovered in 1922. It had cumulative gas production of 2.4 TCF until 1994 (Levey et al., 1994). The field produces gas from the Oligocene-age Frio Formation, a thick, fluvially deposited sand-shale sequence that is a prolific producer along a regional depositional trend. This trend was a result of major progradation, during an 8-m.y. period, into the downdip expanded side of the Vicksburg Fault (Galloway et al., 1982).

Stratton field is within the middle Frio interval, which is aggradational in nature. Aggradation rates alternated over time; during periods of low aggradation, laterally stacked fluvial-splay channel systems developed, whereas during periods of high aggradation, vertically stacked channel-splay systems developed (Kerr and Jirik, 1990). Individual channel-fill sequences are 3–10 m (10–30 ft) thick; commonly they amalgamate into multilateral belts as much as several kilometers (miles) wide. Seismic reflection records suggest that most faults in the area extend through the lower Frio, but not into the middle Frio (Hardage et al., 1996).

One of the reservoir intervals, called the F37 interval, consists of crosscutting, meandering-river-channel sandstones and associated floodplain shales. A 3D seismic time slice of this reservoir interval (at approximately 1.64 s, according to Levey et al., 1994) suggests that the interval is complex, but it does not indicate directly that there is more than one channel sandstone (Fig. 6.30). However, pressure data and well-log correlations point to at least three mutually isolated, good-reservoir-quality (greater than 20% porosity and 10–92-md permeability) channel sandstones with different pressure regimes within a 1 mi^2 area (detected

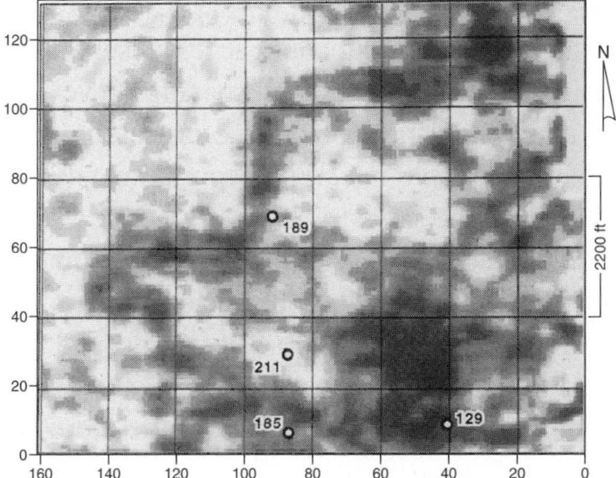

Fig. 6.30. A 3D horizontal seismic or seiscrop time slice through the F37 meandering-river reservoir in Stratton field. Note the meander bends shown by the blue/dark color, which represents low amplitudes. Locations of four wells are shown. After Hardage et al. (1996). (Reprinted with permission of AAPG, whose permission is required for further use.)

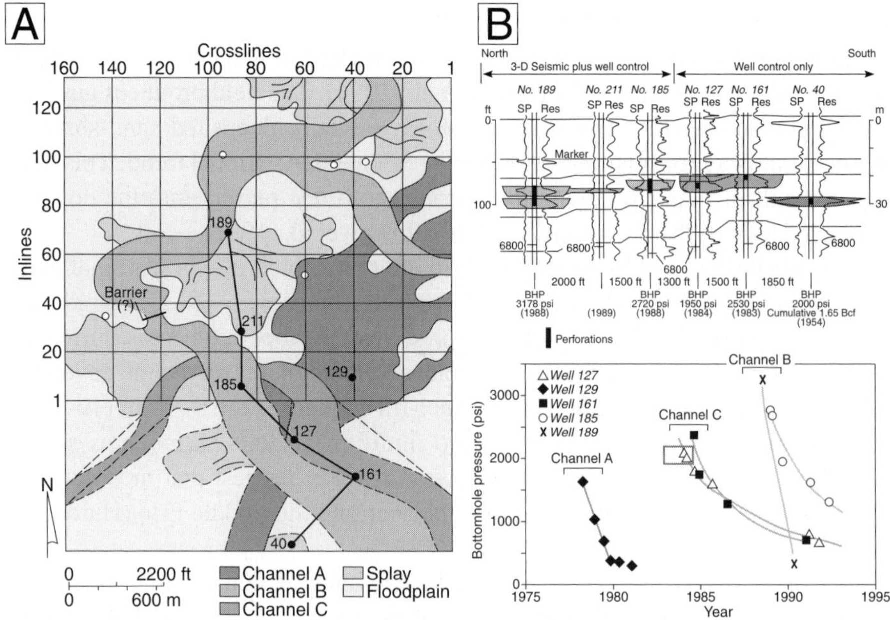

Fig. 6.31. Interpretation of the meandering-river system from the 3D seismic time slice in Fig. 6.30. Three different channel sandstones have been mapped, on the basis of a well log stratigraphic cross-section and different pressure profiles in the three sandstone bodies (B). The location of the cross-section is shown in (A), and some of the wells are also shown in Fig. 6.30. After Hardage et al. (1996). (Reprinted with permission of AAPG, whose permission is required for further use.)

in wells spaced 40–80 acres apart) (Fig. 6.31). Reservoir assessment has indicated that wells with 40-acre spacing could access reserve additions that would not otherwise be captured with greater well spacing (Levey et al., 1992).

Sanchez (2003) applied a variety of seismic-processing techniques to a shallower stratigraphic interval (B46 interval, at approximately 1.24 s, according to Levey et al., 1994) within the same 3D seismic volume (Fig. 6.32). The objective was to evaluate the techniques' ability to improve imaging and resolution capabilities for this fluvial system. Petrophysical evaluation was performed on gamma-ray, neutron–porosity, density–porosity, sonic, and resistivity logs from 21 wells in the field to determine interval porosity, V_{shale}, S_{w}, S_{g}, and net pay thickness. As is the case with the F37 interval, this horizon has virtually no structural complexity.

Spectral-decomposition time slices of the B46 interval at different seismic frequencies were discussed in Chapter 2 (Figs. 2.22 and 2.23). The time slices illuminate different features at different frequencies, so they are a suitable technique for improved imaging of subtle reservoir features.

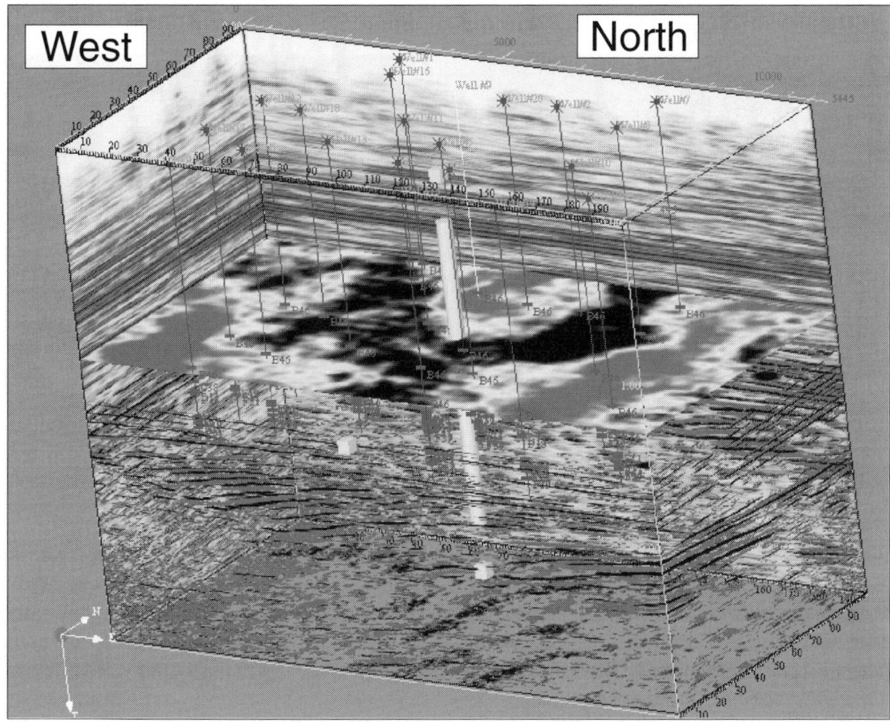

Fig. 6.32. Seismic time slice at the B46 level in the 3D seismic survey of Stratton field. Note the good well control (vertical blue/black lines). (Reprinted with permission of M.E.N. Sanchez.)

A seismic time slice on the flattened B46 interval shows an apparent meander loop highlighted by relatively high seismic amplitude (Figs. 6.32 and 6.33A). Variations in amplitude (indicated by color differences) are an expression of variations in lithology and/or fluid content.

Another display (Fig. 6.33B) is an acoustic-impedance time slice of the same horizon shown in Fig. 6.33A. Seismic-reflection amplitudes are a result of relative differences in acoustic impedance of adjacent rock layers. Acoustic impedance is the product of rock density and velocity and therefore is a measurable physical rock property. Acoustic impedance data can be converted directly to lithologic or reservoir properties, such as porosity and net pay. In this case (Fig. 6.33B), the seismic volume was inverted through a model-based inversion algorithm to produce an acoustic-impedance volume, from which the time slice was generated. It shows greater detail (and compartmentalization) than is depicted by the seismic-amplitude slice (Fig. 6.33A).

Figure 6.33C displays a variance slice of the same horizon. Coherence (or its inverse, variance), mentioned in Chapter 2, is a numerical measure of lateral variation or similarity of a seismic time event. Variations in the gray scale in

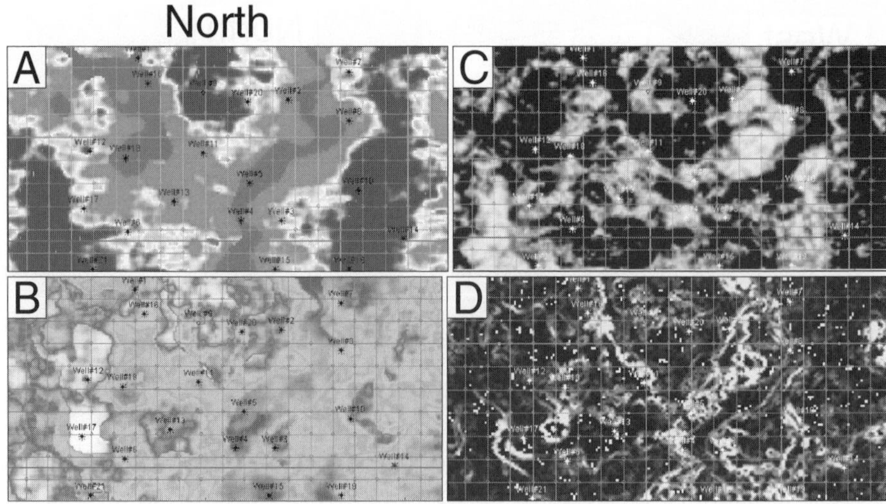

Fig. 6.33. (A) Seismic time slice through reservoir interval B46. The same time slice is shown in Fig. 6.32. (B) Acoustic-impedance display of the same time slice. (C) Coherency-cube display of the same time slice. (D) Acoustic-impedance coherency-cube display of the same time slice. See text and Sanchez (2003) for details of how these displays were generated. (Reprinted with permission of M.E.N. Sanchez.)

this example (Fig. 6.33C) are a measure of variations in lithologic properties. As is the case with the acoustic-impedance time slice, this slice also implies considerable lithologic variation.

The final display (Fig. 6.33D) is a combination of the variance attribute applied to the acoustic-impedance volume. This image provides additional detail.

Based on results from the above processing schemes, Sanchez (2003) combined the acoustic impedance, variance, spectral decomposition, amplitude envelope, and instantaneous frequency of seismic traces to generate a volume-based attribute analysis using an artificial neural network to predict porosity and resistivity. This was done over a 0.34-s interval with its top at the B46 horizon and its base approximately 0.09 s above the F37 horizon. The analysis was based on several data control points from the 21 wells in the field. A cross-plot of predicted versus actual porosity of a series of stratigraphic intervals in several wells within this 0.34-s seismic window shows an excellent correlation ($r = 0.976571$) (Fig. 6.34A). Maps of porosity and resistivity over the interval clearly highlight zones of high porosity and resistivity (Fig. 6.34B and C). Sanchez's (2003) work shows that such seismic-attribute analyses can highlight drilling targets and thereby reduce drilling risk.

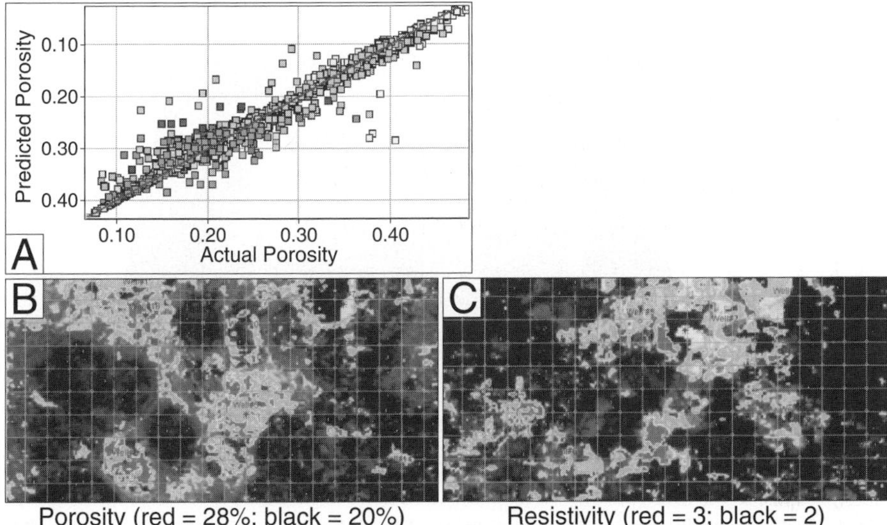

Fig. 6.34. (A) Neural-network-derived crossplot of actual (horizontal axis) versus predicted (vertical axis) porosity of different stratigraphic intervals in 18 test wells (different colored data points). $S = 0.976571$; $Error = 0.0139535$. **(B) Map view of the distribution of predicted porosity in the area shown in Fig. 6.33. (C) Map view of the distribution of predicted resistivity in the area shown in Fig. 6.33. Both maps (B) and (C) were derived from neural-network analysis of calibrated seismic attributes. After Sanchez (2003).**

6.4 Incised-valley-fill deposits and reservoirs

6.4.1 Processes and deposits

An incised valley forms when a river has cut into its own floodplain or underlying strata sufficiently that even at flood stage, flow does not overtop the banks (Fig. 6.35). The formerly active floodplain is left abandoned to serve as interfluves (Posamentier, 2001). Valley incision can occur as a result of base-level fall, tectonic tilting of an alluvial plain, and/or decrease in fluvial discharge to form underfit streams. During the stage in which valley incision of the underlying substrate occurs, fluvial sediment is transported downstream, beyond the confines of the valley. Filling occurs mainly during the turnaround and rise of the base level (Fig. 6.36). The fill of a valley can be quite complex, both laterally and vertically. Laterally, the same meandering river processes that we described earlier in this chapter occur in incised valleys, so the resulting deposits are equally complex. The horizontal transect of a valley fill also varies in the downstream direction, with primarily fluvial deposits upstream and more marine-influenced deposits downstream (Bowen and Weimer, 2003, 2004). Vertically, the ideal valley-fill sequence consists of a basal fluvial lag, which grades upward into fluvial sandstones, and these are overlain by estuarine sandstones

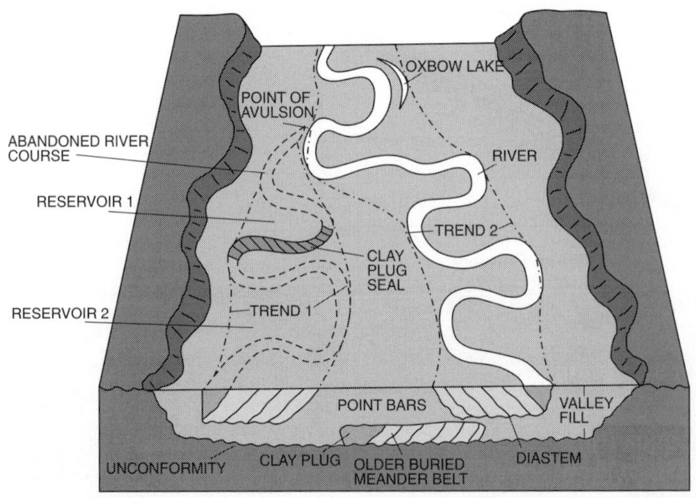

Fig. 6.35. Cartoon illustrating the distribution of fluvial sediments in an incised valley. The valley has been incised into preexisting sediments, which are shown in brown/light gray. The valley floor is an unconformity. The locations of point-bar deposits vary laterally and vertically as a result of lateral migration of the meander bends over time (Fig. 6.25). (Figure courtesy of R. Weimer.)

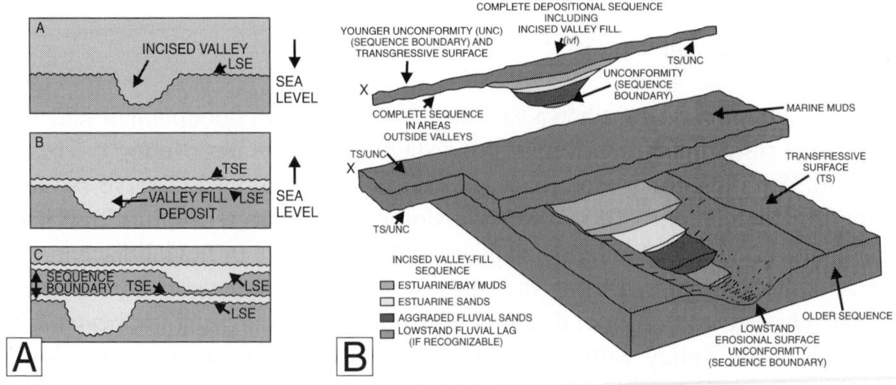

Fig. 6.36. (A) Sequence of events in the incisement and filling of a valley as a result of fluctuations in sea level. During falling sea level the valley is carved, and during rising sea level the valley is backfilled. (B) Ideal sequence of strata that fill a valley. The lowermost fill is fluvial lag. As sea level rises and begins to finger into the valley, a marine-influenced estuary forms and estuarine sediments, reworked from the underlying fluvial substrate, are deposited. Upon complete filling of the valley, open-marine shale is deposited over the top. After Weimer (1992). (Reprinted with permission of AAPG, whose permission is required for further use.)

and mudstones (Fig. 6.36). However, the vertical sequence will exhibit a preponderance of fluvial strata upstream and more estuarine strata downstream, closer to the marine environment. Figure 6.2C is an example of a modern incised valley on the west coast of the North Island of New Zealand. This valley is near the marine shoreline, and at low tide, as is the case in the picture, river flow dominates in this reach of the valley. However, during high tide, the lower reaches of the valley are inundated by tidal waters, and estuarine processes rework the river-derived sediment. The tidal effect diminishes progressively upstream, in the inland direction.

6.4.2 Reservoir examples

6.4.2.1 Sooner Unit, Colorado

The Sooner Unit in the Denver Basin of Colorado produces oil from the upper Cretaceous D Sandstone (Sippel, 1996; Montgomery, 1997). Maps based on well control and on a 3D seismic-reflection survey were discussed in Chapter 1 and shown in Fig. 1.10. The reservoir sandstone was deposited within an incised valley. The map that was based solely on well control showed an interconnected sandstone as thick as 8 m (25 ft). The 3D seismic survey revealed much greater complexity, with several compartments. Recognition of the compartmentalized nature of the fluvial sandstone resulted in redesign of waterflood and infill-drilling programs, which led to production of an additional million barrels of oil at reduced operating cost.

6.4.2.2 Sorrento field, Colorado

Sorrento field is located in the southeast corner of Colorado (Mark, 1998). It is part of a trend of Morrowan (Lower Pennsylvanian) incised-valley-fill and related deposits that extends through eastern Colorado, southwestern Kansas, and northwestern Oklahoma (Bowen and Weimer, 2003, 2004). More than 200 MMBO and 8 TCFG have been produced from this trend. The incised valleys formed and filled during periods of sea-level lowstand and early rise (Fig. 6.36).

The reservoir sandstone at Sorrento field forms an elbow-shaped body with net sand at least as thick as 17 m (50 ft) (Fig. 6.37) (note that less than 75 API units are cut off on a gamma-ray log) (Mark, 1998). Conventional well logs display a good sandstone response, with little apparent complexity (Fig. 6.38). However, results of gamma-ray-, neutron-, density-, and sonic-log values calibrated to cores indicate significant vertical variability within the reservoir sandstone. The basal sandstone is a fluvial channel sandstone; it is overlain by reworked estuarine sandstone, and that is topped by a marine shale (compare this succession with the idealized succession in Fig. 6.36). The reworked sandstone has lower permeability than does the coarser-grained fluvial sandstone

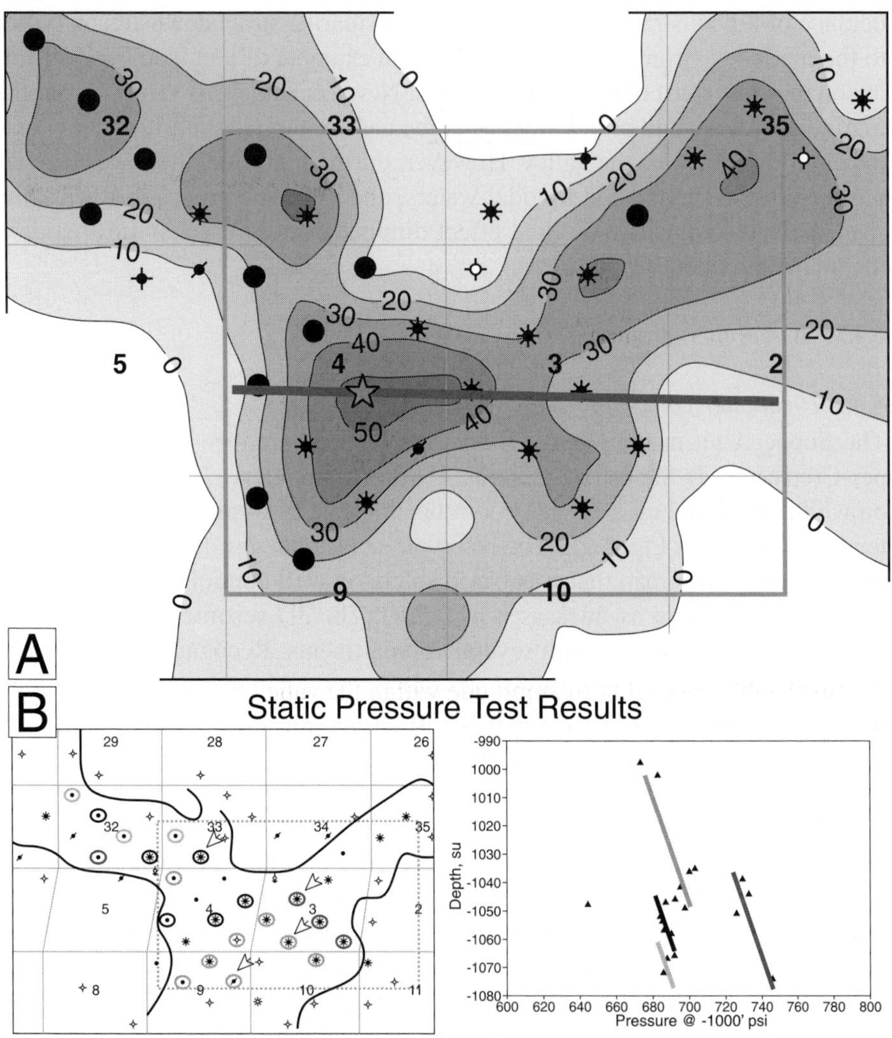

Fig. 6.37. (A) Isopach map of sandstone thickness in Sorrento field. Note the curved shape of the reservoir, indicative of a meandering river deposit (within an incised valley). The thickest sand occurs in the bend in the reservoir, which probably corresponds to a point-bar deposit. (B) Results of static reservoir-pressure tests conducted in wells from the field. The plot of pressure versus depth shows four different trends, indicating that the reservoir is divided into a series of at least four reservoir compartments. The locations of the wells in which the pressure tests were run are also shown. The wells are color coded to show which of the trends are from which wells. Note that each trend occurs mainly within separate parts of the field, thereby confirming the presence of geographically distributed compartments. Sorrento field, in southeast Colorado, is another incised-valley-fill reservoir. After Mark (1998).

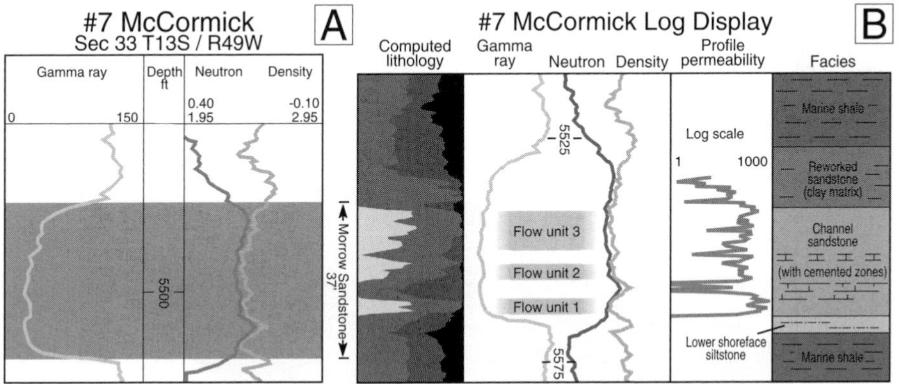

Fig. 6.38. (A) Typical log for a well in Sorrento field. The gamma-ray log suggests a rather uniform sandstone. However, as is shown in (B), this well was cored, and that revealed fluvial and marine-reworked sandy facies. Also, calcite streaks occur that are very impermeable (see permeability profile). Thus, the sand is very highly stratified. After Mark (1998). (Reprinted with permission of Rocky Mountain Association of Geologists.)

(Fig. 6.38). Permeability "tight streaks" are the result of carbonate cement that has differentially cemented fluvial channel sandstones.

A 3D seismic survey revealed the outline of the master incised valley as well as considerable variability in amplitude within the valley fill (Fig. 6.39A). Well and core control confirmed the complexity of the internal fill (Fig. 6.39B). To correlate individual seismic-reflection events to sedimentary facies, each facies was identified from well and core control and tied to the corresponding reflection on a line-by-line basis (Fig. 6.40). Though a time-consuming process, this approach to detailed mapping of a 3D seismic volume provided 3D maps of individual flow units (Mark, 1998). Static pressure tests confirmed that the four mapped channel sandstones (flow units) that fill the incised valley are mutually isolated (Fig. 6.37). This local compartmentalization is partly a result of the tight carbonate-cemented zones, which formed as caliche horizons when channels were open between depositional periods. The compartmentalization is also partly a result of the complex facies distribution in the field. An S-wave seismic survey also was acquired over part of the field and revealed greater internal complexity and compartmentalization than was imaged by the P-wave survey (Fig. 2.30) (Blott et al., 1999). This level of characterization proved useful for designing an improved infill-drilling program in this field.

6.4.2.3 Southwest Stockholm field, Kansas

Stockholm field is another example of a Morrowan-age incised-valley-fill reservoir that lies along the same general trend as does Sorrento field (Fig. 6.41) (Tillman and Pittman, 1993). It, too, takes the shape of a meandering river system. Production is from the Stockholm Sandstone, which is underlain by a lower

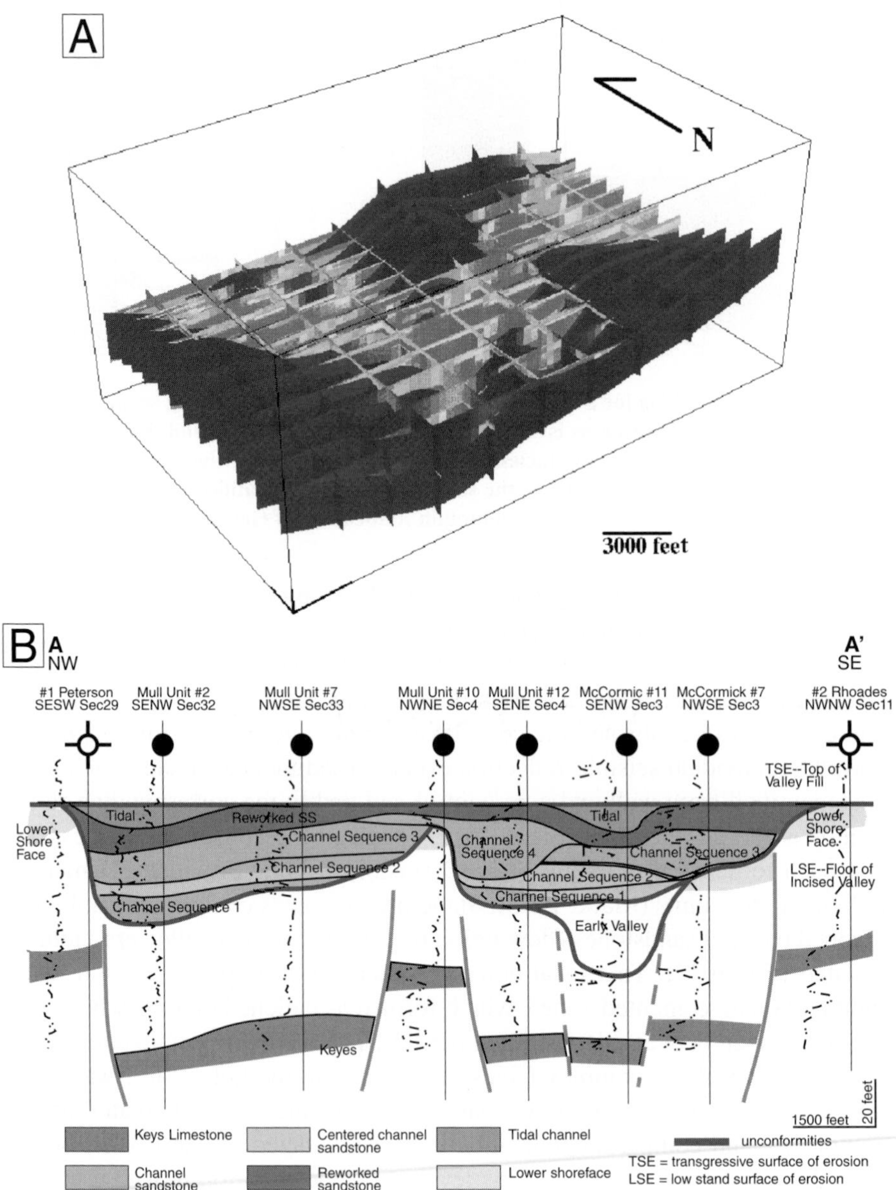

Fig. 6.39. (A) A 3D seismic volume, showing an outline of Sorrento field's incised valley and also the field's internal complexities, which are illustrated by variations in seismic-amplitude (colors). (B) Structural cross-section showing the association of faults to the valley and the distribution of different channel sandstones mapped from well and seismic control. After Mark (1998). (Reprinted with permission of Rocky Mountain Association of Geologists.)

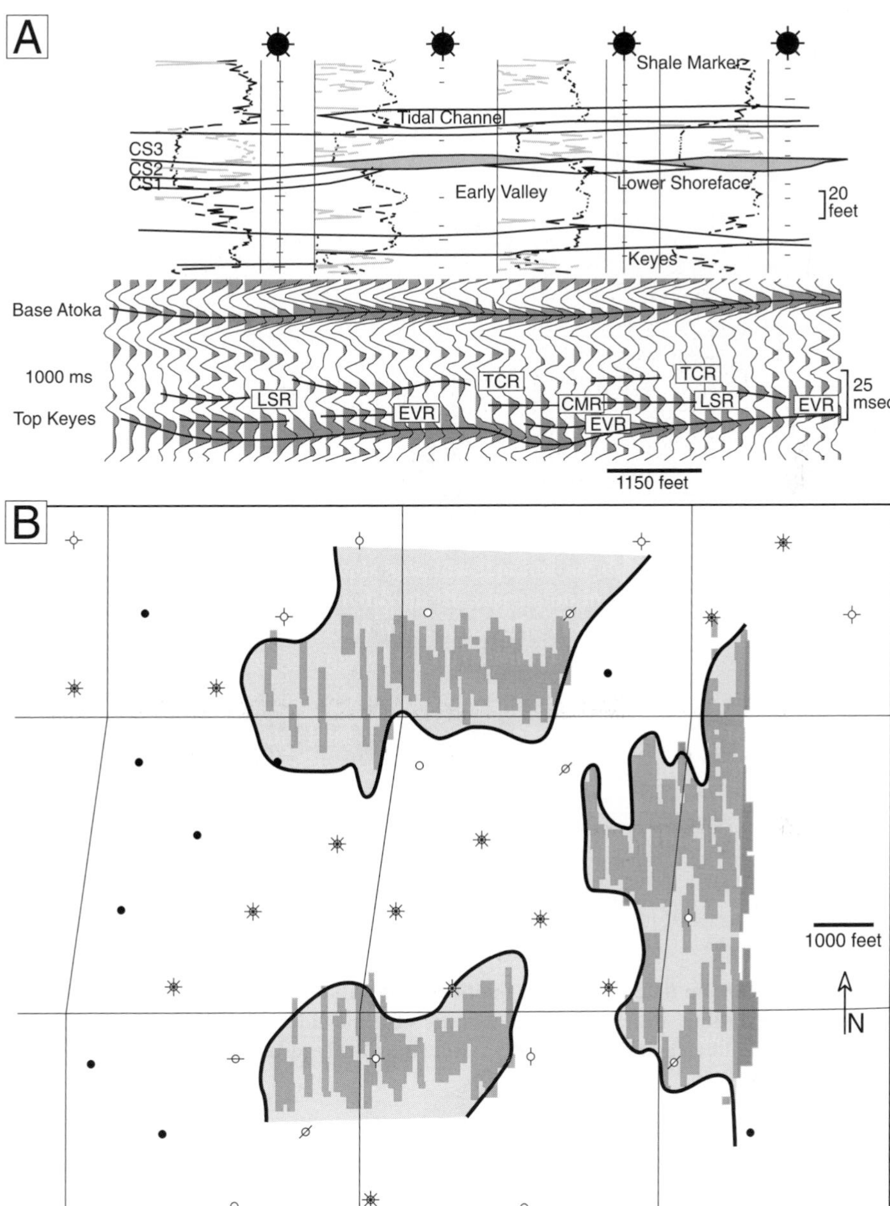

Fig. 6.40. (A) Vertical seismic profile from the 3D seismic data volume, showing the different reflections and their interpretation based on well and core control. Each of these reflections was mapped on a line-by-line basis, in order to map the distribution of sedimentary facies in the field. (B) Example of one of the mapped facies. Vertical bars indicate the presence of that particular seismic reflection on each vertical seismic line. After Mark (1998). (Reprinted with permission of Rocky Mountain Association of Geologists.)

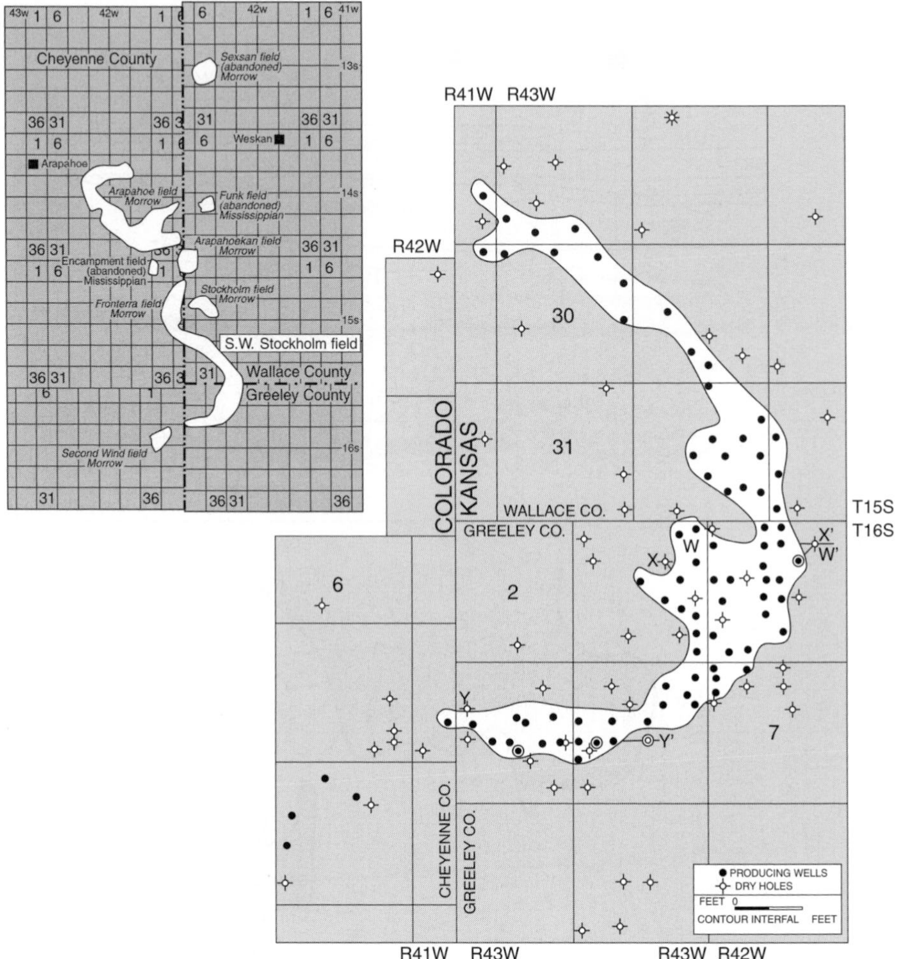

Fig. 6.41. Stockholm field (inset) and Southwest Stockholm field, Kansas. Note the reservoir's curved outline, which is very diagnostic of a meandering river (within an incised valley). After Tillman and Pittman (1993). (Reprinted with permission of Penn Well Books.)

Morrow limestone into which the valley has been incised (Fig. 6.42). Conventional 2D seismic-reflection profiles reveal some internal complexities but cannot resolve detailed heterogeneity as effectively as that resolved by Mark (1998) and Blott et al. (1999) for Sorrento field. Although the conventional well logs of the Stockholm Sandstone appear similar from well to well, core descriptions reveal both estuarine and fluvial facies in vertical and lateral juxtaposition (Fig. 6.43). Permeability of the estuarine facies is nearly an order of magnitude lower than that of the fluvial facies, because of smaller pore-throat sizes (Table 6.3). This example demonstrates that when one is comparing and correlating sandstone intervals with conventional well logs, significant lateral and

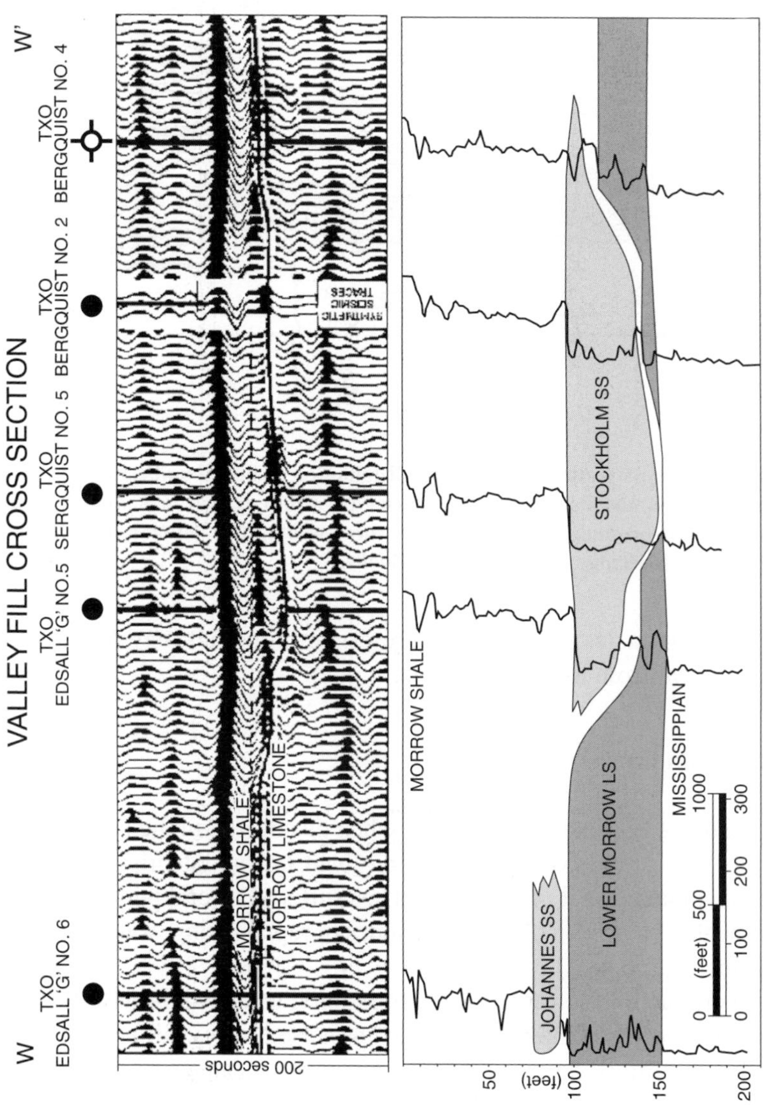

Fig. 6.42. A 2D seismic profile and corresponding interpretation of the Southwest Stockholm field's incised-valley-fill reservoir. The valley was incised into the underlying lower Morrow Limestone. The surface between the limestone and sandstone is an unconformity. After Tillman and Pittman (1993). (Reprinted with permission of Penn Well Books.)

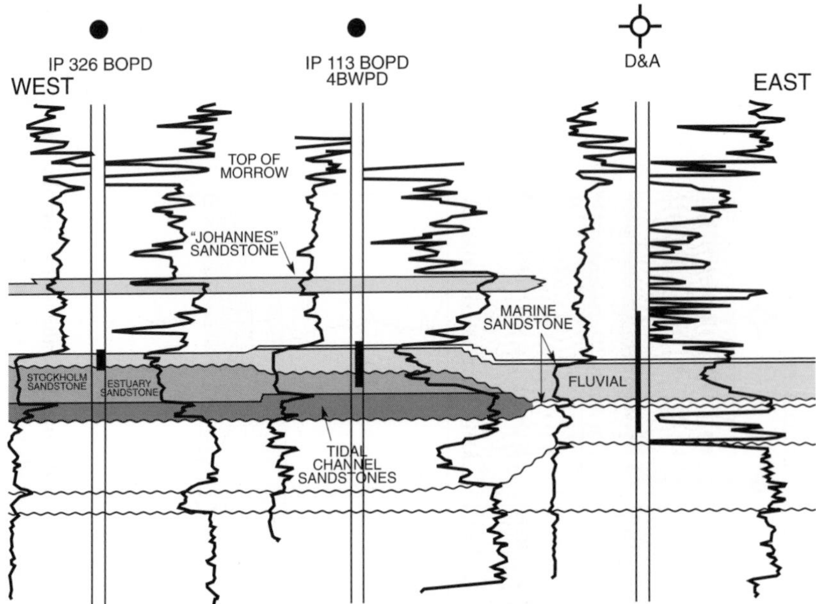

Fig. 6.43. Three-well cross-section, showing the distribution of fluvial and estuarine facies within the Stockholm Sandstone. Although the facies are different, they look similar on the gamma-ray logs. Oil is produced from two of the three wells. After Tillman and Pittman (1993). (Reprinted with permission of Penn Well Books.)

vertical variations in facies and associated reservoir properties may be revealed that are not apparent from log shapes alone.

6.5 Combination fluvial reservoirs

Glenn Pool field is a very mature field that, when discovered in the early 1900s, paved the way for a rich, colorful, and long-lived history of oil and gas exploration in Oklahoma (Fig. 6.44). The field has suffered several periods of production decline that have been offset by episodes of repressuring, waterflooding, and new drilling (Fig. 6.44).

Glenn Pool production is from the Desmoinesian (Pennsylvanian) Bartlesville Sandstone, which has been interpreted to be an incised-valley-fill system composed of a lower braided-river facies and an upper meandering-river facies (Figs. 6.45 and 6.46) (Kerr et al., 1999; Ye and Kerr, 2000). Braided-river sandstones are laterally continuous, with few shale drapes or discontinuities, whereas meandering-river sandstones are finer-grained, lenticular, and laterally discontinuous (Fig. 6.46). Porosity and permeability are much better in the braided-river facies than in the meandering-river facies (Fig. 6.46). Core description and analysis from one well illustrates some of the complex com-

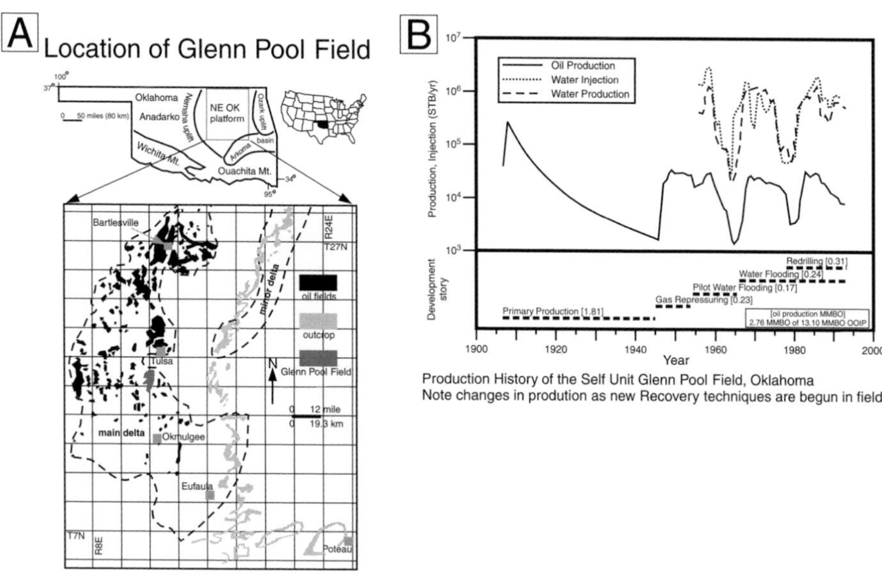

Fig. 6.44. (A) Location and (B) production history of Glenn Pool field in Oklahoma. Also shown in (A) are the surficial distribution of oil fields (black) and outcrops (yellow/bright gray) of the Bartlesville Sandstone. Various enhanced-recovery techniques have been used over the years to reverse this field's production declines, as is shown in (B). After Kerr et al. (1999) and Ye and Kerr (2000). (Reprinted with permission of AAPG, whose permission is required for further use.)

Table 6.3 Reservoir properties of Southwest Stockholm field, Kansas.

	Permeability (md)	Porosity (%)	R_{35} (microns)
Fluvial sandstones			
Average	703	16.3	19
Range	129–1890	11.7–20.6	11.2–33.5
Estuarine sandstones			
Average	80.8	14.4	7.1
Range	50–111	11.5–17.8	6.0–8.4

partmentalization that has affected waterflood performance (Fig. 6.46) from the Glenn Pool field. In this well, the presence of oil stain and bleeding oil is confined to the upper meandering-river facies, indicating that injected water has selectively swept the lower, more-permeable braided-river facies and bypassed the less-permeable meandering-river facies (see Fig. 6.47 and compare it with Fig. 6.25). Therefore, even in this very mature field, one opportunity to increase reserves would be to selectively waterflood the meandering-river facies.

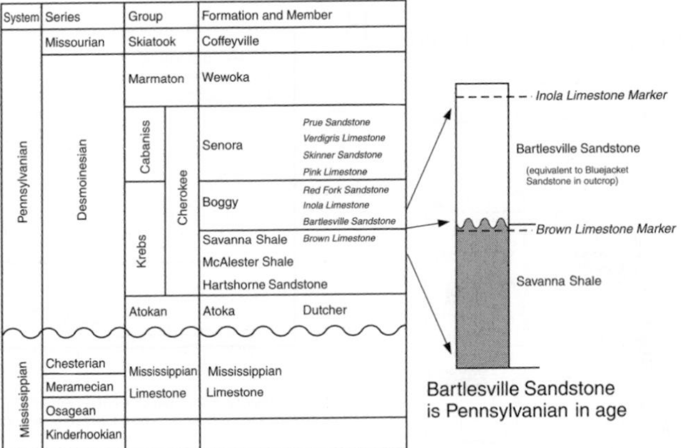

Fig. 6.45. Geologic column showing the Cherokee (Pennsylvanian) age of the Bartlesville Sandstone, which comprises the Glenn Pool field. After Kerr et al. (1999) and Ye and Kerr (2000). (Reprinted with permission of AAPG, whose permission is required for further use.)

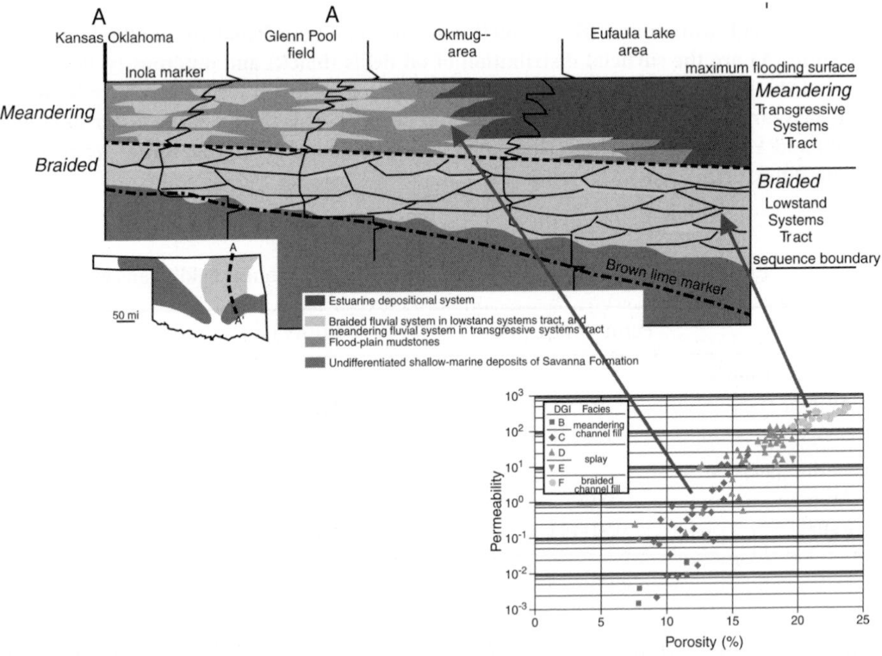

Fig. 6.46. Cartoon illustrating the stratigraphic distribution of braided- and meandering-river deposits that comprise Glenn Pool field. The Savanna Formation (orange) underlies the valley fill. The braided-river deposit sits atop the unconformity and underlies the meandering-river deposit. A porosity versus permeability plot shows that the braided-river facies is of better reservoir quality than is the meandering-river deposit, because the braided-river deposit is coarser grained. After Kerr et al. (1999) and Ye and Kerr (2000). (Reprinted with permission of AAPG, whose permission is required for further use.)

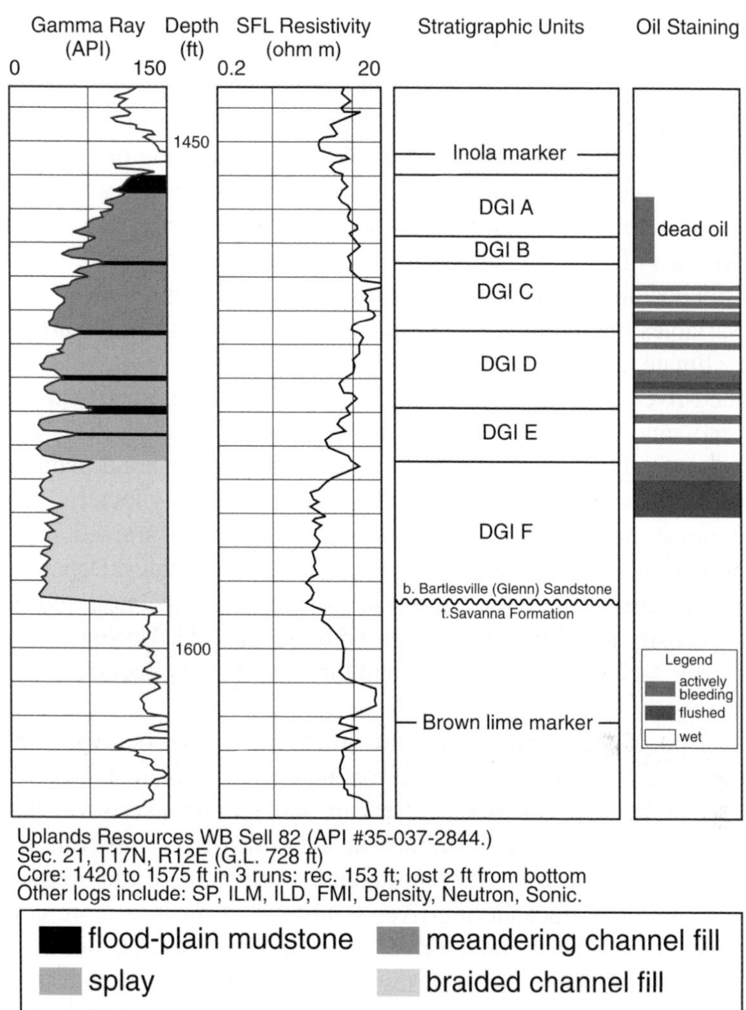

Fig. 6.47. Gamma-ray and resistivity logs from a cored well of the Bartlesville Sandstone. The yellow/bright-gray interval comprises a braided-river deposit. The orange/middle-gray interval comprises a meandering-river deposit. The pink/dark-gray interval is a transitional interval. The right-hand column depicts oil staining in the well, with green/middle gray being zones where oil was actively bleeding when the core was brought to the ground surface. Blue/black zones contained oil stain but did not bleed (because most of the oil had been flushed during a prior waterflood). White zones are water-bearing. Note that most of the actively bleeding intervals are in the transition and meandering-river deposits, because the permeability is lower in these strata and water selectively bypassed these zones. After Kerr et al. (1999) and Ye and Kerr (2000). (Reprinted with permission of AAPG, whose permission is required for further use.)

6.6 Summary

There are different types of fluvial deposits and reservoirs. The two end-member depositional types are braided-river and fluvial-river deposits. A third type, incised-valley fill, can contain either or both of these end members within the confines of the valley. In addition, fluvial deposits near the mouths of the valleys may become reworked by estuarine and tidal processes, which ultimately produce a different set of reservoir properties. The geometry, size, and reservoir characteristics of each fluvial type depend upon transportational, depositional, and postdepositional (diagenetic) processes that are controlled by several external variables, including geographic location, sediment source areas (provenance), climate, and degree of tectonic activity.

Braided-river deposits tend to be relatively coarse-grained and consist of gravel and sand, with little to no mud. Because of this, the beds tend to be laterally continuous over much or all of the width of the braid plain, although the presence of some shale beds may disrupt the continuity locally.

By contrast, meandering-river deposits tend to be finer-grained, more lenticular, and partially or completely encased in floodplain shales. Depending upon the deposit's degree and type of postdepositional compaction and cementation, its porosity and permeability can be quite variable. However, in general, braided-river facies are more porous and more permeable than are meandering-river facies.

As a result of these differences, fluvial reservoirs can be expected to have quite varied performances. Any reservoir-management plan should include an evaluation of the type of fluvial reservoir and its characteristics. For example, sweep efficiency will be higher in a braided-river reservoir than in a meandering-river reservoir. Also, horizontal wells may be more efficient in a set of discontinuous meandering-river sandstones than in a more continuous and interconnected set of braided-river deposits. Seismic-reflection techniques, as well as well-log, core, and well-test analyses, all can be used to adequately define the type of fluvial reservoir and predict the recovery performance and efficiency of that reservoir.

Chapter 7

Eolian (windblown) deposits and reservoirs

7.1 Introduction

Eolian (windblown) sandstone deposits can form excellent oil and gas reservoirs. Wind is a very effective agent for sorting sand-size, silt-size, and clay-size grains, both vertically within the air column and horizontally in the downwind direction. Therefore, at any given eolian depositional site, the deposit will tend to be fairly well-sorted by grain size. Good sorting often leads to good reservoir quality (Chapter 5).

Arid deserts are one of the two major types of environments in which eolian sands are deposited. Often, windblown sands are deposited within an arid intermontane basin. In such a setting, the source of the sand may be alluvial fans that fringe the mountain range and that are subjected to frequent winds (Fig. 7.1). Also, intermittent rivers that flow only during wet periods transport fluvial sediment to the wadi basin (a wadi is a desert channel or dry wash through which an intermittent river flows; the term also applies to the closed desert basin, where the wadi terminates, that is filled with fluvial sediments deposited by the intermittent river). Such fluvial deposits can be reworked later into eolian deposits. Inland sand-sea deposits often are quite thick and areally extensive (Fig. 7.2). They form part of a larger nonmarine system of alluvial-fan, braided-river, sand-sea, and playa-lake environments. They sometimes occur on a coastal plain leading downslope to a marine environment (Fig. 7.1).

Coastal zones on the landward side of beaches are the second major type of environment in which eolian sands accumulate (Figs. 7.2 and 7.3). Along most coastlines, winds blow toward the shore ("onshore winds"). These winds can pick up beach sand and transport it behind the beach, forming sand dunes away from the influence of the surf zone. Vegetation may grow on the dune surface, thus stabilizing the dune and preventing it from migrating farther inland. Many elongate coastal beaches are bordered on their landward side by equally elongate dune deposits, which potentially make good oil and gas reservoirs.

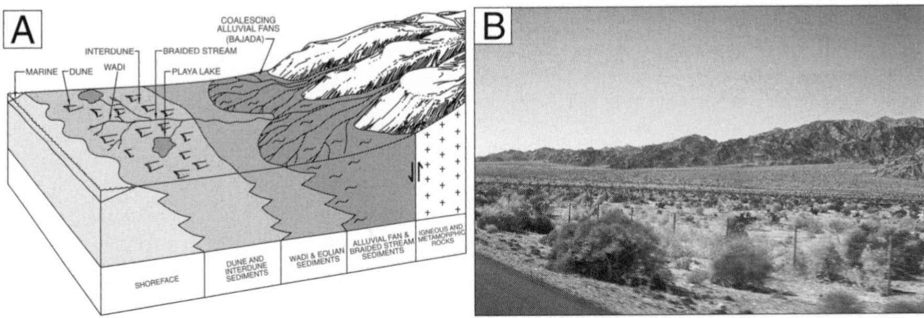

Fig. 7.1. Alluvial and eolian depositional system with various sedimentary environments. The photograph shows an alluvial fan at the base of a mountain. Alluvial fans often provide the source material for eolian deposits. Diagram after Davidson et al. (2002). (Reprinted with permission of Prentice-Hall, Inc.)

Fig. 7.2. An inland sand sea in an arid, intermontane basin. The large dunes are asymmetric in shape, reflecting their downwind migration from left to right.

The largest eolian reservoirs are those that were deposited as sand seas. Coastal dune sands also can form reservoirs, but they tend to be only one of a set of productive facies within a shoreface or barrier-island succession of strata (Chapter 8). In this chapter, the processes of sand transport and deposition are discussed, and reservoir examples are provided.

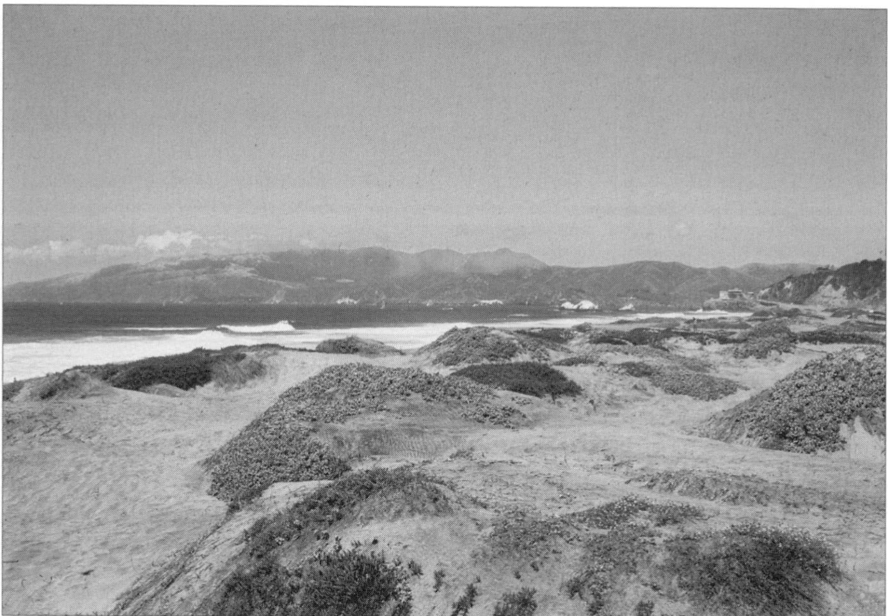

Fig. 7.3. A coastal sand dune field. To the left is the Pacific Ocean with a surf zone. Landward of the surf zone is a sandy beach. The dune sand is sourced from the beach by onshore winds. Vegetation on the crests of dunes helps to stabilize them and prevent further landward migration.

7.2 Processes and deposits

The processes by which sand is transported in the eolian environment were mentioned briefly in Chapter 3 (Fig. 3.36). During a wind storm, sand, silt, and clay deposits can be eroded and the grains swept into the air. Dust devils, which are wind funnels that form when wind crosses a depression on the ground surface, are a common phenomenon in wind-prone areas. Dust devils often erode grains of sediment and entrain them into the air mass (Fig. 7.4A). As grains become entrained, they tend to merge into a sheetlike body that migrates downwind (Fig. 7.4B). The larger and heavier sand-size grains move by saltation, or in suspension if the wind is strong enough, and the smaller grains are carried farther skyward in suspension within the flow. Although air is much less viscous than water, the transport processes in air are similar to the transport of grains within a river (Fig. 3.37). Dust and sand storms can be generated and can migrate very quickly during periods of strong winds. They can be destructive when they impinge upon populated areas (Fig. 7.4C) by depositing sedimentary layers in open-window cars and homes and by abrading buildings and other structures.

On a smaller scale, an individual sand dune moves (migrates) in the downwind direction when wind carries sand grains up and over the crest of the

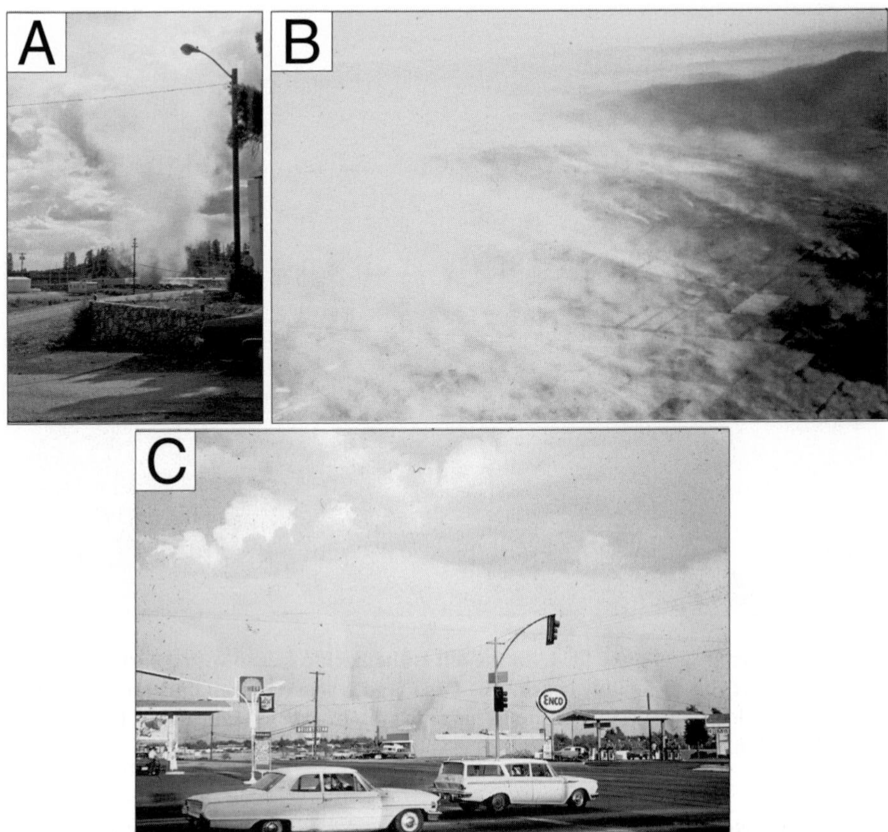

Fig. 7.4. The grain sizes of eolian deposits are mainly sands and silts. (A) Formation of a 'dust devil' in a trailer park. A strong wind has blown over a depression to form a local 'cyclonic' wind that is capable of lifting sediment (if sediment was on the floor of the depression) upward into the air mass, where it then moves in a downwind direction. (B) A dust (silt) storm lifting sand and silt from a farmland area and transporting it in the air. Note that the trails of sand and silt arise from a series of local point sources, which suggests that a series of local dust devils lifted the particles into the air. (C) A sandstorm approaching a street intersection in Phoenix, Arizona. Such storms darken the sky and partially fill automobiles with sand or silt if the windows have been left open during the storm.

windward side of the dune and down the leeward side (also called the dune's avalanche face or slip face (Fig. 3.36). Processes by which eolian sediments are transported and deposited as laminae on the leeward side of dunes include: grainflow (grains avalanche down the steep leeward face of a migrating dune), grainfall (grains fall out of suspension in the dune's upper leeward slope), and wind-ripple migration (wind blows sand on the dune's leeward face, but transverse to the direction in which the dune is migrating) (Hunter, 1977).

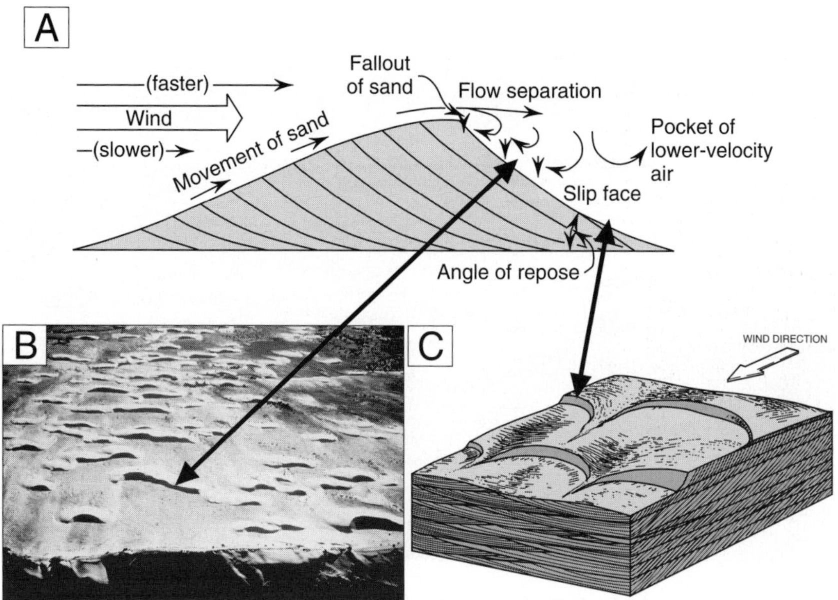

Fig. 7.5. Illustration of how sand dunes migrate. (A) The wind is blowing from left to right. Individual sand grains move up the windward face of the dune and over the crest of the dune to the leeward side or slip face. Cross-beds form internally within the dune as a result of the downwind and lateral movement of the sand. In side view, it is possible to determine the direction of wind transport by noting the direction of dip (downward) of the cross-beds. (B) The leeward side of the dune is the downwind side. (C) In an orientation perpendicular to the wind direction, the internal sedimentary structures may be more lenticular than they are in the direction parallel to the wind. After Weber (1987). (Reprinted with permission of Society for Sedimentary Geology, SEPM.)

Thus, large-scale cross-bedding is a predominant feature of eolian sandstones (Fig. 3.36, Figs. 7.5 and 7.6). Depending on the variability of wind direction and the orientation of the sandstone with respect to the observer, cross-beds may be inclined in a unidirectional manner or they may dip at varying angles (Figs. 7.6 and 7.7). Also, a typical dune avalanche face may become less steeply inclined and may be concave-upward from the crest to the base of the dune, so that cross-bed angles decrease downward along the dune's leeward face. Walker (1980) has provided the following statistics on nine Jurassic–Permian eolian sandstones of the western United States:

Range of maximum thickness (m)	40 to 70
Range of thickness of cross-bed sets (m)	15 to more than 30
Range of dip of cross-beds (°)	15 to 35

Eolian deposits are not the only strata that exhibit cross-beds. Thus, to demonstrate that cross-bedded strata are eolian in origin requires additional ev-

Fig. 7.6. Eolian cross-bed sets along the side of Canyon DeChelley in Arizona. Wind direction was from left to right, as determined by the orientation of the cross-beds. Note that horizontally stratified beds separate cross-bed sets. These horizontal beds are either deflation surfaces or fluvial/playa deposits that can act as barriers to vertical communication of cross-bed sets in an analog reservoir.

idence. Other features that might occur in association with sand dunes include raindrop prints, animal tracks, ripple marks on the avalanche faces, with crests parallel to the dip of the face (wind ripples), and horizontally bedded playa and fluvial deposits of intermittent streams.

Playa lakes and sabkhas are both features that are formed by intermittently occurring collections of shallow, usually saline water trapped on the floor of a desert or semi-arid region (playa lake) or in a supratidal coastal environment (sabkha). In both situations, water collects and then evaporates, leaving evaporite-saline minerals, tidal-flood, and eolian deposits. Water also seeps beneath the surface to form a shallow water table. These features develop during relatively wet climatic spells. During dry spells, dunes may migrate over the playa deposits. Also, wind deflation may expose sediment above the water table, forming a horizontal deflation surface (Friedman and Sanders, 1978).

Such climatic cycles have been related to eustatic glacial–interglacial periods, which created eolian sandstone deposits such as the Pennsylvanian-Permian Tensleep Sandstone (discussed in Section 7.3.5). According to Carr-Crabaugh et al. (1996), each parasequence (conformable succession of genetically related beds or bed sets) is bounded at the base by an unconformity, upon which sits eolian sandstone overlain by reworked eolian sandstone, which

Fig. 7.7. (A) Eolian sandstone at a quarry for building stone. Note the parallel nature of the slipface cross-bed laminae. (B) Eolian sandstone with somewhat more diverse cross-bed angles than in (A), indicating slightly more variable wind directions.

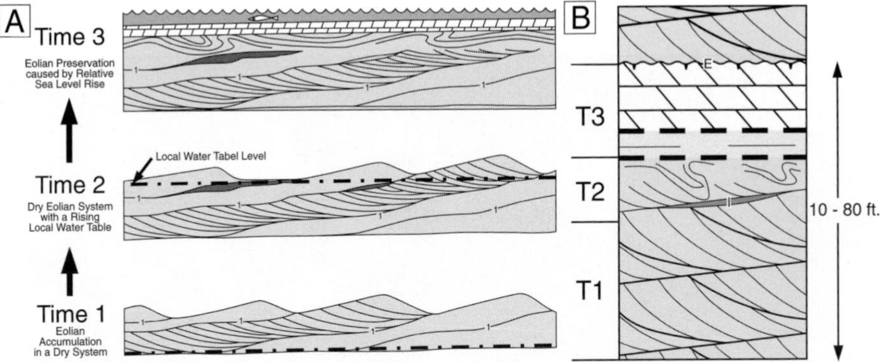

Fig. 7.8. (A) Schematic diagrams illustrating the accumulation, migration, and preservation of the Tensleep Sandstone parallel to the direction of wind transport. During Time 1, dunes accumulate and migrate (from left to right) on a dry surface. The water table is well below the ground surface. First-order bounding surfaces are highlighted. During Time 2, sea-level rise drives the water table upward and causes it to flood the interdune area. During Time 3, continued rise in sea level floods the dunes, reworking their tops. Dunes are preserved, especially if sea level continues to rise, as marine sediments are deposited atop them. (B) Typical vertical sequence resulting from one sea-level-rise cycle. The erosion surface capping the marine strata are a result of erosion as the next cycle of falling-stage begins. After Carr-Crabaugh et al. (1996). (Reprinted with permission of GCSSEPM.)

itself is overlain by a marine sandy dolomite or dolomitic sandstone. Eolian dunes are deposited during glacial lowstands of sea level (Fig. 7.8A). Later, at the onset of an interglacial period, sea level rises, the dunes become submerged, and the dune tops avalanche down the steep dip slope. With continued marine incursion, the dunes eventually are capped by marine strata and thus are preserved (Fig. 7.8B).

In contrast, Ciftci et al. (2004) identify the same facies, but instead place the base of the parasequence at the base of the dolomites and dolomitic sandstones as a flooding surface. In either case, horizontally bedded strata may separate eolian cross-bed sets of sand dunes (Fig. 7.6). In reservoirs, these horizontal beds can act as vertical barriers to communication between individual cross-bed sets (reservoir facies), particularly if the horizontal beds contain fluvial muds or evaporitic playa deposits.

Also, variations in grain size and thickness of strata deposited by grainflow, grainfall, and wind-ripple processes result in variations in the degree of cementation during burial, and ultimately result in fine-scale variations in permeability (Fig. 5.5). Thus, a truly realistic reservoir model for eolian sandstones requires very detailed permeability measurements taken with a minipermeameter, rather than more widely spaced core-plug measurements (Fig. 5.7). Prosser and Maskall (1993) have determined that thin grainflow laminae and beds (less than 1 to 6 cm thick) in Leman Sandstones from Auk field, in the North Sea, have permeabilities in the range of 13–960 md (based on minipermeameter readings), whereas millimeter- to centimeter-thick wind-ripple laminae have permeabilities in the range of 0.75–489 md. The authors claim that core-plug measurements at the standard sampling interval of 30 cm (1 ft) cannot accurately depict a permeability profile. Rather, measurements should be made with a minipermeameter at an interval spacing of one measurement for every lamina that is in a profile perpendicular to sedimentary dip and one measurement every 15–30 cm (0.5–1 ft) for a profile that is parallel to the long axis of the laminae.

7.3 Reservoir examples

Eolian reservoirs typically are complex. Complexities arise from their fine-scale cross-stratification at various angles, the occurrence of related fluvial, playa, and sometimes marine facies, and the existence of differential diagenesis, all of which result in a high degree of permeability anisotropy. Examples from reservoirs (and related outcrops) are provided next to illustrate the complexities and the hydrocarbon-production issues associated with eolian reservoirs.

7.3.1 Leman Sandstone gas reservoirs, North Sea

Several fields in the southern North Sea produce gas from the Leman Sandstone of the Permian Rotliegend Group (Weber, 1987; Abbots, 1991). This sandstone was deposited along an extensive intracratonic, semi-arid basin. Fields typically occur as faulted, northwest–southeast-elongate structures (Fig. 7.9). The Leman Sandstone is overlain by the Zechstein evaporites, which comprise an excellent topseal to the reservoir sandstone (Fig. 7.10). Gas is thought to be derived from underlying Carboniferous coal measures. The most common sedimentary facies within the Leman Sandstone are cross-bedded eolian sandstones, fluvial sandstones, sabkha and/or playa mudstones, and reworked eolian sandstones. Eolian sandstones comprise the reservoir and fluvial and sabkha/playa deposits are nonreservoir facies. Reworked sandstones exhibit much lower porosity and permeability than do the underlying dune sandstones and are not important reservoir intervals. Dipmeter logs are particularly useful for differentiating the facies in wellbores (Fig. 7.10).

The largest of these gas fields is the Leman field, which is considered a giant field with an EUR of 11.5 TCF of gas. The Leman Sandstone in this field is 180–270 m (540–810 ft) thick (Weber, 1987). Cross-bed sets average 4.5 m (13.5 ft) in thickness. In cores, the cross-bed sets represent the leeward or avalanche slopes of transverse dunes with remarkably uniform dips (Fig. 7.7). Well productivity indicates good lateral continuity and connectivity between

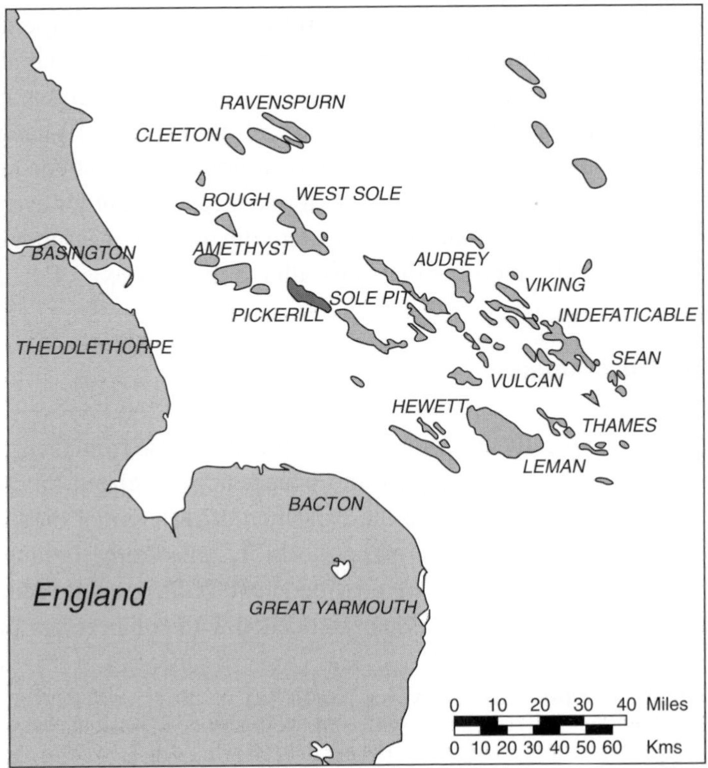

Fig. 7.9. Eolian gas fields in the southern North Sea. The Leman field, Pickerill field, and Rough field are discussed in the text. The producing sandstone is called the Rotliegend Sandstone, which is Permian in age. The source of the gas is underlying coals from Pennsylvanian-age fluvial–deltaic deposits. Overlying the Rotliegend Sandstone is the Permian Zechstein Salt, which forms a top seal to the reservoirs. After Abbots (1991). (Reprinted with permission of Geological Society of London.)

wells. Average length of cross-bed sets in analog outcrops of the DeChelly Sandstone, Arizona (Fig. 7.6) is 900 m (2700 ft) (Weber, 1987). This length is longer than well spacing in Leman field, thereby explaining the good well-to-well connectivity.

A common problem in Leman Sandstone reservoirs is the complexity of permeability within facies. For example, grainfall, grainflow, and wind-ripple laminae and beds exhibit different grain sizes. In addition, interdune sands tend to be horizontally laminated and to have a bimodal grain-size distribution as a result of wind deflation. Such variations in grain size translate into differential diagenetic cementation within the laminae (Fig. 5.5), and ultimately, to lamina-scale variations in permeability, as mentioned above.

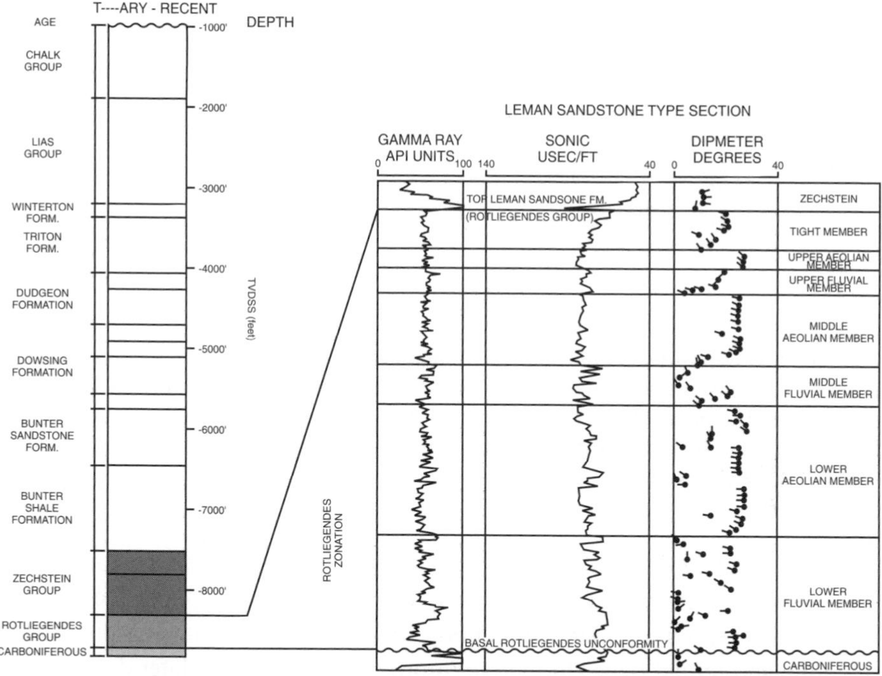

Fig. 7.10. Carboniferous–Cretaceous stratigraphy in the North Sea. The Leman Sandstone forms the reservoir in many gas fields in the southern North Sea (Fig. 7.9). Although there is not much variation in the character of the gamma-ray logs, dipmeter logs allow differentiation of eolian and fluvial facies. After Abbots (1991). (Reprinted with permission of Geological Society of London.)

7.3.2 Rough gas field, North Sea

Another field, the Rough gas field, shows other characteristics that are typical of Leman Sandstone reservoirs (Ellis, 1993). This field is composed of three units. The lowest unit comprises eolian dune and interdune sandstones interbedded with fluvial sheetflood conglomerates and reworked eolian sandstones. The unit is 7–12 m (20–35 ft) thick and fines upward from conglomerates at the base to eolian and sheetflood sands, and then to fluvial sandstones and conglomerates at the top. The middle unit is a 15-m (45-ft)-thick, low-permeability unit consisting of fluvial conglomerates and sandstones with minor eolian deposits. The uppermost unit consists of 8–15 m (25–45 ft) of eolian and interdune sandstones interbedded with fluvial sheetflood sandstones (mainly toward the base). Secondary dolomite cement is prominent in the upper part of this zone. In all three units, porosity and permeability are highest in the eolian facies and lowest in fluvial facies. This is illustrated in Fig. 7.11, which shows that areas of eolian deposition within the field correspond with high-permeability areas, and

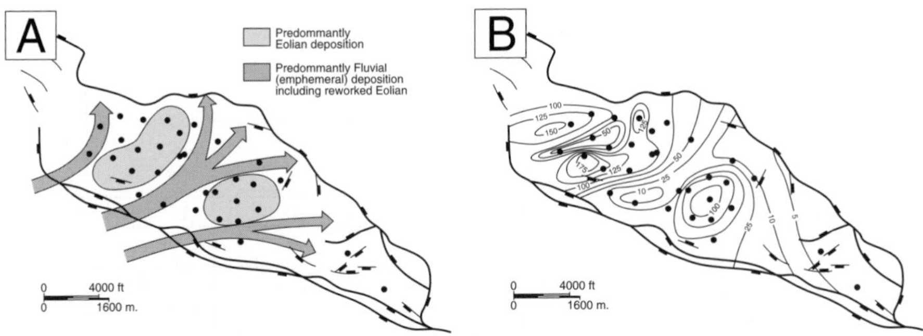

Fig. 7.11. (A) Depositional model for the Rough field, North Sea. Eolian deposits are separated by fluvial (wadi) deposits that were derived from the west and southwest. (B) Map of average well-test permeability distribution within the Rough field. Note the correspondence between areas of high permeability (as high as 100 md) and dominance of eolian deposits. After Ellis (1993). (Reprinted with permission of Geological Society of London.)

areas of fluvial deposition correspond with low-permeability areas. For that reason, wellbore flow-profile studies have shown close correspondence between injectivity/productivity and sedimentary facies. Also, the mapped distribution of eolian and fluvial facies has led to improved placement of later development wells for optimal productivity.

7.3.3 Pickerill field, North Sea

Pickerill field illustrates the effect of structure on yet another Leman Sandstone reservoir in the North Sea (Fig. 7.12). Pressure data for the discovery and first delineation wells indicate that the field is subdivided into at least two distinct compartments. The northwest-trending fault zone, which separates wells 1 and 2 from wells 3–5 and over which well 6 is positioned, is the major fault zone. Core from wells outside this zone exhibit the typical Leman Sandstone cross-beds, whereas core from well 6 is highly faulted and fractured and has infills of cement (Fig. 7.13).

In Pickerill, conventional well logs do not readily differentiate eolian from fluvial sandstones (Fig. 7.13A), although dipmeter logs show the same patterns as those illustrated in Fig. 7.10. The higher sonic-velocity and density values near the tops of reservoir sandstones are a result of the presence of high-density evaporite cement that was precipitated within sandstone pore spaces when the transgressing Zechstein sea submerged the dunes.

Pickerill field was discussed in Chapter 5 as an example of seismic detection of porous zones (Figs. 4.44–4.47). The additional detail provided by 3D seismic data not only shows the highly faulted nature of the field, but also the compartmentalized nature of its porous zones.

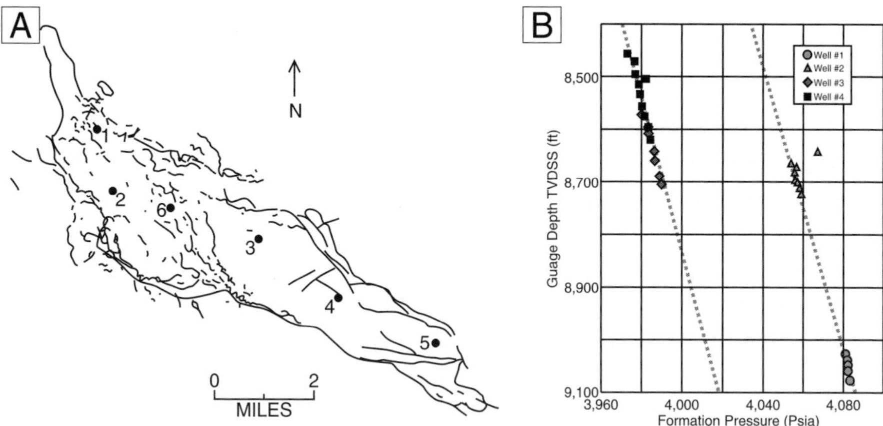

Fig. 7.12. (A) Outline of Pickerill field, southern North Sea. The wavy lines are faults detected from 2D seismic data. A 3D seismic amplitude map is shown in Fig. 5.47. (B) Formation pressure trends for wells 1–4 in (A). Note the two trends, one for the northwest part of the field (wells 1, 2) and the other for the central-southeast part of the field (wells 3, 4). Well 6 occurs over a highly faulted zone in the field.

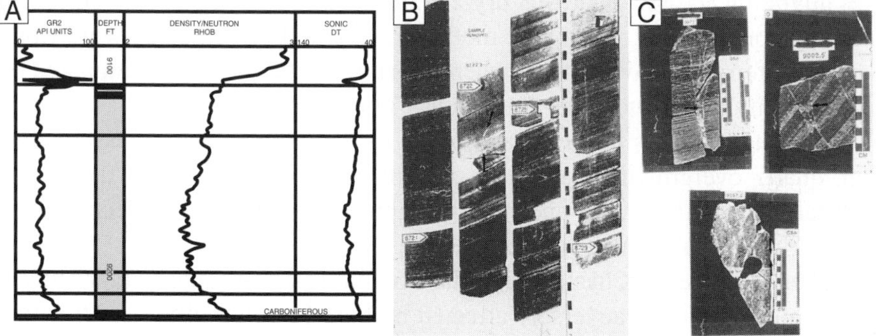

Fig. 7.13. (A) Gamma-ray, density-neutron, and sonic log tracks for a well in Pickerill field, North Sea. The gamma-ray log is nondescript. The higher density and lower transit time of the upper part of the eolian sand is a result of concentration of evaporite cement near the top of the sandstone. (B) Typical eolian cross-bedding in core from one of the other wells in the field. (C) Pieces of core from well 6 (Fig. 7.12) showing a number of cement-filled faults and fractures.

7.3.4 Painter Reservoir field, Wyoming

Painter Reservoir field is one of several fields that produce oil from the Jurassic Nugget Sandstone in southwestern Wyoming and northeastern Utah, USA. (Tillman, 1989). The field is an asymmetric anticline located near the edge of the Absaroka Thrust plate. Original oil in place has been estimated at 138 MMBO, with more than 30 MMBO having been produced during the time interval

1977–1989 (more recent figures are not available). The Nugget Sandstone was deposited primarily by eolian processes in a coastal to inland sand sea that extended from southern Wyoming to southeastern and northern Utah.

The Nugget Sandstone in Painters Reservoir field is very fine- to coarse-grained, moderately to well-sorted sand that was deposited in dune and interdune environments. Individual dune beds range from being laminated to thin-bedded and dip as much as 25°. Thicker strata (greater than 1 cm or 0.4 in thick) are grainflow deposits, and laminae less than 1 cm thick are wind-ripple laminae. Dunes (cross-bed sets) range in thickness from less than 0.3 m (1 ft) to almost 15 m (50 ft). Some dune intervals in cored wells are composed entirely of grainflow laminae and thin beds, whereas other intervals are composed of interstratified grainflow and wind-ripple laminae. The toesets of the dunes (the toeset is the forward part of a dune's advancing frontal slope) display upward-increasing dip and are concave-upward.

Interdune deposits or sand-sheet facies are composed of beds ranging in thickness from less than 0.3 m (1 ft) to greater than 10 m (30 ft). Interdune facies are subdivided into two subfacies. The "dry" subfacies is moderately sorted, horizontally laminated to ripple-laminated sandstone that formed by wind ripples migrating across a relatively dry, flat interdune area. The "damp" interdune facies is composed of burrowed, mottled, wavy and/or contorted wind-ripple laminae inferred to have formed when moisture in the sediment stimulated organic activity and reduced the competency of the sand.

Diagenetic features in these sandstones include compaction, pressure solution, quartz overgrowths, carbonate cement, dissolution molds of quartz and feldspar, and grain-coating and pore-bridging illite (Chapter 3). Tectonic features include both open and closed fractures.

Although diagenesis has reduced the original reservoir quality (particularly permeability, which is adversely affected by pore-bridging illite) and fractures and structure exert some influence, production characteristics are controlled mainly by depositional factors. Porosity and permeability values vary with facies (Table 7.1). Dune facies have better reservoir quality than interdune facies. The ratio $K_v : K_h$ ranges from 3:1 to 7:1 among the four facies, because of their well-laminated nature. As is the case in most eolian reservoirs, a strong directional permeability occurs in the dune facies at Painter Reservoir field, as determined from dipmeter logs (after removal of structural dip; Chapter 2). The direction of maximum permeability parallels the axis of the dune crests (the elongation axis of the dunes' lee facies), which is northwest to southeast (Fig. 7.5C). Directions of reduced permeability are a result of the laminated nature of the sandstone, which results in lower $K_v : K_h$ values across laminae than along their lengths (Fig. 7.5C). This permeability anisotropy was substantiated by production history during a period of pressurization of the gas cap to elevate field pressure.

Table 7.1 Relationship between sedimentary facies and reservoir quality in Painter Reservoir field. After Tillman (1989)

Facies	Porosity (%) average	Horizontal permeability (md)		Vertical permeability (md)	
		Average	Range	Average	Range
Dune					
Avalanche	14.5	22.5	0.2–1450	7.5	0.19–631
Mixed	12	6.7	0.04–363	1.6	0.06–275
Interdune					
Dry	9.9	1.8	0.06–120	0.36	0.04–30
Damp	9	0.49	0.04–10	0.22	0.04–10

7.3.5 Tensleep Sandstone, Wyoming, USA

The Tensleep Sandstone provides an excellent example of applying outcrop studies to improve subsurface reservoir characterization.

7.3.5.1 Location and outcrop characteristics

The middle Pennsylvanian-Lower-Permian Tensleep Sandstone is composed of eolian and marine strata that were deposited in a coastal-plain setting that extended throughout central and north-central Wyoming (Ciftci et al., 2004). Age-equivalent eolian deposits extend much farther, into other parts of Wyoming and parts of Colorado and Utah. These eolian sandstones comprise one of the most important hydrocarbon-producing formations in the US Rocky Mountain region. Permeability anisotropy, discussed below, sometimes results in oil recoveries as low as 15%.

The Tensleep is divided into lower and upper intervals across the Bighorn Basin. The upper interval contains the greatest proportion of eolian reservoir facies. Grainflow and wind-ripple laminations dominate the eolian sandstones. These facies are of differing thicknesses, continuity, grain size, packing, and sorting, which collectively give rise to very complex stratification and permeability anisotropy.

The main depositional factor that affects fluid flow in a Tensleep reservoir is the nature of bounding surfaces. The bounding surfaces form a fourfold hierarchy: marine-to-eolian surfaces (mentioned earlier in this chapter), and first-order, second-order, and third-order surfaces (Figs. 7.14 and 7.15) (Carr-Crabaugh et al., 1996; Ciftci et al., 2004). First-order bounding surfaces represent the migration and accumulation of the main dune bedforms. They generally dip less than 1° and are laterally extensive. Compartments between these surfaces are 3–20-m (9–60-ft) thick, large-scale tabular-planar, eolian cross-bedded sandstones. Lower-permeability interdune deposits sometimes occur

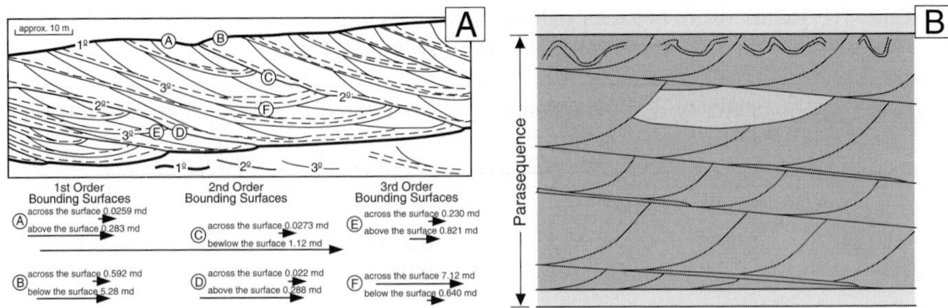

Fig. 7.14. (A) Hierarchy of first-, second-, and third-order bounding surfaces. Note the permeability of 0.0259 md at the first-order bounding surface and the permeability of 5.28 md in sandstone below the surface. The surface itself is a low-permeability barrier to fluid flow across it. Similar differences are noted for the other bounding surfaces shown in the figure. Modified from Carr-Crabaugh et al. (1996). (B) A parasequence according to Ciftci et al. (2004). This parasequence differs from that of Carr-Crabaugh (Fig. 7.8B), in that the top of their parasequence is above the marine strata, and the top of the parasequence here is at the base of the marine strata. (A – Reprinted with permission of GCSSEPM. B – Reprinted with permission of AAPG, whose permission is required for further use.)

Fig. 7.15. Outcrop of the Tensleep Sandstone showing first- and second-order bounding surfaces. Wind direction was from right to left. See Fig. 7.14 for estimates of permeability at and across the bounding surfaces. (Photo provided by N. Hurley.)

at the bounding surfaces. Second-order bounding surfaces lie between first-order surfaces and separate bundles of eolian cross-bed sets. Along a second-order bounding surface, cross-beds transition from tabular-planar near the top to troughs toward the base of a set. Third-order bounding surfaces are the reactivation surfaces that bound eolian cross-bed sets and that were caused by fluctuations in wind direction or speed. At all scales, permeability differences exist across the boundaries (Fig. 7.14).

7.3.5.2 Outcrop 3D geologic model

Ciftci et al. (2004) have eloquently documented scales of depositional compartmentalization in 3D space from a series of northeast–southwest-oriented

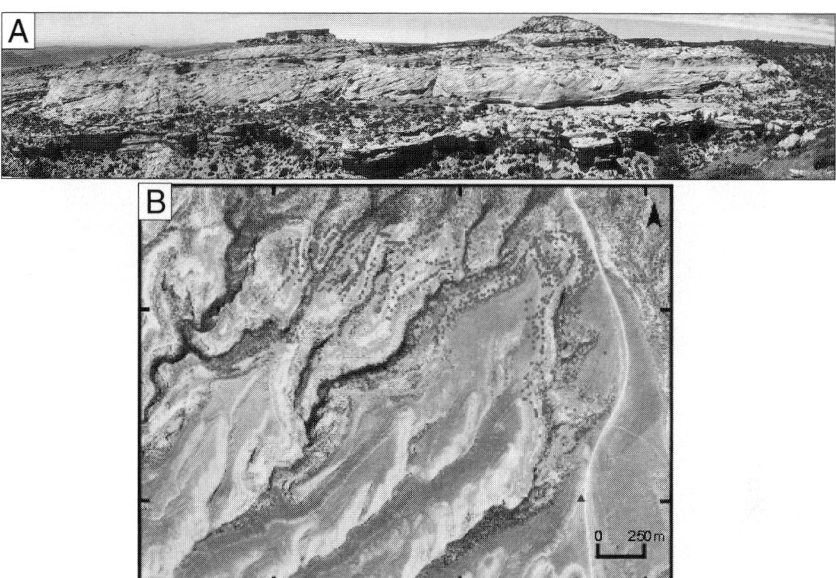

Fig. 7.16. (A) Long-distance photo of the same cliff face of the Tensleep Sandstone shown in Fig. 7.15. (B) Aerial photograph showing a number of parallel canyons. Each canyon wall exposes the Tensleep Sandstone. Several stratigraphic sections were measured in the series of canyons and positioned in 3D space using GPS. Locations of measured sections are shown by the green*/gray and red/dark-gray dots. Because the beds are flat-lying (horizontal), they could be correlated from canyon to canyon, giving a 3D geologic model of the Tensleep Sandstone across the area shown on the map. After Ciftci et al. (2004). (Reprinted with permission of AAPG, whose permission is required for further use.)

canyons that superbly expose as much as 61 m (200 ft) of the Tensleep (Fig. 7.16). The stratal features in several of these canyons were mapped and positioned (using GPS) and then features between canyons were correlated. Those data provided a 3D geologic model over an area of 4.04 km^2 (1000 acres). Outcrop subdivision of parasequence and first-order bounding surfaces identified several different zones that are correlative between canyons and that provided the basis for construction of a 3D geologic model (Fig. 7.17). Each zone represents a discrete reservoir compartment bounded by permeability baffles. Lithologies and other properties vary among the zones (Table 7.2).

Ciftci et al. (2004) geologic model was used to simulate vertical well performance. A series of templates with different well spacings designed to drain areas ranging in size from 0.04–0.65 km^2 (10–160 acres) was developed and

---

*The indicated color is for a CD which contains all of the figures in color.

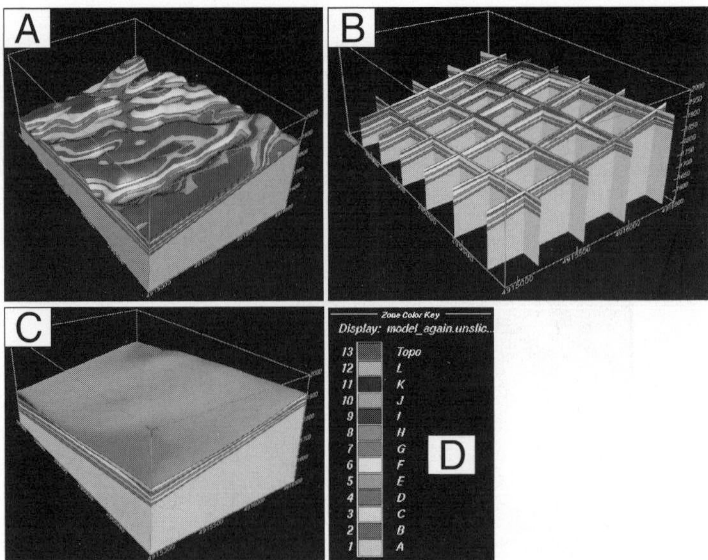

Fig. 7.17. (A) Thirteen mapped horizons in each canyon are shown by the color display. (B) Fence diagram correlation of the 13 horizons with ground topography removed. (C) Filling the 13 horizons on the 3D model. (D) Color code of the 13 measured horizons. After Ciftci et al. (2004). (Reprinted with permission of AAPG, whose permission is required for further use.)

Table 7.2 Properties of zones in the Tensleep outcrops*

Zone	Thickness (m)	Lithology	Facies	Porosity (%)	Permeability (md)
L		Mudstones, dolostones, evaporates	Evaporite flat	Topseal in reservoirs	
I–J–K	0–26	Cross-bedded ss	Eolian	Similar to B–F	
H	3–10	Sandstones and dolostones	Interdune	4.8	
G	up to 12	Cross-bedded ss, but with fine-scale bedding planes		Lower than others, due to abundance of wind-ripple laminae	
F	6–28	Cross-bedded ss, but deformed near the top	Eolian, marine, reworked near top	17.7	420
E	thins to 0	Cross-bedded ss	Eolian		
C–D	2–16	Cross-bedded ss	Eolian	15.7–16.5	167–216
B	12	Dolomitic ss	Marine	5.7–12.1	4.6–54
A	Base is beneath ground surface				

*Reservoir quality values are averages based on similar facies in a nearby Tensleep field. Boundaries between zones are confined to parasequence and first-order bounding surfaces.

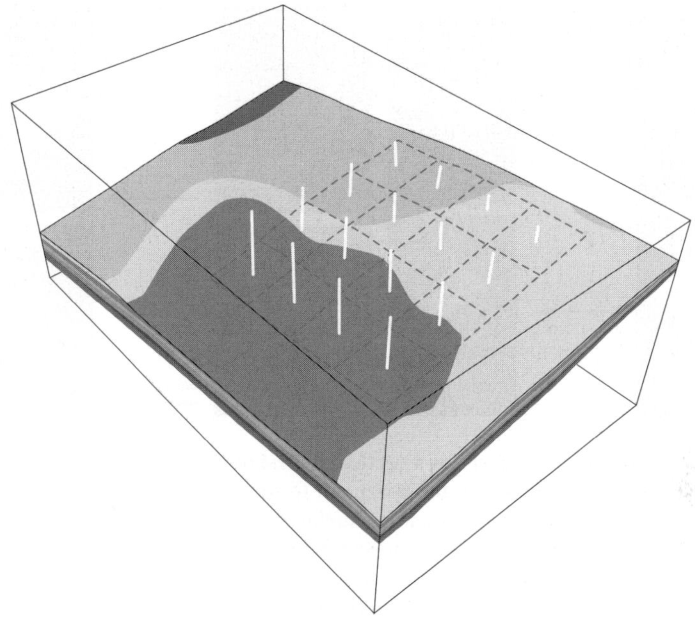

Fig. 7.18. The 0.08-km^2 (20-acre) geologic model showing the vertical well template. Dashed lines represent drainage areas of wells. After Ciftci et al. (2004). (Reprinted with permission of AAPG, whose permission is required for further use.)

simulated to evaluate the effect of bounding surfaces on fluid flow (Fig. 7.18). Under each scenario, expected or ideal drainage volume was calculated and compared with simulated drainage within the same volume of the model (Fig. 7.19). Differences between simulated and ideal drainage volumes were attributed to the negative effect of bounding surfaces.

Drainage-volume simulations from the geologic model demonstrated that vertical wells are efficient enough to effectively drain eolian reservoir intervals between first-order bounding surfaces (Fig. 7.20). Even the 0.65-km^2 (160-acre) case received almost 95% of the ideal drainage volume. This simulated result contrasts with much lower production levels in Tensleep oil fields, which typically are developed from vertical wells. Thus, it would appear that second-order bounding surfaces also influence well performance. This was verified by lower drainage efficiencies from vertical-well simulations that included second-order bounding surfaces as no-flow boundaries between first-order surfaces (Fig. 7.20). Even at 0.04-km^2 (10-acre) well spacing, only 42% of the ideal volume was drained. At the 0.65-km^2 (160-acre) well spacing, only 10% of the volume was recovered. According to these simulations, in order to economically drain a reservoir interval that has second-order bounding surfaces as transmissibility barriers, well spacing must be reduced to 33 m (100 ft), which is uneconomic in the Tensleep and most other eolian reservoirs. Horizontal-well

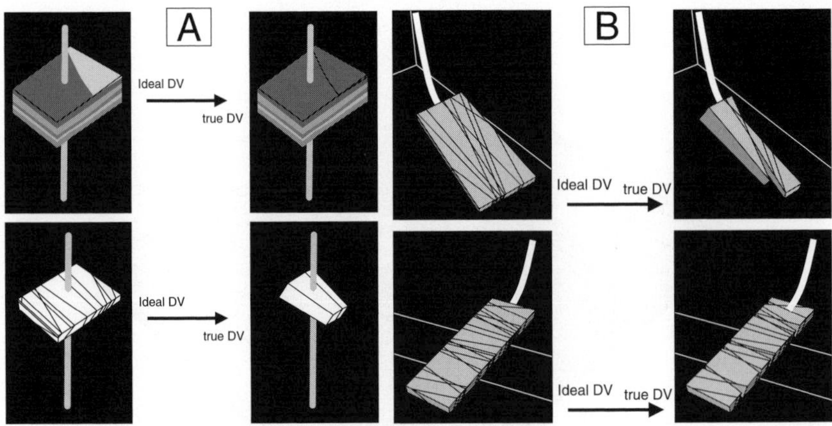

Fig. 7.19. (A) Volumetric analysis for a vertical well, showing the ratio of true drainage volume (DV, or volume contacted by the borehole) to the ideal drainage volume for a single vertical well with only first-order bounding surfaces and also with second-order bounding surfaces. (B) Volumetric analysis for a horizontal well, showing the ratio of true drainage volume (DV, or volume contacted by the borehole) to the ideal drainage volume for a single horizontal well with only first-order bounding surfaces and also with second-order bounding surfaces. After Ciftci et al. (2004).

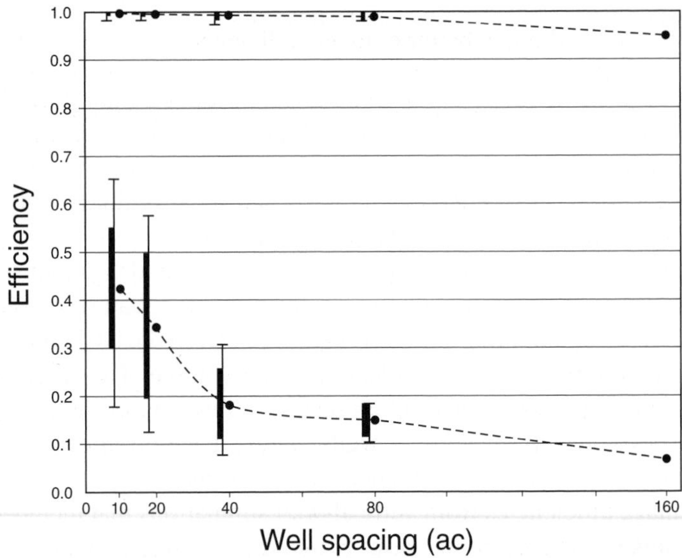

Fig. 7.20. (A) Average volumetric analysis carried out for vertical wells at different spacings and with first-order bounding surfaces (and compartments). Efficiency is the ratio of the true drainage volume to the ideal drainage volume for the well at different well spacings. (B) Average volumetric analysis carried out for vertical wells at different spacings and with second-order bounding surfaces (and compartments). After Ciftci et al. (2004). (Reprinted with permission of AAPG, whose permission is required for further use.)

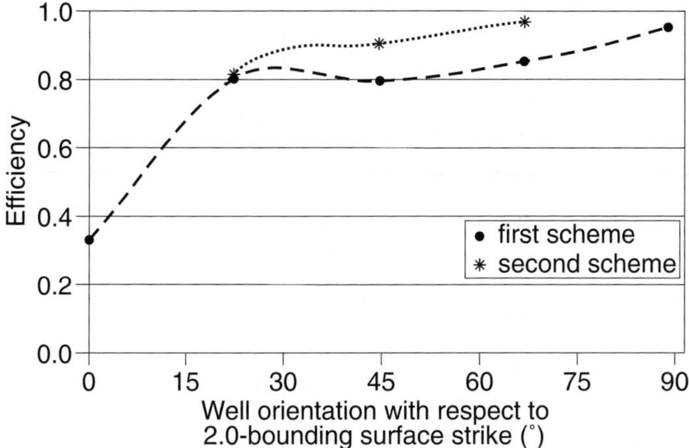

Fig. 7.21. (A) Two separate volumetric analyses for horizontal wells. The ratio of true drainage volume (DV, or volume contacted by the borehole) to the ideal drainage volume for a single horizontal well with second-order bounding surfaces and compartments defines the efficiency. The horizontal axis is the well orientation with respect to the orientation of the bounding surfaces. Note that drainage efficiency increases as the orientation of the well approaches the dip direction (90°) of the bounding surfaces. After Ciftci et al. (2004). (Reprinted with permission of AAPG, whose permission is required for further use.)

simulations provided much better drainage efficiencies (Fig. 7.21). The outcrop studies indicated that the horizontal spacing of second-order bounding surfaces in the dip direction is 33 m (100 ft). The most important factor in obtaining high drainage volumes was found to be the orientation of the horizontal well with respect to the second-order bounding surfaces. Wells oriented parallel to the dip direction of second-order bounding surfaces (or perpendicular to the long axis of dune crests; Fig. 7.5C) produced the maximum efficiency by crossing several second-order bounding surfaces. With 33-m spacing of second-order bounding surfaces, a 150-m-long horizontal reach would cross through five compartments within an eolian interval that was bounded vertically by first-order bounding surfaces and bounded horizontally by second-order bounding surfaces (Fig. 7.14). In this study, third-order bounding surfaces were not included in simulations. However they also can complicate fluid-flow paths in a reservoir.

7.3.5.3 Application to Tensleep subsurface reservoirs

Because the simulations indicated the influence of bounding surfaces on reservoir performance, it is worthwhile to develop criteria for identifying such surfaces using subsurface data. Although core is the ideal material from which to differentiate bounding surfaces, dipmeter (and borehole-image) logs also provide a means of differentiating them (Fig. 7.22) (Carr-Crabaugh et al., 1996).

In addition, the importance of bounding surfaces should be recognized when constructing well-log cross-sections. An illustrative example is provided in

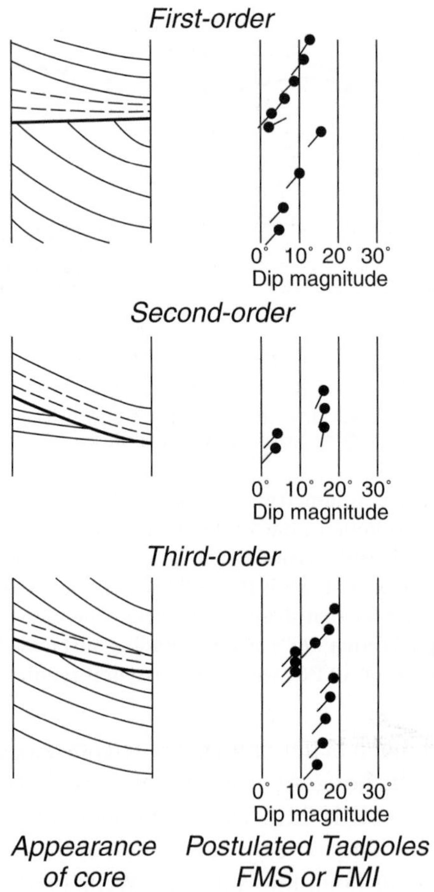

Fig. 7.22. Different core and dipmeter patterns for first-, second-, and third-order bounding surfaces in eolian strata. After Carr-Crabaugh et al. (1996). (Reprinted with permission of GCSSEPM.)

Fig. 7.23. The correlation scheme shown in the lower figure probably could not have been determined without the application of dipmeter or borehole-image logs to identify the inclined bounding surfaces.

The outcrop model simulation results were tested in Byron field, a 5.6-km (3.5-mi)-long, 2.4-km (1.5-mi)-wide producing area within a doubly plunging anticline that has 180 m (600 ft) of structural closure (Hurley et al., 2003). In 1992, a horizontal well had been drilled there that trended northeasterly, or roughly perpendicular to the axis of the eolian dune crests and parallel to the known major fracture orientation. The reason for orienting the well that way was to intersect as few of the fractures as possible, in an attempt to minimize water cut. The horizontal portion of the well was drilled on an uphill slant and stayed within a 6-m (20-ft)-thick stratigraphic interval for its final 105 m (315 ft)

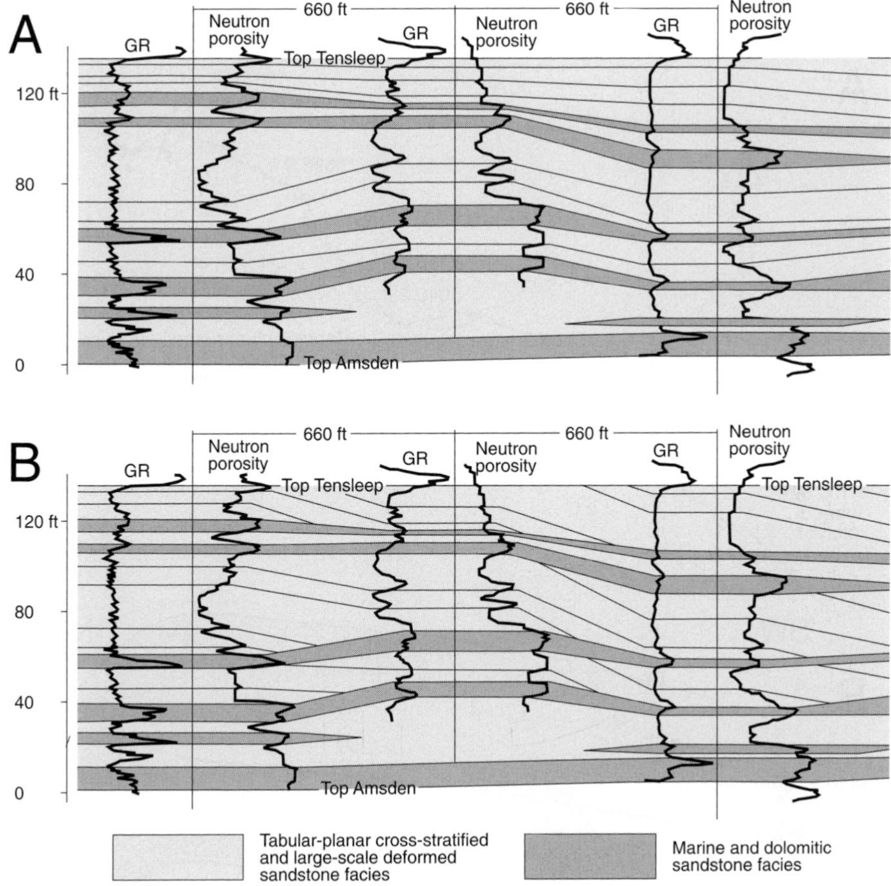

Fig. 7.23. Well-log correlations showing alternate correlations of first-order bounding surfaces in Byron field, Wyoming. (A) A conventional layer-cake correlation pattern. (B) The dips of first-order bounding surfaces, which are permeability baffles or barriers within each sandstone interval. After Hurley et al. (2003). (Reprinted with permission of AAPG, whose permission is required for further use.)

(Fig. 7.24). Formation Micro-Scanner (FMS) log interpretations from this well revealed two open fracture sets: the dominant northwest-trending set with fracture spacing of 2.3 m (7.5 ft) and a northeasterly trending set with fracture spacing of 0.9 m (3 ft). Formation Micro-Imager (FMI) measurements used to construct cumulative dip and vector plots (Chapter 2) also revealed that the wellbore had crossed at least five compartments that were bounded by second-order surfaces (Figs. 7.24 and 7.25), with an average spacing between compartments of 19 m (63 ft). On the basis of the outcrop model simulation, the horizontal well appears to have been drilled in an optimal orientation with respect to stratigraphic (second-order bounding-surface) compartmentalization.

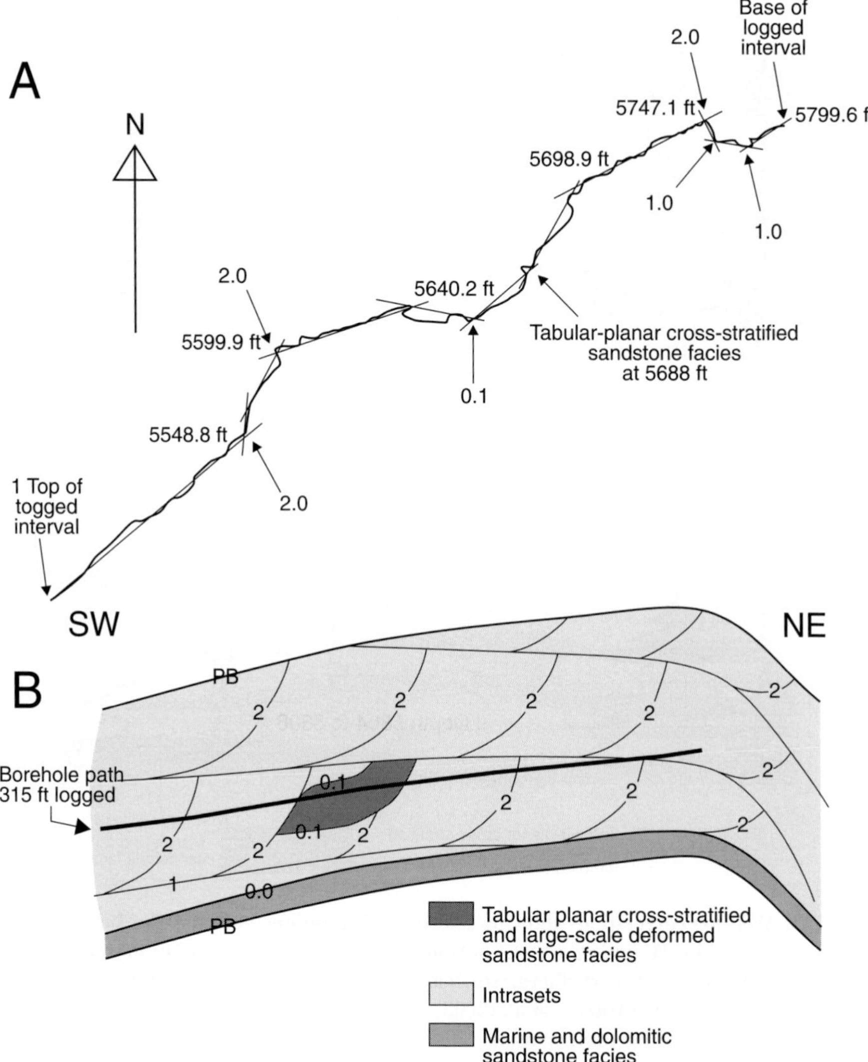

Fig. 7.24. (A) Vector plot showing dip domains interpreted from FMS bedding-plane dip directions in the Lindsay #3H horizontal well, Byron field, Wyoming. The plot is read from the deepest stratigraphic depth at the upper right to the shallowest depth at the lower left. Different straight-line trends define different second-order bounding surfaces and compartments. (B) Schematic diagram of the inferred relationship between the borehole path and the first- and second-order bounding surfaces.

In terms of fractures, the well was drilled perpendicular to the orientation of the main fracture set and crossed approximately 46 of these fractures. Undoubtedly, many of the secondary fractures also were crossed, even though that set is roughly parallel to the orientation of the wellbore.

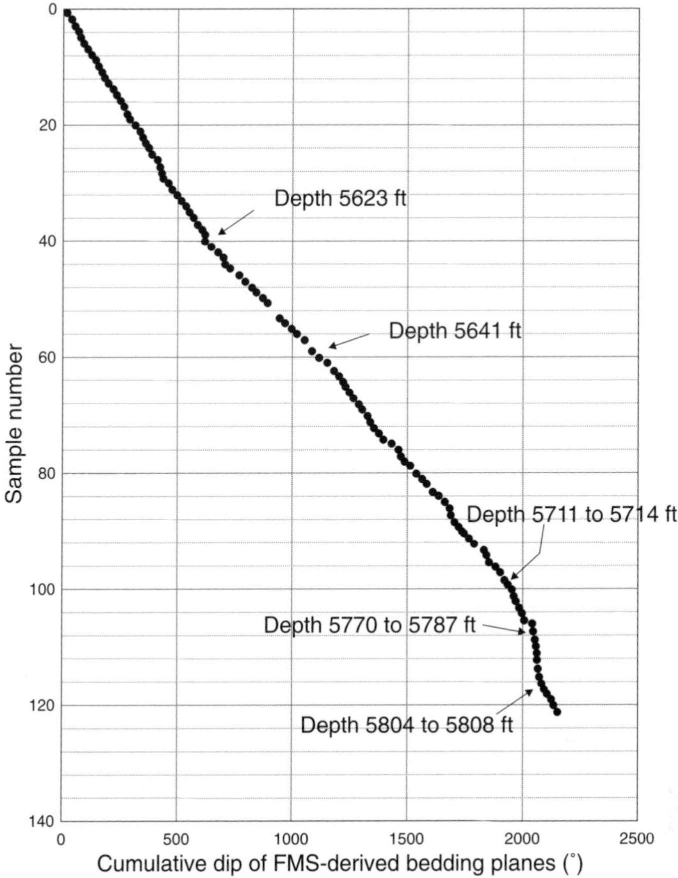

Fig. 7.25. Cumulative dip plot of FMS-derived bedding planes in the Lindsay #3H horizontal well, Byron field. Inflection points between dip domains are interpreted to be bounding surfaces. After Hurley et al. (2003). (Reprinted with permission of AAPG, whose permission is required for further use.)

7.4 Summary

The focus of this chapter has been on reservoirs, with only a secondary emphasis on description of outcrops. That is because the unique, fine-scale-stratification characteristics of eolian deposits that affect their reservoir performance have been very well documented from the reservoirs themselves. Because of the likelihood of stratigraphic compartmentalization and permeability anisotropy resulting from bounding surfaces, it is very important that eolian reservoirs be characterized in detail. In addition to the effects of bounding surfaces, variations in cementation within laminae of different grain sizes result in small-scale variations in porosity and permeability, which are difficult and expensive to measure and document. That fact further emphasizes the importance of detailed reservoir characterization.

Chapter 8

Nondeltaic, shallow marine deposits and reservoirs

8.1 Introduction

The continental shelf is defined as "that part of the continental margin that is between the shoreline and the continental slope (or, when there is no noticeable continental slope, [is at] a depth of 200 m)" (Jackson, 1997, p. 138). Globally, the average width of the continental shelf is 75 km and the average gradient is 1.7 m/km (0.1°) (Fig. 8.1). However, the shelf's width and slope vary considerably from area to area. For example, the continental shelf off the east coast of the United States is broad and relatively flat, whereas the continental shelves off the west coasts of South America and West Africa tend to be much narrower (Fig. 3.29).

Depositional processes and their resulting deposits vary significantly. Where major rivers debouch into the ocean, deposits are very different from those where major rivers are not present (Fig. 3.3). Deposition at the mouth of a major river generally leads to a deltaic shoreline and an adjacent continental shelf that are quite different from an "interdeltaic" (i.e., a nondeltaic) shoreline and shelf. This chapter discusses nondeltaic shoreline to shelf-edge deposits and reservoirs, and the next chapter discusses deltaic shoreline-shelf deposits and reservoirs.

8.2 Shallow marine processes and environments

Nondeltaic shorelines receive their sediment supply from a variety of sources. For example, sediment may be deposited temporarily at the mouths of rivers and then picked up and moved alongshore by waves and currents to form beaches (Figs. 3.39C and 8.2). Also, sediment can be picked up from the offshore seafloor and moved landward, again to form beach deposits. In general, there is a net transport of sand onshore, from the offshore seafloor to the shoreline.

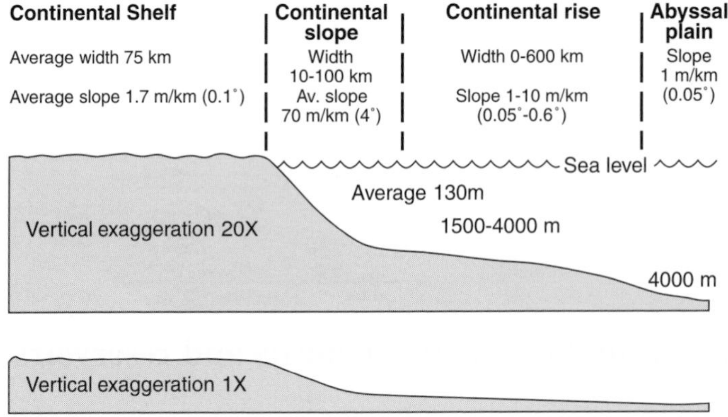

Fig. 8.1. Physiography and dimensions of the continental shelf, slope, rise, and abyssal plain. (Source of figure unknown.)

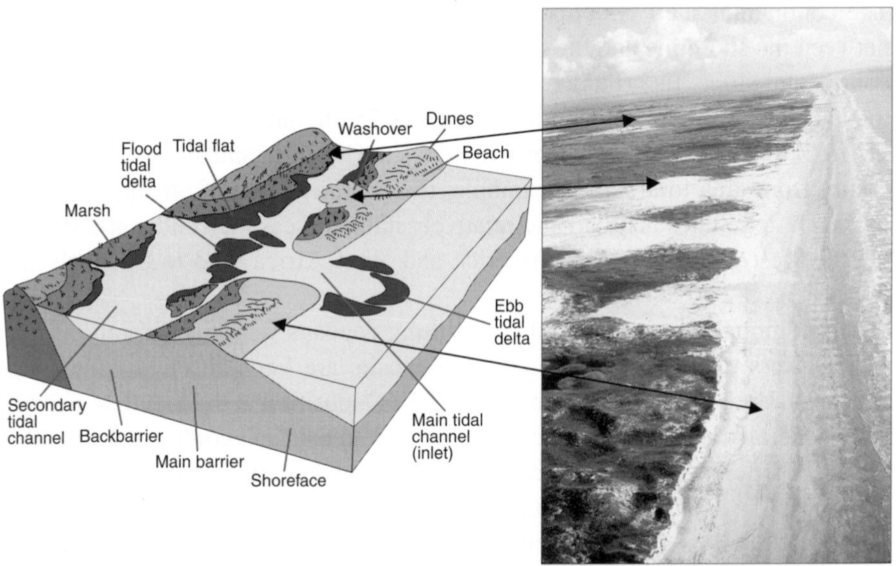

Fig. 8.2. Sandy beach on the Texas Gulf Coast, with washover fans overlying swamp and lagoonal organic-rich muds. The fans formed during a hurricane on the south Texas coast when onshore winds blew the sand inland from the beach. The beach sand and the washover sands can form excellent reservoirs when they are buried. Inset is a block diagram of a barrier island system, which is discussed in the text.

This net landward transport occurs because of the landward movement of water by wind-generated waves and nearshore currents.

Water particles tend to move in a circular fashion in the open ocean (Fig. 8.3A). However, as a wave moves landward, it begins to "feel" the seafloor

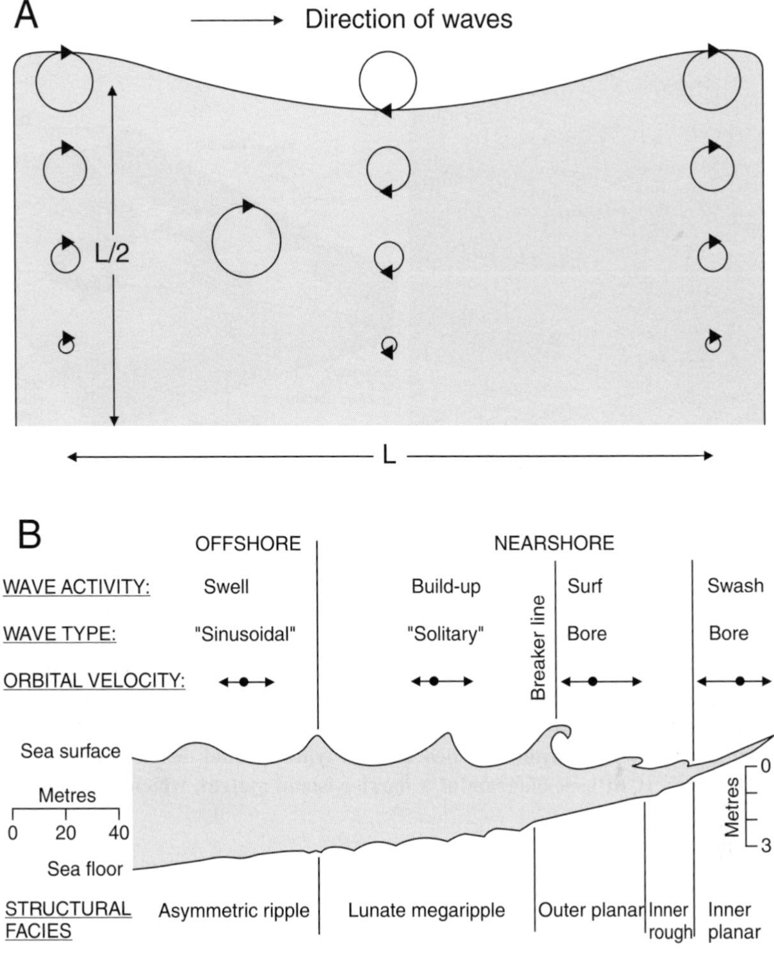

Fig. 8.3. (A) Movement of a wave involves circular or orbital movement of particles of water from the water surface downward to a depth that is half the length of the wave ($L/2$). Orbital diameters diminish progressively downward from the water surface to the depth $L/2$, beneath which there is no associated particle motion. (B) Offshore to nearshore marine environments, showing the wave motion and orbital velocities as the wave approaches the shore. When the wave reaches a water depth of $L/2$, the wave begins to "feel" the bottom and the water particles' circular motion tends toward elliptical motion. Associated frictional effects decrease the velocity of the wave near the seafloor relative to that at the water surface, until the wave becomes unstable and "breaks", creating surf and swash zones. Source of figure is unknown.

at a water depth that is approximately one-half its wavelength. As the wave continues to move landward, frictional forces convert the circular orbital pattern of its water particles to an elliptical pattern (Fig. 8.3B). Friction reduces the landward motion of the near-bottom water particles relative to that of particles closer

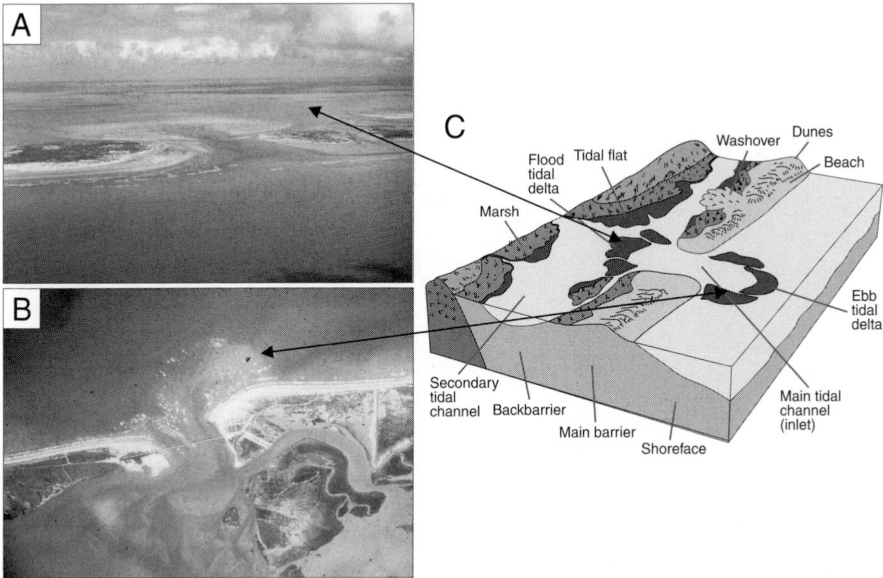

Fig. 8.4. (A) Photograph of a flood-tidal delta. Note the delta shape. The ocean is to the bottom and lagoons are toward the top of the picture. Land is also in that direction. The flood-tidal current apparently is strong enough to transport sediment through the tidal channel and into the landward lagoon. (B) Photograph showing a beach, lagoon, and ebb-tidal channel. Ebb-tidal deltas form when the outgoing tides are sufficiently strong to transport sand seaward, beyond the beach. Both types of tidal deltas can form good, but small reservoirs. (C) Block diagram of a barrier island system, which is discussed in the text.

to the sea surface, so the latter outrun the former and the wave breaks, producing surf. The breaking wave moves up the beach as swash, then back down the beach as backwash. These different modes of water-particle motion result in transport and deposition of coarser-grained sediment in a landward direction. The outcome is a variety of sedimentary structures, the characteristics of which are dependent upon water velocity and sediment grain size (Fig. 3.38).

Tides also play a significant role in the patterns of nearshore sedimentation. Small flood-tidal deltas may form on the landward sides of open tidal channels if the flood tide is significantly stronger than the ebb tide (Fig. 8.4A). Conversely, ebb-tidal deltas may form on the seaward side of tidal channels if the ebb tide is stronger than the flood tide (Fig. 8.4B). Internal sedimentary structures on a tidal flat or tidal delta often exhibit bidirectional or "herringbone" cross-beds that are indicative of alternating tidal-current movement (Fig. 3.42).

Associated with shoreface environments are the offshore or open-shelf environment (seaward) and the foreshore, swash, beach, and dune environments (landward) (Fig. 8.5). Because water depth increases progressively in an offshore direction, sediments tend to become finer-grained in that direc-

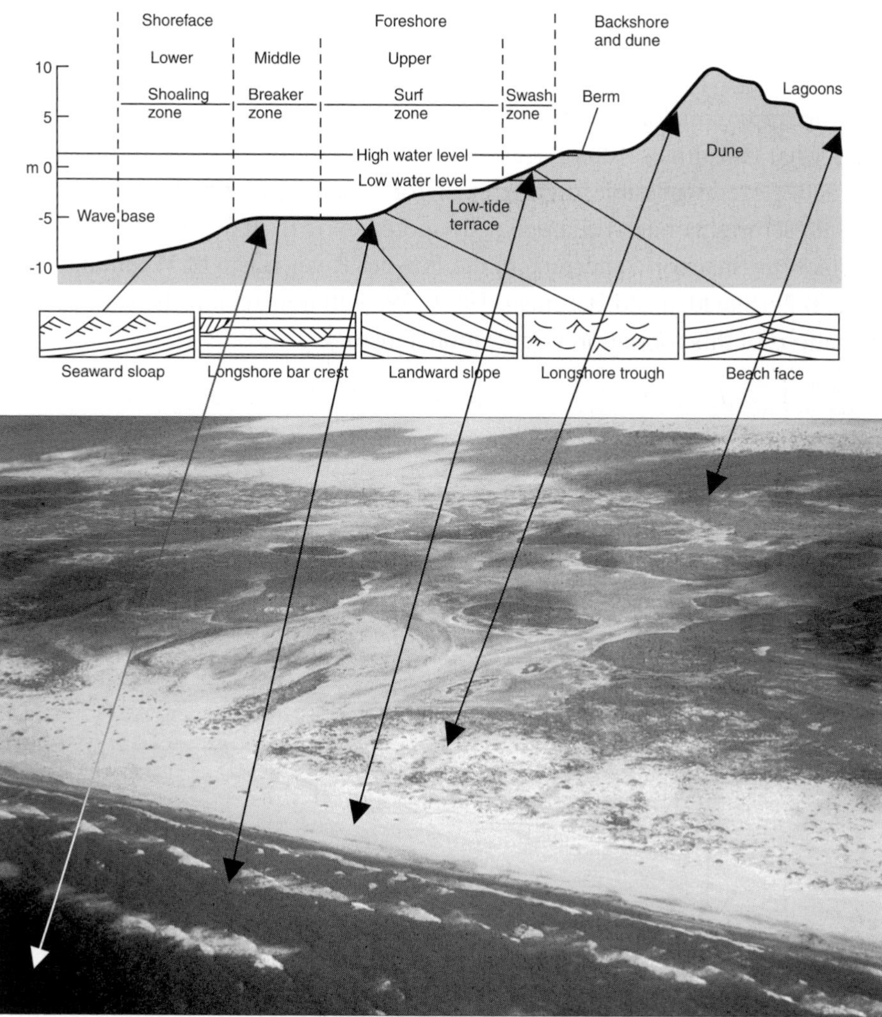

Fig. 8.5. Photograph showing a modern foreshore and backshore/lagoonal deposit. The shoreface is beneath the water level in this photo. The various sedimentary structures of sandstones that are diagnostic of the different environments are shown in the diagram. After Berg (1986).

tion (Fig. 8.5). High-energy, upper-shoreface deposits typically are composed of well-sorted sand. The sand content and grain size decrease toward the shelf edge and the mud content increases. As a result of the differing energy levels of waves and currents and the varying types of substrate, diverse organisms occupy the various parts of the shoreface to open-shelf profile. Trace fossils of the *Skolithos* and *Cruziana* ichnofacies are particularly diagnostic of certain specific parts of that profile (Fig. 3.60).

8.3 Shallow marine deposits

During approximately the past 25 years, considerable debate has taken place over what constitutes “shoreface” and “shelf” deposits and how those deposits formed. Several possible origins exist for “shallow marine” sandy deposits, and those origins have been debated vigorously (Fig. 8.6). An example is the Cretaceous Shannon Sandstone in the Powder River Basin of Wyoming, USA. (Fig. 8.7) (Snedden and Bergman, 1999). We will now review the various interpretations of this formation’s depositional history.

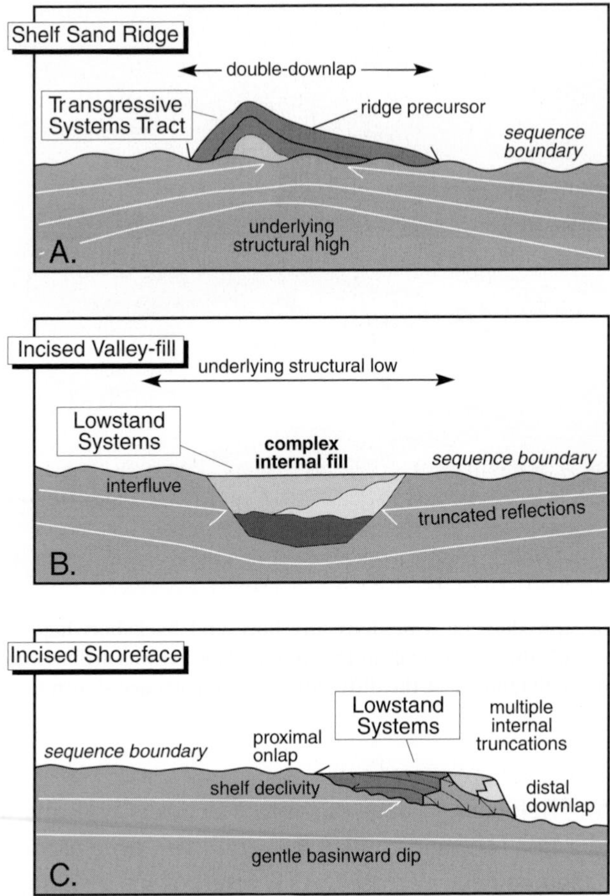

Fig. 8.6. Differing stratal patterns of the three types of shallow marine sand bodies discussed in the text: (A) shelf sand ridge, (B) marine-dominated incised valley fill, and (C) incised shoreface. After Snedden and Bergman (1999). (Reprinted with permission of Society for Sedimentary Geology, SEPM.)

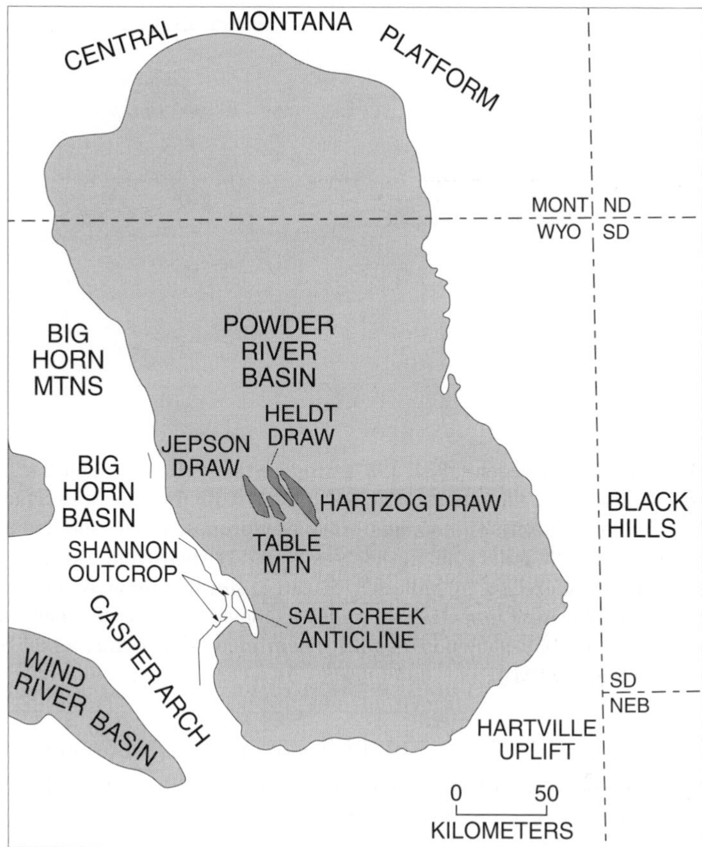

Fig. 8.7. Powder River Basin, Wyoming and Montana, showing the location of Hartzog Draw and related fields (green*/dark-gray), which produce hydrocarbons from the Cretaceous Shannon Sandstone. Note the elongate character and orientation of the various Shannon Sandstone fields. After Tillman and Martinsen (1984). (Reprinted with permission of Society for Sedimentary Geology, SEPM.)

8.3.1 Offshore bars or sand ridges

In the 1970s and 1980s, workers interpreted most linear sandstones deposited within the Cretaceous Western Interior Seaway to be shelf ridges or bars that were disconnected from the paleoshoreline and deposited some distance onto the continental shelf (Berg, 1975; Hobson et al., 1982; Tillman and Siemers, 1984; Turner and Conger, 1981; Beaumont, 1984; Slatt, 1984; Tillman and Martinsen, 1984, 1987; Gaynor and Scheihing, 1988; Gaynor and Swift, 1988) (Figs. 8.6A and 8.8). Such "shelf sandstones" often are isolated from sandy shoreline deposits by muddy shelf deposits, so the sand was thought to have

*The indicated color is for a CD which contains all of the figures in color.

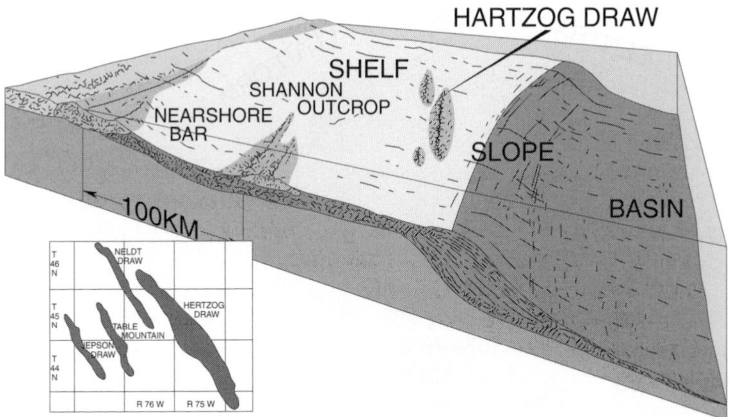

Fig. 8.8. Tillman and Martinsen (1984, 1987) proposed an "offshore shelf-bar" model for the Shannon Sandstone in the outcropping Salt Creek Anticline and subsurface Hartzog Draw field of Wyoming (inset). Tillman and Martinsen proposed that sand had been transported approximately 100 miles (160 km) across a muddy shelf from a contemporaneous north–south-trending shoreline, by obliquely oriented, storm-generated currents that deposited or reworked the sand into elongate ridges on the middle to outer shelf. Inset shows an enlargement of the fields shown in Fig. 8.7. (Reprinted with permission of Society for Sedimentary Geology, SEPM.)

been sourced from nearby deltas. This theory of origin was most popular prior to development of sequence stratigraphy concepts related to sea level fluctuations (i.e., a static sea level was invoked). However, subsequent arguments have been raised against this shelf-sandstone model. They include the facts that (1) depositional discontinuities exist that sometimes are overlooked in shelf-sandstone interpretations, (2) a satisfactory explanation is lacking for the transport of large quantities of sand across muddy paleoshelves and the reworking of this sand into ridges during static sea level, and (3) shoreface trace-fossil assemblages may be present (Bergman, 1994).

8.3.2 Shoreface parasequences and successions

Considerable research in the late 1980s and the 1990s (e.g., Leggitt et al., 1990; Van Wagoner et al., 1990; Walker and Eyles, 1991; Pattison and Walker, 1992; Bergman, 1994; Hart and Plint, 1994; and many papers in Van Wagoner and Bertram, 1995), coupled with the acceptance of sequence stratigraphy concepts, has resulted in general agreement that at least some, if not most, of the linear sandstones deposited within the Cretaceous Western Interior Seaway actually are shoreface deposits that formed during periods of fluctuating sea level (Fig. 8.6C).

A common interpretation is that the shoreface surface was incised during a stage of falling sea level, and sands were deposited in that declivity. Then, with

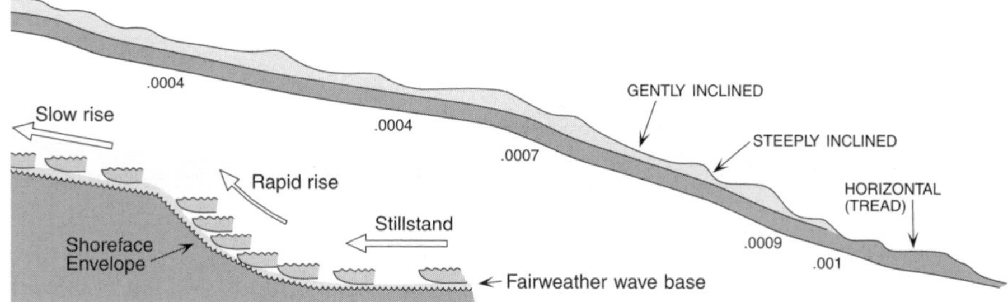

Fig. 8.9. Schematic illustration of a profile tilted to make each individual bench (tread) horizontal. Inset (bottom left) shows the development of a shoreface envelope by the lateral and upward migration of individual shorefaces. After Walker and Eyles (1991). (Reprinted with permission of Society for Sedimentary Geology, SEPM.)

a subsequent rise in sea level, the same sand was reworked into an elongate deposit parallel to the trend of the paleoshoreline (Fig. 8.6C).

According to Walker and Eyles (1991), the rate of sea level rise controls the final configuration, extent, and continuity of the shoreface sandstones. Walker and Eyles used detailed maps of key surfaces and deposits in the Cretaceous Cardium Formation, in Alberta, Canada, from which they recognized the following sequence of events: (1) uplift of the basin floor in the paleolandward direction, which caused a relative lowering of sea level and a basinward shift in the shoreline, (2) an extended period of subaerial erosion, during which time the major unconformity surface was formed, (3) subsequent downwarp of the basin floor in the paleolandward direction, causing a continued marine transgression, and (4) periods of relative sea-level stillstand during overall transgression, which generated smaller, horizontal, incised shorefaces, called "Treads", upon which shoreface sands prograded (Fig. 8.9).

During periods of rapid sea-level rise, shoreface deposits would backstep stratigraphically in the landward direction in a relatively steep manner, whereas during periods of slower rise, individual shoreface sands would prograde over a shallower-gradient slope while backstepping across the slope surface (Fig. 8.9).

The basic stratigraphic unit of a shoreface sequence is called a parasequence, which is defined as a "relatively conformable succession of genetically related beds or bedsets bounded by marine-flooding surfaces or their correlative surfaces" (Van Wagoner et al., 1990, p. 8; Fig. 8.9). The significance of parasequences within a sequence stratigraphic framework is discussed in more detail in Chapter 11.

Figure 8.10 shows a seaward-prograding parasequence, with its various facies from the foreshore to the offshore shelf environments (left to right). Corresponding gamma-ray log patterns are also shown. The progressive decrease in sandiness in the seaward direction is reflected in the shalier appearance of the gamma-ray log patterns in that direction. During progradation, shallower-water facies are deposited atop deeper-water facies (Fig. 8.11). Prograding

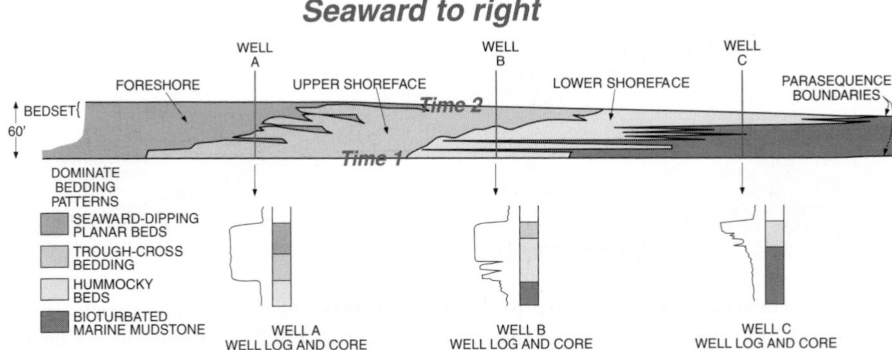

Fig. 8.10. Illustration of a parasequence. It is a conformable succession of genetically related beds or bedsets bounded by marine-flooding surfaces or their correlative surfaces. A seaward-prograding parasequence is shown, with its various facies from foreshore to upper and then to lower shoreface and finally to offshore shelf (left to right). Corresponding well-log patterns are also shown. After Van Wagoner et al. (1990). (Reprinted with permission of AAPG, whose permission is required for further use.)

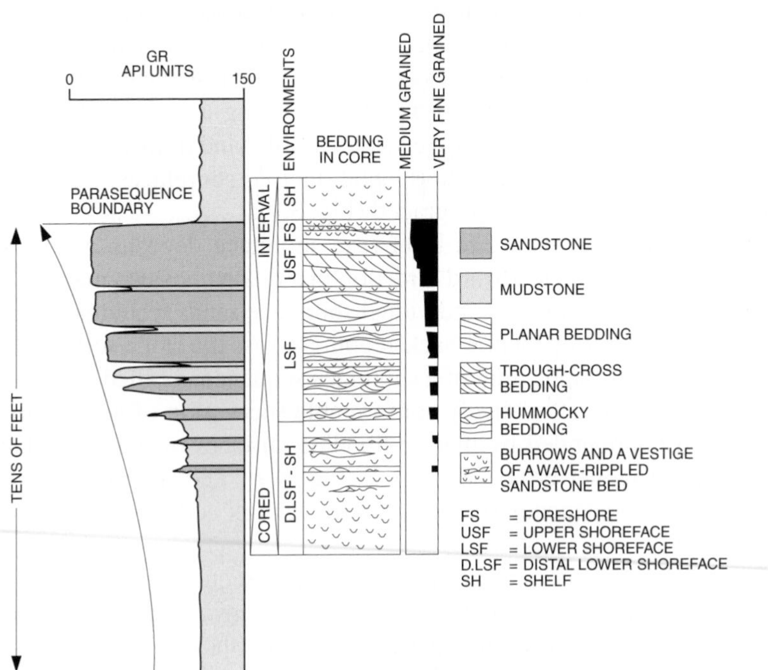

Fig. 8.11. Characteristics of a upward-coarsening parasequence deposited in a sandy beach environment. Modified from Van Wagoner et al. (1990).

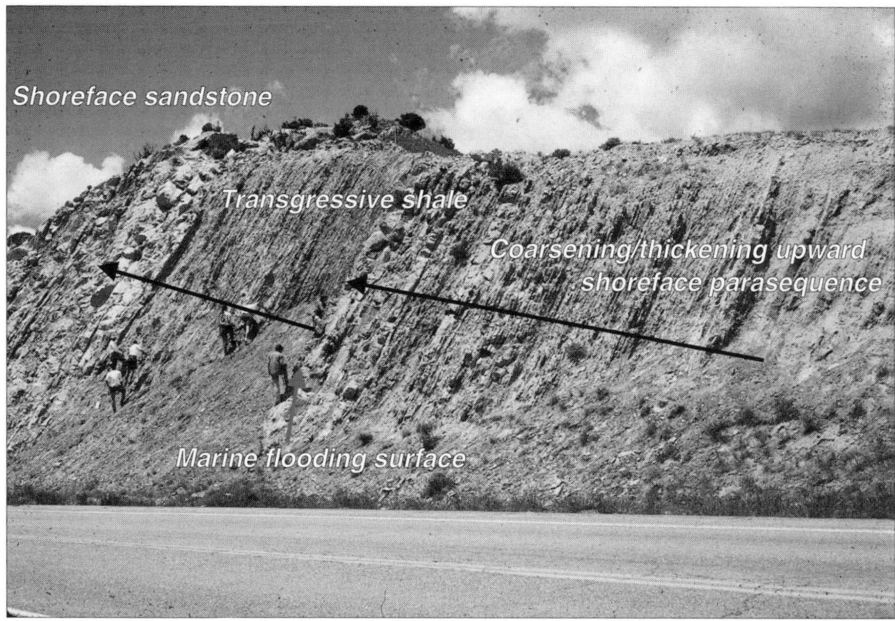

Fig. 8.12. Photograph of an outcrop of two Cretaceous parasequences, considered to be analogous to Terry Sandstone parasequences, discussed in the text. The beds dip steeply, and the stratigraphic top is toward the left. The lower sequence (toward the right of the photo; right-hand red/black arrow) consists, from the base toward the top, of mudstone, interbedded mudstone and sandstone, and sandstone, representing offshore, lower-shoreface, and upper-shoreface strata, respectively. The overlying thick, darker colored shale is a transgressive marine shale. The sandstone surface upon which this shale sits is a relatively coarse-grained, transgressive lag surface (TSE) that contains wood fragments and shark teeth. The transgressive shale is laterally continuous for a long distance and can provide an excellent topseal for an analog subsurface reservoir. Overlying the transgressive marine shale first is mudstone, then interbedded sandstone and mudstone, and finally sandstone, representing another parasequence (left-hand red/black arrow) as a result of progradation.

parasequences thus can be recognized by an upward increase in (1) grain-size and sandiness, (2) higher-energy sedimentary structures, and (3) higher-energy trace fossils. These features normally can be recognized in cores and well logs as well as in outcrops (Fig. 8.12).

Another variation of the shoreface model is the stranded lowstand shoreline associated with a forced regression (Posamentier et al., 1992). According to this model, a zone of sedimentary bypass, subaerial exposure, and possible fluvial erosion may result from an abrupt seaward migration of shallow water and shoreline facies during a relative lowering of sea level – that is, during a forced regression (Fig. 8.13). Two major features of forced-regression deposits are

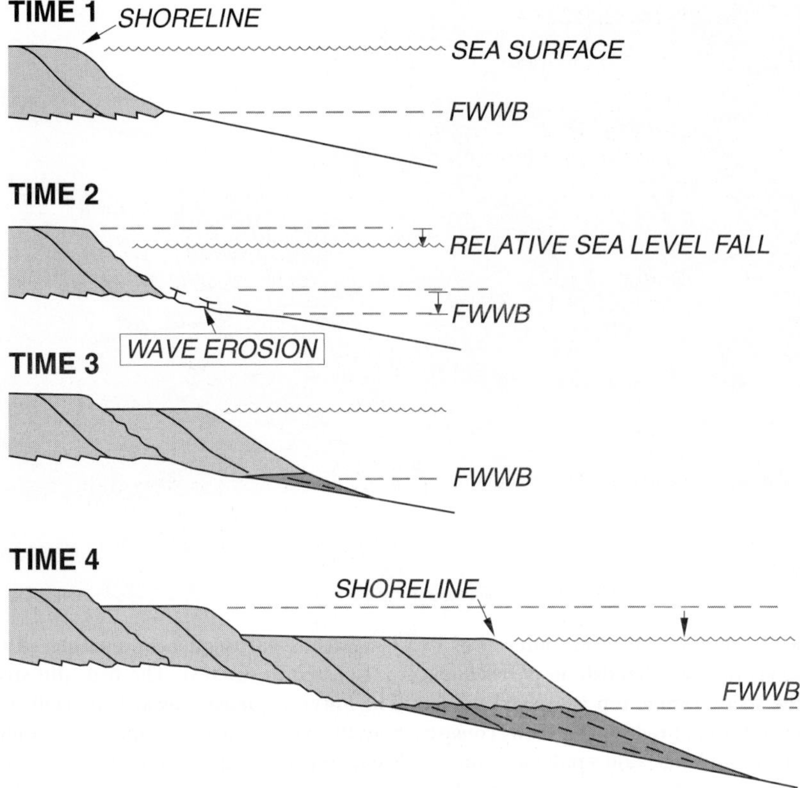

Fig. 8.13. Posamentier et al. (1992) have interpreted rapid drops in relative sea level to be "forced regressions". Such drops result in a basinward shift of sedimentary environments and resultant facies, leading to deposition of coarser-grained sediment seaward of finer-grained sediment at the same apparent stratigraphic horizon. Also, an erosional surface forms in the landward direction and grades seaward into a gradational vertical rock succession. FWWB is fair weather wave base. After Posamentier et al. (1992). (Reprinted with permission of AAPG, whose permission is required for further use.)

(1) the occurrence of coarser-grained, more proximal facies seaward of finer-grained, more distal facies (Fig. 8.13) and (2) a basal erosional surface upon which landward strata rest, grading seaward into a gradational contact (Figs. 8.13 and 8.14).

8.3.3 Marine-dominated, incised-valley-fill deposits

Incised-valley-fill deposits were discussed in the previous chapter. Such deposits also tend to be linear. Some sandstones have been interpreted to have been deposited in incised valleys that were carved during a stage of falling sea

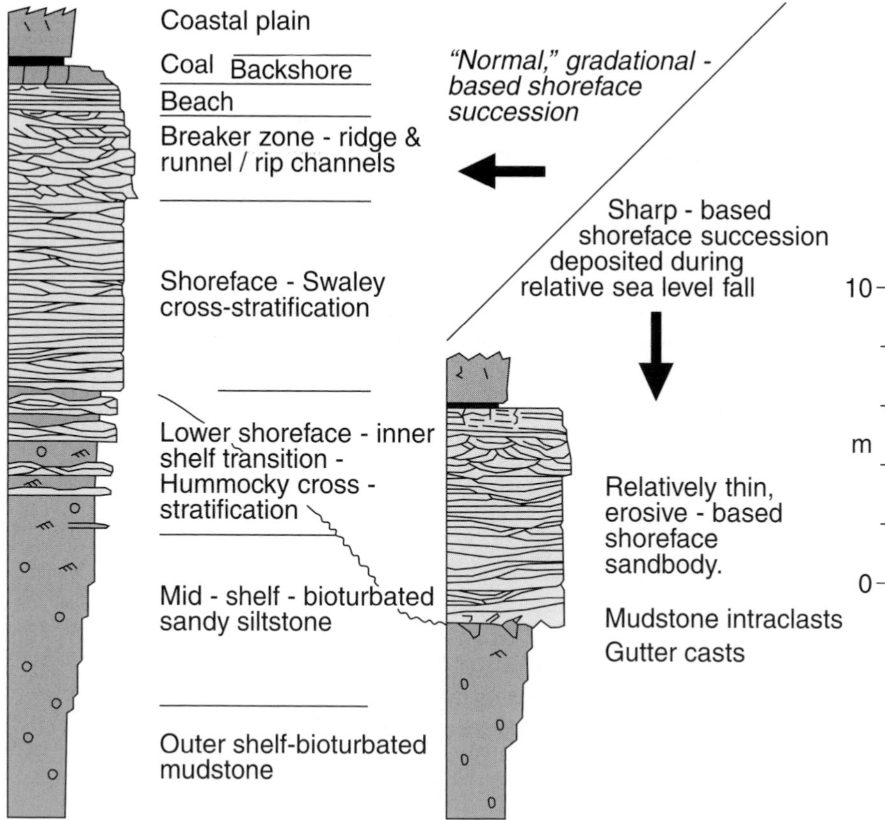

Fig. 8.14. During a relative-sea-level drop of the type shown in Fig. 8.13, the unconformity or erosional surface will be expressed as a sharp-based sandstone in the landward direction, but it may be gradational in the seaward direction. After Walker (1980). (Reprinted with permission of Geological Association of Canada.)

level. These incised valleys then filled mainly with estuarine strata during the subsequent rise in sea level (Fig. 8.6B).

8.3.4 Significance of the origin of deposits

Irrespective of their origin, nondeltaic sandstone sequences can extend laterally for long distances (Fig. 8.2), thus making them potentially excellent reservoirs in the subsurface. However, it is important to note that the internal connectivity of reservoir sandstones differs among the three proposed models and can result in different fluid-flow barriers and flow paths in reservoir analogs (Fig. 8.15). Thus, it is critical that we thoroughly evaluate the data for such a reservoir and select the most likely model for reservoir characterization and ultimate performance.

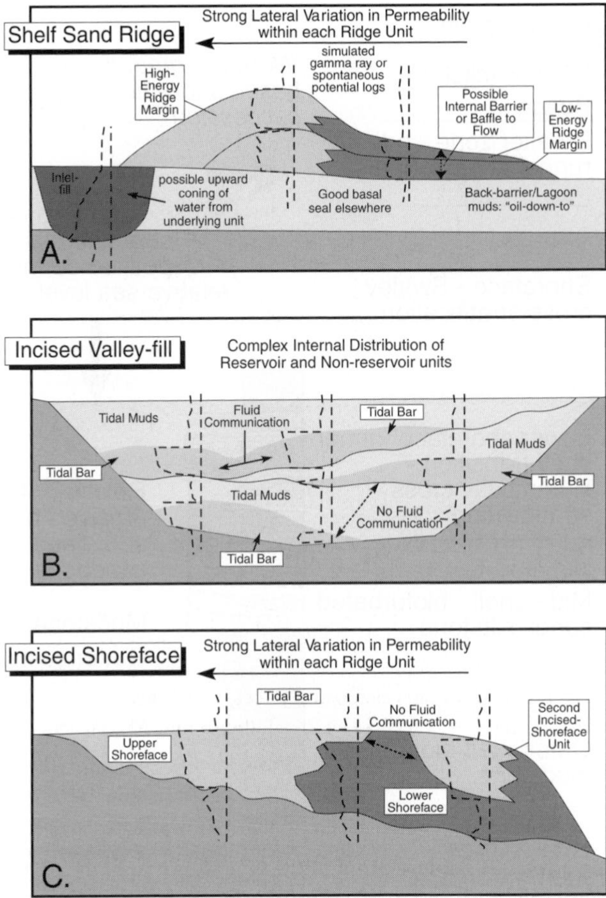

Fig. 8.15. Different log responses and production aspects of the three shallow marine sandstone types shown in Fig. 8.6. After Snedden and Bergman (1999). (Reprinted with permission of SEPM, Society for Sedimentary Geology.)

8.4 Shallow marine reservoirs

8.4.1 The puzzle of Hartzog Draw field

Hartzog Draw field (Figs. 8.7, 8.8, and 8.16) is the largest Shannon Sandstone field in the Powder River Basin. It is approximately 35 km long and as much as 5 km wide. It was discovered in 1975, and primary oil production peaked in 1977 at about 39,000 BOPD, with water injection commencing in 1981 (Sullivan et al., 1997). The field produces from depths of 2,750–3,000 m, out of a 30-m-thick Shannon Sandstone interval (net sand is approximately 20 m thick). Maximum production during waterflood was about 19,000 BOPD in 1987, and production has been on a steady decline since that time.

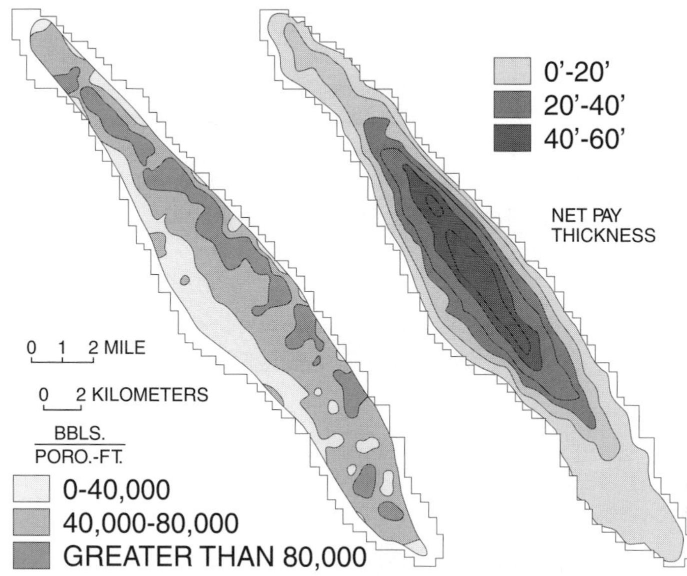

Fig. 8.16. The yellow–orange (right) image is a net-pay isopach map of the Shannon Sandstone reservoir at Hartzog Draw field. The thickest sandstone is not in the center of the field but is displaced somewhat toward the northeast (upper right). The green (left) image shows early production in the field, in barrels/porosity-feet, which also was greatest along the paleoseaward (northeast) flank of the field. After Tillman and Martinsen (1987). (Reprinted with permission of Society for Sedimentary Geology, SEPM.)

8.4.1.1 Hartzog Draw field as an offshore sand ridge (shelf bar) (1984–1987)

Tillman and Martinsen (1984, 1987) proposed an "offshore shelf-bar" model for the Shannon Sandstone in the outcropping Salt Creek Anticline and subsurface of Hartzog Draw field in Wyoming (Fig. 8.8). They proposed that sand was transported from a contemporaneous north–south-trending shoreline for approximately 160 km (100 mi) across a muddy shelf. The sand was carried by obliquely oriented storm-generated currents that deposited or reworked it into elongate ridges on the midshelf and outer shelf. Tillman and Martinsen (1984) identified and interpreted the following facies: (1) a basal, bioturbated shelf siltstone deposited on the open shelf, which is the "background shelf facies" (Fig. 8.8), (2) parallel- to ripple-laminated, interbar sandstone facies with occasional sandy storm deposits, and (3) cross-bedded, coarser-grained sandstones representing higher-energy deposition along the flanks (margin) and crest of the ridge.

Cumulative production (to 1980) and net pay thickness both were greater along the eastern flank of the field than along the western side (Fig. 8.16). Thus, Gaynor and Scheihing (1988) concluded that currents (either wave or tidal) had reworked sand on the paleo-oceanward side of sand bars, thereby providing

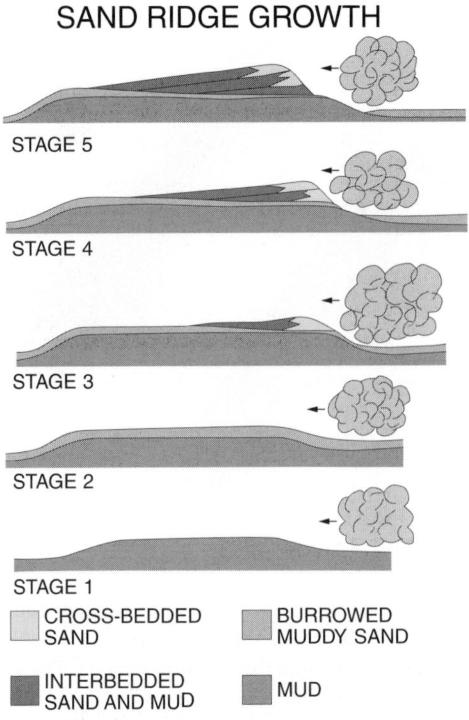

Fig. 8.17. The asymmetry of production shown in Fig. 8.16 was explained by D. Swift (personal communication, 1990) to be a result of cleaner, better-sorted and thus more porous and permeable sandstones on the seaward flank of the offshore ridges. The cleaner sandstones resulted from greater storm wave reworking of sands in the seaward direction. This origin is illustrated in the figure, which shows progressive sand ridge growth by first reworking and winnowing of sand from the shelf substrate (Stages 1 and 2) and then by progressive winnowing and buildup toward the seaward flank of the ridge, as a result of storm-generated currents (Stages 3–5). In this manner, the ridge deposit not only would be better sorted, cleaner, and coarser in the paleoseaward direction, it also would be better sorted, cleaner, and coarser vertically upward. Picture provided by D. Swift.

better-sorted and coarser-grained sands, and thus, more porous and permeable sands on that side (Fig. 8.17).

8.4.1.2 Hartzog Draw field as a lowstand shoreface deposit (1993–1994)

Walker and Bergman (1993) and Bergman (1994) reinterpreted the Shannon Sandstone in Wyoming (with emphasis on the Hartzog Draw–Heldt Draw fields) as shoreface deposits (Fig. 8.18). Their cross-sectional model suggests that the Shannon Sandstone comprises a series of stacked, lowstand shoreface sandstones that prograded from west to east and subsequently were partially truncated by transgressive erosion during several time periods. The mapped elongation of the sandstone is a combination of deposition along a long, linear

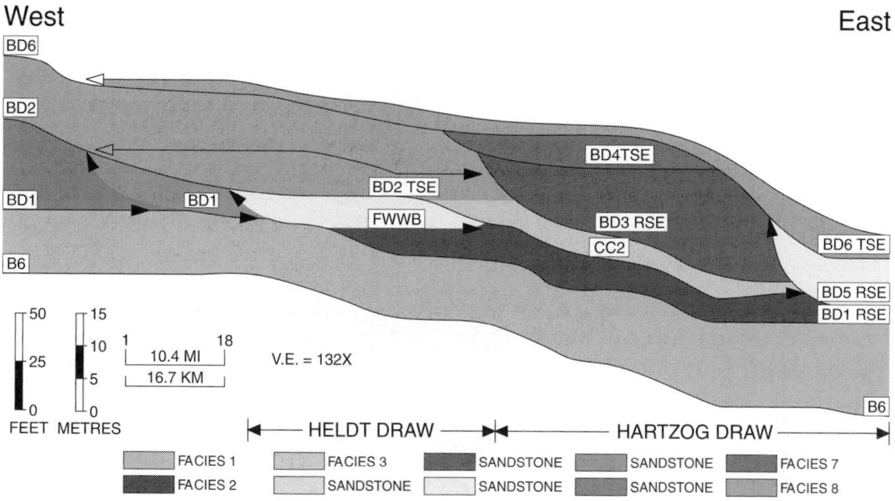

Fig. 8.18. Walker and Bergman (1993) and Bergman (1994) reinterpreted the Shannon Sandstone in Wyoming as shoreface deposits. This cross-sectional model indicates that the Shannon Sandstone comprises a series of stacked, lowstand shoreface sandstones (shown by different colors) that prograded from west to east and subsequently were partially truncated by transgressive erosion during several time periods (see Fig. 8.9 and the text for a diagrammatic illustration of the processes involved). The open-headed arrows represent onlapping markers, and closed-headed arrows represent truncation. BD1 – regressive surface of erosion. BD2 – erosional transgressive surface. CC2 – correlative conformity. BD3 – regressive surface of erosion. BD4 – erosional transgressive surface. BD5 – regressive surface of erosion. BD6 – erosional transgressive surface. TSE – transgressive surface of erosion. RSE – regressive surface of erosion. FWWB – fair weather wave base. CC2 – correlative conformity. After Bergman (1994). (Reprinted with permission of Society for Sedimentary Geology, SEPM.)

shoreline and transgressive erosion of an originally more extensive shoreface deposit. The end result is an internally complex set of individual shoreface sandstones. Their model is similar to that proposed for the Cretaceous Cardium Formation by Walker and Eyles (1991), as discussed above (Fig. 8.9).

8.4.1.3 Hartzog Draw field as a tidal sand-bar deposit associated with incised valley fill (1997)

Sullivan et al. (1997) interpreted the Shannon Sandstone at Hartzog Draw field as a series of stacked, upward-coarsening tidal bars associated with southeast-trending incised valleys (Fig. 8.19). They identified three major, unconformity-bounded depositional sequences: (1) the "Copenhagen Blue" sequence, comprising distal tidal-bar deposits at the base of which is a regional unconformity that separates it (the Copenhagen Blue sequence) from underlying offshore mudstones of the Cody Shale; (2) the "Crimson Red" sequence, which is the main reservoir interval and is dominated by high-reservoir-quality, proxi-

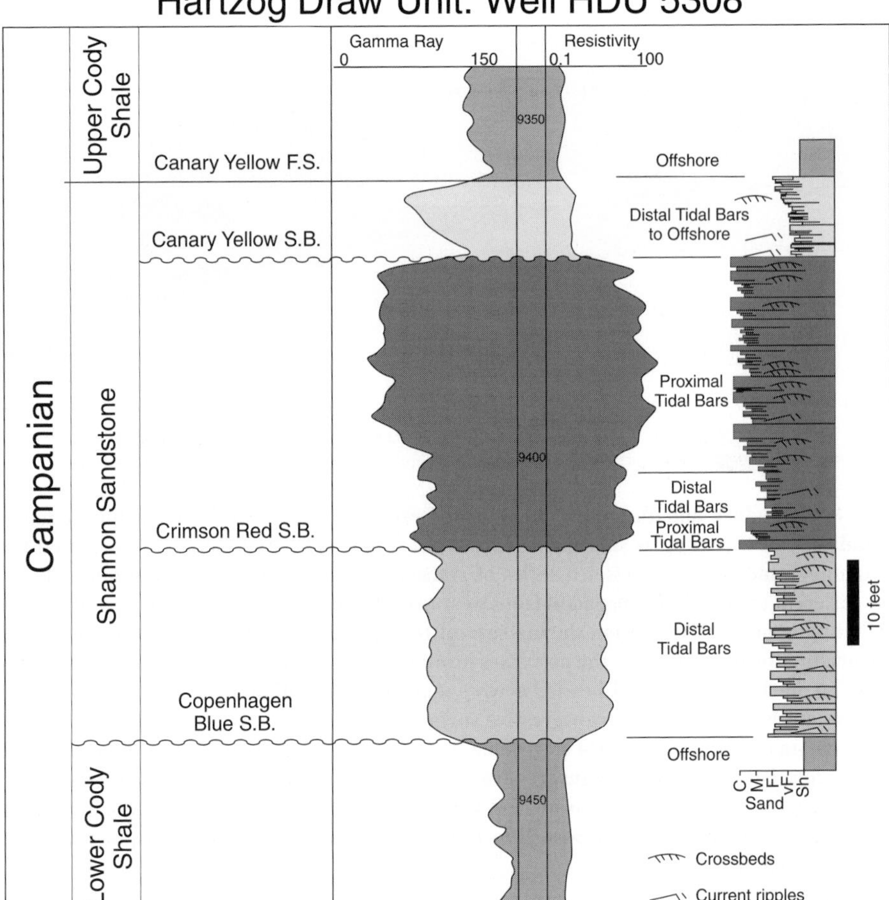

Fig. 8.19. Sullivan et al. (1997) reinterpreted the Shannon Sandstone at Hartzog Draw field to be a series of stacked, upward-coarsening tidal bars. Thickness trends are interpreted to be the product of variable erosion associated with southeast-trending incised valleys. Three major, unconformity-bounded depositional sequences have been identified: (a) the lowermost "Copenhagen Blue" sequence is composed of distal tidal-bar deposits sitting atop offshore mudstones of the Cody Shale; (b) the "Crimson Red" sequence is the main reservoir interval and is dominated by high-reservoir-quality, proximal tidal-bar sandstones; (c) the highly incised, overlying "Canary Yellow" sequence is composed of distal tidal bars and offshore mudstones. After Sullivan et al. (1997). (Reprinted with permission of Society for Sedimentary Geology, SEPM.)

mal tidal-bar sandstones; and (3) the highly incised, overlying "Canary Yellow" sequence, whose boundary forms the trap at Hartzog Draw field by juxtaposing the reservoir sandstones of the Crimson Red sequence with the distal tidal bars and offshore mudstones of the Canary Yellow sequence. Table 8.1 summarizes

Table 8.1 Reservoir quality of Shannon Sandstone at Hartzog Draw field

Facies	Porosity (%)	Sand/shale (%)	Permeability (md)
Proximal tidal bar	12.9–13.6	80–100	13.6–15.9
Distal tidal bar	7–11	10–90	2.4–8.7
Offshore	3.5–5.3	0–10	0.5–1.7

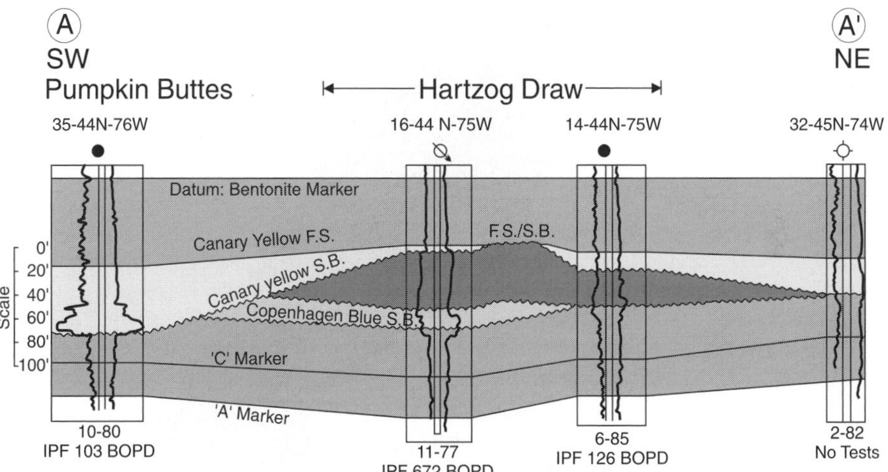

Fig. 8.20. A southwest–northeast stratigraphic cross-section through Hartzog Draw and adjacent Pumpkin Buttes fields, illustrating the three sequences developed within the Shannon Sandstone lowstand sequence set. Both the Copenhagen Blue and Crimson Red sequences have been highly dissected by the Canary Yellow sequence boundary, giving rise to the thickest section of high-reservoir-quality Crimson Red strata within Hartzog Draw field. F.S. – flooding surface. S.B. – sequence boundary. After Sullivan et al. (1997). (Reprinted with permission of Society for Sedimentary Geology, SEPM.)

the reservoir properties of proximal and distal tidal-bar facies and offshore facies. Proximal tidal-bar facies have the highest net sand content and the best porosity and permeability, whereas the offshore facies have the lowest net sand content, porosity, and permeability.

A southwest–northeast stratigraphic cross-section through Hartzog Draw and adjacent Pumpkin Buttes fields illustrates the stratigraphic relationships of the three lowstand sequences developed within the Shannon Sandstone (Fig. 8.20). Both the Copenhagen Blue and Crimson Red sequences have been highly dissected by the Canary Yellow sequence boundary, so that the Crimson Red sequence is thickest, has the best reservoir quality, and the lowest remaining reserves. The high-reservoir-quality facies have been swept effectively by waterflood, so there are low remaining reserves. In this instance, an area of moderate-quality facies has not been well swept, so it contains the most wa-

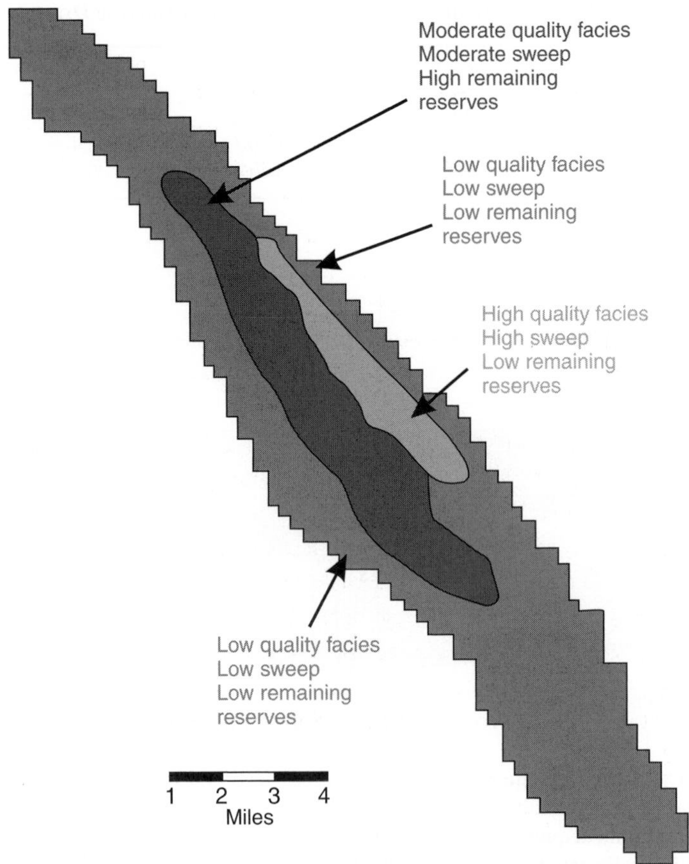

Fig. 8.21. Diagram showing the distribution of remaining reserves and expected recoveries by facies (as of 1997). The red/dark-gray area comprises relatively low-quality facies (mainly Canary Yellow) with low remaining reserves. The purple/gray area comprises high-reservoir-quality facies (mainly Crimson Red) that have been effectively swept by waterflood, so the remaining reserves are low. The blue/black area comprises moderate-quality facies with moderate sweep efficiency, so it contains the most waterfloodable reserves (about 60% of the total remaining reserves). The cross-section of Fig. 8.20 is perpendicular to the trend of reserves shown in this figure. After Sullivan et al. (1997). (Reprinted with permission of SEPM, Society for Sedimentary Geology.)

terfloodable reserves (about 60% of the total remaining reserves). In Fig. 8.21, the high-quality facies that have been swept by waterflood are the good-quality Crimson Red facies. The moderate-quality facies that have high remaining reserves are the southwest part of the Crimson Red facies, some Canary Yellow facies, and probably also some Copenhagen Blue facies.

8.4.2 Terry Sandstone, Denver Basin, Colorado

The Terry Sandstone in Hambert-Aristocrat field provides an example of a series of stacked shoreface sandstones separated vertically by shales (labeled B–G in Fig. 8.22) (Slatt, 1997). The Terry Sandstone produces both oil and gas from the field, which is located on the east flank of the Denver Basin, Colorado, USA. Production was initiated in 1972 with 320-acre spacing for gas wells and 40-acre spacing for oil wells. Cumulative production through August 1993 was 24.8 BSCFG and 1 MMSTBO. Permeability generally averages 1 md or less and rarely exceeds 10 md in individual sandstones, because these fine-grained sandstones have a high degree of cementation (Figs. 8.23 and 8.24).

As was the case for many Cretaceous sandstones in the Western Interior Seaway, the Terry Sandstone was interpreted in the 1980s to be offshore shelf bars in the Spindle field area of Colorado (Porter and Weimer, 1982) and in the Antelope-LaPoudre field area (north of Hambert-Aristocrat field; Siemers and Ristow, 1986). Later, Slatt used detailed mapping and applied sequence stratigraphy concepts and concluded that the Terry Sandstone in the Hambert-Aristocrat field (and likely also in the other Terry Sandstone fields) comprises a series of retrogradational shoreface parasequences formed in a manner similar to that of the Cardium Formation (Fig. 8.9) (Slatt, 1997).

The sequence stratigraphy of the Terry Sandstone in Hambert-Aristocrat field is discussed in a later chapter. For the purposes of describing the reservoir facies, we will discuss only Units A and B here.

Unit A is a burrowed mudstone deposited on an open shelf environment. Overlying Unit A is shoreface sandstone B. In the western (paleolandward) part

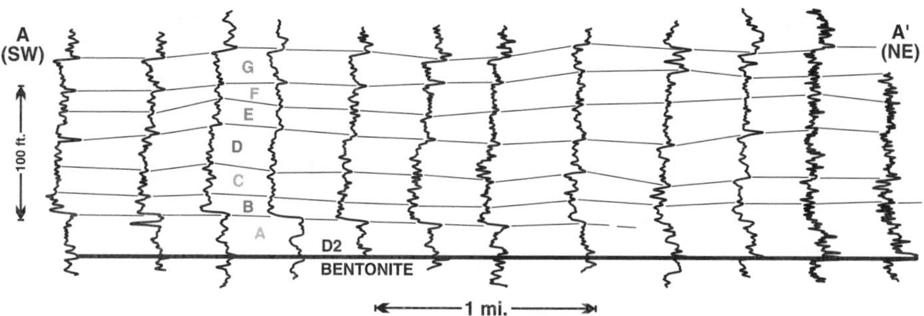

Fig. 8.22. Southwest-to-northeast stratigraphic cross-section across Hambert-Aristocrat field. A prominent bentonite bed forms the stratigraphic datum. Six shoreface intervals (B–G), each separated by a flooding shale, have been identified as comprising the Terry Sandstone. Interval A is shelf shale. Note the sharp upper surface of interval A toward the southwest (paleolandward) and the gradational contact toward the northeast (paleoseaward), which is typical of shoreface parasequences (Figs. 8.8, 8.10, 8.13, and 8.14). After Slatt (1997). (Reprinted with permission of SEPM, Society for Sedimentary Geology.)

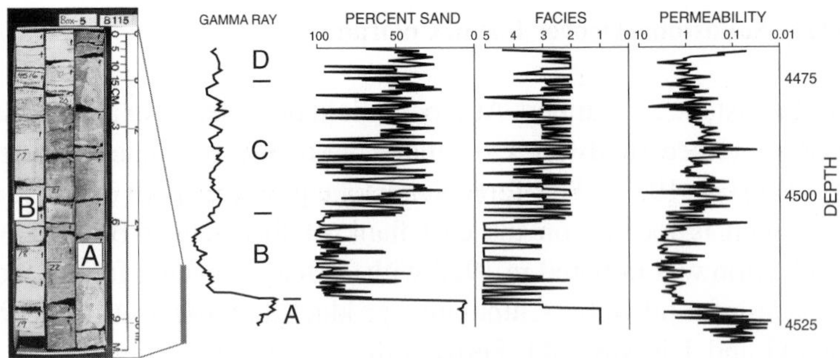

Fig. 8.23. Core and a gamma-ray log through three parasequences (B–D; see Fig. 8.22 and underlying shelf facies A). The positive relationships among the percentage of sand, the facies, and permeability are shown graphically by the profiles. The contact between A and B is quite sharp, as seen on both the log and in the core. Also, Parasequence B clearly exhibits a sandier gamma-ray log response, reflecting the sandier nature of upper shoreface strata. Parasequences C and D exhibit a shalier overall log response, with alternating thin sandstones and shales reflecting the overall muddier middle to lower shoreface environment of deposition. Overall, in this well, the Terry Sandstone becomes finer-grained (muddier) upward from the base of Parasequence B, reflecting an overall temporal deepening of water during deposition. Facies 1 – Bioturbated shelf mudstone. Facies 2 – Burrowed to bioturbated sandy mudstone. Facies 3 – Burrowed to bioturbated muddy sandstone. Facies 4 – Planar- to cross-bedded sandstone. Facies 5 – Ripple-bedded sandstone. After Slatt (1997). (Reprinted with permission of SEPM, Society for Sedimentary Geology.)

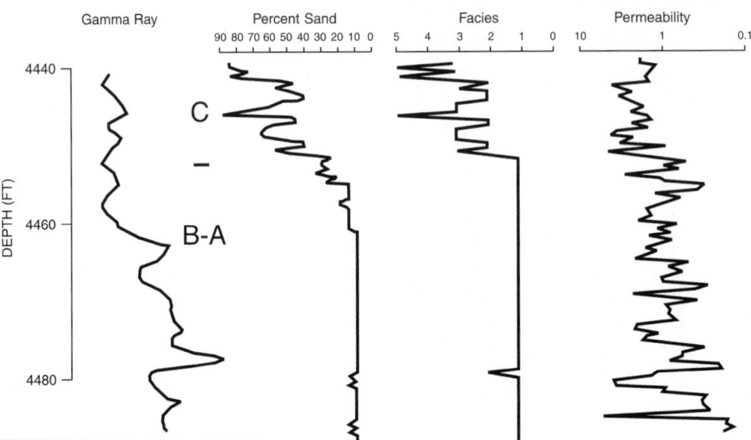

Fig. 8.24. A gamma-ray log, and profiles of sand content, facies, and permeability for another cored well in the Hambert-Aristocrat field area. The vertical profile differs from that of the interval shown in Fig. 8.23 in that it becomes sandier upward. The contact between A and B is gradational, and the strata are bioturbated mudstones. Parasequence C is composed of bioturbated sandy mudstone to muddy sandstone, interpreted to have been deposited in a lower shoreface setting. After Slatt (1997). (Reprinted with permission of SEPM, Society for Sedimentary Geology.)

of the field, Unit B has an erosional base upon which a net sand of about 100% with an average permeability of 7–8 md was deposited in an upper shoreface setting (Fig. 8.23). To the east, in the paleoseaward direction, the well log contact between Unit A and Unit B is gradational rather than erosional (Fig. 8.24). Thus, it is difficult to place any boundary between the open-shelf Unit A and the middle- to lower-shoreface Unit B. Figure 8.25 shows the general area where the transition from blocky to gradational log response of Units A and B occur. Net sand in Unit B seaward of the boundary between the two log patterns is less than 30% and permeability is less than 1–3 md.

The seaward progression of facies, and the associated net sand and permeability, conform to that of a typical seaward-fining shoreface sequence (Fig. 8.6), which has a sharp erosional base resulting from lowstand erosion in the nearshore region and a gradational base resulting from progradation in the seaward direction (Figs. 8.10, 8.13 and 8.14). Although Unit B is a tight-gas sandstone of low permeability, it illustrates the more general case of a seaward reduction in permeability within a single parasequence, as a result of the seaward reduction in sand content (and presumably of grain size).

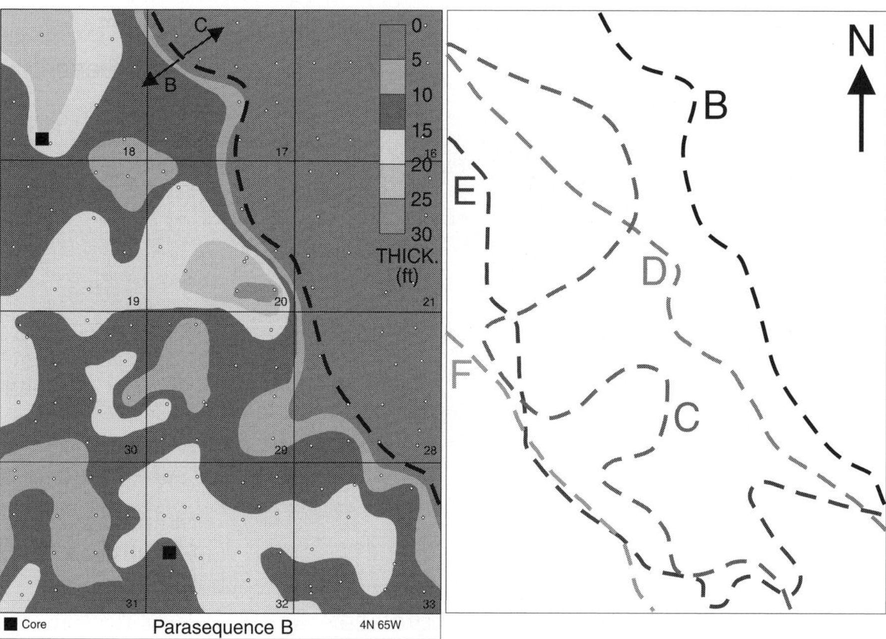

Fig. 8.25. Map of part of the Hambert-Aristocrat field, showing the positions of the boundary between blocky sandstone and gradationally coarsening-upward sandstone of each parasequence. Modified from Slatt (1997). (Reprinted with permission of Society for Sedimentary Geology, SEPM.)

8.5 Barrier island deposits and reservoirs

8.5.1 Complex processes and deposits

Barrier island deposits are quite complex because of the variety of processes involved in their formation (Fig. 8.26). On the larger time scale, they are affected by sea-level fluctuations. In the shorter time frame, they are the result of interactions of shallow-water waves (Fig. 8.3) and ocean currents that impinge upon the shoreline and stack sand along a beach front (Figs. 8.2 and 8.27). Sand also is deposited landward of the beach to form sand dunes (Fig. 7.3). Intense, storm-generated waves and currents, including those resulting from hurricanes, can throw sand into the back-barrier area as washover deposits (Fig. 8.2). Tides also have an effect by producing strong tidal currents that can cut tidal channels and ebb- or flood-tidal deltas on either side of the channel (Figs. 8.4, 8.26 and 8.28). It is not surprising, then, that subsurface barrier island reservoirs are complex and present challenges to successful, economic oil and gas production.

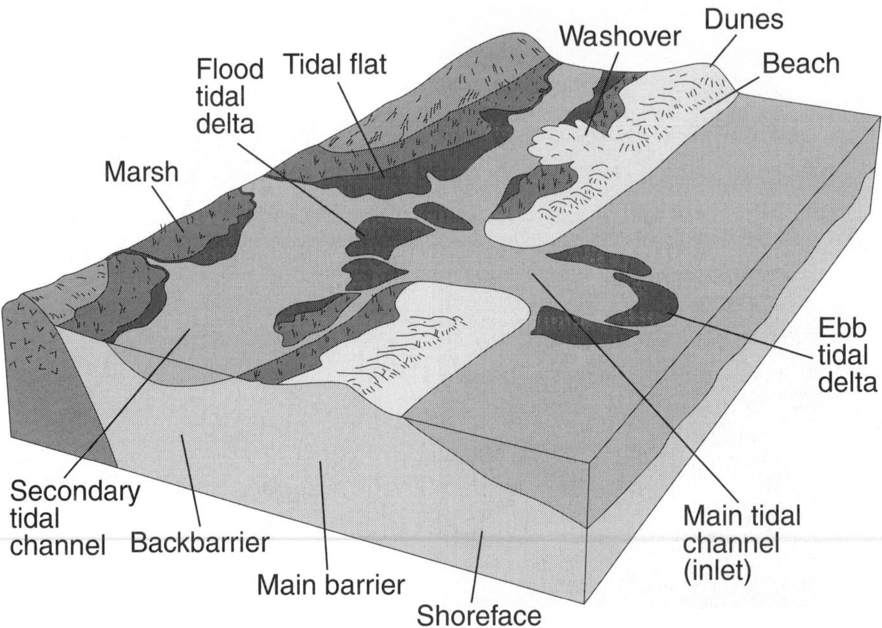

Fig. 8.26. Block diagram model of a barrier-island system, showing the complex nature of the deposit. A barrier island system consists of a number of different component parts, as discussed in the text. After Walker (1980). (Reprinted with permission of Geological Association of Canada.)

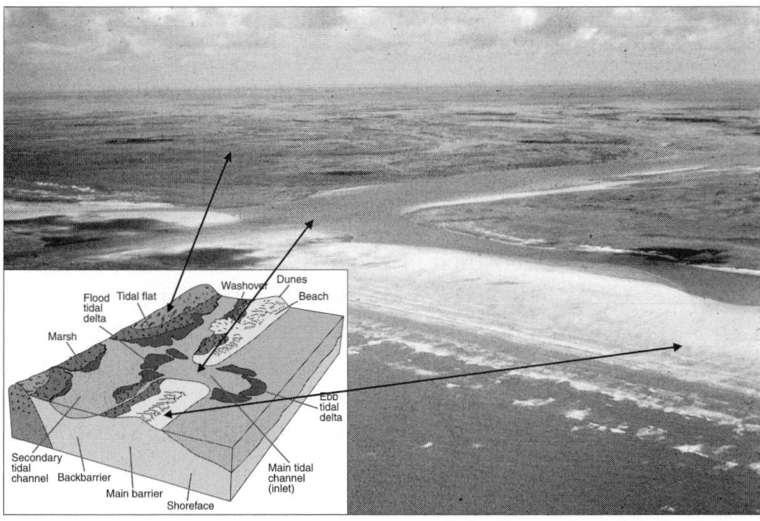

Fig. 8.27. A modern barrier sand, tidal channel, and lagoon from the Texas Gulf Coast. When they are buried in the subsurface, the barrier sand and the tidal channel sands can be good reservoirs. The lagoonal deposits can form hydrocarbon source rocks if they are organic rich, as is often the case in lagoons. The inset is a block diagram of a barrier island system, which is discussed in the text.

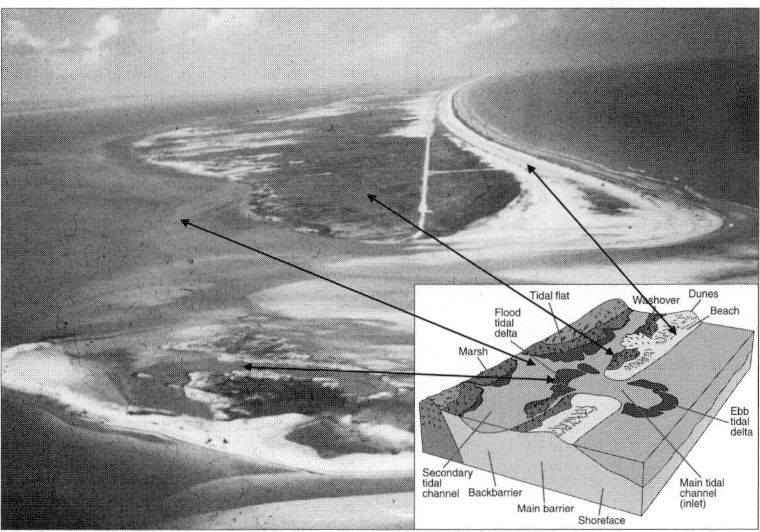

Fig. 8.28. A flood-tidal delta and a very long beach-sand deposit. The flood-tidal delta formed when incoming tides moved sand progressively inland, beyond the beach and through the tidal channel. Both the flood-tidal delta and the beach sand can be good reservoirs when they are buried in the subsurface. Block diagram is of a barrier island system, which is discussed in the text.

8.5.2 Bell Creek and Recluse fields, Montana and Wyoming, USA

Bell Creek and Recluse fields are two mature and well-studied fields that occur within a series of Cretaceous Muddy Sandstone reservoirs on either side of the joint Wyoming and Montana state border (Fig. 8.29) (Berg, 1986). Because these fields have been so heavily drilled, they provide excellent examples of the complexities of geology and production in barrier island reservoirs. Both fields were discovered in 1967. By 1977, annual production from 195 wells in Bell Creek field was 8.8 MMBO, for a 10-year cumulative production of 86.3 MMBO. Annual production during the same time period reached 8.4 MMBO from 58 wells in Recluse field, for a 10-year cumulative production of 20.8 MMBO (Berg, 1986).

The northeasterly elongation of Bell Creek field (Fig. 8.30) suggests that it might be a beach deposit, and the mapped bulge on the northwest side of the field is interpreted to be a flood-tidal delta (the ocean was to the southeast during the time of deposition) (Berg, 1986). A core from one of the wells in the

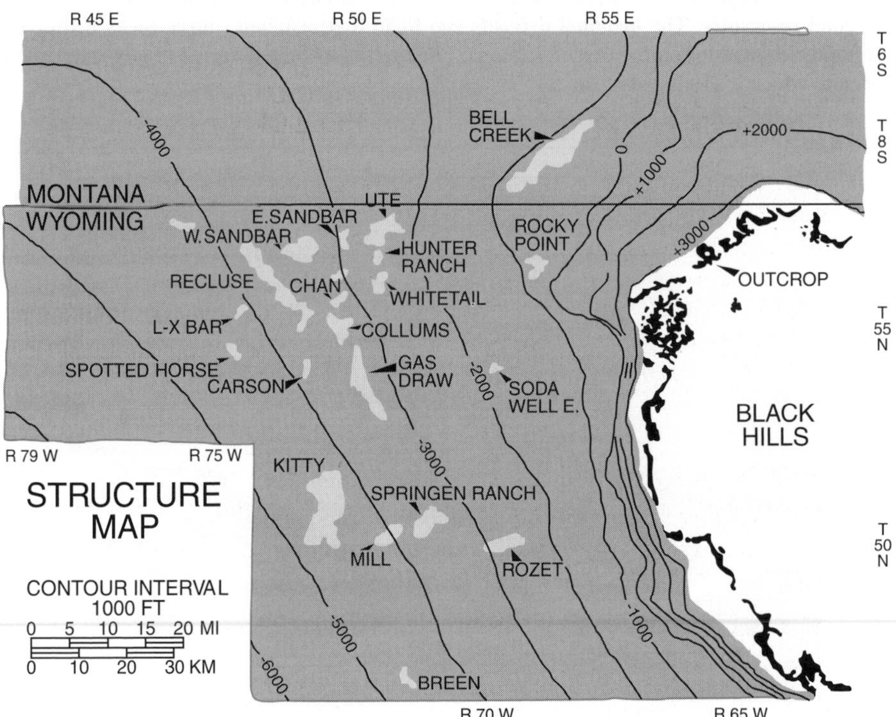

Fig. 8.29. Locations of Bell Creek and Recluse fields in the Powder River Basin of Wyoming and Montana. Several other Cretaceous fields are also shown in yellow/bright-gray. After Berg (1986). (Reprinted with permission of AAPG, whose permission is required for further use.)

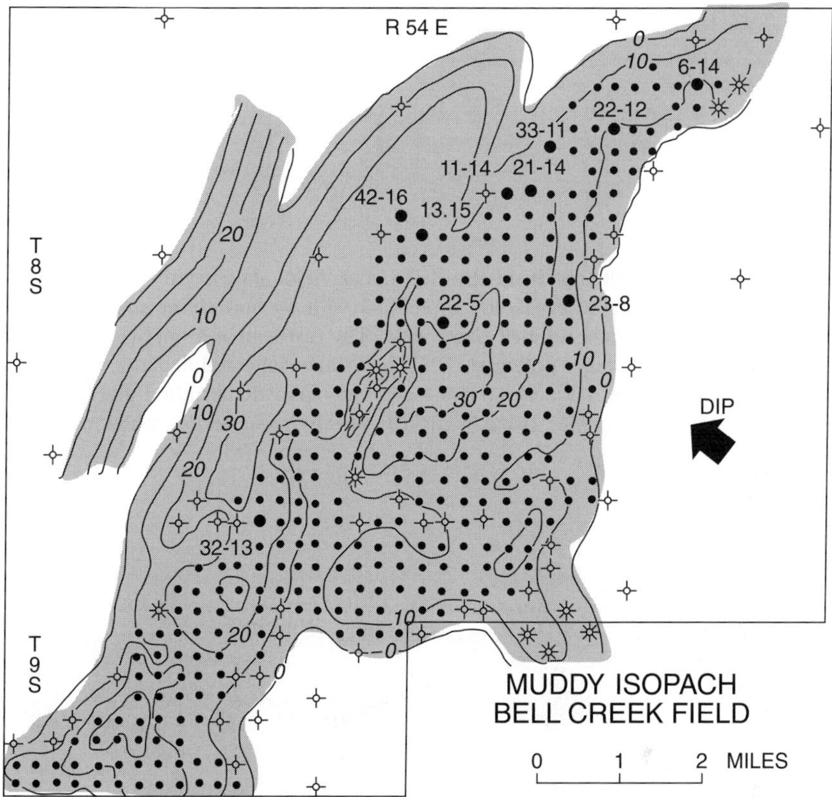

Fig. 8.30. An isopach map of the Muddy Sandstone, which comprises the Bell Creek field, Montana. Locations of wells are shown by black dots. This field has been very heavily drilled. After Berg (1986). (Reprinted with permission of AAPG, whose permission is required for further use.)

field contains the following facies: (1) 1.5 m of very fine-grained, massive eolian sandstone, (2) 1.5 m of fine-grained, laminated beach and upper-shoreface sandstone, (3) 3 m of massive to laminated fine-grained middle-shoreface sandstone, and (4) 0.9 m of shaly, highly bioturbated lower-shoreface mudstone. This sequence is identical to that of a modern Galveston Island, Texas, USA barrier island. Six different zones, sitting atop an unconformity surface, have been recognized by correlating core to the log responses (Fig. 8.31A). In specific zones, thick barrier-bar sandstones might grade laterally into thinner-bedded, shalier lagoonal facies (Fig. 8.31B). Discontinuities between wells are a result of shaly lagoonal deposits that have compartmentalized the reservoir sandstones (Fig. 8.32).

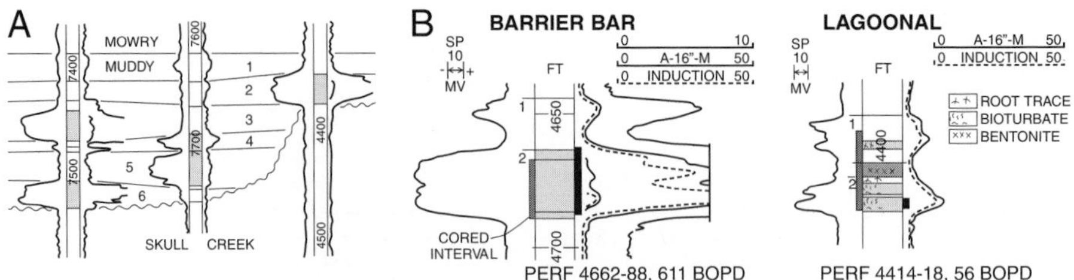

Fig. 8.31. (A) Three well logs from the Bell Creek field, showing the six intervals that have been identified. The intervals are in contact with an underlying unconformity surface. (B) Two well logs from the Bell Creek field. The reservoir is a barrier or beach sand. It produced 611 BOPD initial potential (IP). The thinner lagoonal deposits produced only 56 BOPD initial potential (IP). These two facies vary laterally within a given interval, as shown in (A). After Berg (1986). (Reprinted with permission of AAPG, whose permission is required for further use.)

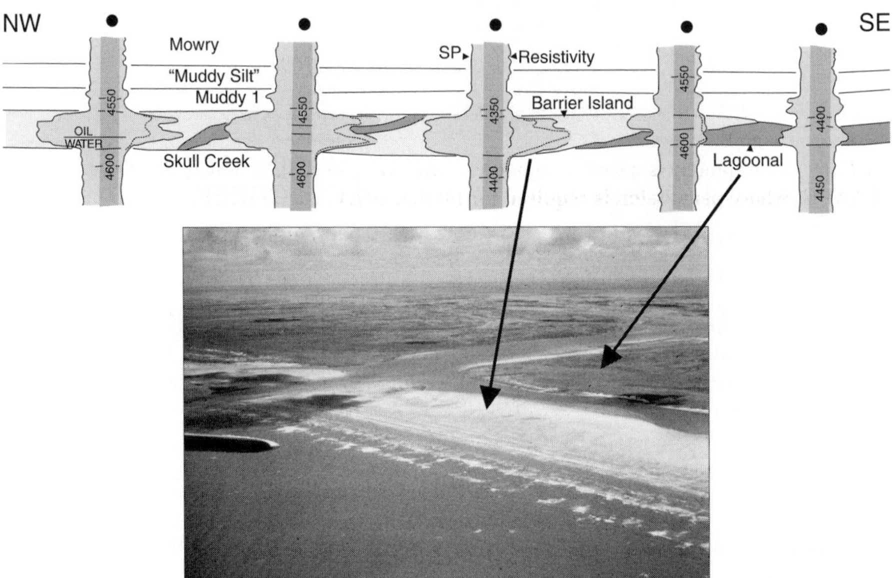

Fig. 8.32. The upper figure is a cross-section across Bell Creek field. The main reservoir is composed of thick barrier-island sandstones. The shales penetrated in wells are lagoonal shales. Lagoonal shales are interpreted between wells 42-16 and 13-15 and between wells 13-15 and 22-5. The reason for this interpretation is that the sands in these three wells are isolated, therefore there must be a shale barrier between them. Inset shows the different facies in a modern Texas Gulf Coast setting. After Berg (1986). (Reprinted with permission of AAPG, whose permission is required for further use.)

The orientation of Recluse field is approximately perpendicular to that of the Bell Creek field (Fig. 8.33) (Berg, 1986). Although the Recluse field is

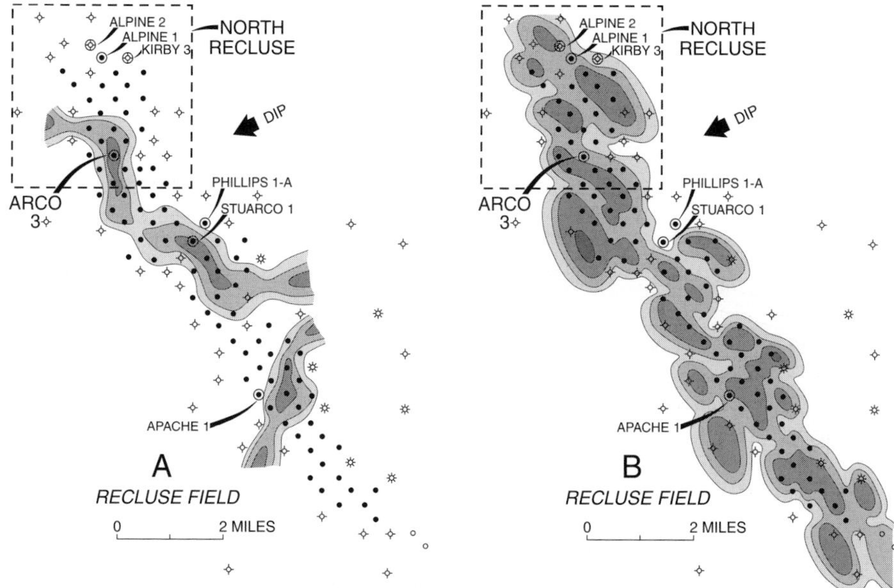

Fig. 8.33. Isopach thickness maps are shown for the Recluse field intervals A and B. Note the large number of wells in the field. The Recluse B zone has been mapped as a series of partially to completely isolated sand bodies. After Berg (1986). (Reprinted with permission of AAPG, whose permission is required for further use.)

composed of an elongate sandstone, as is Bell Creek field, sandstone bodies appear to be less continuous in Recluse field, suggesting compartmentalization in Recluse. Such compartmentalization also is suggested by the presence of water in wells that are higher structurally than are oil wells in the same field (Fig. 8.34). Sandstones in Recluse field vary from being fine- to medium-grained (Fig. 8.35). The former exhibit lower porosity and permeability than do the latter, because of the grain-size effect discussed in Chapter 5.

8.6 Summary

Shallow marine environments, from the shoreline to the shelf edge, are complex and result in complex deposits. In turn, complex deposits translate into complex reservoirs. To maximize reservoir performance, it is imperative to understand the type of shallow marine deposit that makes up the reservoir. That is not an easy task, as is exemplified by the various interpretations that have been assigned to the linear sandstones of the U.S. Cretaceous Western Interior Seaway. These sandstones, in both outcrop and subsurface reservoirs, have been interpreted to be offshore shelf bars or ridges, shoreface bodies, and tidally influenced incised-valley fill. Interpreting the type of deposit is not merely an academic exercise, it is essential because each of these different types of sandstone

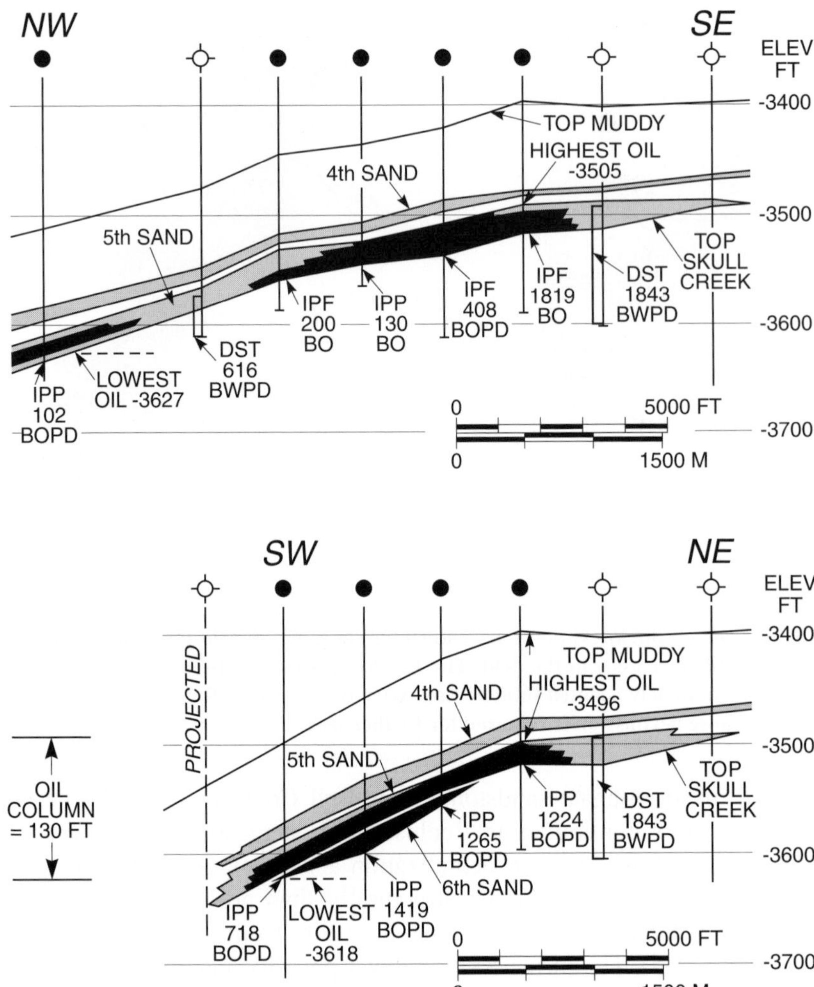

Fig. 8.34. Structure cross-sections, with updip toward the right. In the upper figure, the IPs (initial potential flow rate) show more oil in structurally updip wells. However, the uppermost Davis 3 well contains water (BWPD – barrels of water/day). Water above oil usually indicates some barrier to flow between wells. A similar situation occurs in the lower cross-section, where the structurally higher well contains water and the structurally lower wells are oil bearing. After Berg (1986). (Reprinted with permission of AAPG, whose permission is required for further use.)

bodies is characterized by different geometries and degrees of compartmentalization.

Barrier island deposits provide a particularly challenging reservoir-characterization problem. Because of the variety of sedimentary processes that can influence barrier island formation, several different sandstone and shale geome-

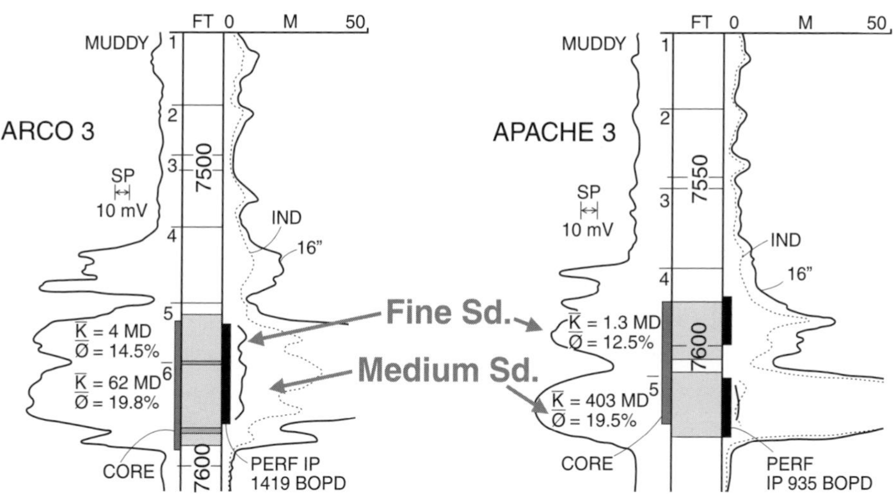

Fig. 8.35. Two logs from Recluse field are shown. In both logs, the lower sand is medium-grained and the upper sand is fine-grained. Note on the figure the different porosities and permeabilities. The fine sands have lower porosities and permeabilities than do the medium sands, as would be expected from the discussion in Chapter 5. The wells IP'd at 1,419 and 935 BOPD, which are good wells. After Berg (1986). (Reprinted with permission of AAPG, whose permission is required for further use.)

tries and trends can occur. That variation in geometries can lead to the potential for a high degree of compartmentalization that is difficult to predict. Again, depositional-geometry prediction and well placement are facilitated by an understanding of the nature of the deposit and how it was formed.

Chapter 9

Deltaic deposits and reservoirs

9.1 Introduction

The shoreline is the transition zone that separates nonmarine processes and environments from marine processes and environments. In this zone, there are both marine and nonmarine influences. The nonmarine influence increases in the landward direction, and the marine influence increases toward the shore zone. The previous chapter discussed processes and deposits that occur in the coastal zone and on the open continental shelf, where major rivers are absent, and there is no significant, recurrent sediment source. By contrast, small to large rivers flowing to the ocean deposit their sedimentary load in the shore zone and, over time, they may form a deltaic deposit (Fig. 9.1).

A deltaic deposit's morphology and stratigraphy depend on several factors, such as the sediment load of the source river, the volume of the river and its drainage area, the topography over which the sediment travels on its way to the coast, and the nature and intensity of nearshore marine processes that act to rework and disperse the sediment once it reaches the coast. Also, sediment can be supplied to the shore zone from offshore and alongshore sources unrelated to a river. This chapter describes the variety of deltaic deposits and discusses the processes responsible for those deposits' architecture and stratigraphy. Examples of deltaic hydrocarbon reservoirs also are presented to demonstrate how important it is to be understand deltaic architecture if they are to be managed optimally.

9.2 General deltaic processes, environments (physiographic zones), and types

When sediment carried by fresh river water reaches the shoreline, the mixture extends into the ocean as a sediment plume (Fig. 9.2; also see Fig. 10.6). If

Fig. 9.1. Areal view of Juneau, Alaska, which is built on a delta.

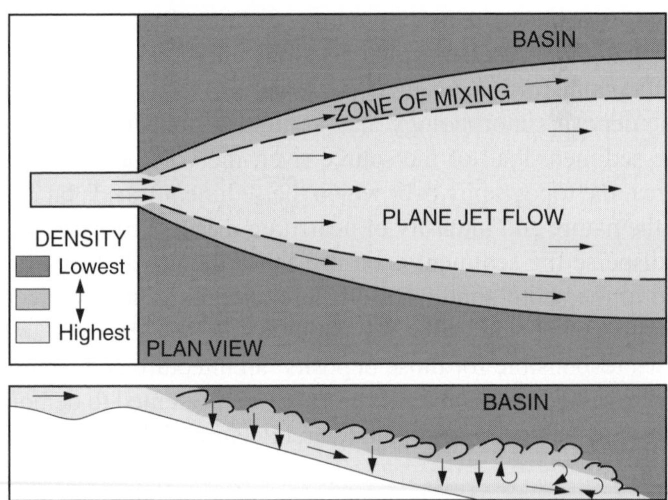

Fig. 9.2. Flow behavior of river-borne sediment into the marine shore zone. Sediment suspended in freshwater moves across the denser seawater surface as a plane jet. If the sediment concentration is sufficient for the density of the sediment–freshwater plume to be greater than that of seawater, the flow will sink to the bottom and move offshore under the influence of gravity as a hyperpycnal flow. Modified from Fisher et al. (1969). (Reprinted with permission of AAPG, whose permission is required for further use.)

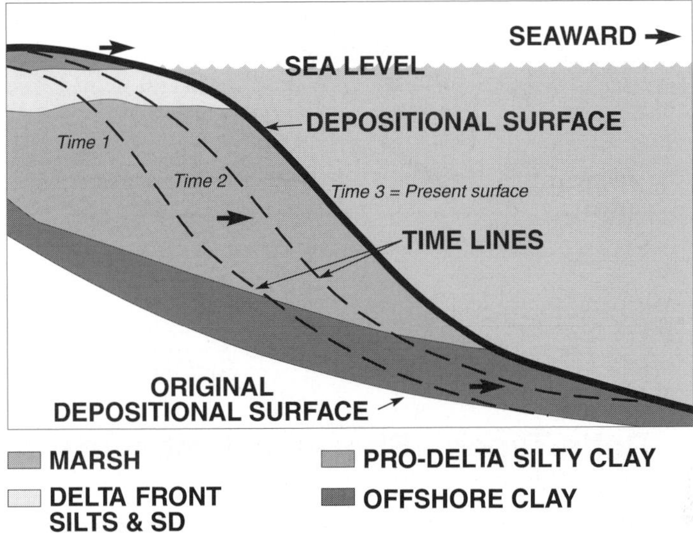

Fig. 9.3. Process of a delta's seaward progradation over time (from Times 1 through 3). Time lines form clinoforms and cross-cut facies boundaries because the facies migrate seaward during progradation. Generalized facies are shown. At any one vertical position through the delta, such as where a vertical well might be drilled, grain size of the delta strata will increase upward. Modified from Scruton (1960). (Reprinted with permission of AAPG, whose permission is required for further use.)

the sediment–freshwater mixture is denser than seawater, the mixture will sink through the water column and flow along the seafloor. If the mixture is less dense than seawater, the sediment will ride atop the seawater column and will settle slowly to the seafloor over a large area. In either case, with continued deposition of sediment on the seafloor, the floor builds upward and moves seaward, and the deposit progrades or migrates in a seaward direction over time (Fig. 9.3). Traditionally, deltas have been classified according to a tripartite division. If there are no marine influences acting to disperse the river-borne sediment laterally as it enters the marine environment, a prograding "river-dominated" delta will form. However, if the coast is in an area in which waves, currents, and/or tides are present (Fig. 9.4), the sediment can be dispersed in different directions, depending upon the directions of the waves, currents, and tides. If waves dominate the shore zone, the resulting delta is called a "wave-dominated" delta, and if tides are dominant, the resulting delta is called a "tide-dominated" delta.

In reality, rivers, tides, waves, and currents, in varying proportions, all play a role in the ultimate distribution of deltaic sediment along different coastlines. Bhattacharya and Walker (1992) illustrated a six-fold subdivision of deltas as a function of the respective influences of waves, tides, and rivers (Fig. 9.5). Also, the influence of rivers, waves, and tides may vary from place to place along a given shoreline, giving rise to a very complex facies distribution, both strati-

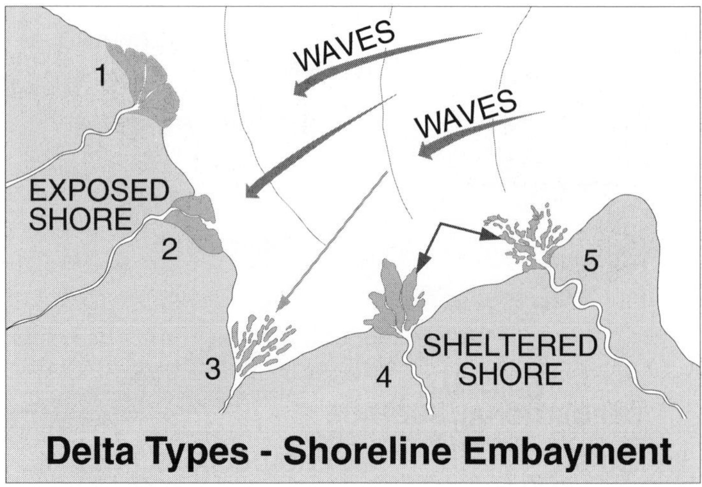

Fig. 9.4. Illustration of the types of deltas that form as a result of the interactions of rivers, waves, and tides with the configuration of the shoreline. Deltas 1 and 2 are wave-dominated deltas that receive the direct impact of waves impinging upon the river-derived deltaic sediment. Delta 3 is a tide-dominated delta that is impacted by the narrow embayment and resultant focused tidal energy. Deltas 4 and 5 are river-dominated deltas that are protected from the effects of open ocean waves and tidal currents. (Source of figure is unknown.)

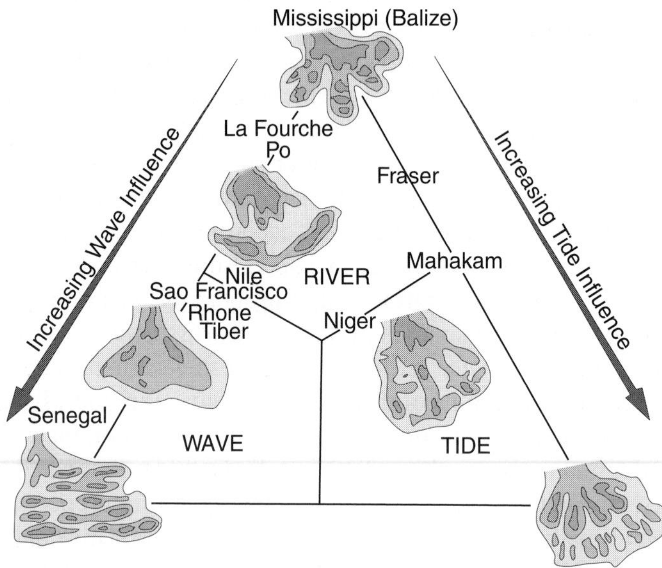

Fig. 9.5. Tripartite classification of deltas, with six general styles based on relative influences of rivers, waves, and tides. After Bhattacharya and Walker (1992). (Reprinted with permission of Geoscience Association of Canada.)

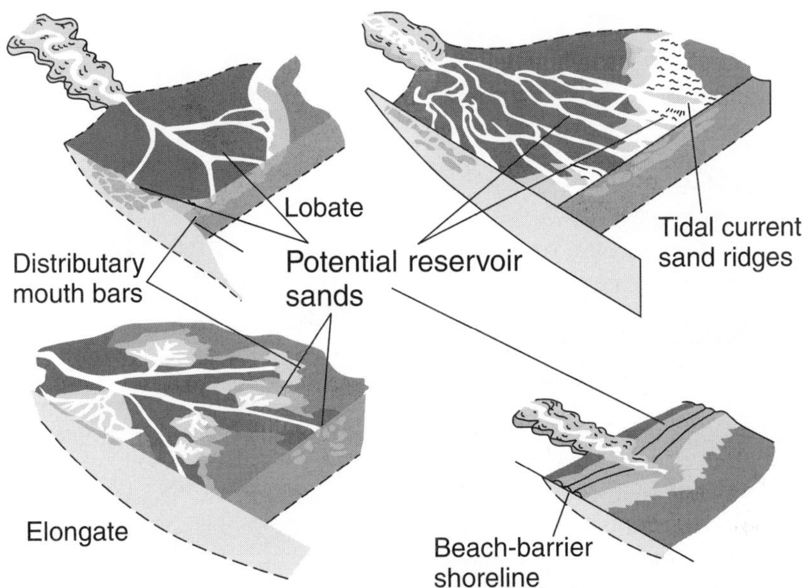

Fig. 9.6. General geometries of deltas and potential reservoir sands in each delta. Modified from Reading (1986). (Reprinted with permission of Blackwell Scientific.)

graphically and laterally, within one delta system. For example, the Danube River's delta system, deposited onto the northwestern margin of the Black Sea, consists of a northern lobe resembling a river-dominated delta and southern lobes resembling wave-dominated deltas (Bhattacharya and Giosan, 2003). With deltaic reservoirs of hydrocarbons, it is essential to determine the type of deltaic deposit in order to maximize reservoir development and production. Errors are easily made in interpreting types of deltas in the subsurface environment, where we have only scattered wells and limited cores or image logs from which to identify depositional processes and environments. Nevertheless, the geometry, size, and orientation of reservoir sandstones and shale barriers differ among the different types of deltas (Fig. 9.6), and we must understand these components fully to develop deltaic reservoirs successfully.

Deltaic coasts are composed of three principal zones, based on their bathymetry, slope gradient, and resultant sedimentary facies. In a river-dominated delta (Fig. 9.7), (a) the delta plain is composed largely of freshwater and brackish-water muds, sands, and peats, (b) the delta front is composed of sands that decrease in grain size with increasing water depth, and (c) the prodelta is composed mainly of mud with minor sand. Farther seaward, prodelta muds grade laterally into shelf muds. In the delta-front environment sufficient sand can accumulate in the form of distributary mouth bars and channels and crevasse splays that reservoirs form upon burial. Natural levees are associated with both trunk and distributary channels.

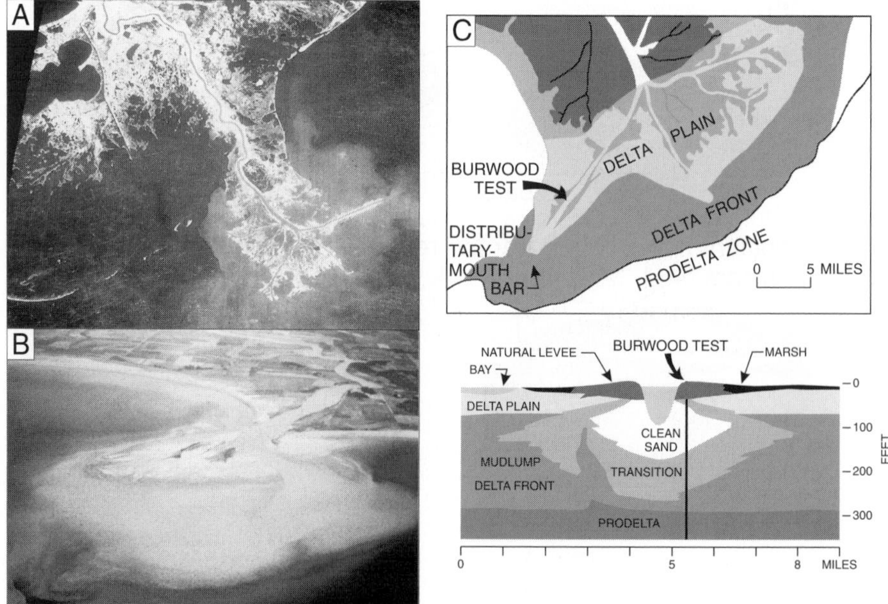

Fig. 9.7. A river-dominated delta. (A) Suspended sediment plume from aerial infrared photograph. The Mississippi River delta, a modern birdsfoot delta. (B) Suspended sediment plume emerging from a river mouth. (C) Lateral and vertical distribution of sedimentary facie characteristic of river-dominated deltas. Modified from Fisk (1961). (Reprinted with permission of AAPG, whose permission is required for further use.)

9.3 River-dominated delta deposits and reservoirs

9.3.1 Processes and deposits

The Mississippi River delta (Fig. 9.7) has long been considered the "type" river-dominated delta (Fig. 9.5). Strong waves and currents do not impinge upon its protected shoreline, so sediment deposited at and near the shore zone is not reworked or dispersed laterally (Figs. 9.4 and 9.7). With time, and sufficient accommodation space between the sea surface and seafloor to accept sediment (see Chapter 11 for more details on accommodation), such a delta will prograde seaward (Fig. 9.3), as will the delta zones (Fig. 9.7). By that process, surfaces representing instants in geologic time crosscut boundaries between sedimentary facies (see the time lines in Fig. 9.3). However, over time, progradation and vertical aggradation lead to a progressive shallowing of the water depth, to the point at which the main river(s) feeding sediment to its delta will shift toward deeper water (toward lower points on the seafloor), thereby causing "delta-lobe switching". During the past 4600 years, there have been no fewer than eight periods of lobe switching on the Mississippi River delta (Fig. 9.8).

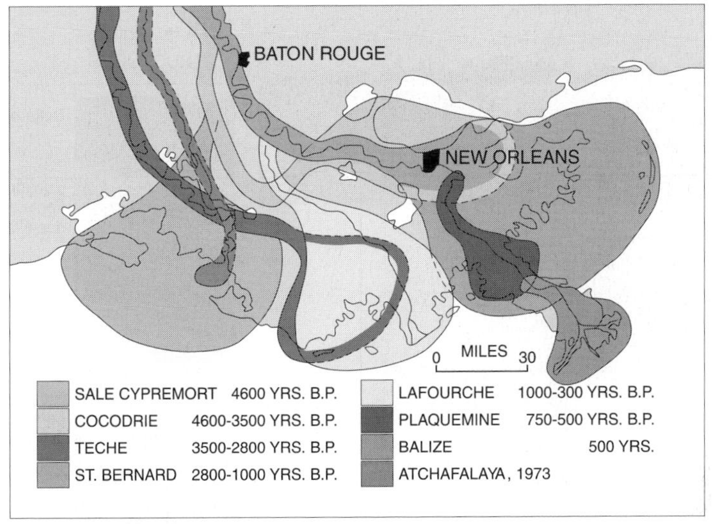

Fig. 9.8. Growth patterns of the Mississippi River delta's lobes during the past 4600 years. Modified from Kolb and Van Lopik, Fig. 2, in Shirley (1966). (Reprinted with permission of Houston Geological Society.)

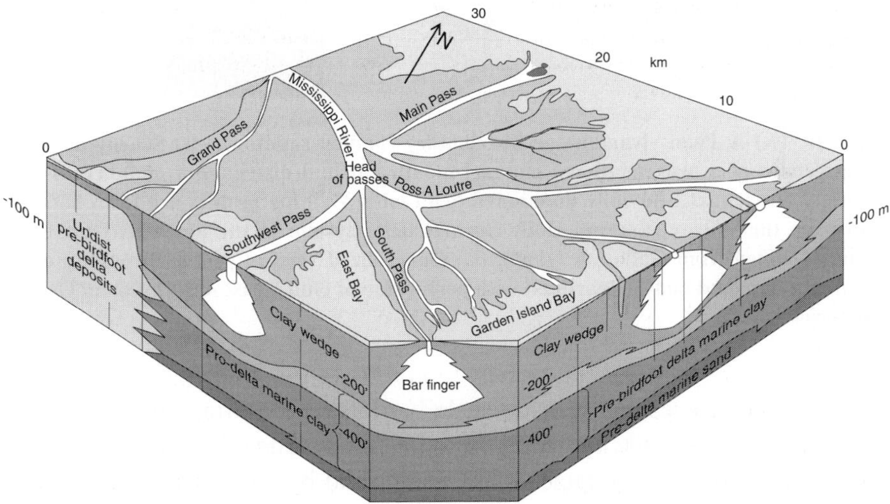

Fig. 9.9. Schematic illustration of the distribution of facies and environments in the modern Mississippi birdsfoot delta. Modified from Fisk (1961). (Reprinted with permission of AAPG, whose permission is required for further use.)

The main facies of a river-dominated delta, and their distribution in space, are shown in Fig. 9.9. Sandy reservoir facies are deposits of distributary channels and distributary mouth bars. Interdistributary bays, marshes, and lagoons

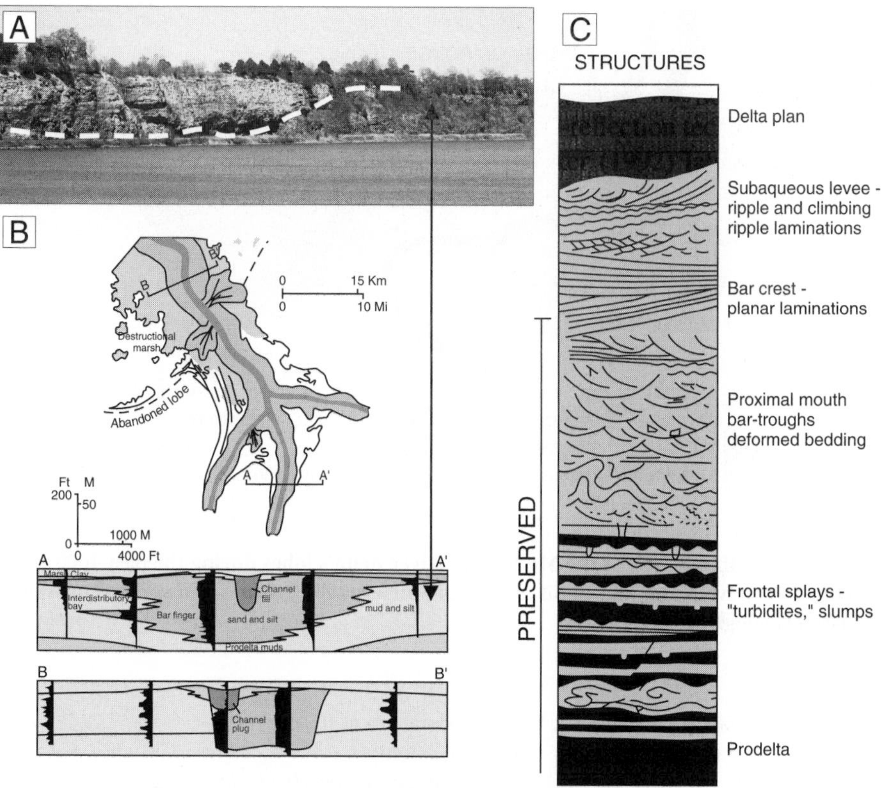

Fig. 9.10. (A) A Pennsylvanian-age distributary-channel sandtone. (B) Schematic facies cross-sections of a progradational channel mouth bar and distributary channel-fill sand bodies of an elongate, fluvially dominated delta lobe. Well log profiles are from borehole data from the Wilcox (Eocene) Holly Springs delta system of the northern Gulf Coast Basin. Modified from Galloway (1968). (C) Generalized vertical profile through a channel-mouth-bar sand body. (Reprinted with permission of Gulf Coast Association of Geological Societies.)

separate sandy facies and provide shale barriers in subsurface reservoirs, and sometimes, they provide hydrocarbon source rocks. The different facies can be recognized in outcrop by facies relationships and by a distinctive coarsening-upward grain size with a characteristic assemblage of sedimentary structures (Fig. 9.10). In the subsurface, these same features can be identified in cores and on well logs, including a lateral thinning and fining log response away from the channel axis (Fig. 9.10).

In Prudhoe Bay field, reservoir sandstones were deposited in the river-dominated delta's distributary channels and as distributary mouth bars (Fig. 9.11) (Tye and Hickey, 2001). To evaluate formative processes, a 1000-ft (300-m) horizontal well was drilled into a distributary mouth-bar reservoir sand,

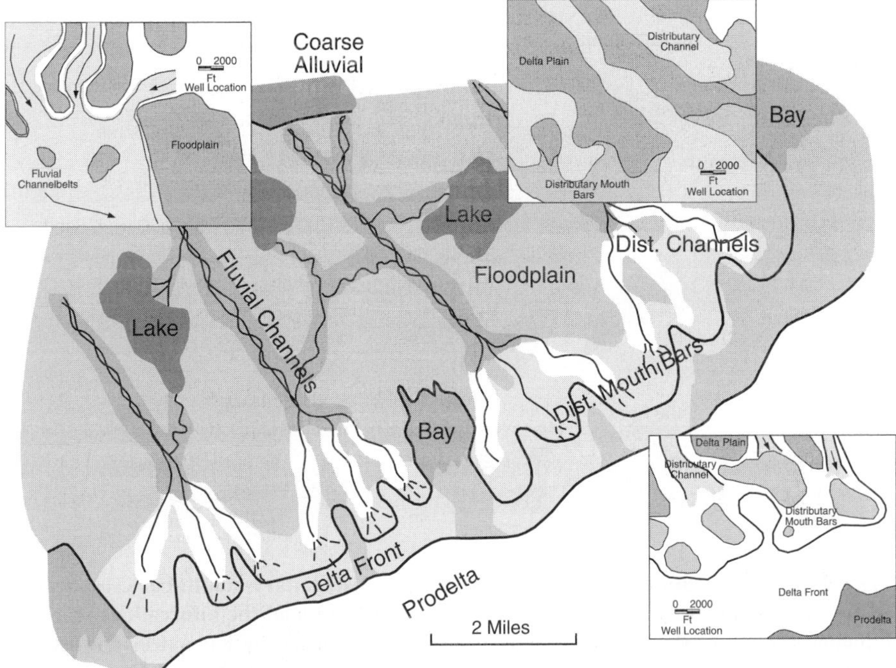

Fig. 9.11. Schematic paleogeographic reconstruction of Romeo facies associations, Prudhoe Bay field, as they existed during deposition. Insets are facies-association maps for three Romeo stratigraphic layers, from proximal (upper left) to distal (lower right). After Tye et al. (1999). (Reprinted with permission of AAPG, whose permission is required for further use.)

and three cores were obtained. Seven sandy lithofacies were identified in the core; all are moderately to very well sorted and fine to very fine grained. Six of the lithofacies exhibit average horizontal permeabilities of 12–40 md. The seventh, best-sorted lithofacies averages 129 md. The origin of these lithofacies was determined by studying modern river-dominated delta processes and sediments. Tye and Hickey (2001) believe that sediment plumes emanating from distributary channels during flood stage deposited the best-sorted and most-permeable sands (the seventh lithofacies) at the apex or most-proximal part of the distributary mouth bars, as well as along the proximal-bar margin (Fig. 9.12). Finer-grained, more poorly sorted sands and muds were deposited more distally, under lower flow-regime conditions.

Over a long period of time, repeated bifurcation of distributary mouth channels during progradation led to development of a broad, strike-parallel sheet composed of numerous mouth bars. However, across a single bar, there will be discontinuities in facies and permeability.

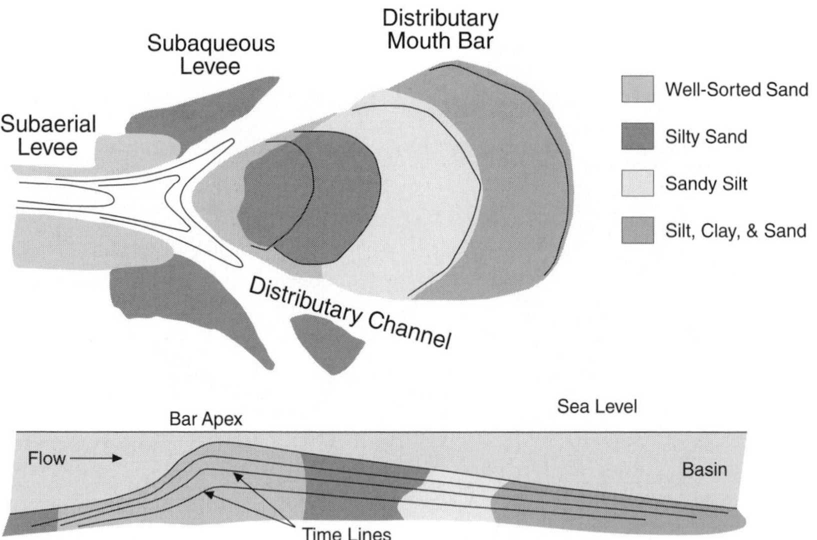

Fig. 9.12. Typical sediment distribution pattern and distributary-mouth-bar cross-section. Distributary-mouth channels typically bifurcate as shown, and the bifurcations segregate lithofacies by concentrating the coarser-grained sediment on the bar apices downstream of the bifurcation point and along bar margins. Modified from Tye and Hickey (2001). (Reprinted with permission of AAPG, whose permission is required for further use.)

9.3.2 Reservoir example: Prudhoe Bay field

General geology of Prudhoe Bay field, Alaska, in the river-dominated deltaic deposits of the ancient Sagavanirktok River, was discussed in Chapter 6. The most prolific reservoir rocks in the field are of fluvial, braided-river origin. The lower stratigraphic interval, called the Romeo Zone (Figs. 9.11–9.13), is composed mainly of poorer reservoir-quality, river-dominated delta-front and prodelta deposits (Figs. 6.18, 6.20, and 9.11) that in recent years have been targeted for production, mainly from horizontal wells (Tye et al., 1999; Tye and Hickey, 2001). Productive deltaic facies are distributary-channel and distributary mouth-bar sandstones (Figs. 9.11, 9.13, and 9.14). The depositional history depicts southeastward progradation of low-gradient, fine-grained fluvial systems that extended about 100 km (60 mi) from their source area. Individual fluvial channels were separated laterally by muddy floodplains. In the downcurrent direction, fluvial channels bifurcated and formed a delta plain and associated distributary mouth bars (Fig. 9.11).

River-dominated deposition resulted in significant lateral and vertical facies changes, which caused a high degree of lateral and vertical heterogeneity in the resulting reservoir. Past reservoir characterization studies of this delta complex had correlated sandstone and shale units on the basis of lithostratigraphic

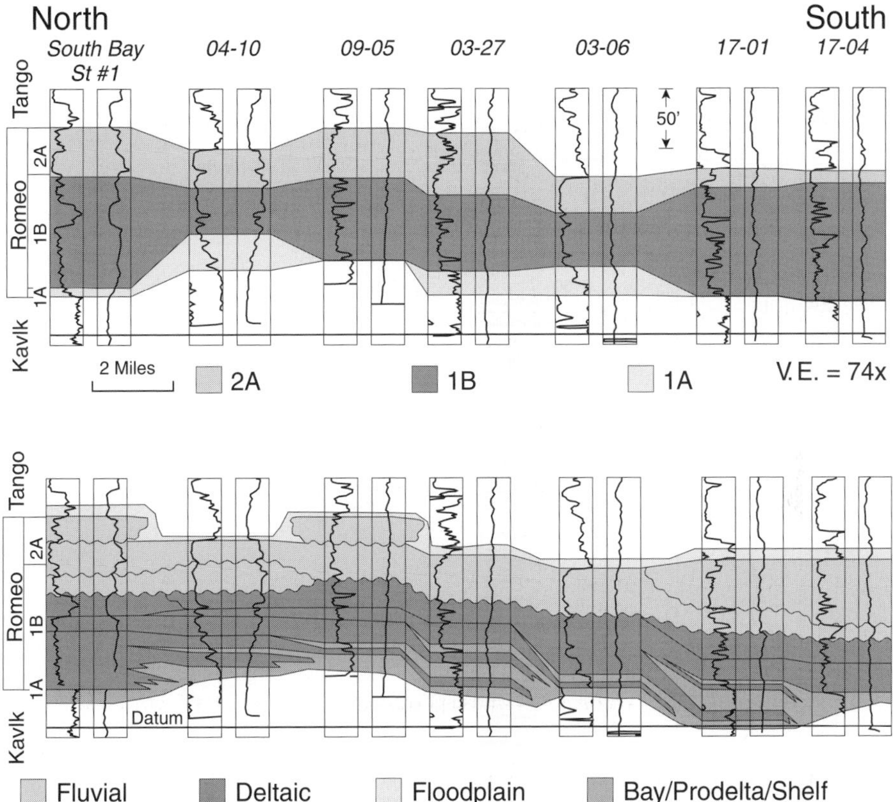

Fig. 9.13. Comparison of prestudy lithostratigraphic correlations (upper cross-section) and poststudy sequence stratigraphic correlations and interpretations of the Romeo interval (lower cross-section). In the lithostratigraphic cross-section, only the interval 1A is discontinuous. In the sequence stratigraphic cross-section, downlap patterns are evident by the gently dipping nature of the basal intervals. The poststudy interpretation shows the stratigraphy is not tabular; instead, sand-prone deltaic wedges interfinger downdip with bay and prodelta/shelf shales. After Tye et al. (1999). (Reprinted with permission of AAPG, whose permission is required for further use.)

features. However, Tye et al. (1999) were better able to match well correlations with production history using a sequence- or time-stratigraphic correlation scheme (discussed in Chapter 11) (Fig. 9.13). Prior to this sequence stratigraphic interpretation, the gently dipping nature of these reservoir sandstones was unrecognized and led to improper correlations.

Insights gained into the formation of distributary mouth bars led to improved recovery in the Prudhoe Bay field. Detailed information on the lateral attributes of the mouth bars (including permeability), gained from the horizontal, cored well, rather than from interpretations of standard vertical cores, provided the

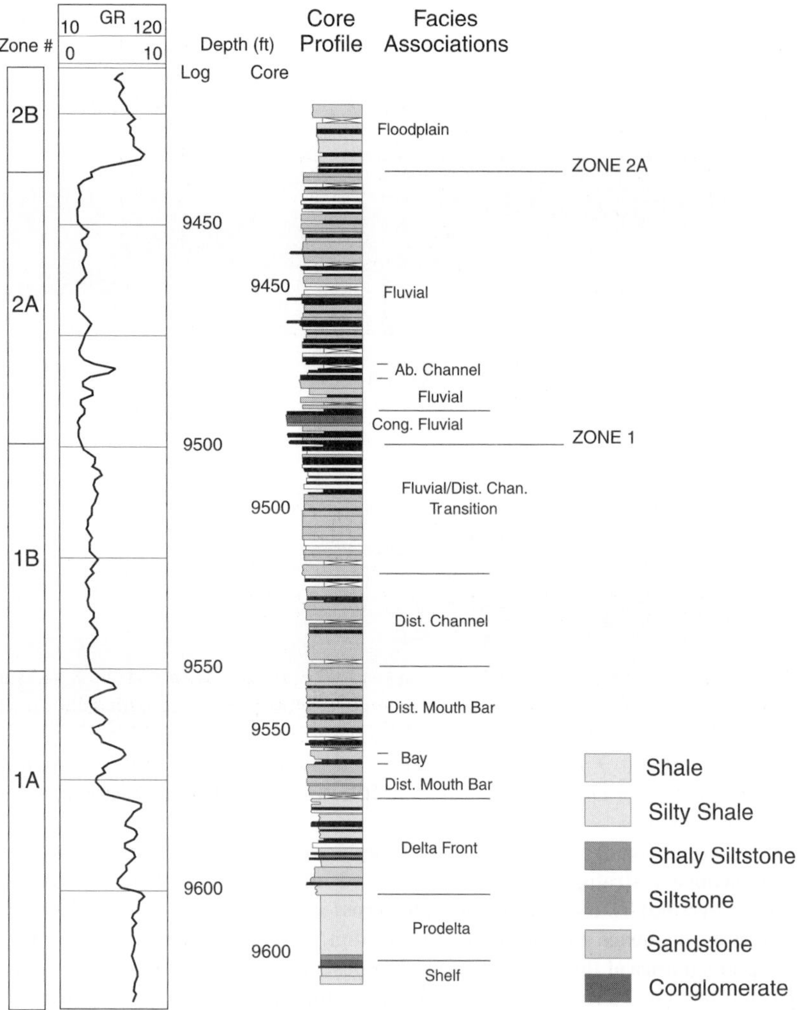

Fig. 9.14. Typical stacking pattern and gamma-ray response of Romeo lithofacies and facies associations in well 07-06. After Tye et al. (1999). (Reprinted with permission of AAPG, whose permission is required for further use.)

knowledge that a sinusoidal, horizontal well path would contact more high-permeability lithofacies than would a conventional well.

Shales also were found to be particularly important to production and fluid flow in Prudhoe Bay field's deltaic facies. Some shales are continuous over areas larger than that of the well spacing, but other shales are less continuous because of erosion by an overlying sandstone or by stratigraphic pinch-out (Fig. 9.15). Continuous shales create vertical flow barriers and help to focus a waterflood. By restricting perforations in horizontal wells to deltaic sandstones, developers avoided competition for water between these lower-permeability sandstones

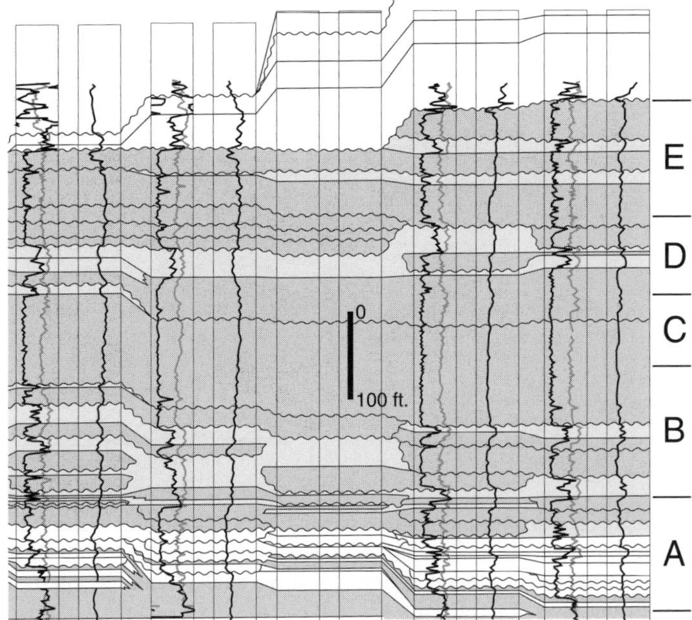

Fig. 9.15. Part of a north–south wireline log correlation across the Sadlerochit Sand at Prudhoe Bay field. Tan is floodplain facies. Yellow*/bright-gray is fluvial/distributary channel facies. White is distributary mouth bar/delta front and prodelta/shelf facies. A – Romeo interval, B – Tango interval, C – Victor interval, D – X-ray interval, and E – Zulu interval. Modified from Tye et al. (1999). (Reprinted with permission of AAPG, whose permission is required for further use.)

and overlying higher-permeability fluvial sandstones (Figs. 9.16 and 9.17) (Tye et al., 1999). One innovative application of the understanding of shale continuity was applied to a vertical well that had maintained a steady output in excess of 300 BOPD between May 1992 and January 1993. During that time, water production increased, indicating that oil production in that well would decline considerably over time (Fig. 9.18). A shale interval at 8950 ft was mapped as a sealing shale, which suggests that oil might be trapped in an underlying deltaic sandstone. A sidetrack horizontal well was drilled as an unconventional completion into the sandstone, and it led to new production of approximately 400 BOPD (Fig. 9.18).

A second example of the value of understanding shale continuity is presented in Fig. 9.19. The horizontal well 11-30 was drilled beneath a bay shale that acted as a shield from the overlying gas cap. A 894-ft (272-m)-long horizontal section was drilled entirely within the targeted sandstone, of which 475 ft (144 m) was determined to be net pay. As of 1999, production had stabilized at 1,600 BOPD,

* The indicated color is for a CD which contains all of the figures in color.

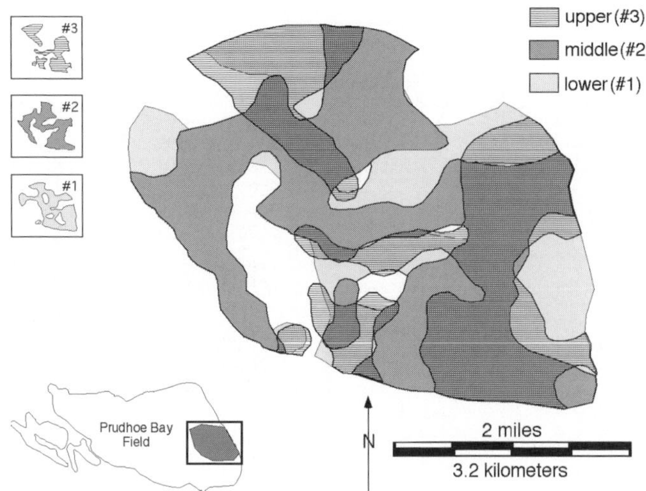

Fig. 9.16. Generalized distribution of three stratigraphically distinct shales in the eastern part of Prudhoe Bay field. In places, all three shales are absent as a result of nondeposition or erosion, whereas in other locations all three shales separate adjacent reservoir layers. Only with this level of stratigraphic detail can gravity drainage, waterflood, and enhanced-oil-recovery depletion mechanisms be effectively managed. After Tye et al. (1999). (Reprinted with permission of AAPG, whose permission is required for further use.)

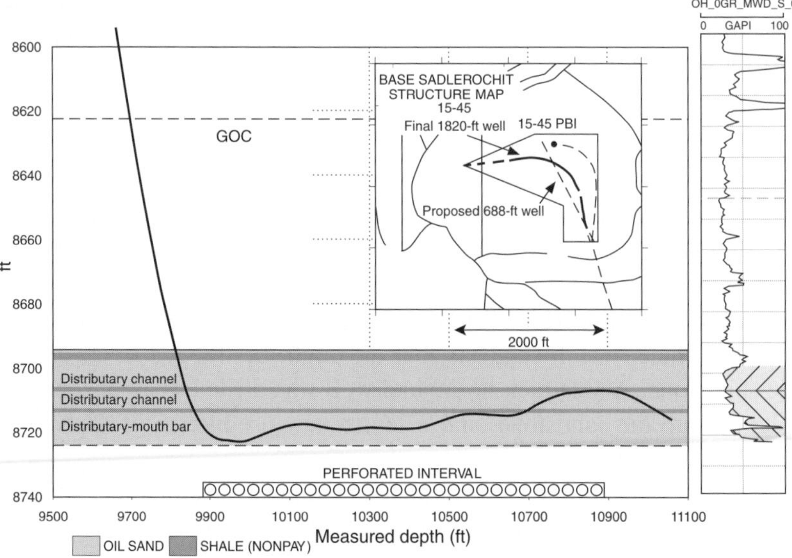

Fig. 9.17. Horizontal well drilled within distributary-channel and mouth-bar facies separated by shales. The shales compartmentalize the sandstones and shield against gas-cap encroachment. After Tye et al. (1999). (Reprinted with permission of AAPG, whose permission is required for further use.)

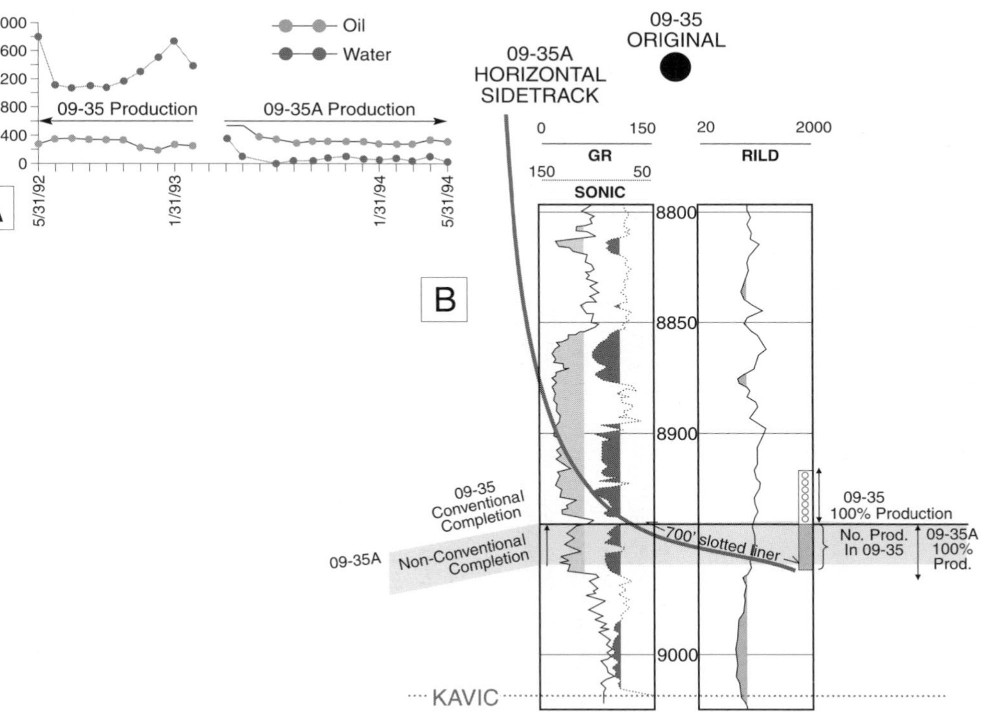

Fig. 9.18. Wireline logs for the original well 09-35, and a horizontal sidetrack well 09-35A. Well 09-35A was designed to produce from the basal Romeo sandstones without competition from stratigraphically higher perforations, and to use the shale at 8,936 ft (2,725 m) as a barrier from slumping water. The stippled pattern on the gamma-ray log shows the productive sandstones. The graph in the upper left compares oil and water production from the two wells, with a slight increase in oil rate and a dramatic decrease in water production from well 09-35A. Modified from Tye et al. (1999). (Reprinted with permission of AAPG, whose permission is required for further use.)

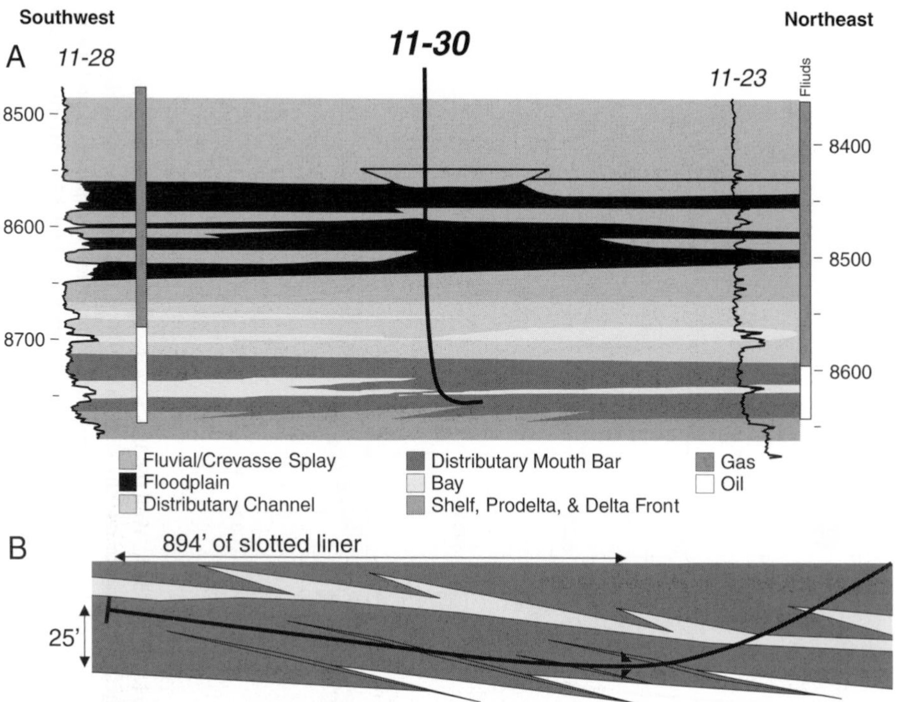

Fig. 9.19. Schematic cross-section illustrating facies-association interpretations between wells 11-28 and 11-23 and the predrill strategy for placing well 11-30. The horizontal trajectory of 11-30 is perpendicular to the cross-section in A. Wellbore path through the distributary-mouth-bar sandstones beneath a continuous bay shale is shown in B. Modified from Tye et al. (1999). (Reprinted with permission of AAPG, whose permission is required for further use.)

with a low gas/oil ratio (Tye et al., 1999). Detailed characterizations and applications such as these are increasing the life of this giant field.

9.4 Wave-dominated deltas

9.4.1 Processes and deposits

Along a coast that is unprotected from waves and ocean currents, sediment that reaches the shoreline is dispersed laterally (Fig. 9.4). If sufficient sand is supplied to the shoreline, over time the shore zone will accrete laterally as well as prograde basinward (Fig. 9.6). The end result is a fan-shaped or arcuate body of sand, such as the Nile Delta, which is areally extensive and quite thick (Fig. 9.20). Progradation of sandy beach ridges creates a vertical stratigraphic sequence that becomes progressively sandier and coarser-grained from the base

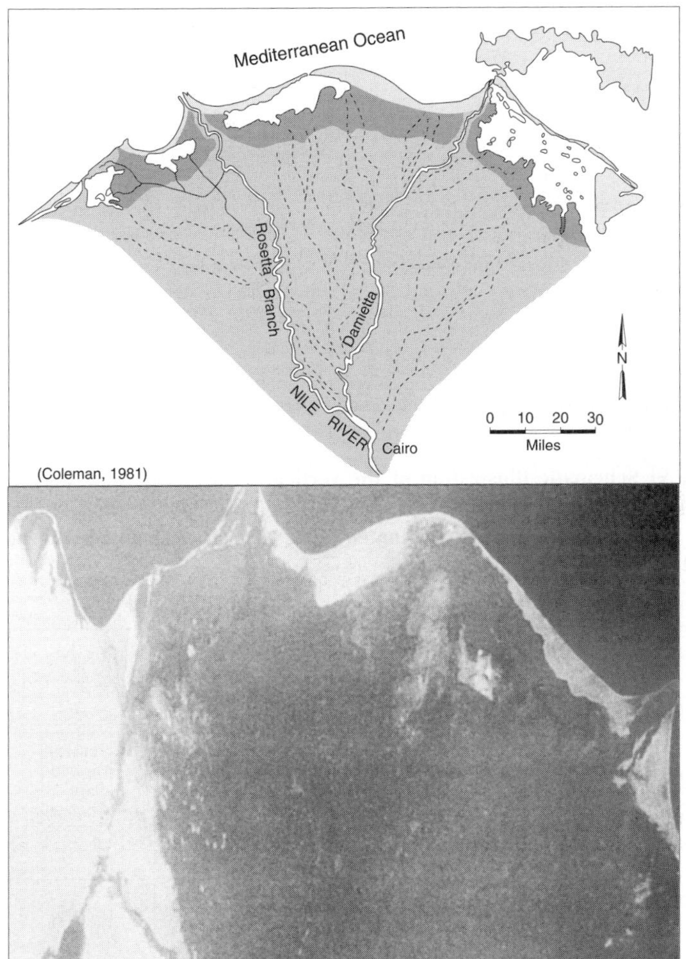

Fig. 9.20. Generalized geology of the Nile fan. Modified from Coleman et al. (1981). Lower photo is a satellite image of the Nile fan.

to the top (Fig. 9.21). Porosity and permeability also increase upward because of the grain-size effect discussed in Chapter 5 (Fig. 9.22).

In addition to becoming coarser-grained and sandier from base to top, the vertical sequence of sedimentary structures exhibits a preponderance of wave- and current-derived cross-bedding, broken shell fragments, and other indicators of high-energy processes (Fig. 9.21). This vertical stratigraphy is quite distinct from that of a river-dominated delta (compare Figs. 9.10, 9.14, and 9.21), and it provides the means for differentiating the two types of deltas in subsurface core. A subsurface well log cross-section of a wave-dominated delta deposit will exhibit greater lateral continuity of sandy beach ridges (Figs. 9.6 and 9.21) than does the river-dominated delta deposit (Figs. 9.6 and 9.10).

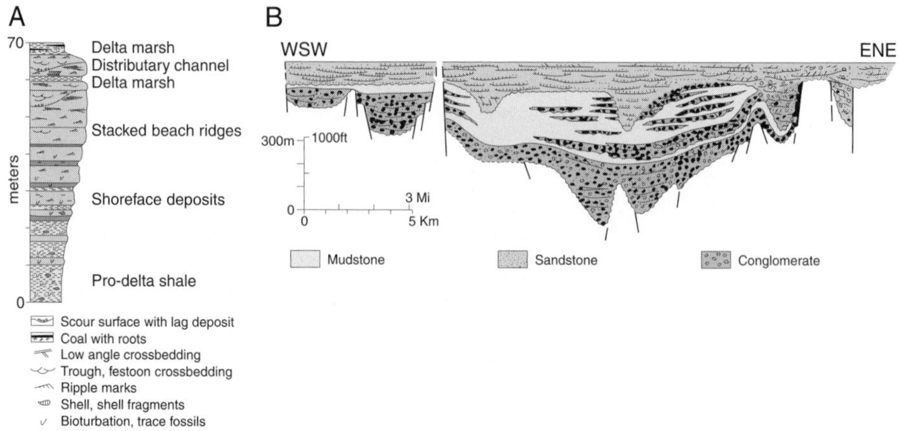

Fig. 9.21. (A) Schematic illustration of the vertical stratigraphy and sedimentary characteristics of a wave-dominated delta deposit. After Miall (1980). (B) Generalized lateral lithofacies distribution of a wave dominated delta deposit. (Reprinted with permission of Geological Association of Canada.)

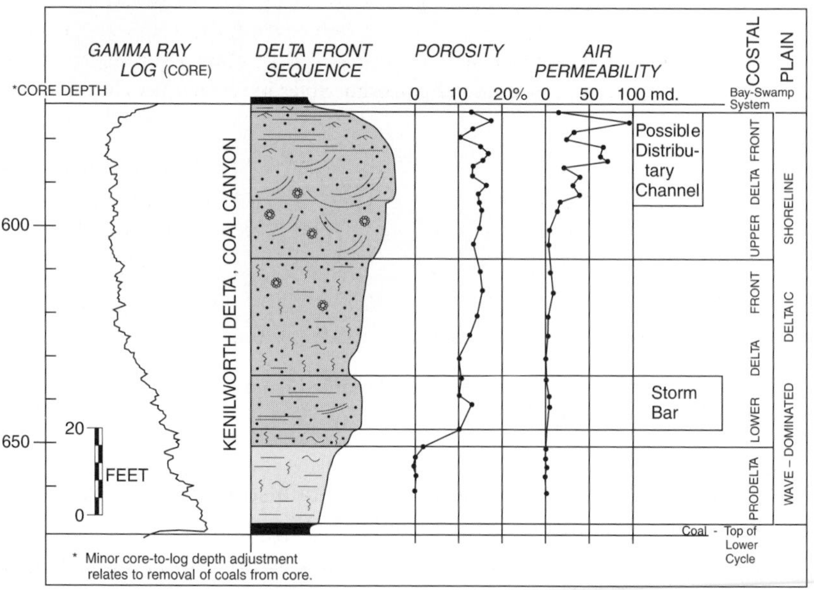

Fig. 9.22. One depositional cycle of the wave-dominated Kenilworth delta, Blackhawk Formation, Carbon County Utah. Modified from Balsley (1980). (Reprinted with permission of AAPG, whose permission is required for further use.)

Bhattacharya and Giosan (2003) have differentiated symmetric wave-influenced deltas from asymmetric ones. The distinction is based on the relative importance of fluvial discharge of sediment at the coast and on the degree

Fig. 9.23. (A) Aerial photo showing deflection of sand toward the lower left by waves as the sand reaches the river mouth. (B) Aerial photo showing shoreline deposits of part of a prograding wave-dominated delta deposit. Progradation is in the direction of (a) and waves move shoreline sands in the direction of (b) as the sand reaches the river mouth. (C) Aerial photo showing the seaward progression of paleoshorelines in the manner shown in (B).

and direction of lateral dispersal of sand when the river-borne sediment reaches the coast. In areas where there is fluvial input of sediment but little or no net longshore sediment transport, an arcuate to cuspate, symmetric delta forms (Fig. 9.6). Beach ridges at distributary mouths are more or less equally distributed on both sides of the mouth. However, if there is sufficient longshore drift of sediment once it reaches the coast (Fig. 9.23A), the updrift side will consist of a beach ridge plain of sand trapped by the river, and a downdrift side will consist of elongate sandy ridges that can be separated by mud-filled troughs (Fig. 9.23B and Fig. 9.23C). Examples of symmetric and asymmetric wave-influenced deltas are shown in Fig. 9.24, and a general model for asymmetric, wave-dominated deltas is provided in Fig. 9.25. Bhattacharya and Giosan (2003) also point out that both symmetric and asymmetric wave-influenced deltas can form within a single delta system; for example, the symmetric Rosetta lobe and the asymmetric Damietta lobe of the Nile Delta (Fig. 9.20).

9.4.2 Reservoir example: Budare field

The Budare field, in the eastern Venezuela Basin, provides a reservoir example of a mature wave-dominated deltaic complex (Hamilton et al., 2002). The field has produced 95 MMBO of an estimated 443 MMSTBO since its discovery in 1954. A sustained 6-year decline during the early 1990s reduced daily

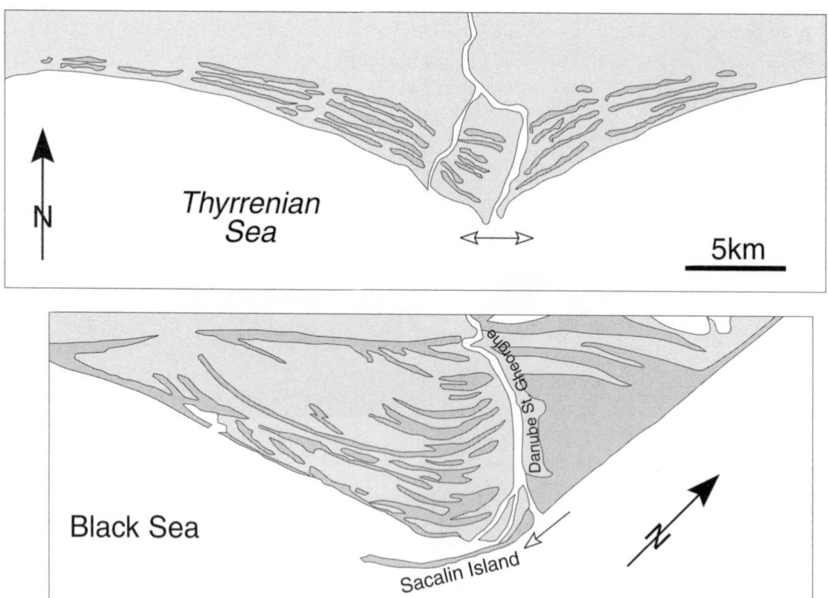

Fig. 9.24. Morphology of a symmetric, wave-dominated delta, Tiber Delta, Italy (top figure) and an asymmetric, wave-dominated delta from St. Gheorghe lobe of the Danube delta, Romania (bottom figure). Asymmetry is the result of downcurrent redistribution of sediments by waves, once the sediments reach the shoreline, and upcurrent blocking of sand transported toward the river by longshore drift. Top figure after Bellotti et al. (1994). Bottom figure after Gastescu (1992). (Reprinted with permission of Society for Sedimentary Geology, SEPM.)

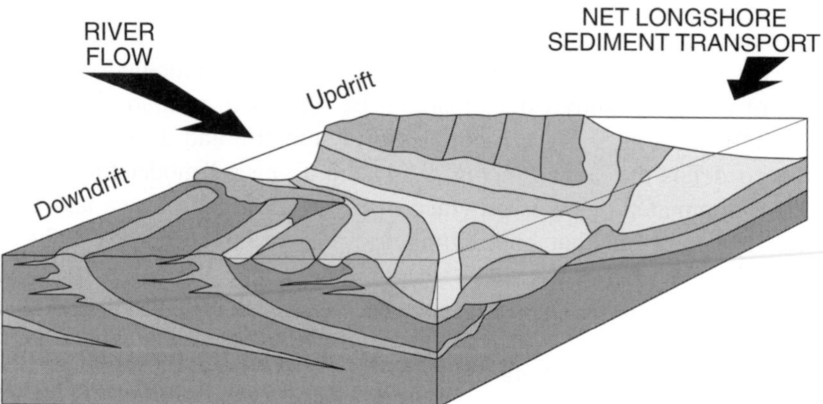

Fig. 9.25. Schematic diagram illustrating the inferred 3D facies architecture of an asymmetric delta. Sandy barrier bar complexes and associated prodelta muds form in the downcurrent portion of the delta. The upcurrent side of the delta comprises a sandy beach ridge plain. Modified from Bhattacharya and Giosan (2003).

production to 3,000 BOPD. Detailed reservoir characterization resulted in a sustained 4-year increase in production, to 16,000–17,000 BOPD, culminating in incremental recovery of more than 24 MMBO. The reservoir characterization included integration of geologic, 3D seismic, and engineering data through the life of the field, coupled with a study of analog outcrops.

The main producing zones in Budare field are the Tertiary-age Merecure and Oficina reservoirs, which were deposited as a fluvial and wave-dominated delta complex. The Merecure interval is divided into zones A, B, and C, with zone A being the wave-dominated delta interval (only zone A will be discussed further). The Merecure A zone is 30–37 m (100–120 ft) thick and is bounded at the top and base by laterally continuous, easily correlative transgressive marine shales (maximum flooding surfaces, as defined in Chapter 11). Net sandstone thickness ranges from 12 to 29 m (40–95 ft). Sandstone geometries are both depositional-dip-elongate in the north–south orientation, and depositional-strike-elongate in the west–east orientation (Fig. 9.26). These two trends reflect the two dominant facies, distributary channels (the north–south trend) and distributary mouth bars and thinner strandplain or beach-ridge deposits (the west–east trend) (Fig. 9.27). Both distributary-channel and mouth-bar facies exhibit a blocky, sand-rich log character, whereas thinner, serrate, and subtly upward-coarsening log patterns

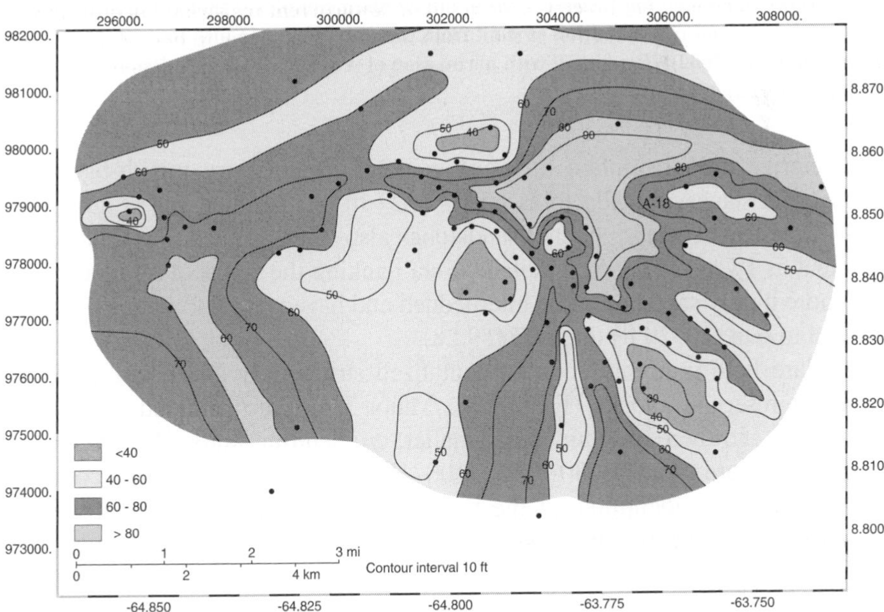

Fig. 9.26. Gross sandstone thickness map, Merecure unit A, Budare field, Venezuela. Note the change in orientation of sandstones from north–south in the southern part of the field to east–west in the northern part of the field. After Hamilton et al. (2002). (Reprinted with permission of AAPG, whose permission is required for further use.)

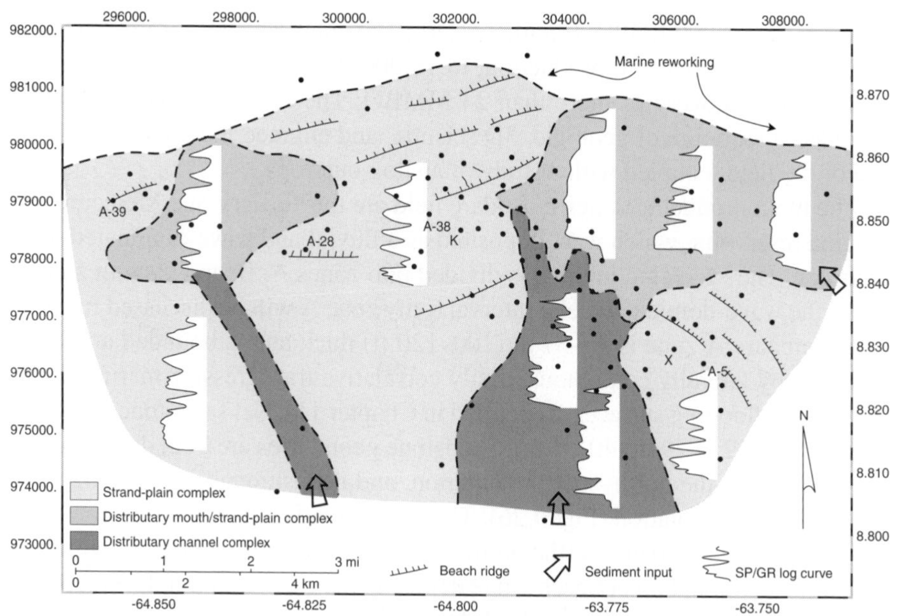

Fig. 9.27. Log facies, Merecure unit A, Budare field, Venezuela. Blocky well log facies represent distributary channel and channel mouth bar facies, and thin, serrate, and subtly upward-coarsening log facies represent strandplain complexes that resulted from marine reworking and redistribution of sand from the distributary mouth bar complex. After Hamilton et al. (2002). (Reprinted with permission of AAPG, whose permission is required for further use.)

characterize the strandplain facies (Fig. 9.27). Distributary mouth-bar sandstones that stack vertically to as much as 27 m (90 ft) thick act as individual flow units. Distributary channel sandstones also stack vertically to thicknesses as great as 17 m (55 ft). Strandplain strata flanking the distributary mouth bars are more highly sand and shale interbedded and have typical flow units ranging from 3 to 6 m (10–20 ft) thick (Fig. 9.28).

Budare field is highly compartmentalized structurally and contains 18 individual reservoir blocks (Fig. 9.29). Major faults are oriented northwest–southeast, and there are numerous smaller, crosscutting faults. Throws on the larger faults are typically 30–46 m (100–150 ft). Of the 18 isolated compartments, five were identified on the basis of variable oil–water contact levels, five were identified from the lowest-known-oil level, two were ascertained from pressure or production data, and the remaining six were identified from data in other intervals in addition to Merecure A. For example, a vertical well in one compartment recorded a pressure of 1,112 psi in 1991; a horizontal stepout well in 1993 crossed a fault, and the measured pressure in the adjacent compartment was 1,636 psi (Fig. 9.30). Pressure measurements recorded by another pair of wells drilled across the same fault, but into a different compartment, were

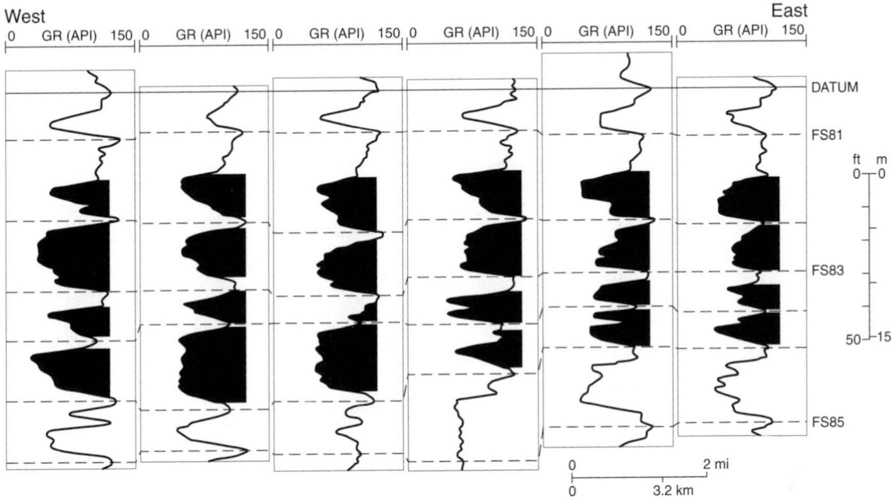

Fig. 9.28. East–west cross-section of Merecure unit A, Budare field, Venezuela, illustrating the bedding architecture of the strandplain and associated facies. Note the good long-distance continuity of the facies. After Hamilton et al. (2002). (Reprinted with permission of AAPG, whose permission is required for further use.)

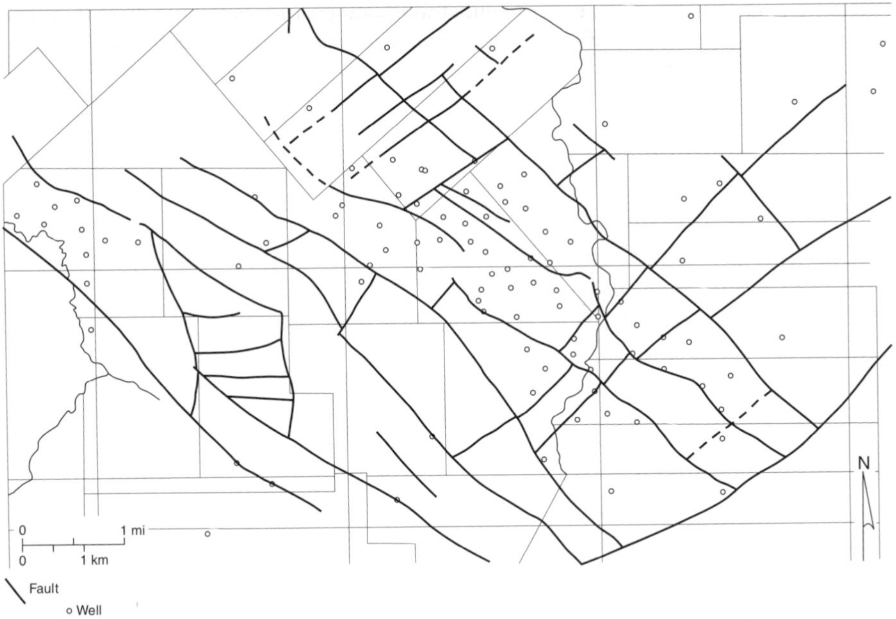

Fig. 9.29. Structure and faulting patterns in the Merecure unit A, Budare field, Venezuela. Note the high degree of potential structural compartmentalization of the field. After Hamilton et al. (2002). (Reprinted with permission of AAPG, whose permission is required for further use.)

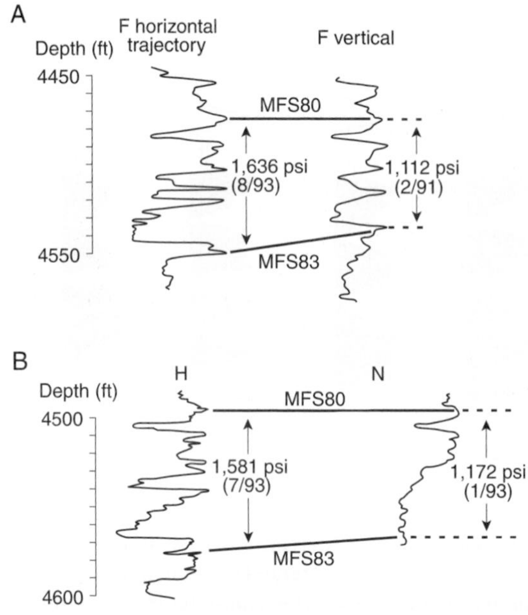

Fig. 9.30. Pressure data for wells in the Merecure unit A, Budare field, Venezuela. In both cases A and B, pressure differentials between two wells (vertical F and horizontal F in A; vertical H and N in B), are due to a sealing fault between the wells. After Hamilton et al. (2002). (Reprinted with permission of AAPG, whose permission is required for further use.)

1,172 and 1,581 psi (Fig. 9.30); these two wells were drilled 6 months apart. Several additional examples of fluid behavior related to compartmentalization are provided by Hamilton et al. (2002).

The key to successful reservoir characterization of the Budare field was identification of the degree and types of heterogeneities and compartments of the Merecure A and other reservoir intervals. That analysis resulted in highgrading the earlier estimated OOIP to 622 MMSTB and inspired location of numerous drilling scenarios – which led to calculated reserve additions of much as 52.8 MMSTB.

9.5 Tide-dominated deltas

9.5.1 Processes and deposits

In an embayed coastline, waves and tides can interact closely, depending upon the configuration of the embayment and the orientation of the incoming waves (Fig. 9.4). Tidal ranges (from normal low tide to normal high tide) are classed as microtidal (less than 2 m), mesotidal (2–4 m), and macrotidal (more than 4 m). The macrotidal range has the most pronounced effect on tidal delta deposits, in

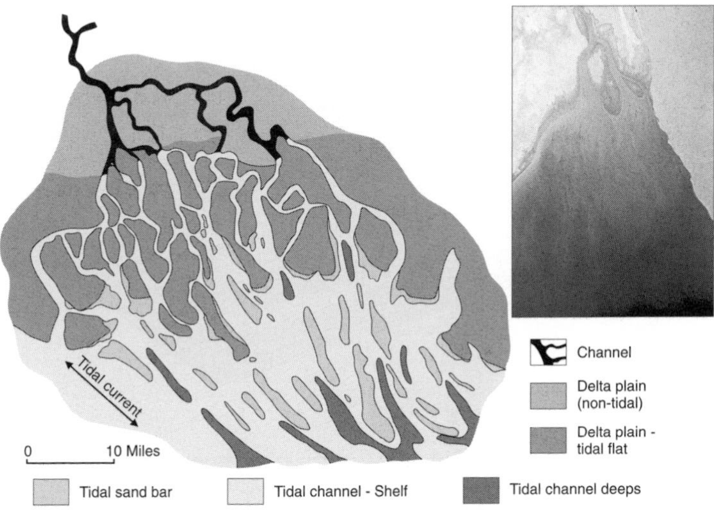

Fig. 9.31. Schematic illustration of the modern Gulf of Papua's tide-dominated delta. After Fisher et al. (1969). Note the elongate orientation of the subtidal sand bars. Source of inset photo is unknown. (Reprinted with permission of Gulf Coast Association of Geological Societies.)

part because of the relatively greater area of the shore zone surface that can be reworked during a single cycle.

Particularly in narrow embayments, tidal energy can build progressively landward, giving rise to a very large tidal range. Thus, the tide-dominated delta can be a very high-energy environment, and the sediments will be relatively coarse grained. Generalized environments of deposition and resultant facies are shown in Fig. 9.31. In this model, the tidally influenced delta plain is muddy. Sand is confined to tidal channels and offshore tidal-sand ridges, which are aligned parallel to the direction of the tidal currents.

As is the case with river-dominated and wave-dominated deltas, the vertical stratigraphy of tide-dominated deltas offers some unique characteristics that provide a means for identifying and differentiating them. Such characteristics may include cross-bedding, including bidirectional herringbone cross-bedding, and the presence of shell hash, burrows, and well-sorted sands. The elongate nature of the tidal channels and of offshore tidal sand ridges leads to lenticular and laterally discontinuous sand bodies in the depositional strike orientation, but also to good lateral continuity of sand bodies in the depositional dip direction (Fig. 9.32).

9.5.2 Reservoir example: Lagunillas field

Eocene tide-dominated delta strata in Lagunillas field, Maracaibo Basin, western Venezuela, serve as an example of a tide-dominated-delta reservoir. Here,

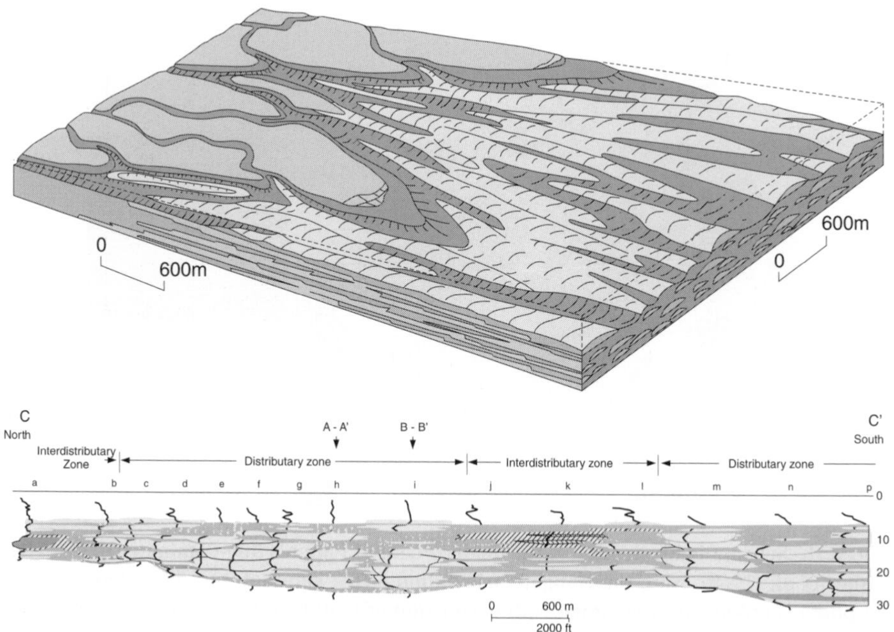

Fig. 9.32. Schematic 3D depositional model of the Eocene-frac tide-dominated delta (top figure), and strike-oriented log-facies cross-section of one of the reservoir intervals (bottom figure), Lagunillas field, Maracaibo Basin, western Venezuela. After Maguregui and Tyler (1991). (Reprinted with permission of Society for Sedimentary Geology, SEPM.)

deltaic sandstones are of low permeability (average 10 md) but contain more than 5 BBOOIP (Maguregui and Tyler, 1991). As a result, recovery efficiency is about 14%. These reservoir sandstones are termed "Eocene frac" reservoirs because they are highly cemented sandstones that require fracture stimulation to facilitate oil flow into the wellbore. When they are artificially fractured, their production increases from about 100 BOPD to 500 BOPD. In addition to being low-permeability sandstones, they have lateral discontinuity in the depositional strike direction (Fig. 9.32) that is an additional cause of low recovery efficiency.

The four dominant facies within the reservoirs are: estuarine distributary-channel sandstones, tidal sand-ridge sandstones, prodelta/shelf facies, and tidal-channel facies.

Estuarine distributary-channel sandstones (Fig. 9.33) are characterized by medium- and small-scale trough cross-stratification at the base, and by wavy laminations, ripples, and flaser bedding at the top. Individual sandstones averaging 2.6 m (8 ft) thick and stack to form a single estuarine distributary-channel deposit.

Tidal sand-ridge sandstones (Fig. 9.34) average 2 m (6 ft) thick and are characterized by an upward increase in grain size, good grain sorting, small-scale trough cross-bedding in their upper part, and substantial burrows.

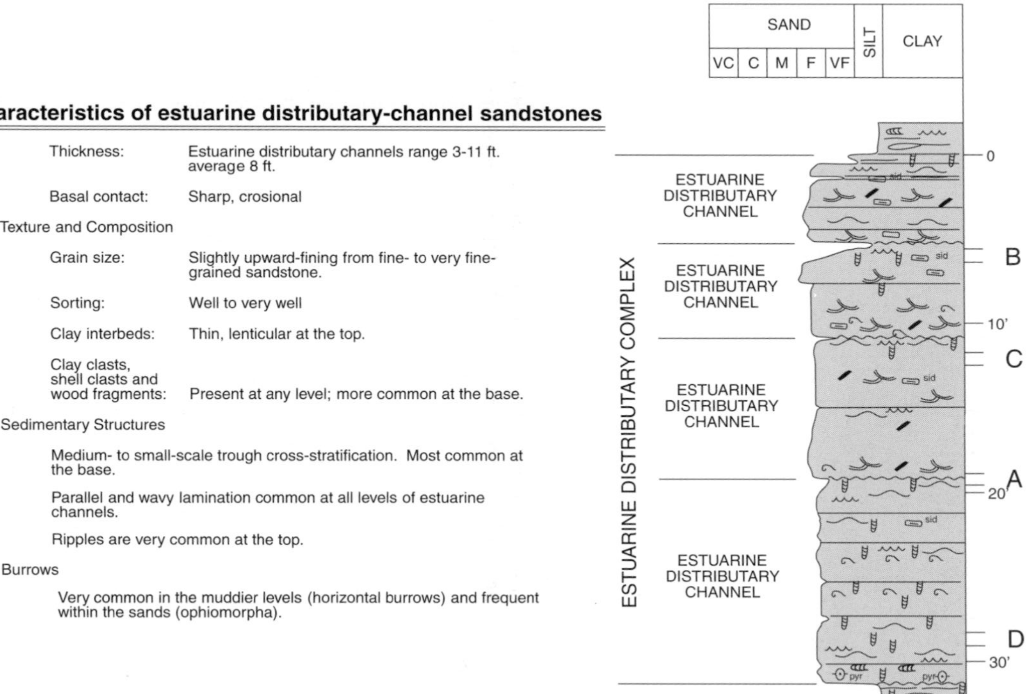

Characteristics of estuarine distributary-channel sandstones
Thickness: Estuarine distributary channels range 3-11 ft. average 8 ft.
Basal contact: Sharp, crosional
Texture and Composition
Grain size: Slightly upward-fining from fine- to very fine-grained sandstone.
Sorting: Well to very well
Clay interbeds: Thin, lenticular at the top.
Clay clasts, shell clasts and wood fragments: Present at any level; more common at the base.
Sedimentary Structures
Medium- to small-scale trough cross-stratification. Most common at the base.
Parallel and wavy lamination common at all levels of estuarine channels.
Ripples are very common at the top.
Burrows
Very common in the muddier levels (horizontal burrows) and frequent within the sands (ophiomorpha).
SAND
VC
C
M
F
VF
SILT
CLAY
ESTUARINE DISTRIBUTARY COMPLEX
ESTUARINE DISTRIBUTARY CHANNEL
ESTUARINE DISTRIBUTARY CHANNEL
ESTUARINE DISTRIBUTARY CHANNEL
ESTUARINE DISTRIBUTARY CHANNEL
0
B
10'
C
A
20'
D
30'
sid
pyr

Fig. 9.33. Sedimentary characteristics and stacking pattern of the estuarine distributary channel sandstone. Individual sandstones exhibit subtle upward-fining vertical trends and a decrease in scale of sedimentary structures. Erosive basal contacts commonly separate fine-grained sandstones at the base from very fine-grained sandstones at the top of the underlying channel. Lagunillas field, Maracaibo Basin, western Venezuela. After Maguregui and Tyler (1991). (Reprinted with permission of SEPM, Society for Sedimentary Geology.)

Characteristics of tidal sand-ridge sandstones

Thickness: Range 3-12 ft. average 6 ft.

Basal contact: Transitional with marine shales.

Texture and Composition

Grain size: Upward-coursening from siltstone/very fine-grained sandstone at the base to upper fine-grained sandstone at the top.

Sorting: Well to very well.

Clay interbeds: Very common and gradually decreasing toward the top. Their relative content defines the upward-coursening, thickening trendl.

Clay clasts and wood fragments: Common in the upper half.

Shell clasis: Mainly in the sandier levels of the upper part. Frequently in high concentrations at the top.

Sedimentary Structures

Wavy lamination and ripples are dominant in the lower two thirds of the facies.

Small-scale trough cross-stratification is usually present in the upper one-third facies.

Burrows

Most common within the lower half or at the very top of the facies. Frequently muddy sandstones are homogenized by bioturbation. Ophiomorpha traces are common in sand-rich levels.

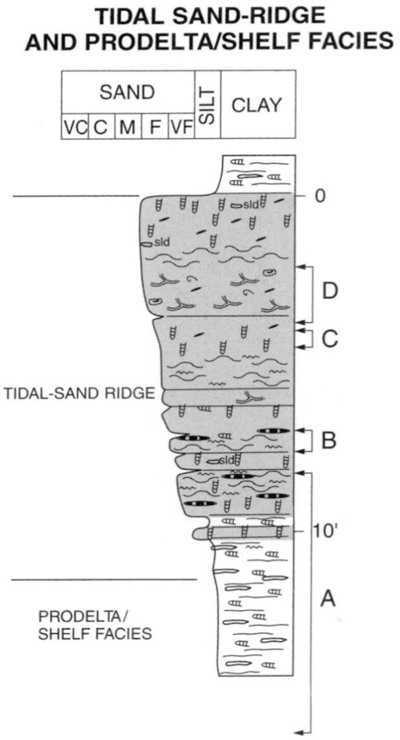

TIDAL SAND-RIDGE AND PRODELTA/SHELF FACIES

SAND
VC C M F VF
SILT
CLAY
0
10'
D
C
B
A
sld
TIDAL-SAND RIDGE
PRODELTA/
SHELF FACIES

Fig. 9.34. Sedimentary characteristics and stacking pattern of tidal sand-ridge sandstones. Sandstone intervals coarsen- and thicken-upward from a gradational basal contact with prodelta/shelf facies. Tops of the sandstone ridges exhibit well developed, small-scale trough cross-stratification, Lagunillas field, Maracaibo Basin, western Venezuela. After Maguregui and Tyler (1991). (Reprinted with permission of SEPM, Society for Sedimentary Geology.)

Prodelta/shelf facies are composed of highly laminated, burrowed shales intercalated with very thin lenses of siltstone and very-fine-grained sandstone. This facies thickens seaward and overlies tidal sand-ridge facies.

The tidal-channel facies was not cored, so its sedimentary characteristics are unknown. On well logs, this facies has a low net-sand content, which allows it to be differentiated from distributary channel sandstones. Well logs show a sharp base, a serrated pattern of sandstone/shale interbedding, and a fining-upward grain-size trend. Interpreted areal distribution of the four facies is shown in Fig. 9.35.

Five complete tide-dominated delta cycles comprise the reservoir. Elongation of the sandstone bodies is in the depositional–dip (west–east) direction. Cycle 1, the lowermost cycle, is composed, from the base upward, of unit A, which consists of a series of stacked tidal sand-ridge sandstones, and unit B, which con-

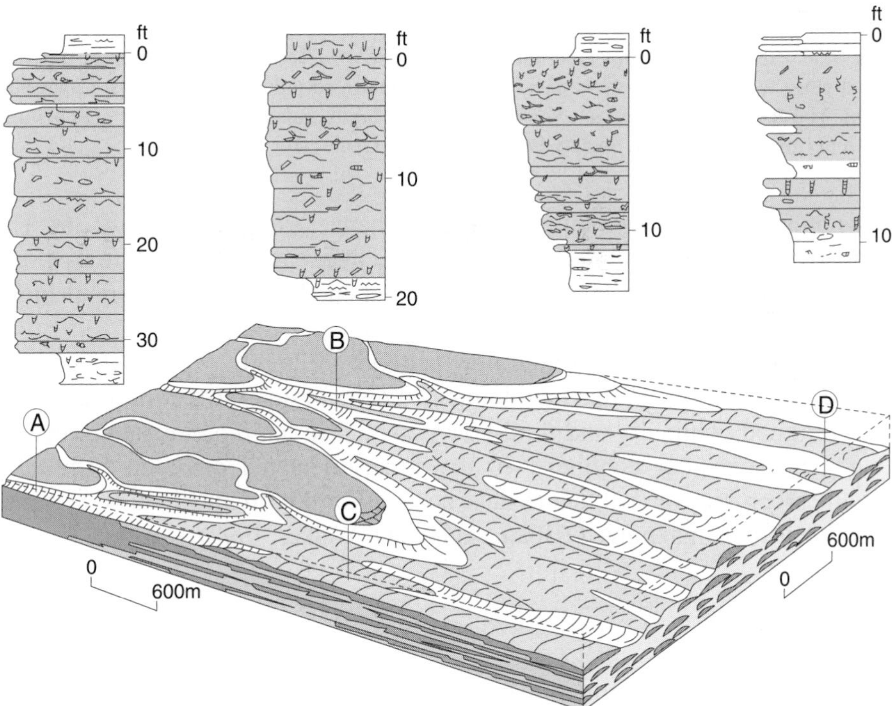

Fig. 9.35. The same schematic 3D depositional model as that shown in Fig. 9.32, but in this figure, sedimentary facies, and their positions within the model, are illustrated. A is a typical vertical profile of an estuarine distributary complex. B is a seaward 'transitional' facies with greater reworking by tidal currents. C is a proximal tidal sand ridge facies. D is a more distal, thinner-bedded sand ridge facies, deposited farther from the estuary, where the sediment supply is limited and tidal currents are weaker. Lagunillas field, Maracaibo Basin, western Venezuela. After Maguregui and Tyler (1991). (Reprinted with permission of Society for Sedimentary Geology, SEPM.)

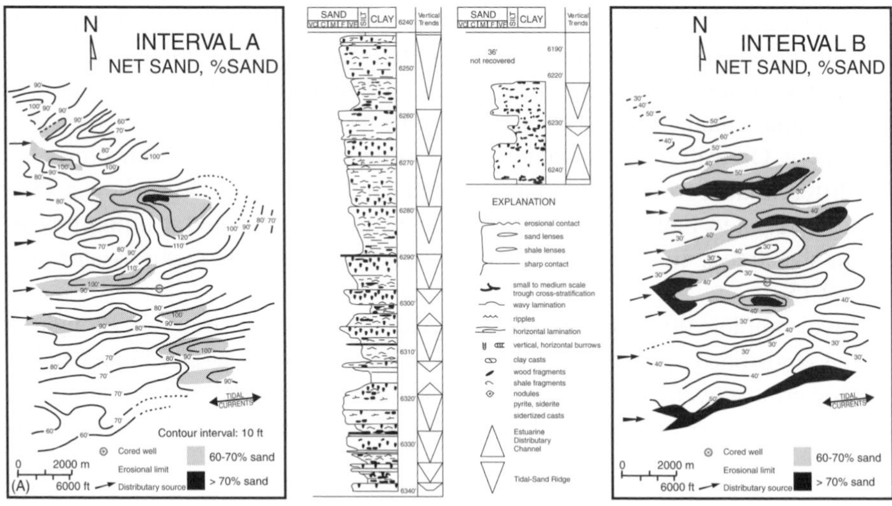

Fig. 9.36. Net-sand maps for reservoir intervals A and B. Trends of elongation of the tidal sand ridge facies are east–west, parallel to the interpreted paleo-tidal currents, Lagunillas field, Maracaibo Basin, western Venezuela. After Maguregui and Tyler (1991). (Reprinted with permission of SEPM, Society for Sedimentary Geology.)

sists of at least some estuarine distributary-channel sandstones (Fig. 9.36). The geometries of unit A and B sandstones show a linear pattern parallel to the tidal-current direction. Other cycles exhibit vertical and lateral depositional patterns similar to those of the four facies described above.

Reservoir continuity and fluid-flow patterns are highly dependent upon depositional processes in this tide-dominated delta system. Reservoir sandstones exhibit good continuity and fluid-flow potential in the dip-elongate direction, but they have poor continuity in the strike-orientated direction. Advanced hydrocarbon-recovery strategies must account for this architectural style if production is to be maximized (Bhattacharya and Walker, 1992).

9.6 Summary

Globally, deltas often contain major oil and gas reservoirs. The geometry, size, and internal architecture of deltas are functions of many variables related to the delta's mode of formation. A tripartite classification of deltas, into river-, wave-, and tide-dominated deltas, has been a standard for many years. However, even within each of these delta types, the distribution of properties can vary considerably depending on the delta's depositional history. With regard to reservoir performance and optimization, perhaps the most significant difference in delta properties is in orientation and continuity of sand (reservoir) and shale (barrier) trends. Reservoir quality also varies according to the facies within the delta.

To maximize hydrocarbon production, it is not sufficient to merely classify the reservoir as a delta. A complete understanding of the characteristics and variations of an individual delta's reservoir is required for proper well placement and reservoir management.

Chapter 10

Deepwater deposits and reservoirs

10.1 Introduction

The deepwater depositional system is the one type of reservoir system that cannot be easily reached, observed, and studied in the modern environment. The study of deepwater systems requires many different remote-observation techniques, each of which can provide information on just one part of the entire system. As a consequence, the study and understanding of deepwater depositional systems as reservoirs has lagged behind that of the other reservoir systems, whose modern processes are more easily observed and documented. Geoscientists use an integrated approach to study deepwater systems, working in interdisciplinary teams with multiple data types, including outcrop studies, 2D and 3D seismic-reflection data (both shallow- and deep-resolution data), cores, log suites, biostratigraphy, and well-test and production information. These data sets are routinely incorporated into computer reservoir models to simulate reservoir performance.

10.1.1 Definitions

The term "deep water" is used in two ways. First, in the geologic sense, deep water refers to sediments that have been transported under gravity-flow processes and deposited in the marine environment, beneath storm-wave base, from the slope to the floor of a basin. Sediment gravity-flow processes also are operative in lakes and in cratonic basins in which water depths may exceed 300 m. Unless otherwise stated, in this book the term "deepwater systems" refers to marine-sediment gravity-flow processes, environments, and deposits. Other authors have used different terms for describing deepwater processes and deposits, such as "turbidite systems" (Mutti and Normark, 1987, 1991), "turbidite system complexes" (Stelting et al., 2000), and "submarine fans" (Bouma et al., 1985). Deepwater deposits and reservoirs occur in basins that are currently both onshore and offshore.

Second, the engineering definition of deep water refers to offshore reservoirs that have been drilled in modern water depths – specifically, to depths more then 500 m. This definition of deep water is used by drilling engineers to describe the depth of the water column through which the drill string must extend before the bit reaches the seafloor (i.e., the "mud line"). In this chapter, deep water refers to water depths from 500 m to 2,000 m, and ultradeep water is any water deeper than 2,000 m.

10.1.2 Global deepwater resources

Exploration and production in deep and ultradeep water have expanded greatly during the past 15 years, to the point at which they are now major components of the petroleum industry's annual upstream operating budget. Unique challenges in deepwater exploration and production include the facts that: (1) drilling is occurring in progressively deeper waters, as exploration continues to move farther offshore (basinward); (2) deepwater reservoirs are geologically complex; and (3) leading-edge technology is required for exploring for, and developing deepwater reservoirs. To recoup the large financial outlay needed to meet these challenges, operators must strive to reduce cycle time (time between discovery and first production) and to maintain operations at maximum efficiency, with reasonable cost.

After the first deepwater drilling (engineering definition) in the late 1970s, 38 giant discoveries (more than 500 MMBOE recoverable) have been made (Figs. 10.1 and 10.2) (Pettingill and Weimer, 2001). Although worldwide the total number of giant fields that have been discovered in recent decades has leveled off, the discovery rate of deepwater giants is rapidly increasing. Deepwater giant reserves are approximately 66% oil, compared with 36% oil for all giants found in the same time period (Figs. 10.1 and 10.2). .

Deepwater discoveries account for less than 5% of the current worldwide total oil-equivalent resource, although this amount is increasing steadily (BP, 2000). Gas exploration in deep water is immature, because of current infrastructure and economic limitations, but it is also destined to become a major focus in the future.

By the end of 2003, approximately 78 BBOE of total resources had been discovered in deep water, in 18 basins on 6 continents (Fig. 10.2). Most deepwater resources have been found in the northern Gulf of Mexico, Brazil, and West Africa (Fig. 10.2). This total consists of 43 million barrels (MMBBL) of oil and condensate and approximately 180 trillion cubic feet (tcf) of gas. Deepwater resources constitute about 85% of total discovered resources; ultradeep water has about 15%. More than half of this total has been discovered since 1995. Only about 31% of the total resources are developed or are currently under development, and less than 5% of the total has been produced, underscoring the play's immaturity (Fig. 10.1).

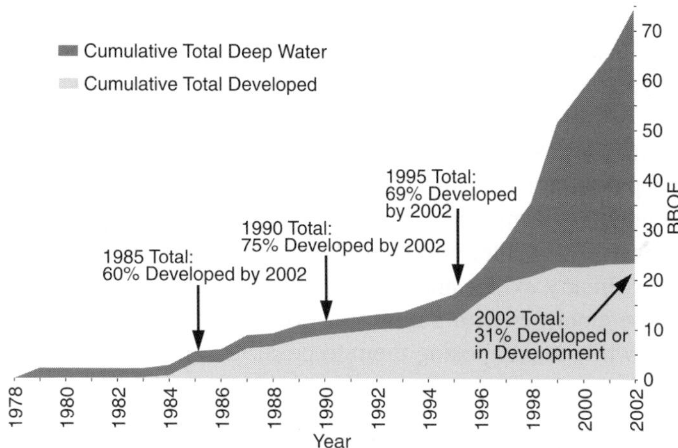

Fig. 10.1. Deepwater discovered resources versus time as of the end of 2002. Resources (BBOE) are based on deepwater reservoirs (present water depths of 500–2,000 m). (Reprinted with permission of AAPG, whose permission is required for further use.)

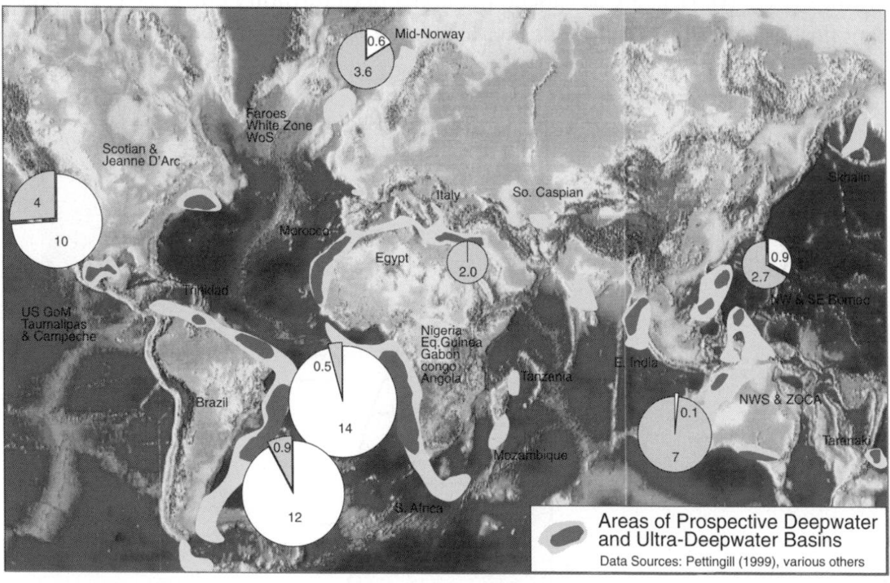

Fig. 10.2. Total discovered deepwater (more than 500 m) recoverable petroleum resources per region, announced as of mid-2001. These resources include producing reserves, those in development, and technically recoverable resources for which development has not been sanctioned. Major prospective basins are also shown. Green*/white – oil, red/gray – gas. Total discovered – 78 BBOE, of which 48 billion barrels is oil and 180 tcf is gas. Updated from Pettingill and Weimer (2001). (Reprinted with permission of AAPG, whose permission is required for further use.)

* The indicated color is for a CD which contains all of the figures in color.

The global deepwater exploration success rate was about 10% until 1985, but since then it has averaged approximately 30% because of remarkable success in the Gulf of Mexico and West Africa (BP, 2000). Exploration success rates have been highest in West Africa and lowest in Asia. In the lower Congo Basin, the success rate has exceeded 80% during the past few years.

Although participation in the global deepwater play initially was restricted to a few large major companies, progressively smaller companies have become involved. Generally, smaller companies are exploring in areas where (1) major infrastructure already exists, and consequently they are able to operate, and/or (2) they can be a partner with a limited working interest, thereby limiting their financial risk while still exposing them to possible high rewards.

10.2 Sedimentary processes operative in deep water

The first real recognition of deepwater (geologic definition) processes and deposits evolved from a classic paper by Kuenen and Migliorini (1950), who described "graded beds" from laboratory flume experiments and outcrop observations (Fig. 10.3). They advanced the concept of turbidity currents as an important process by which sediment is transported from shallow water to deep water

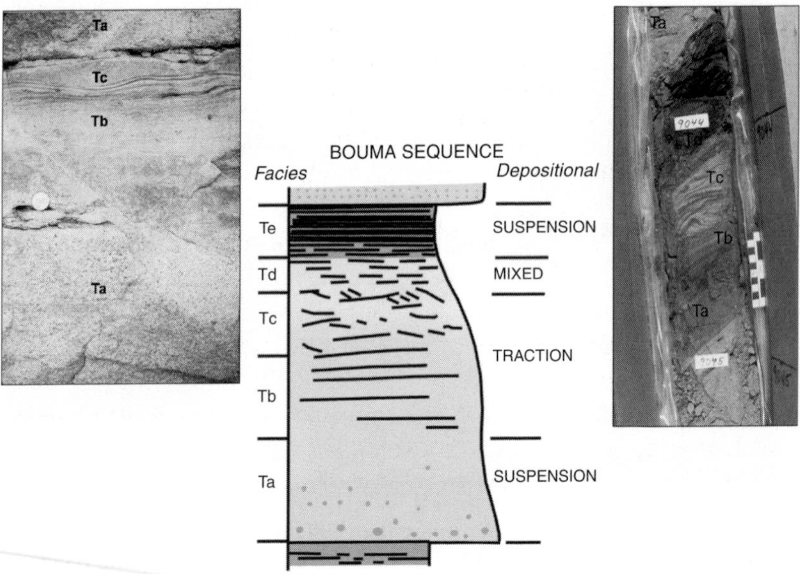

Fig. 10.3. Bouma (1962) sequence in an outcrop, consisting of Ta (massive to size-graded sand), Tb (parallel-laminated sand), and Tc (ripple-laminated sand). Bouma Td (massive siltstone to mudstone), and Te (claystone) have been weathered. The Ta is normally considered to be the "graded sand" interval, but normally, grain size decreases progressively upward from Ta to Te. (Core photograph supplied by C. Jenkins, personal communication, 2003.)

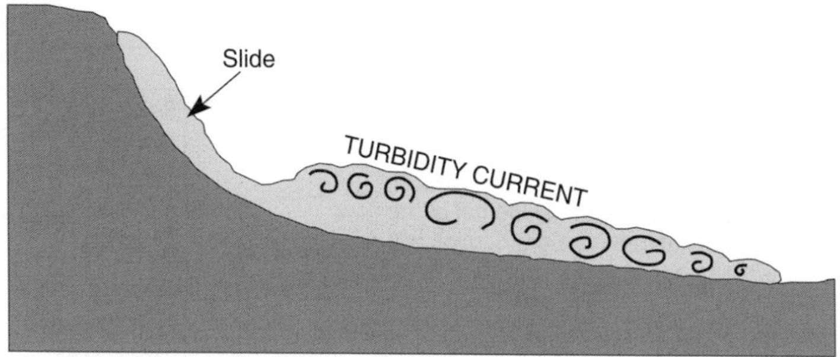

Fig. 10.4. Schematic illustration of the origin of a turbidity current from an upslope slide. Modified from Morris (1971). (Reprinted with permission of SEPM, Society for Sedimentary Geology.)

(Fig. 10.4). The first indirect evidence that this process occurred was periodic breakages of deep-ocean (200–3,500 m water depth) seafloor communications cables in the early and middle 1900s off Grand Banks of Newfoundland, offshore eastern Canada (Heezen and Ewing, 1952); the Magdalena Fan, offshore Colombia (Heezen, 1956); and the Congo River and continental slope (Heezen et al., 1964). After researchers acknowledged this important process, they eventually attributed the cable breaks to erosion by high-velocity, bottom-hugging turbidity currents.

Turbidity currents require only seawater and a small volume of sedimentary particles, relative to the total volume of water, to move downslope under the influence of gravity. Because fluid is the dominant component, the flow becomes turbulent and remains turbulent as it travels downslope. Many recent studies have shown that, in addition to turbidity currents, a variety of types of sediment gravity flows distribute sediment into deep water (Fig. 10.5). These flows behave according to the interactions of the individual particles in the flow. At low sediment concentrations, turbulent, fluidal flow predominates. With higher concentrations of sediment, the grains interact more frequently, and different mechanisms keep the grains moving within the flow (Fig. 10.5).

Sediment gravity flows have been documented to have traveled hundreds of kilometers in deep-ocean basins (Walker, 1992). These sediment gravity flows all have one thing in common – they originate within the marine environment. For example, a sediment gravity flow can be generated by sediment being dislodged from the upper continental slope during an earthquake (Fig. 10.4). These types of flows are called "ignitive" flows, which implies that they experience instantaneous generation. Another group of flows, called "nonignitive", or more specifically, hyperpycnal flows, comprises flows that originate when mixtures of sediment and river water are discharged into the marine environment from the river mouth during flood stage (Fig. 10.6). The density of the freshwater plus the grains is not sufficient for the flow to sink beneath denser seawater, so,

Sediment Support Mechanisms

Fluid turbulence
-random motion of fluid in eddies

Hindered settling
-sediment begins to settle out of the flow, space required for a grain to fall makes water move upward, providing a lift force

Dispersive pressure
-interaction of grains with one another, rattling of grains against each other, happens when shear occurs

Matrix strength
-cohesion, usually provided by fines

Increasing sediment concentration

Fig. 10.5. The various sediment support mechanisms that are operative in the deep marine environment and that lead to deposition of most sediment in deep water. The support mechanisms are related to the volume of grains relative to the volume of grains plus water in the flow. As the concentration of grains increases, the flow transitions from turbulent, dilute flow to laminar, cohesive flow. (Figure provided by D. Pyles, personal communication, 2002.)

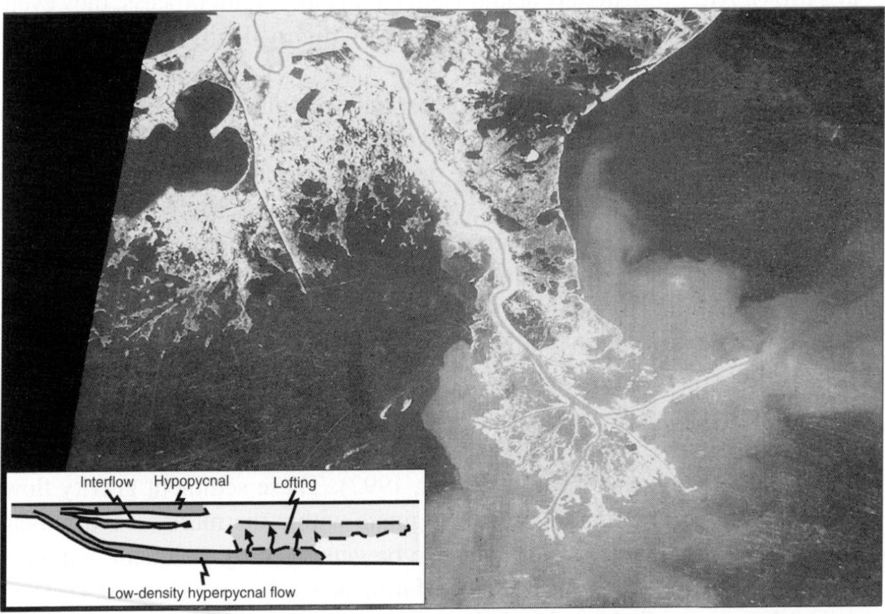

Fig. 10.6. Infrared aerial photograph of a fine-grained sediment plume emanating from the mouth of a Mississippi River distributary channel. Inset shows the types of flows that can be generated when river-borne sediment plumes enter the marine environment. Hyperpycnal turbidity currents are generated when the concentration of suspended sediment in freshwater exceeds 42 kg/m^3. After Mulder et al. (2003). (Reprinted with permission of Geological Society of London.)

instead, the flow floats across the sea surface until the particles slowly disperse through the water column. If a very high concentration of particles reaches the marine environment, the flow's density can exceed that of seawater, and the flow will sink to the seafloor and move downslope in a manner similar to that of "ignitive" flows. The critical density for this behavior is thought to be 42 kg/m^3 (Mulder et al., 2003).

10.3 Depositional models

Pioneering work by Bouma (1962), Mutti and Ricci Lucchi (1972), and Normark (1978) provided early geologic models for submarine fans and their component strata. Walker (1978) attempted to combine models into a comprehensive submarine-fan model composed of a feeder canyon, a proximal suprafan lobe, and a more distal lobe fringe, all sitting on a basin-plain deposit (Fig. 10.7).

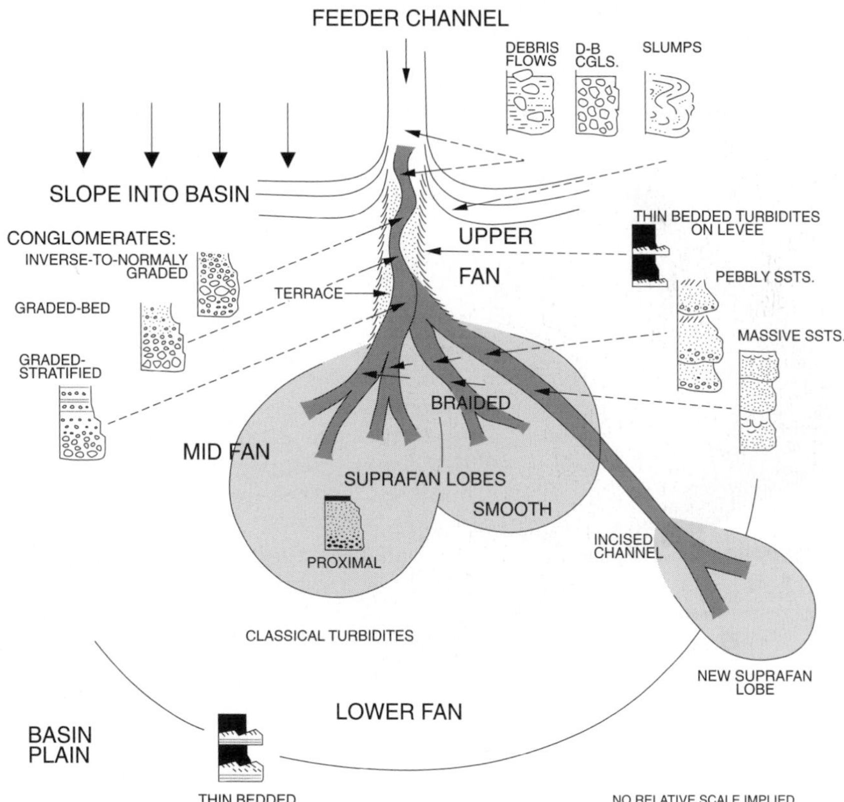

Fig. 10.7. Generalized submarine-fan model of Walker (1978). (Reprinted with permission of AAPG, whose permission is required for further use.)

According to this model, the grain size of the sediments decreases progressively seaward, which suggests that the potential for oil and gas reservoirs also diminishes in a seaward direction. Although this model was the standard for many years, widespread use of 2D and 3D seismic-reflection technologies have revealed that the model was too simplified. Walker (1992) later recanted his comprehensive-fan model on these grounds and stated that one model could not fit all deepwater systems. This statement has been verified by many studies.

A major breakthrough in our understanding of deepwater deposits came with the development of, and rapid improvement in, 3D seismic-reflection data. These data, along with parallel advances in water-bottom imaging technology, such as GLORIA sidescan sonar, allowed geoscientists to characterize detailed stratigraphic and facies relationships in 3D space. As a result, geoscientists have developed verifiable insights into the complex processes and deposits of deepwater environments.

Many slope systems are very muddy. The importance of channels as sand conduits for bypass across the slope and into the basin floor probably was not fully appreciated until about 15 years ago, when researchers recognized that large volumes of sand occur downdip of muddy slope systems (e.g., Angola and the northern Gulf of Mexico) (Fig. 10.8). Many slope channels are marked by evidence of sediment bypass (such as coarse-grained lags, traction deposits,

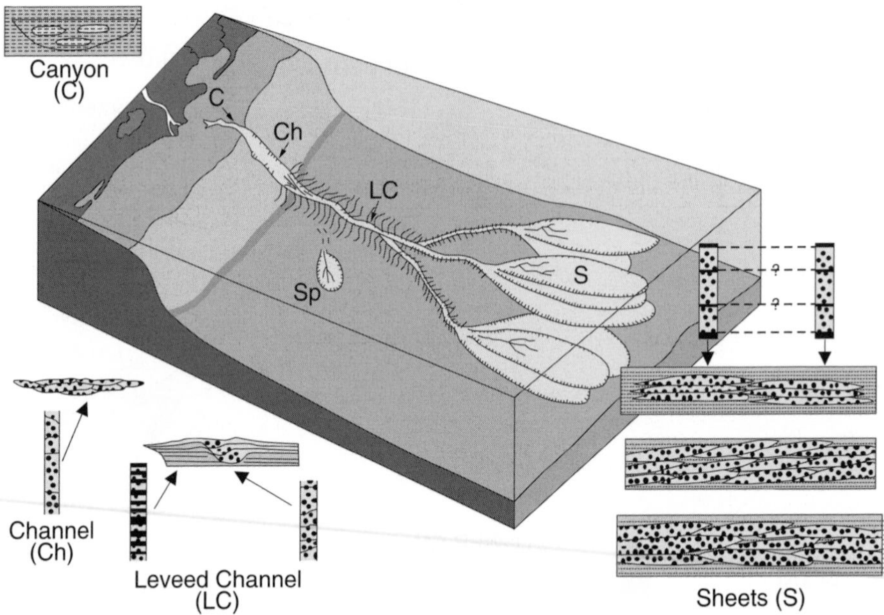

Fig. 10.8. Schematic diagram of the architectural elements of fine-grained deepwater depositional systems. Modified from Bouma (2000). (Reprinted with permission of AAPG, whose permission is required for further use.)

heterolithic deposits of fine-grained tails, and fine-grained levees, in some instances). In such systems, sediment grain size does not decrease progressively seaward. For example, on the Mississippi submarine fan, gravel deposits have been found in deepwater cores 220 km from the present shelf edge (Stelting et al., 1985). On the modern Amazon Fan, there is 5% sand (and 95% mud) on the upper fan, 10–30% sand on the middle fan, 70% sand on the lower fan, and 30% sand on the basin plain (Piper and Normark, 2001). Paleogeographic reconstruction of deepwater sedimentary facies in the Permian Brushy Canyon Formation of west Texas indicates that there is 50% sandstone on the upper slope, 63% sandstone on the lower slope, 76% sandstone at the base of slope, and 93% sandstone on the basin floor (Gardner and Borer, 2000). These systematic, seaward differences are the result of the updip portions of the fan and slope containing the feeder channels and their confined fill (sand), often adjacent to muddier intervals, and the downdip portions containing sheet sands or lobes (Fig. 10.8).

10.4 Architectural elements of deepwater deposits

Mutti (1985) introduced the concept of turbidite elements. Chapin et al. (1994) further developed the concept for Shell Oil Co., to characterize the company's deepwater discoveries in the northern Gulf of Mexico (Fig. 10.9). Chapin et

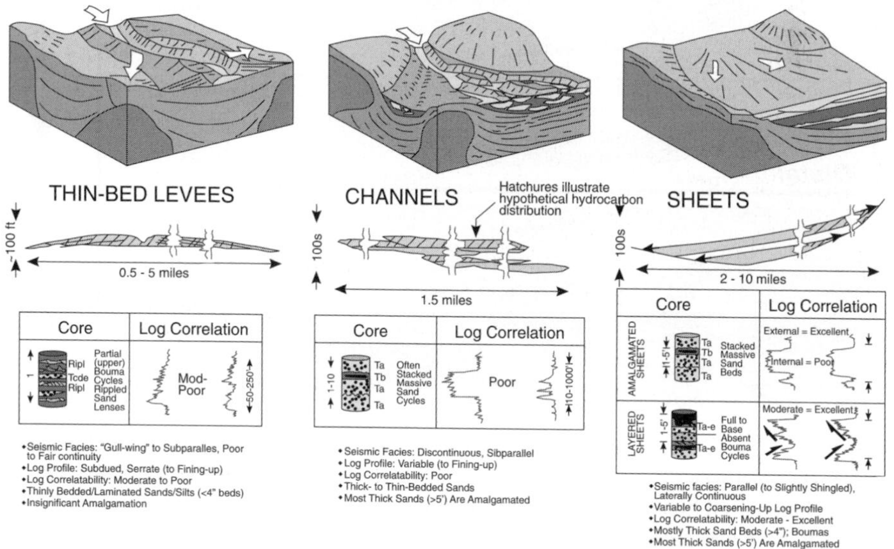

Fig. 10.9. Classification of deepwater architectural elements, with application emphasis on the northern Gulf of Mexico reservoirs. After Chapin et al. (1994). (Reproduced with permission of the GCS–SEPM Foundation.)

al. (1994) emphasized the three main sand-bearing architectural elements (i.e., types of reservoirs): sheets (layered and amalgamated), channels (single and multistory), and thin beds in levee sediments. This descriptive classification of the architectural elements of deepwater systems is commonly used in the oil and gas industry.

The main architectural elements that comprise deepwater depositional systems are: canyons, (erosional) channels, (aggradational) leveed channels, and sheets or lobes (Figs. 10.8 and 10.9). Below, some examples are provided of the characteristics of each element. It is important to note that one should include different types of data in a reservoir characterization, because each type may provide details at a different scale. For example, at the reservior scale, seismic-reflection patterns for the three elements are distinctly different (Fig. 10.10).

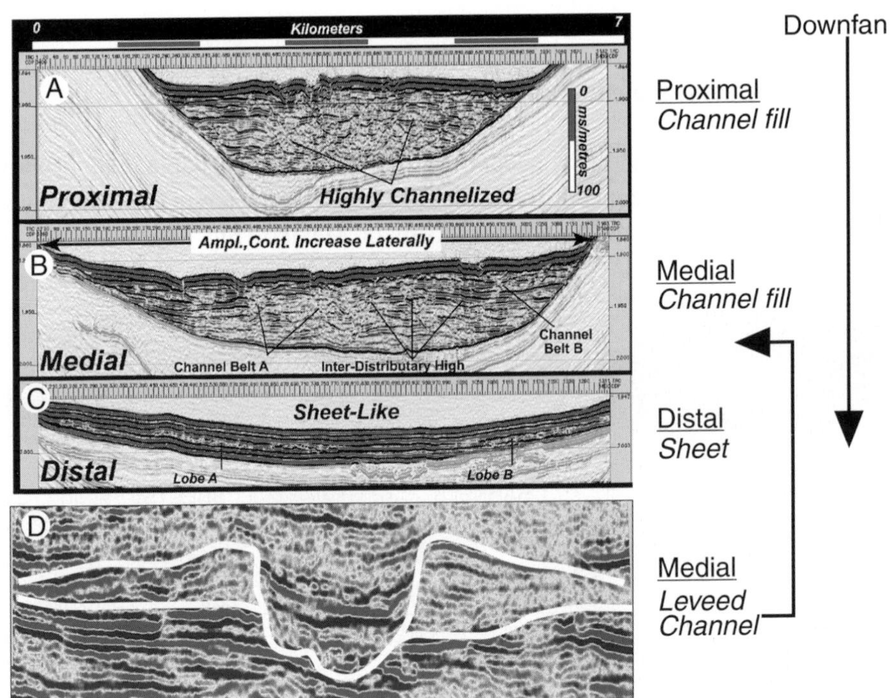

Fig. 10.10. A, B, C are three high-resolution seismic profiles from one shallow intraslope minibasin, northern deep Gulf of Mexico. (A) Proximal and (B) medial profiles cross the upfan channelized systems. (C) A distal profile crosses the sheet deposits. Note that lobes A and B have a slightly mounded appearance among the laterally continuous, sheetlike reflections. The deposits are as large as 50 ms in two-way traveltime. These have laterally continuous reflections that lapout against the side of basins. (D) Seismic profile of a leveed channel complex from the western Gulf of Mexico. A, B, C, after Beaubouef et al. (2003). (Reprinted with permission of the Gulf Coast Section SEPM Foundation.)

10.4.1 Sheet sandstones and reservoirs

Sheet sands and sandstones are considered to be some of the best high-rate, high-ultimate-recovery (HRHU) reservoirs in deep water. This is because they tend to have the simplest reservoir geometries: good lateral continuity, tabular external form, potentially good vertical connectivity, high width-to-thickness ratio (aspect ratio) (greater than 500:1), narrow range in grain size, and few erosional features (Chapin et al., 1994; Mahaffie, 1994). Because of the initial successes with these reservoirs in different parts of the world, the petroleum industry has studied sheet sands and sandstones in great detail to better understand them, and hence, to find and produce hydrocarbons from more of them. In some cases, however, reservoirs that were initially interpreted to be sheet sands were later determined to be amalgamated channel sands.

Sheet sands are deposited from decelerating flows at the termini of channels (Fig. 10.11). Sheet sands reflect the sediments that have bypassed through updip channels (confined flow) and then are deposited in a primarily unconfined, downdip setting. Unlike other deepwater reservoir elements, sheet sands commonly have an areal extent that exceeds the area of the trap. Also, they can extend the entire length of salt or shale minibasins (Fig. 10.12).

Chapin et al. (1994) defined two types of sheet sands: layered and amalgamated (Fig. 10.9). Amalgamated sheets are characterized by high sand content

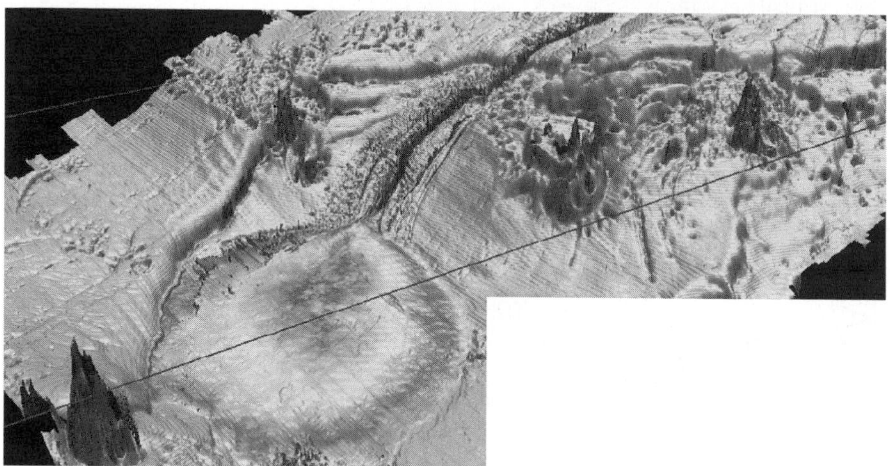

Fig. 10.11. Three-dimensional perspective of an isochron of one depositional lobe with sheet sands draped over seafloor bathymetry. This deposit is in one intraslope minibasin created from shale-deformation on the Nigerian continental slope. The maximum isochron values are shown in red/dark-gray (100 ms). Three distinct areas with sheets are outlined. The vertical "stripes" represent an acquisition "footprint" of the 3D data. After Pirmez et al. (2000). (Reprinted with permission of the Gulf Coast Section SEPM Foundation.)

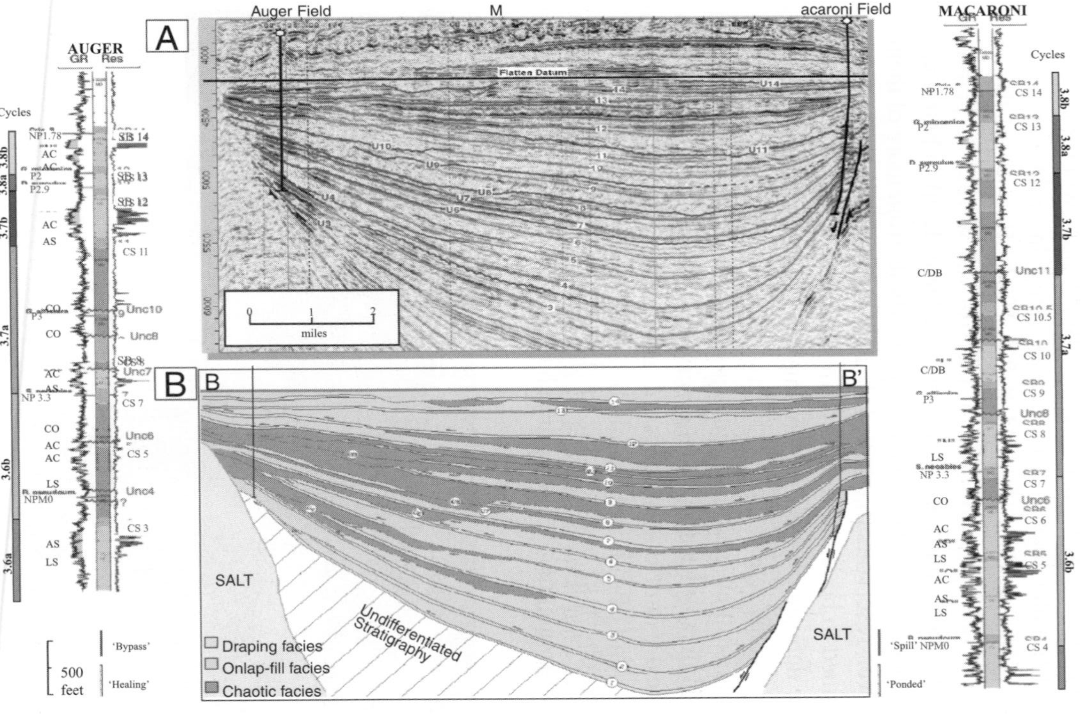

Fig. 10.12. **(A) Flattened seismic profile across the Greater Auger minibasin, showing the relationships between the Auger and Macaroni fields. Multiple sheet sands are interpreted to be present and to extend across most of the basin. (B) Interpreted stratigraphic-fill packages between two wells. Yellow/gray indicates onlapping fill facies (i.e., sand-rich sheets), and the orange/dark-gray areas are channel-fill facies ("bypass facies"). Two different kinds of facies are present. Gamma-ray logs shown from each field illustrate layered sheet (LS) and amalgamated sheet (AS) sands. After Booth et al. (2000). (Reprinted with permission of the Gulf Coast Section SEPM Foundation.)**

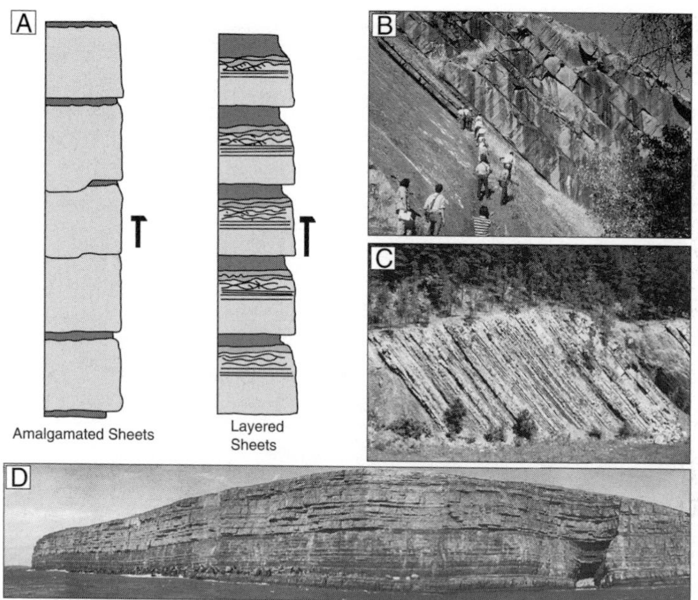

Fig. 10.13. (A) Schematic diagrams of amalgamated- and layered-sheet sands. After Mutti (1985). (B) Amalgamated-sheet sand, California. (C) Layered-sheet sand, Jackfork Group, Arkansas. (D) Layered-sheet sands overlain by amalgamated-sheet sands, Kilcloher Cliff Section, Ross Formation, Ireland.

and amalgamated sand-on-sand contacts. They comprise stacked sandstone beds with few interbedded shales (Fig. 10.13). By contrast, layered sheets are characterized by relatively low net sand content and have interbedded shale and sandstone beds. Sheets exhibit a transition from amalgamated to layered form, in both longitudinal and transverse directions.

Even though sheet sands are laterally continuous, three complexities can render them difficult reservoirs from which to produce hydrocarbons. (1) The surfaces of sheet sands are often traversed by channels that periodically feed sediment to the sheet (Fig. 10.14). Mud may be deposited within these channels, once they are abandoned, and this ultimately gives rise to shale breaks and discontinuities at individual stratigraphic levels. (2) The external form of sheets varies and depends partly on the topography of the seafloor upon which they are deposited (Fig. 10.14). This attribute can often be determined seismically. (3) Shales between sandstones – at both the scale of individual beds (layered sheets) (Fig. 10.13) and of separate, thicker amalgamated- or layered-sheet intervals – can extend at least as far within a basin as the sandstones, giving rise to vertically isolated or compartmentalized reservoir intervals. Further are some examples of reservoirs that demonstrate these complexities.

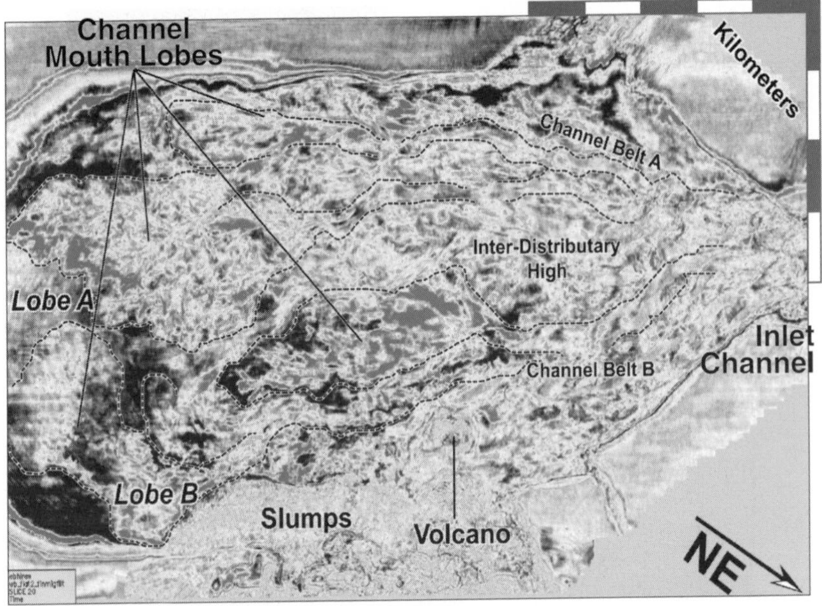

Fig. 10.14. Seismic horizon slice taken 20 ms below the seafloor in one intraslope basin, northern deep Gulf of Mexico. Two distinct upfan channel belts (A, B) to the right (north) change downfan to channel-mouth lobes. Also present are basin margins, a mud volcano, and "slumps". After Beaubouef et al. (2003). (Reprinted with permission of the Gulf Coast Section SEPM Foundation.)

10.4.1.1 Auger field

Auger field is a Gulf of Mexico example of a series of sheet sands within a salt minibasin (Fig. 10.12) (McGee et al., 1994; Bilinski et al., 1994; Booth et al., 2000; Kendrick, 2000; Beaubouef et al., 2003). The field is notable for its extremely high production rates from one sand, the S sand, thus qualifying it as an HRHU reservoir. Although 120 MMBE (millions of barrels of equivalent oil) have been attributed to the S sand, as of the year 2000, seven wells had produced 110 MMBE, as a result of excellent aquifer support and perhaps conservative initial estimates of original oil in place (OOIP) and flow rates.

The S sand consists of a series of layered and amalgamated sheet sands within a combined fault and stratigraphic-pinchout trap. Oil-bearing zones occur stratigraphically beneath water-bearing zones. This compartmentalization is a result of subseismic-scale shales that extend as far as the sands do and vertically compartmentalize them (Fig. 10.15). Though they are subseismic in scale, these shales restrict fluid flow vertically (Kendrick, 2000).

10.4.1.2 Mensa field

A second example of sandstone reservoir complexities is Mensa field in Mississippi Canyon blocks in the northern Gulf of Mexico (Pfeiffer et al., 2000).

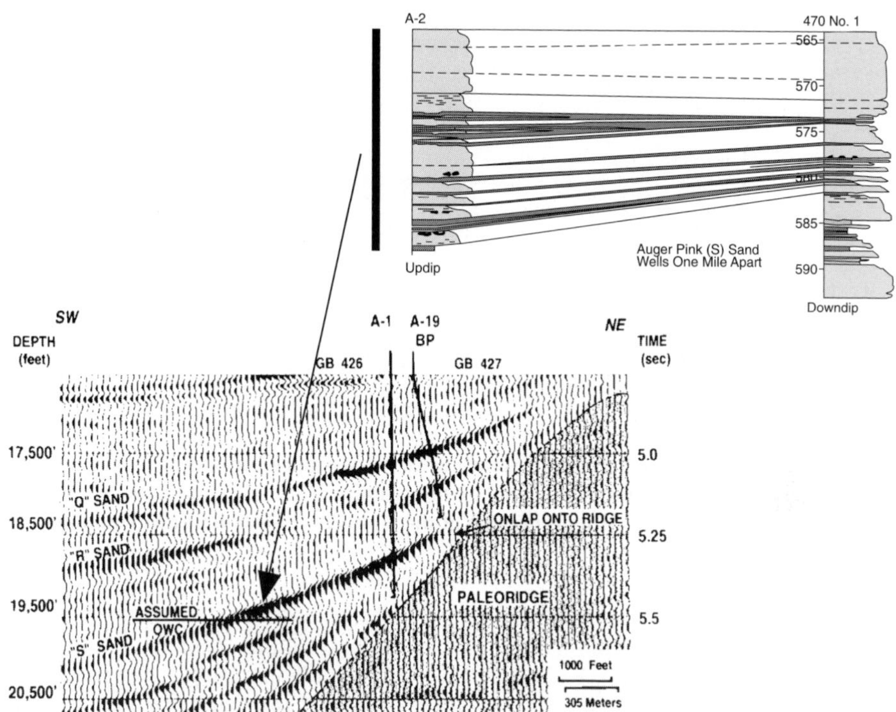

Fig. 10.15. The upper-right figure shows correlation of the S sand in two wells spaced about 1 mi (1.6 km) apart. Sands are in yellow/gray and shales are in brown/dark-gray. Note that there are several continuous (interpreted) shales that separate the sands. The lower figure shows the seismic amplitude corresponding to the S sand. Modified from Bilinski et al. (1994). (Reprinted with permission of the Gulf Coast Section SEPM Foundation.)

Mensa field was discovered in 1985 as a prominent seismic anomaly draping a major turtle structure (an anticline) in an intraslope minibasin (Fig. 10.16). The main pay (the I sand) has a net sand content of almost 90%. Porosity varies from 29 to 32%, and permeability varies from 500 to 2000 md. Original gas in place (OGIP) estimates are 1.3 tcf, with the I sand interval containing 750 bcf.

On the basis of core and seismic data and wireline-log response, the reservoir was initially interpreted to be a homogeneous sand connected downdip to a major aquifer that provided the field's drive mechanism. However, after production began in 1997, pressure measurements did not support the original reservoir model. New reservoir simulations indicated that the reservoir was primarily a depletion-drive reservoir with little to no aquifer support. Newer, higher-frequency seismic data imaged a more complex reservoir architecture. Instead of one laterally continuous reflection associated with one sheet sand, as had been seen previously, three slightly offset stacked sheet sands were revealed (Fig. 10.17). One sheet probably cut into the other, and they may have been separated by a partial permeability barrier between them. Additionally, an

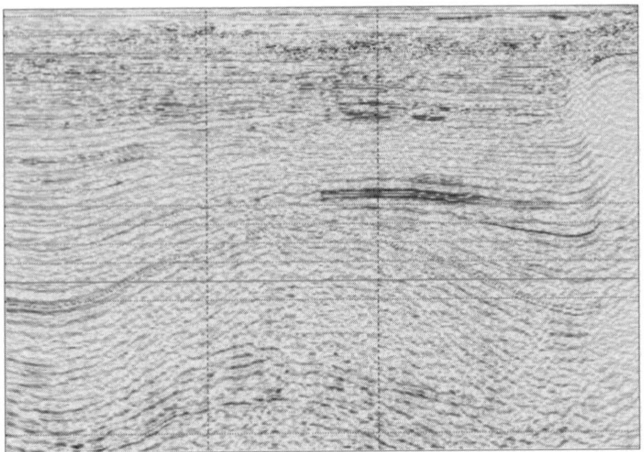

Fig. 10.16. Seismic profile across the Mensa field, Mississippi Canyon 731, northern deep Gulf of Mexico. The reservoir is indicated by the prominent seismic amplitude. After Pfeiffer et al. (2000). (Reprinted with permission of the Gulf Coast Section SEPM Foundation.)

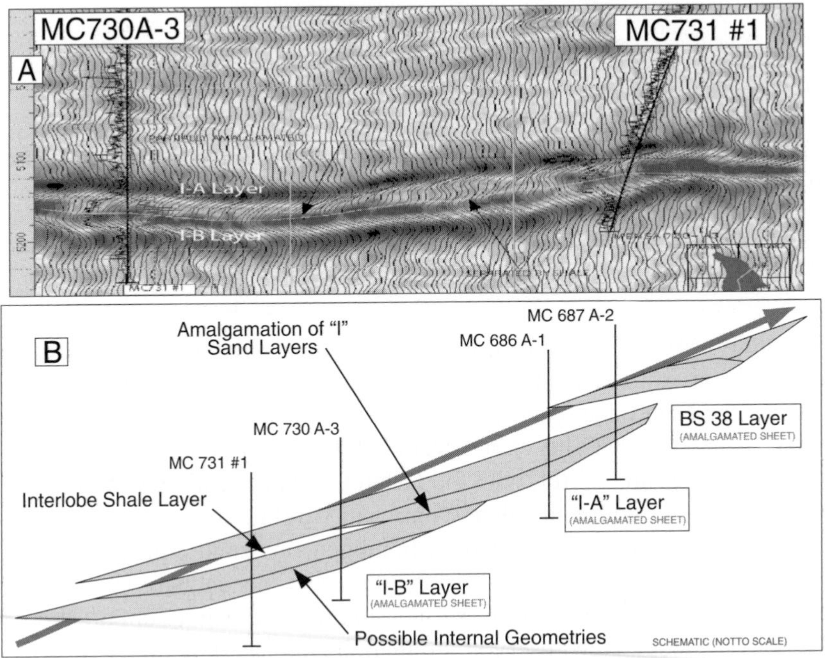

Fig. 10.17. (A) Seismic profile across Mensa field. This profile is from a higher-frequency data set and shows the time-based gamma-ray logs and the two amalgamated-sheet sands, I-A and I-B. (B) Schematic cross-section drawn after the higher-frequency seismic data have been integrated with well information. Additional data defined three discrete sheets, two of which are in pressure communication. After Pfeiffer et al. (2000). (Reprinted with permission of the Gulf Coast Section SEPM Foundation.)

erosional channel that cuts the main pay sand to the west may limit the amount of pressure support that the aquifer can provide.

As is the case in Auger field, Mensa field's reservoir complexities are subseismic in scale. In the end, additional well work indicated that there was sufficient communication between reservoirs for a subsea tieback to be maintained.

10.4.1.3 Ram Powell J sand

A third example of sandstone reservoir complexity is the Ram Powell field in the Viosca Knoll 956 block, in the northern Gulf of Mexico (Rossen and Sickafoose, 1994; Clemenceau, 1995). The Ram Powell field was discovered in 1984, and production began in late 1997. It was one of the earliest discoveries in the northern deep Gulf of Mexico. Production in Ram Powell field is from a series of individual reservoir sands that include all the major deepwater architectural elements: amalgamated sheet/channel (J) sand, channel-levee (L and M) sands, and amalgamated channel (N) sands (Fig. 10.18).

The J sand is a sheet-sand complex overlain by a channel-levee deposit. Core porosity averages 30%, and permeability ranges from 640 to 2,680 md. In-place

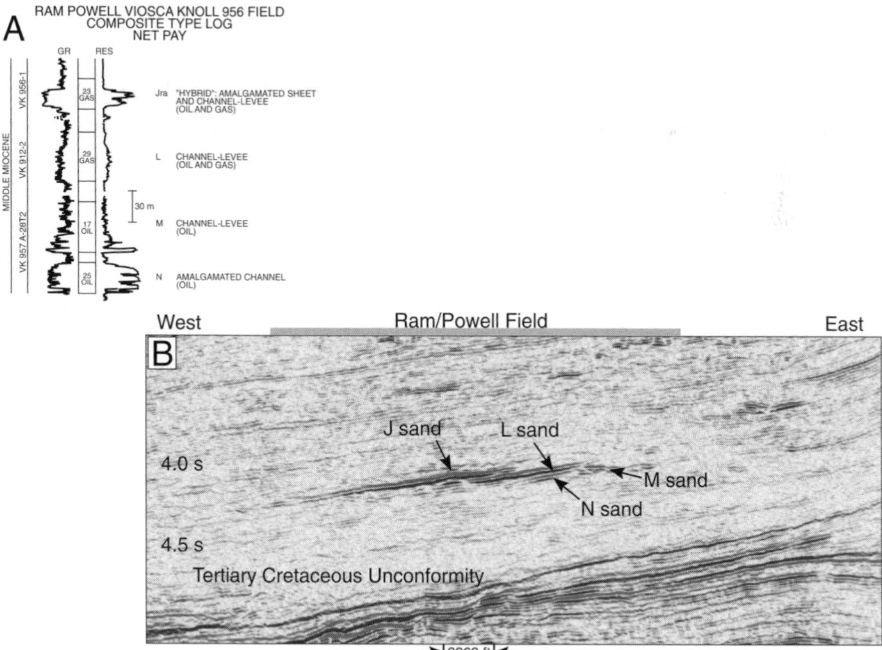

Fig. 10.18. (A) Type well log showing the J, L, M, and N reservoir sands in Ram Powell field, northern Gulf of Mexico. After Craig et al. (2003). Reprinted with permission of AAPG. (B) Seismic profile across the Ram-Powell field, illustrating the J, L, M, and N sands. After Clemenceau et al. (2000). (Reprinted with permission of the Gulf Coast Section SEPM Foundation.)

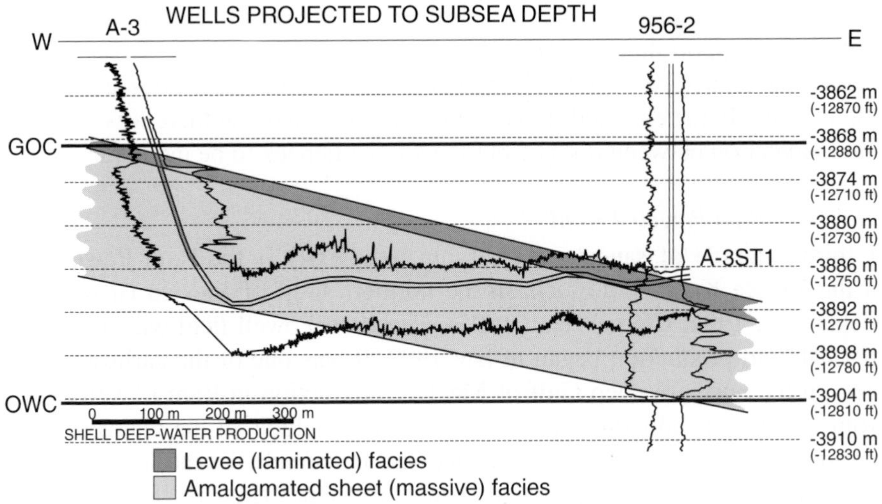

Fig. 10.19. Horizontal well path in the Viosca Knoll 956 A-3 well through the J sand. The J sand consists of a lower amalgamated-sheet facies overlain by a thin-bed levee facies. After Craig et al. (2003). (Reprinted with permission of AAPG, whose permission is required for further use.)

hydrocarbon volumes have been estimated at 80 MMBO and 600 bcf gas. The oil rim is estimated to contain 50 MMBO. At year-end 2001, cumulative production was 29 MMBO and 205.9 bcf gas from three horizontal wells. One well – the Viosca Knoll 956A-3ST1 (Fig. 10.19) – held the Gulf of Mexico production record of 40,900 barrels of oil equivalent per day (BOEPD) for some time.

In 1993, the original J sand development plan consisted of eight vertical wells – six producers from an oil rim and two gas-cap blowdown wells. The well spacing was planned at 1.4 km^2 (340 acres), with expected production of 6,000 BOPD. However, advances in horizontal drilling and completion technology offered the opportunity to produce the oil rim more efficiently. Specifically, reservoir modeling suggested that three horizontal wells, spaced at 3.2 km^2 (800 acres), could drain the oil rim at a rate of 30,000 BOPD, thus resulting in substantial development-cost savings (Fig. 10.19). Drilling revealed some unpredicted features of the reservoir, which resulted in modification of the drilling plan. However, the program was a success.

10.4.1.4 Long Beach Unit, Wilmington field

The fourth example is the Long Beach Unit of the Wilmington field, in southern California, USA (Slatt et al., 1993; Clarke and Phillips, 2003). Wilmington field is the largest of several giant oil fields in the Los Angeles Basin of southern California. The Long Beach Unit comprises the southeastern portion of the field,

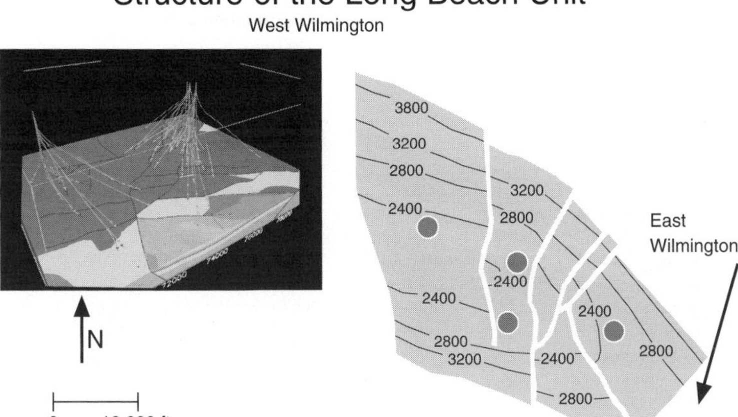

Fig. 10.20. The Long Beach Unit of the Wilmington field is bounded by production on the west and east. West Wilmington began production in 1936, whereas East Wilmington started in 1954. The Long Beach Unit sits atop a large anticline with a structural relief of about 530 m (1,600 ft). Faults compartmentalize the reservoir and the unit is primarily developed from four offshore islands (red/dark-gray dots). Dips are 20° on the north, steepening to 60° on the south. Wilmington field's structure is 11 mi (18 km) long, 3 mi (5 km) wide, and encompasses 13,500 acres. 10% of production comes from under the City of Long Beach, the rest of LBU production is from under Long Beach Harbor faults. Modified from Clarke and Phillips (2003). (Reprinted with permission of AAPG, whose permission is required for further use.)

in part lying offshore and in part underlying the city of Long Beach (Fig. 10.20). Total original oil in place for the Long Beach Unit is 3.8 billion barrels. More than 1,500 wells have been drilled within this unit, through a number of zones and subzones. In the late 1960s, unit-wide oil production peaked at more than 100,000 BOPD.

The Wilmington field is a very mature field, having been discovered in 1936. Development began in the Long Beach Unit in 1965. It went on waterflood soon after, primarily to maintain reservoir pressure in order to reduce subsidence beneath the city of Long Beach. Later geologic study of the most prolific zone in the Long Beach Unit – the Ranger Zone – revealed that more selective perforation and waterflooding might increase production by producing from previously untapped sands that are isolated by laterally continuous shales. Unconsolidated sands, which comprise the Ranger Zone, have porosities averaging about 28% and permeabilities varying from millidarcys to darcys. Additional drilling later proved successful.

In other intervals – called the Tar Zone, the Union Zone, and the Terminal Zone – steam-flood projects were designed and successfully implemented using horizontal wells (Figs. 10.21 and 10.22) (Clarke and Phillips, 2003).

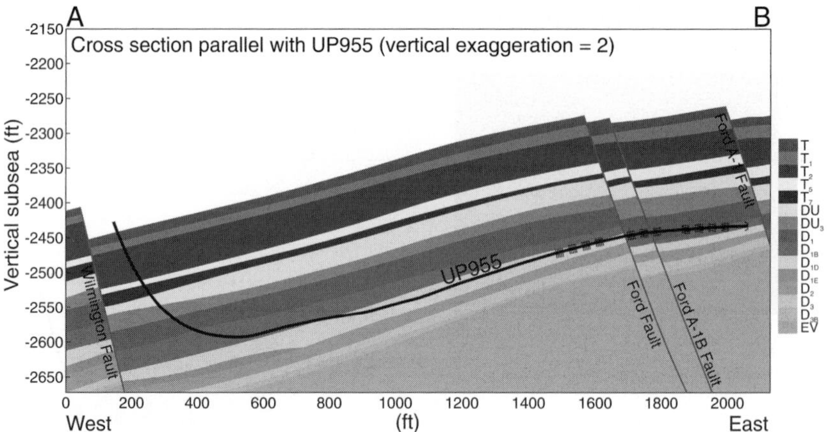

Fig. 10.21. Schematic cross-section showing the horizontal well course of UP 955 through the D1 sand, Long Beach Unit, Wilmington field, California. After Clarke and Phillips (2003). (Reprinted with permission of AAPG, whose permission is required for further use.)

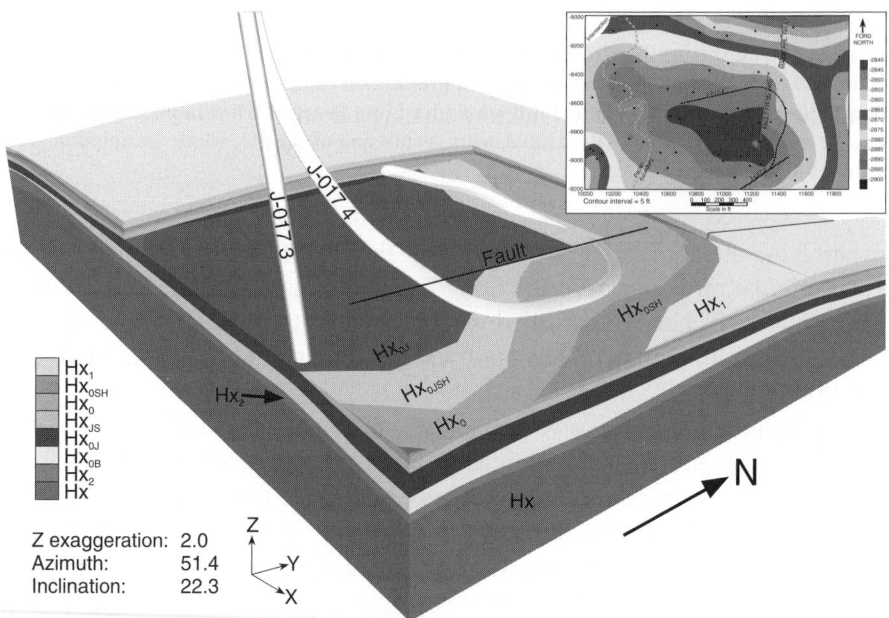

Fig. 10.22. Horizontal well trace in the Hx_0 sandstone of the Long Beach Unit, Wilmington field, California. Note the curved well path. After Clarke and Phillips (2003). (Reprinted with permission of AAPG, whose permission is required for further use.)

Initial oil production from some of the newer horizontal wells has exceeded 600 BOPD, with about 300 BOPD with 80% water cut from average wells. Total unit production in 2003 was 38,000 BOPD. In all these projects, knowledge of the stratigraphy and reservoir characteristics helped developers add hundreds of millions of barrels of new reserves. Of particular importance was recognition of laterally continuous, impermeable shales separating permeable sand packages, and the long-distance continuity of the sands (Fig. 10.21). Some very sophisticated horizontal drilling technologies were employed in these projects (Fig. 10.22).

10.4.2 Canyon and channel-fill sandstones and reservoirs

Deepwater channels have received considerable attention in the petroleum industry during the past decade because of (1) important discoveries in several deepwater basins in which reservoir performance was critical to development decisions and strategies (e.g., Campos Basin, Brazil; offshore Angola; Nile and Mahakam Deltas; northern Gulf of Mexico; West of Shetland Islands; and offshore mid-Norway); (2) the increasing ability of 3D seismic to image the complex internal geometries of channel systems (especially those that are sinuous); and (3) the need to avoid shallow flow problems while drilling channel-fill sediments.

Channels and their fills have been studied for many years, from different perspectives, and using multiple data sets, including data from the modern seafloor, from the shallow subsurface (shallow seismic for shallow-hazards drilling surveys), from deeper-exploration seismic, from reservoirs, and from outcrops. One conclusion that can be reached from the large number of published studies is that most channel fills have at least some unique characteristics that, if they are not properly understood early in the life of a field, can result later in costly production problems.

Deepwater channel fills have been classified into three broad categories: (1) those that originated by erosion of underlying substrate and that have few, or no, associated levee-overbank deposits, (2) those that originated by aggradation of levee-overbank strata to give an intervening depression that serves as a channel for sand and mud transport, and (3) those that formed by a mix of erosional and depositional processes, either contemporaneously or during separate evolutionary phases of the fill (Fig. 10.23) (Mutti and Normark, 1987, 1991; Clark and Pickering, 1996; Morris and Normark, 2000). The fill of erosional channels is sometimes referred to as amalgamated channel sands or large channels, and the fill of aggradational channels is sometimes referred to as leveed channel fill or low-relief channel levees (Mayall and O'Byrne, 2002; Saller et al., 2003). Channels tend to be more erosional updip, because there they have steeper slope gradients that result in higher flow velocities. Farther downdip, channels become mixed erosional–depositional and/or aggradational, as the slope gradient diminishes (Fig. 10.24).

As the name indicates, erosional channels are those in which the depression or conduit has formed by erosion of the underlying substrate. Sandy sediment

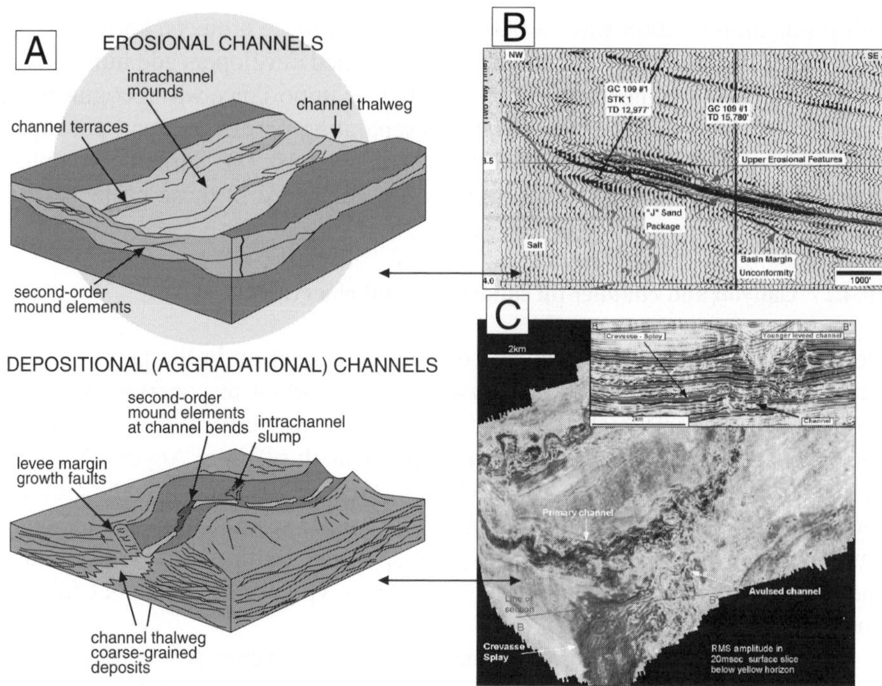

Fig. 10.23. (A) Two end-member types of channel fill: erosional channel fill and depositional or aggradational channel fill. After Clark and Pickering (1996). (B) Seismic profile showing some distinct erosional surfaces that are now filled by shale. After Holman and Robertson (1994). (C) Vertical profile and horizontal seismic horizon slice through a depositional channel. After Mayall and Stewart (2000). (Reprinted with permission of the Gulf Coast Section SEPM Foundation.)

gravity flows are common erosive agents, but the larger canyons and channels must have originated by submarine slides on muddy slopes. The resulting depressions later became conduits for sediment transport. Such slides might be generated in a number of ways, including from (1) seismic activity; (2) instability and slope failure resulting from rapid sedimentation, oversteepening and/or change in pore pressure (particularly during a change in sea level); (3) fine-grained underflows that trigger larger flows, and possibly (4) sudden discharge of gas hydrates (clathrates) upward through slope sediment to the slope floor.

By contrast, aggradational channels are those in which the conduit, or depression, is a result of long-term deposition and buildup of levee and overbank strata parallel and adjacent to the channel. The levee beds tend to be finer grained and thinner bedded than beds deposited in the adjacent channel, and often, levee and channel strata are separated by complex channel-margin facies, which suggests there has been a significant time interval between the deposition of levees and their channel fill.

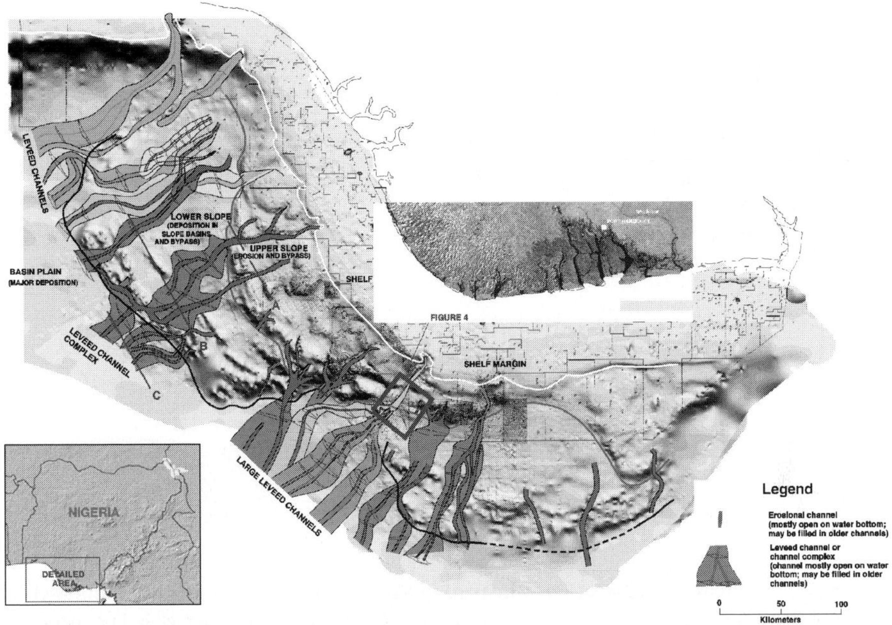

Fig. 10.24. Shaded bathymetry map of offshore Nigeria. Superposed are late Pleistocene slope channels of different ages. The channels are erosional in the upper slope and become less erosional downdip, with aggradational channel-levee or channel complexes downdip. After Mitchum and Wach (2002). (Reprinted with permission of the Gulf Coast Section SEPM Foundation.)

Channel shape varies, from elongate, relatively straight channels to those that are highly sinuous, in much the same manner that the shapes of fluvial systems vary from elongate-braided-distributary systems to meandering ones (Fig. 10.23). Although the origin of deepwater channel sinuosity is hotly debated (and is beyond the scope of this chapter), general observations are that the degree of sinuosity is inversely proportional to slope gradient, and that finer-grained, lower-energy channel fills tend to be more sinuous than do coarser-grained, higher-energy fills.

In addition to downslope changes in channel type and shape, the vertical stacking pattern within channel fill also changes, from more areally widespread, erosional channels near the base, to mixed erosional–aggradational distributary channels upward, to smaller, aggradational channels with prominent levees near the top (Fig. 10.25). This predictable vertical stacking pattern occurs because channels normally backfill during a relative turnaround and early rise in sea level, when the energy, grain size, and volume of flows are all diminishing and the depositional axis is stepping progressively landward. Although the final internal fill of a channel normally is quite complex, channel fill often can be subdivided into an organized, recognizable pattern or hier-

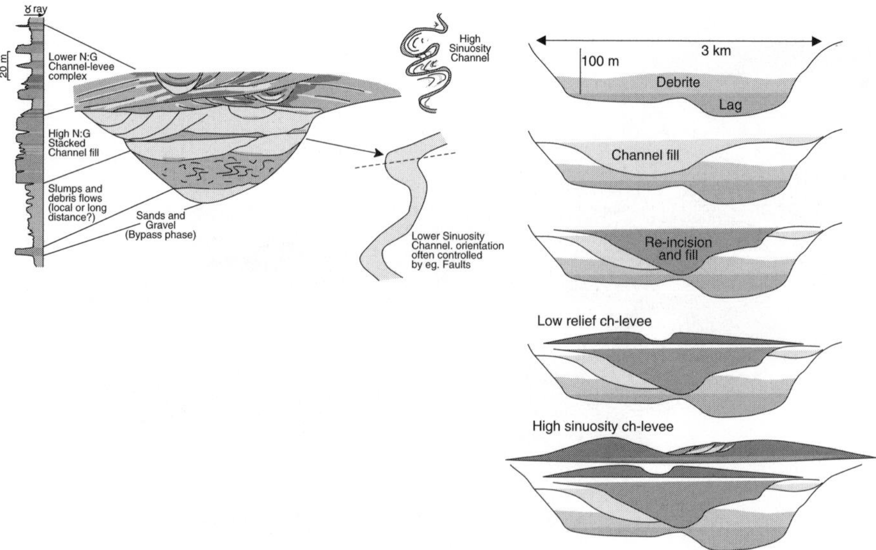

Fig. 10.25. Schematic cross-sections illustrating the sequential development of channel-fill facies. At the base, a large erosional surface is overlain by thin sand-rich lag, which is overlain by shale-rich debrite, which is overlain by thick, amalgamated-channel-fill deposits (with high net:gross value), which are overlain by leveed-channel deposits (with lower net:gross values). An interpreted vertical succession (or gamma-ray log response) is also shown along with the geometries of the channels in plan view (higher sinuosity – channel-levee; lower sinuosity – sand-rich channel fill). Repeated erosion and fill is a common feature in channels. After Mayall and Stewart (2000). (Reprinted with permission of the Gulf Coast Section SEPM Foundation.)

archy of strata (Fig. 10.26) (Gardner and Borer, 2000; Navarre et al., 2002; Sprague et al., 2002).

The composition of the internal fill of channels also varies greatly. Deposits can include gravel, sand, mud, and mixed fill, depending upon a number of factors, including tectonics, climate, and sediment supply (Reading and Richards, 1994; Richards et al., 1998). Channel-fill sediments may consist of a variety of sediment gravity flow deposits, from turbidites to debrites and slide blocks, coupled with hemipelagic suspension fallout. Grain size of the fill generally decreases upward, in accordance with the upward change from more distributarylike channels to smaller leveed channels (Fig. 10.25).

Some example reservoirs are discussed below to demonstrate the degree of variability of channel-fill sandstones.

10.4.2.1 Ram Powell N sand

The Ram Powell N sand is the lowest reservoir sand stratigraphically in the Ram Powell field (Fig. 10.18) (Lerch et al., 1997; Kendrick, 2000; Craig et al.,

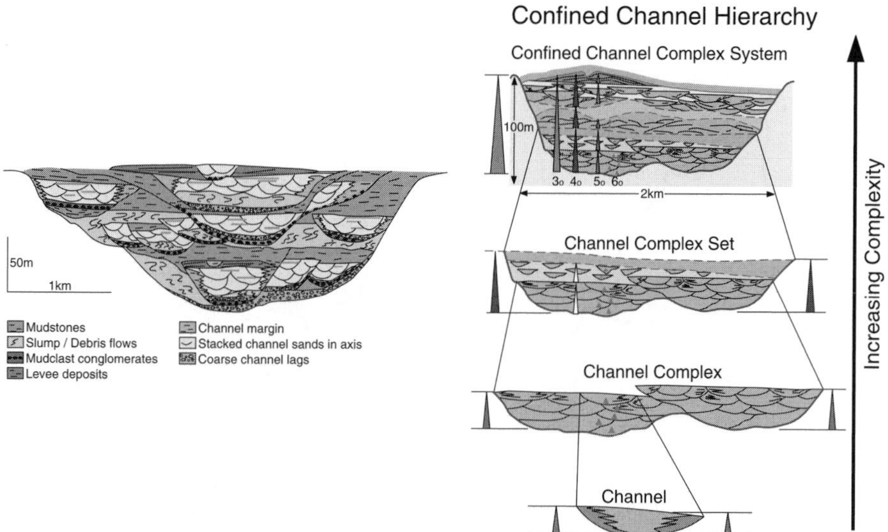

Fig. 10.26. Schematic cross-sections illustrating confined channel hierarchy from a single channel element, through a complex of elements, through a complex set of elements, and finally to a complex system. Multiple channel fills and intervening shale from levee deposits create significant heterogeneity, with many potential flow barriers and baffles. After Sprague et al. (2002). (Reprinted with permission of AAPG, whose permission is required for further use.)

2003). The N sand exhibits a north–south, elongate geometry that reflects the canyon into which the sands were deposited (Fig. 10.27). The reservoir sands are encased in, and capped by, slope shales, and they pinch out updip to form excellent stratigraphic traps on the flank of a subtle, south-plunging anticlinal nose.

Total Ram Powell N sand reserves in 1997 were reported to be 75 MMBOE. Early appraisal wells encountered high net-sand values and thick pay (Fig. 10.27). Reservoir modeling suggested a single oil–water contact and reasonable continuity, indicating that there would be satisfactory sweep efficiency and pressure support. However, the first three development wells encountered multiple fluid contacts updip of the presumed oil–water contact, even though some degree of pressure communication was noted between wells (Fig. 10.27). This indicated a high level of discontinuity and compartmentalization of the reservoir interval. Thus, a horizontal drilling strategy was employed to improve production. Because of the risk of compartmentalization by isolated channel sands, the trajectory of horizontal wells was designed to penetrate multiple sands (Fig. 10.28). One 725-m (2,380-ft) horizontal well produced at a peak rate of 11,681 BOPD.

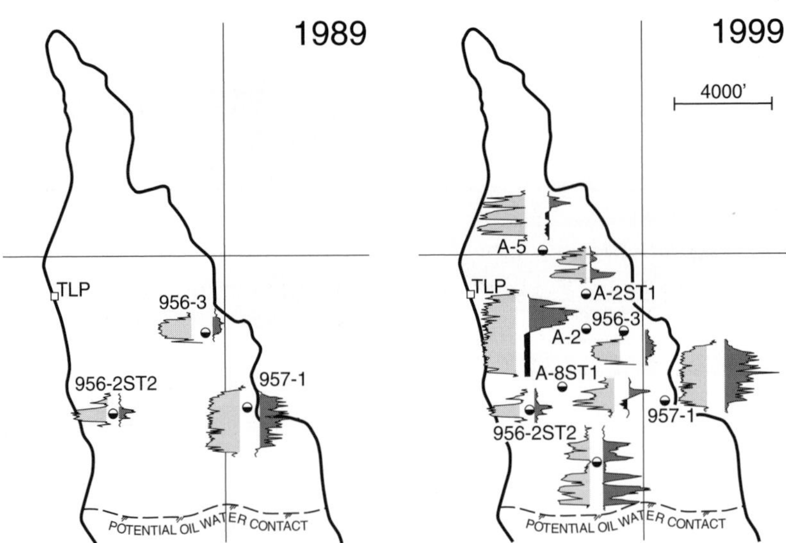

Fig. 10.27. Ram Powell N sand, channel-fill reservoir. The left figure shows the field in 1989. A seismic survey revealed the lenticular outline of the sand body. The first well drilled, the 957-1, was a discovery, with a thick sand (yellow/gray) full of oil (green/dark-gray). The second well, the 956-2ST2, had thinner sand but was oil-filled. The third well, 956-3, was a bit thicker and also oil-filled. On this basis, a series of production wells were drilled. The right-hand figure shows the development drilling results as of 1999. Subsequent wells that were drilled mainly showed partial oil and partial water (blue/black) in the sand. Also, the oil–water contact was at different structural levels in different wells, indicating isolated sands. Thus, the amount of oil in the field turned out to be less than had been calculated originally. After Kendrick (2000). (Reprinted with permission of the Gulf Coast Section SEPM Foundation.)

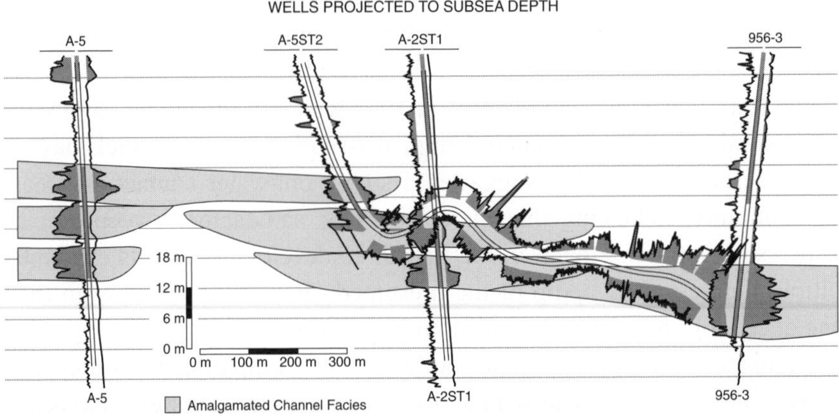

Fig. 10.28. The Ram Powell N sand horizontal drilling program. Shown are four wells in the field, and the distribution of the isolated sand bodies. Horizontal wells are drilled to capture oil from more than one isolated sand body. After Craig et al. (2003). (Reprinted with permission of AAPG, whose permission is required for further use.)

10.4.2.2 Garden Banks 191 field, northern Gulf of Mexico

Garden Banks 191 field was discovered in 1977 by drilling that penetrated a sheet-sand gas reservoir called the "4500-ft" sand (Fugitt et al., 2000).

In 1990, a well tested a deeper seismic-amplitude anomaly, and the "8500-ft" sand was discovered (Fig. 10.29). The 8,500-ft sand is a channel-fill sand reservoir with good vertical connectivity of individual reservoir sands, even though laterally discontinuous shales are common.

During the period 1993–2000, the 8500-ft sand produced 126 bcf of gas from four wells: the A1, A2, A3, and A7 wells (Fig. 10.29). Their combined flow rate of 150 mcfgpd has declined steadily through time. Production performance indicates that the connectivity and continuity of sands is quite variable. For example, RFT pressures show that three individual sandstone intervals (Members 3, 4, and 5) are vertically connected and have behaved as a single sand. The uppermost two members (Members 1 and 2) are connected to each other, but are

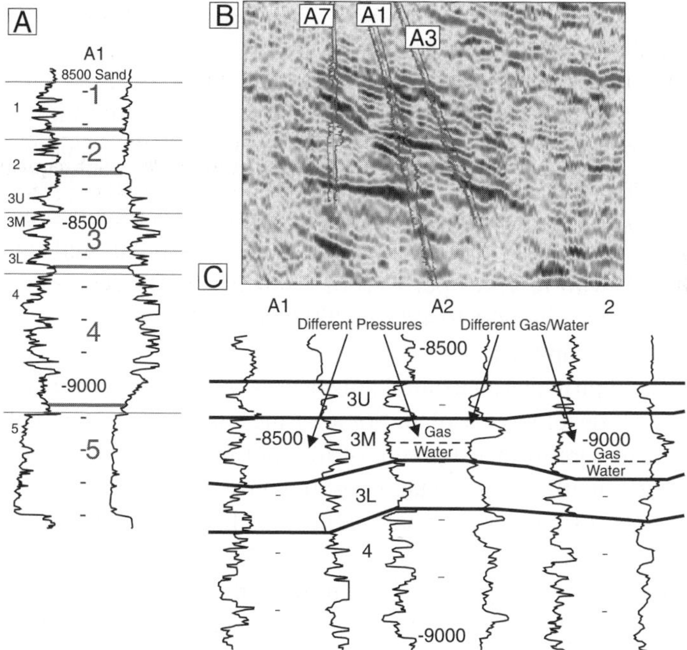

Fig. 10.29. Garden Banks 191 gas field. (A) Well log showing five reservoir intervals in the field, labeled 1–5. (B) Seismic-reflection profile shows the dipping beds, horizontal gas–water contact, and three wells drilled into the reservoir. (C) Interval 3 is subdivided into upper (3U), middle (3M), and lower (3L) units. Different reservoir pressures were encountered in 3M, between wells A1 and A2, and different gas–water contacts were encountered between wells A2 and 2. These features all indicate that the 3M interval is compartmentalized, presumably by the lenticular nature of the channel sandstones. After Fugitt et al. (2000). (Reprinted with permission of the Gulf Coast Section SEPM Foundation.)

separated from the lower members. Member 3 has been subdivided into lower (3L), middle (3M), and upper (3U) intervals. Different reservoir pressures and gas–water contacts within these intervals indicate that there is a level of compartmentalization, presumably as a result of shales separating lenticular sands (Fig. 10.29). Because of these shale breaks, recovery efficiencies would have been reduced considerably if the reservoir were producing oil.

10.4.2.3 Andrew field, UK Sector, North Sea

Andrew field is located in UKCS (UK continental shelf) blocks 16/27a and 16/28, North Sea (Leonard et al., 2000; Jolley et al., 2003). The field was discovered in 1974, approved for development in 1994, and brought online in 1996. Hydrocarbon trapping is by a four-way dip closure over a simple dome structure above an underlying Zechstein salt diapir. Original volumetric estimates in 1996 of 262 MMSTBOOIP were revised upward in 2002 to 315 MMB, and reserves were revised upward from 132 MMBO to 154 MMBO. Gas is estimated to be 280 bcf. More than 70 MMBO were produced in 2.5 years while the field was at maximum flow rates, with an increase in the production rate from 54,500 to 75,000 STBOPD, and an increase in peak production of 1.5 years.

The long time lag between discovery and development was the result of several economic and technical uncertainties that were minimized only after development of horizontal-well technology. Drilling of 11 radial horizontal development wells (down from the 24 originally-planned conventional vertical wells) with high flow rates and low drawdown pressures provided the incentive to proceed in developing the oil rim. Individual horizontal well rates average 10,000 BOPD, with reserves per well of 13 MMBO. The recovery factor also increased from 46% in 1996 to 49% in 2002 and is expected to rise to 53% with additional drilling options.

Andrew field provides an example of success by using a carefully monitored, horizontal-well development plan. Some of the success with this field also can be attributed to careful investigation of the four categories of potential heterogeneities that can affect horizontal wells: the presence of (1) variations in channel geometry, (2) shale barriers and baffles, (3) high-permeability streaks, and (4) faults.

Let us review an example from this investigation for shale barriers and baffles. The performance of a horizontal well through the oil column was simulated under scenarios of both the presence and absence of a laterally continuous shale that might occur across the field (Fig. 10.30). Without a continuous shale, significant gas coning occurred into the heel of the well. With the continuous shale, the seal was sufficient to prevent early coning of gas into the wellbore. The results of this modeling provided the basis for well design. Production logging of individual wells indicated that selective water was underrunning into high-permeability (500-md) sandstones (another of the potential heterogeneities listed above) (Fig. 10.31).

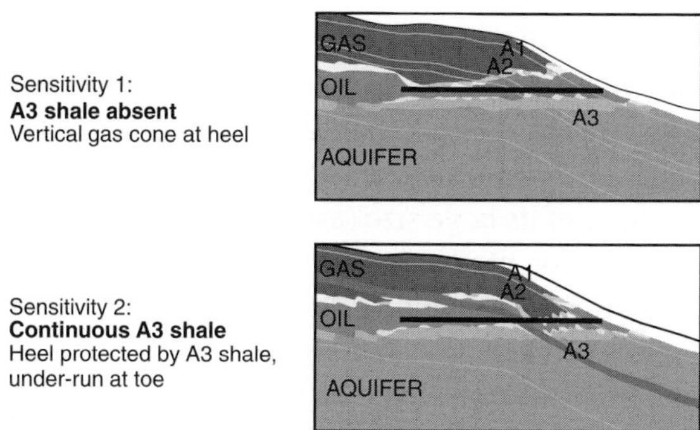

Fig. 10.30. Two horizontal-well simulation models for the Andrews field, showing the effects of the absence (Sensitivity 1) and presence (Sensitivity 2) of a shale barrier between reservoir sands. The horizontal well is drilled into a relatively thin oil rim. The absence of a shale barrier results in gas drawdown and early gas breakthrough into the wellbore. The presence of the shale barrier retards drawdown and allows unaffected oil production from the well. After Jolley et al. (2003). (Reprinted with permission of AAPG, whose permission is required for further use.)

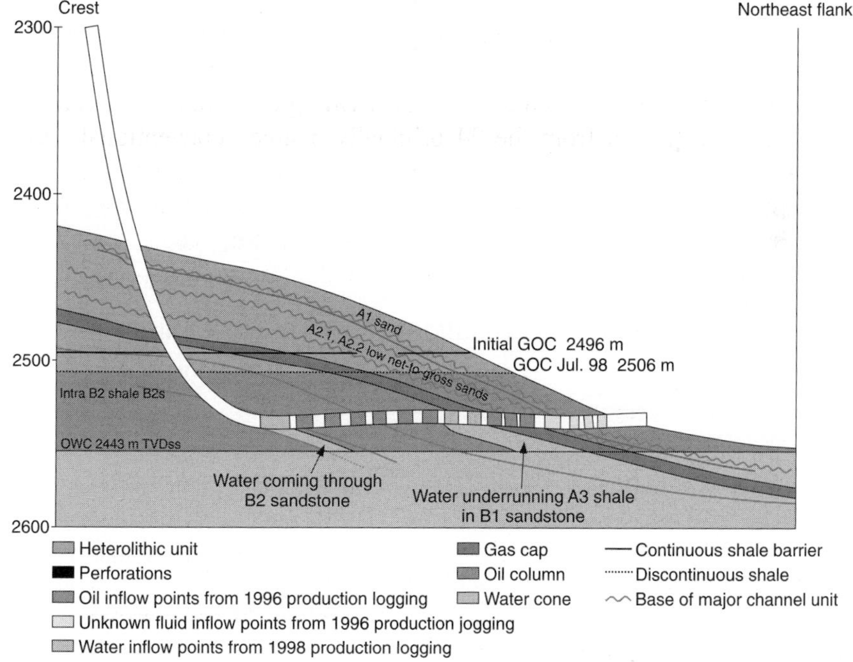

Fig. 10.31. Schematic cross-section along the A03 horizontal well in Andrews field, showing water entry (blue/gray) points along high-permeability (500-md) sandstones, interpreted from an RST log run. The water has underrun the A3 shale in the B1 sandstone. Even though the shale barrier has prevented gas drawdown, different permeabilities in underlying sandstones result in uneven water breakthrough. After Jolley et al. (2003). (Reprinted with permission of AAPG, whose permission is required for further use.)

10.4.2.4 Girassol field, offshore Angola

Girassol field was a major discovery in deepwater (more than 1300 m-deep) Block 17, offshore Angola (Kolla et al., 2001; Beydoun et al., 2002; Navarre et al., 2002). It is an example of a channel-lobe complex that has been investigated intensely because of its large size (estimated to be about 800 MMBOOIP). Although it has not been on production for very long, by mid-2002 the field had reached 200,000 BOPD from eight wells. Girassol field is an example of the value of technologic advancements. A conventional 3D seismic survey acquired and processed in 1996 was considered to be of excellent quality (Fig. 10.32). Later, a 3D high-resolution survey was acquired that nearly doubled the spa-

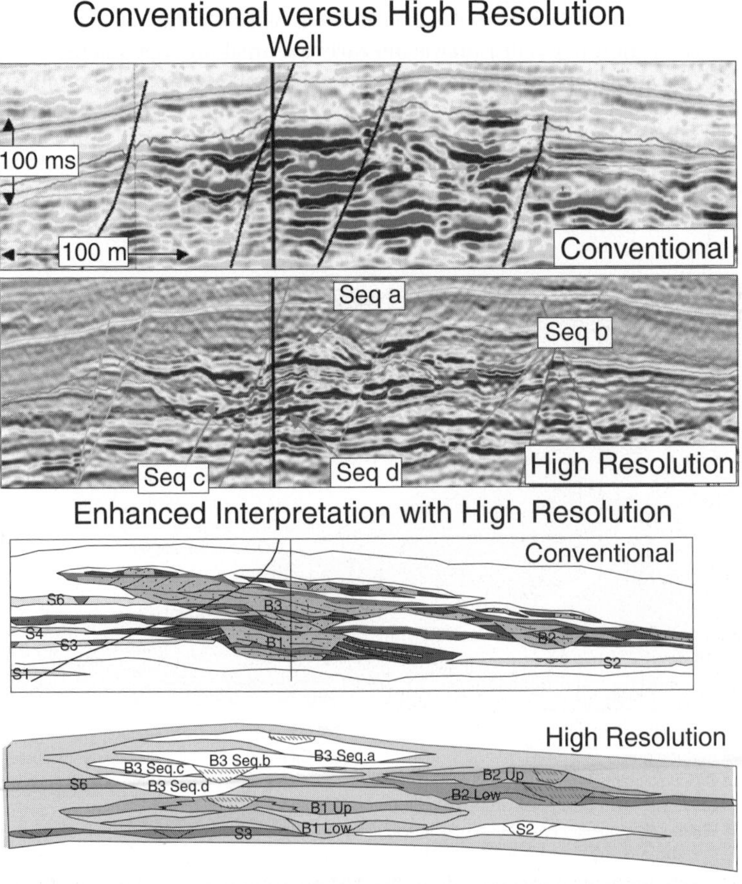

Fig. 10.32. Conventional-resolution versus high-resolution seismic profiles of Girassol field, offshore Angola. The high-resolution seismic profile has provided a considerably better image of the hierarchy of channel fill within the reservoir unit, as shown by the schematic lithofacies interpretations. After Beydoun et al. (2002). (Reprinted with permission of The Leading Edge.)

tial and vertical resolution, thus providing an even greater level of detail than had existed before (Fig. 10.32). Improved design of appraisal and development wells reduced the total number of wells by 25%, compared with the plan that had been developed from the original 3D seismic data.

10.4.3 Levee deposits and reservoirs

For the past decade, most major oil and gas companies have focused on developing channel-fill or sheet-sandstone reservoirs. Much less is known about levee-overbank deposits as potential reservoirs. Levee-overbank deposits consist primarily of muds and thinly bedded (millimeters- to centimeters-thick), laminated sands and sandstones (hereafter termed "thin beds") that form adjacent to sinuous channels (Fig. 10.33). They sometimes exhibit excellent porosity and darcy-range permeability. Thin beds can be ideal stratigraphic traps because of their lateral wedging and thin interbedding of sand and mud. Until

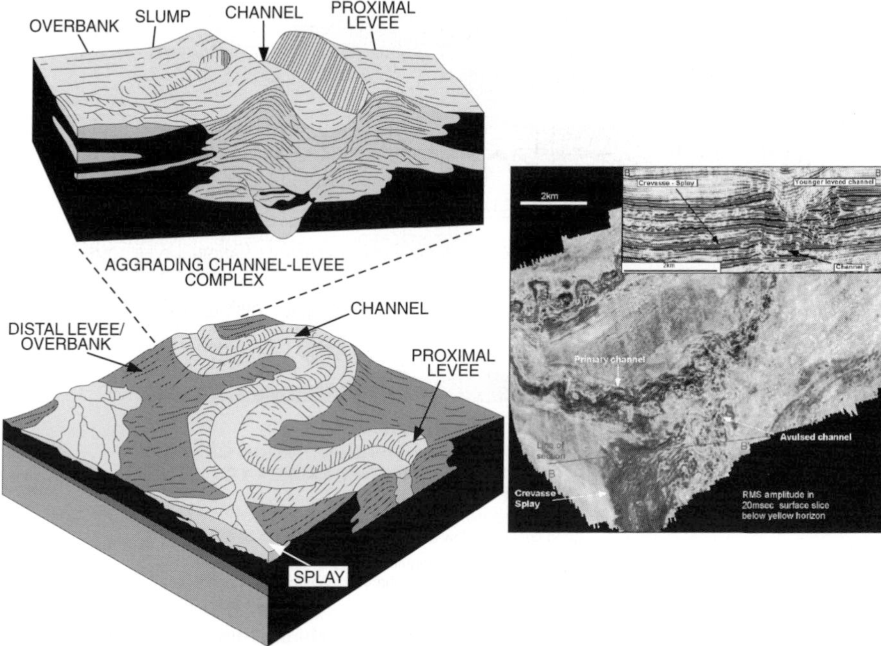

Fig. 10.33. Block diagram of a channel-levee system illustrating the key subenvironments. Modified from Roberts and Compani (1996). Also shown is an RMS amplitude extraction map (20-ms gated window) of a channel-levee system with a prominent Miocene-age crevasse splay, offshore Angola. This crevasse is fan shaped and resulted from an avulsion in the channel. The seismic profile shows the interval from which the attribute was extracted. After Mayall and Stewart (2000). (Reprinted with permission of the Gulf Coast Section SEPM Foundation.)

borehole image logs became more widely used in deepwater settings, potential and discovered thin-bed pay zones were overlooked because, on conventional well logs, they appeared as low-resistivity, low-contrast (shaly) intervals. In recent years, volumetrics in a number of fields have been increased on the basis of recognition of thinly bedded pay from core and borehole image logs.

To date, no comprehensive predictive model for thin-bed reservoirs has been published. A generalized model of a leveed-channel system is shown in Fig. 10.33. The components of this setting include channel-fill, a proximal levee, a distal levee-overbank, slides, crevasse splays, and sediment waves. Of these, the proximal levee and splay deposits, as well as the channel-fill, all can form reservoirs.

Recent work has led to development of well log criteria for differentiating channel-fill and proximal and distal levees, as was discussed in a previous chapter (Fig. 10.34). Trends of bedding-plane dips, determined either from dipme-

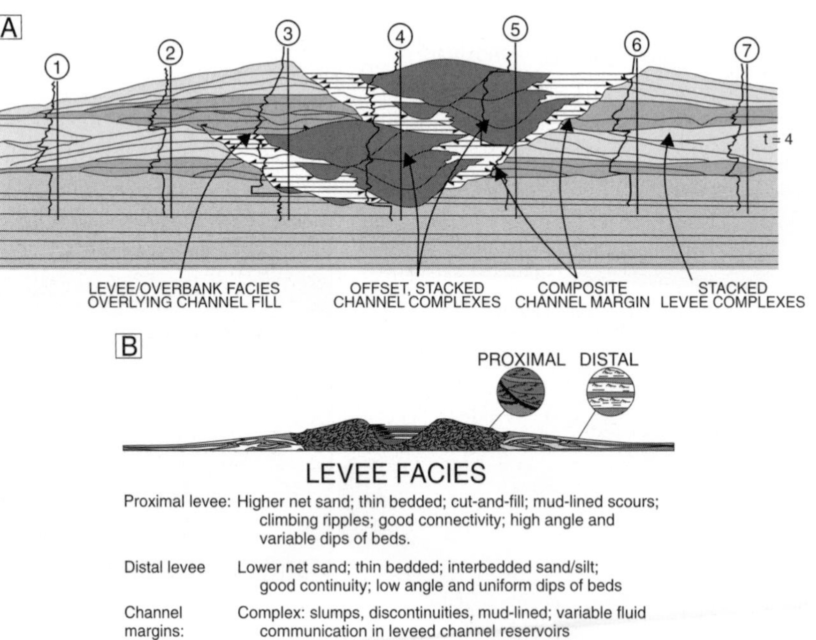

Fig. 10.34. (A) Schematic cross-section across a channel-levee system, and the corresponding gamma-ray or SP logs. Note the vertical decrease in the grain sizes of the levee sediments. This figure illustrates two scales of levees: those outside of the main or master channel (shown in green/bright-gray and orange/dark-gray) and those within the master channel and associated with the smaller channels (shown in yellow/white). Modified from Beaubouef et al. (2003). (B) Similar model for a single channel-levee/overbank deposit, illustrating characteristics of channel margins and the proximal and distal levee. After Slatt et al. (1998). (A – reprinted with permission of AAPG, whose permission is required for further use.)

ter or borehole image logs, are particularly useful for this purpose (Browne and Slatt, 2002). Channel fill exhibits a systematic upward decrease in bed dip. A proximal levee is characterized by a relatively high net-sand content, good vertical connectivity because of commonly occurring erosional scour surfaces, and relatively high and diverse dip angles and orientations. A distal levee/overbank is characterized by lower net sand, greater lateral continuity, but lower vertical connectivity of thin beds, and relatively low and more uniform dip angles and orientations. Also, two scales of levees can be present: those that are associated with a main "master" channel and those that are contained within that channel and associated with smaller, internal channel fills (Fig. 10.34). Slumps are common along channel margins and may compartmentalize or separate the levees from their channel.

Levees form by a portion of channelized sediment gravity flows overbanking or overspilling as they move down the channel axis. "Flow stripping" is the process of separating the components of a sediment gravity flow as it travels within a sinuous channel; it has been called unique to sinuous submarine channels (Piper and Normark, 1983; Peakall et al., 2000). Flow stripping occurs along the outside bends of sinuous channels, where turbidity-current flows accelerate in a way similar to the flow acceleration in fluvial channels along the cutbank margin. High-velocity sediment gravity flows are unable to negotiate the bend, thereby breaching the levee and depositing sediment in the form of splays on the outside of the levee, immediately downcurrent from the bend. Flow velocity normally would be insufficient for all sediment to overtop the levee bend, so the coarser-grained fraction of the flow is transported through the channel and deposited beyond its mouth, whereas the finer-grained sediment is stripped out and transported to an extra-channel or levee location. Through time, the proximal levee receives more sediment than the distal levee does, because the flow decelerates rapidly as the current overtops the proximal levee's banks. The end result is deposition of a wedge-shaped body, with a thick proximal levee and a thinner and shalier distal levee/overbank portion (Fig. 10.35). Gross porosity and permeability also diminish from the proximal to the distal levee/overbank, in the direction of sediment fining and thinning (Fig. 10.36). The channel is thought to fill with sediment after the levees have aggraded – or at least, channel aggradation occurs at a slower rate than does levee aggradation, so that the channel is continually open at the top as it fills from the bottom.

Below are some of the few published examples of levee/overbank reservoirs. Undoubtedly, future discoveries will enhance the economic importance of this significant type of deepwater deposit.

10.4.3.1 Ram Powell L sand reservoir

The Ram Powell L sand is a channel-levee deposit (Fig. 10.18) that is considered to be one of the few purely stratigraphic traps in the northern deep

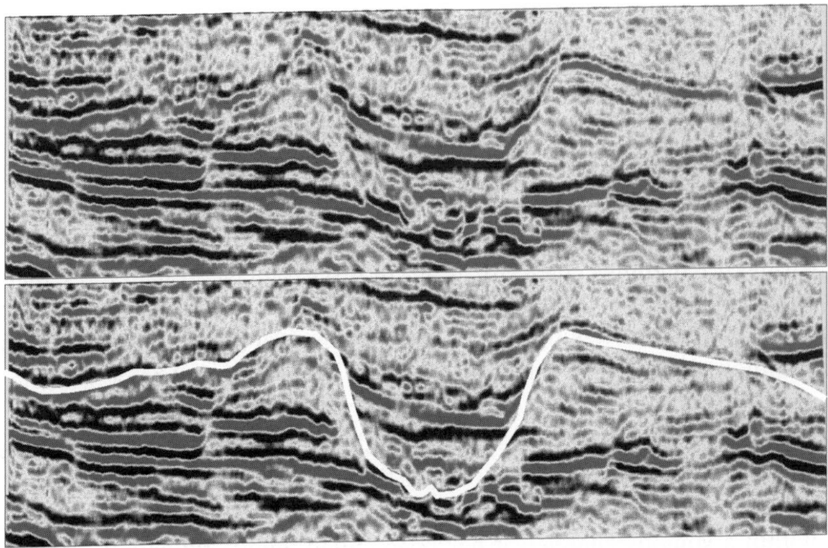

Fig. 10.35. Uninterpreted and interpreted seismic profile of a leveed channel. The master channel and associated levees are outlined in yellow/white. Seismic profile from western Gulf of Mexico.

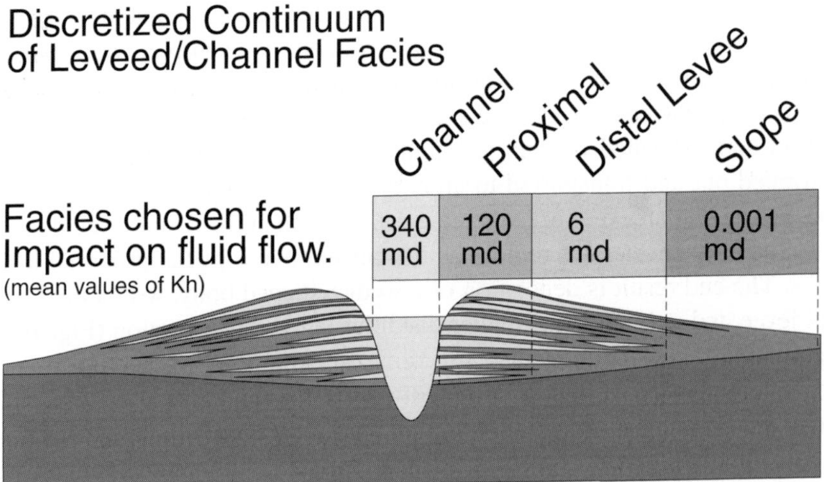

Fig. 10.36. Schematic illustration of the lateral distribution of permeability with distance from the channel. (Diagram provided by C. Jenkins, personal communication, 2003.)

Gulf of Mexico (Clemenceau et al., 2000; Kendrick, 2000). The L sand contains 250–300 bcf of gas. When the L sand was first being developed, five wells were drilled into the channel fill, on the assumption that hydrocarbon-bearing channel sands would be present. All five wells encountered good sands

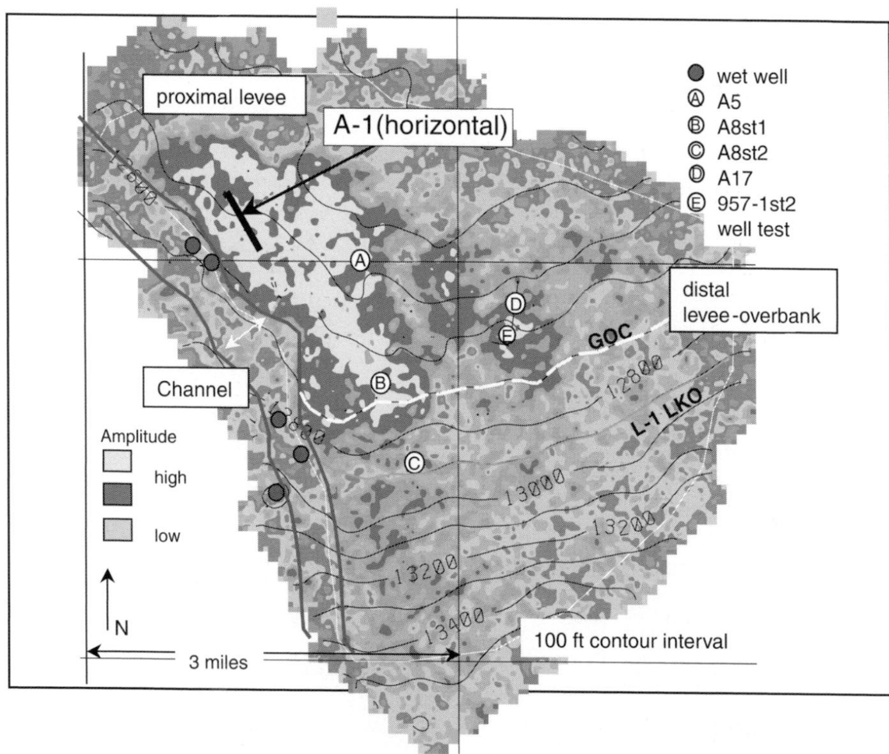

Fig. 10.37. Amplitude extraction map of the Ram-Powell L sand interval superposed on the top L sand structure map, northern deep Gulf of Mexico. Producing levee sands correspond to areas of high amplitude, because of the effects of gas in the reservoir. Note the adjacent channel (outlined by turquoise lines) has low amplitude and no pay. After Clemenceau et al. (2000). (Reprinted with permission of the Gulf Coast Section SEPM Foundation.)

in the 500- to 675-m (1500- to 2000-ft)-wide channel, but they were all water-wet (Fig. 10.37). Wells that were drilled into a high-amplitude reflection zone adjacent to the channel were gas-charged (Fig. 10.37). Near the edge of the high-amplitude zone, a four-day well test flowed 23 MMCFGD and 2700 BCPD (barrels of condensate per day). A strategy was developed to drill a 830-m (2,500-ft) horizontal well parallel to the channel margin, but within proximal levee beds that were identified on the basis of core, dip logs, location relative to the channel, and outcrop analogs. In particular, proximal-levee deposits were identified by a relatively high net-sand content in cores and by relatively high dips on dipmeter logs (Fig. 10.38). Distal-levee beds were identified by a relatively low net sand content in cores and by relatively low and uniform dip patterns on dipmeter logs (Fig. 10.38). The horizontal well performed beyond expectations, with a peak flow rate of 105 MMCFGD and 9600 BCPD (Clemenceau et al., 2000). By year-end 2001, 102.1 bcf gas and 6.5 MMBC had been produced.

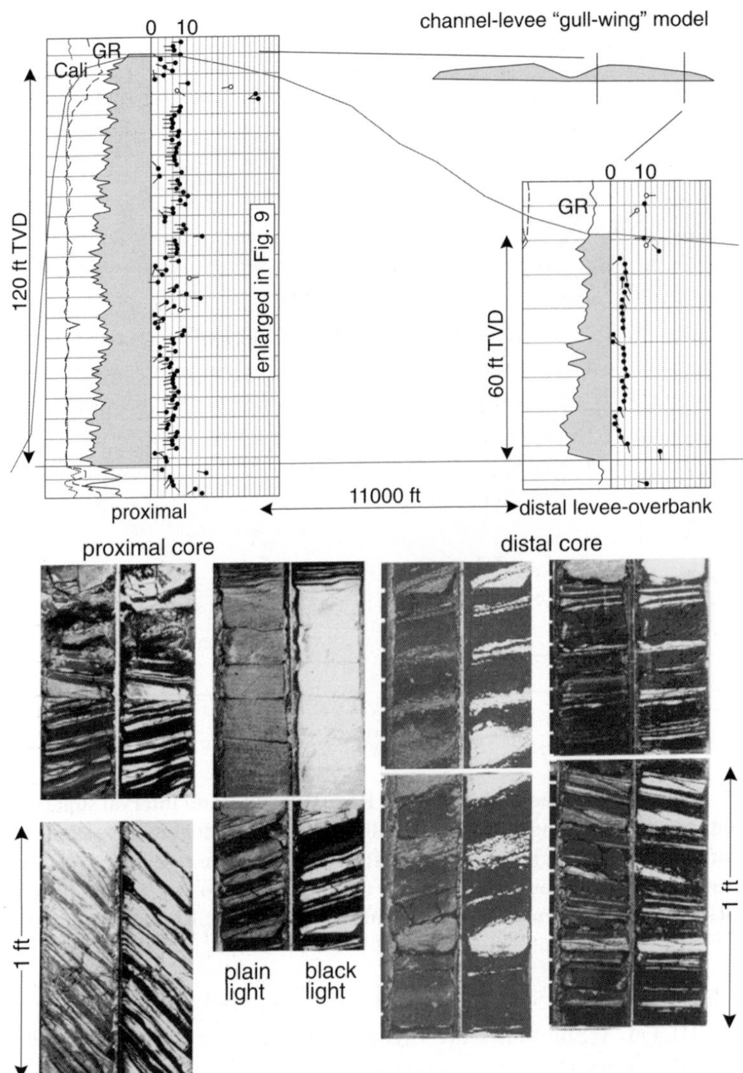

Fig. 10.38. Well logs and core through proximal- and distal-levee strata in the Ram Powell L sand, showing the variations in net sand and dip, which are similar to those described in Fig. 10.34. After Clemenceau et al. (2000). (Reprinted with permission of the Gulf Coast Section SEPM Foundation.)

New information was provided by drilling the horizontal well. The proximal levees comprise four distinct, upward-thinning, thin-bed packages, each package separated from the others by shale beds. At least two of these units have different fluid levels, and pressure data suggest that they act as separate flow units. The post-production data indicate that connectivity occurs across a large area, from the proximal- to distal-levee deposits. Through time, connectivity seems to be increasing as a result of the breakdown of intra-reservoir barriers, as manifested by an increase in pressure followed by flattening of the pressure-decline curve. Pressure tests indicate that the levee and adjacent channel sandstones are not in communication. Most importantly, current estimates indicate that this one well will drain the entire reservoir.

10.4.3.2 The M4.1 sand, Tahoe field, northern Gulf of Mexico

The Tahoe field, located in Viosca Knoll Block 783, was discovered in 1984 (Shew et al., 1994; Shew, 1997; Kendrick, 2000). It was one of the first thin-bed reservoirs discovered and developed in the northern deepwater Gulf of Mexico. Two reservoirs are present; the M4.1 is the deeper and main reservoir. The field is bisected by a prominent channel (Fig. 10.39). The operator (Shell Oil Co.) initially was unsure whether these thin beds could produce at high enough, continuous rates to warrant development. Several outcrop analog studies were undertaken to address this issue, and the 783-4ST2 well was cored and tested within a "shaly" interval (Fig. 10.40). Coring indicated that the reservoir interval consists of a stack of beds and laminae whose individual thicknesses are generally less than 1 cm, with porosity and permeability averages of 27% and 70 md, respectively (and with some beds as high as 500 md) (Fig. 10.40). Nevertheless, the interval produced at the high rates of 29 MMCFGPD and 950 BCPD. Initially, the entire reservoir was developed in January 1994 with one well. A second development phase occurred in 1996 with four additional producing wells.

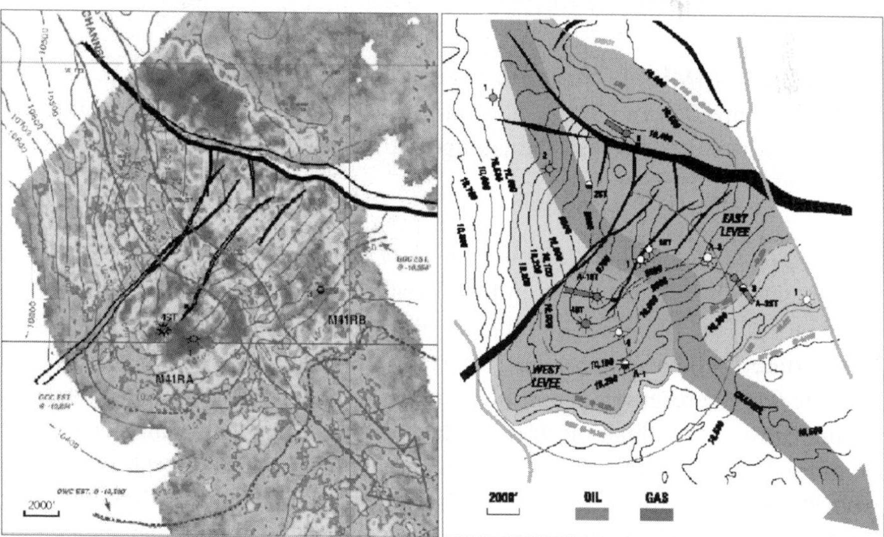

Fig. 10.39. Depth-based seismic profile and interpretation across the M4.1 sand, Tahoe field, northern deep Gulf of Mexico. Note the central channel, which is of relatively low amplitude because it is shale-filled. After Kendrick (2000). (Reprinted with permission of the Gulf Coast Section SEPM Foundation.)

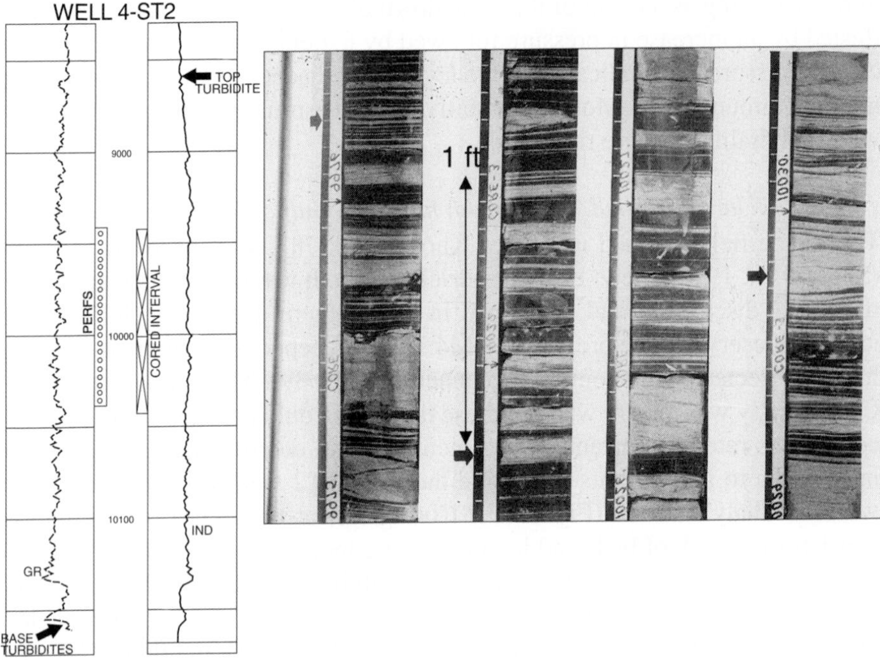

Fig. 10.40. Perforated and cored interval in the 4-ST2 well in Tahoe field, showing the shaly log response and the thin-bed nature of the contained strata. After Shew et al. (1994). (Reprinted with permission of SEPM, Society for Sedimentary Geology.)

More than 17 MMB of equivalent gas and condensate had been produced from four wells through mid-2000. The oil–water contact is shallower in the west levee than in the east levee, indicating that communication does not cross the entire field (Fig. 10.39).

10.4.3.3 Falcon field, northwestern Gulf of Mexico

Falcon gas field is located within East Breaks blocks 579 and 623 in the northwestern Gulf of Mexico (Abdulah et al., 2004). It was discovered in 2001. 3D seismic horizon slices, facies analysis, dipmeter data, and whole core all indicate that the reservoir is part of a channel-levee complex. Gas accumulation occurs within a combination stratigraphic/structural trap that is characterized by high-amplitude, low-impedance seismic anomalies. Channel morphology varies from straight to sinuous. The channel is mud-filled. Flat spots at different structural elevations and on opposite sides of the channel indicate that the channel is a flow barrier. Dipmeter data reveal a pattern of dip orientation away from, and at approximately right angles to, the channel. Two stacked levee sequences dip away from the channel; the upper levee dips 4–6° and the lower levee dips 5–8°.

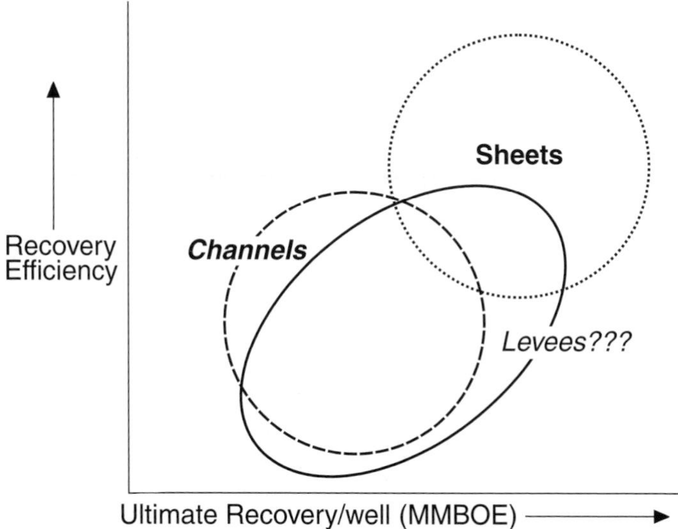

Fig. 10.41. Schematic illustration showing the approximate ultimate recovery/well and recovery efficiency of the three main deepwater architectural elements.

Four facies have been identified: (1) a proximal-medial levee, which is the main gas reservoir, with 84% net thin-bed sand and porosities as high as 37.8%; (2) a distal levee/overbank with 20% net sand; (3) interchannel splays characterized by 50–60% net sand, arranged in coarsening- and thickening-upward, amalgamated-sheet packages; and (4) fine-grained basin-floor facies composed of shales and siltstones that encapsulate the entire sequence. Permeabilities over the entire cored interval range from 0.06 to 6,220 md, and average porosity is 31.4%.

10.5 Summary

This chapter has summarized the important characteristics of deepwater deposits and reservoirs. These reservoirs are quite complex and variable. An understanding of the different architectural elements is critical to hydrocarbon recovery, because the elements exhibit different external geometries, sizes, spatial orientations, and internal sedimentary and stratigraphic features. Because of these differences, the volume of hydrocarbons and the anticipated recovery efficiency will vary by architectural element (Fig. 10.41).

Chapter 11

Sequence stratigraphy for reservoir characterization

11.1 Introduction

Sequence stratigraphy is the study of sedimentary rock relationships within a chronostratigraphic or geologic-time framework. Its basis is identification of stratal surfaces, regional unconformities and their correlative conformities, and relationships among lithofacies and depositional environments, within this chronostratigraphic framework. Sequence stratigraphy differs fundamentally from lithostratigraphy. In sequence stratigraphy, stratal surfaces and rock bodies between the surfaces are defined on the basis of stratigraphic intervals that are time-synchronous, and laterally continuous and regionally correlative (e.g., bentonites or condensed sections), rather than on the basis of the lithologic character of the rocks and their stratigraphic relations (Fig. 11.1). Thus, chronostratigraphic horizons often crosscut lithostratigraphic horizons (Figs. 11.1 and 11.2), and well-log correlations differ substantially between

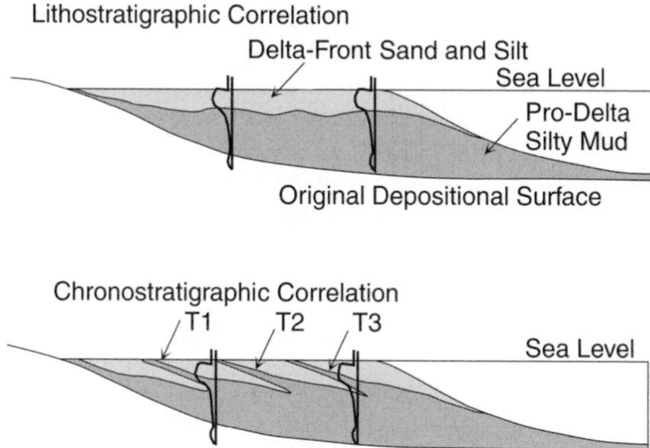

Fig. 11.1. Comparison of a lithostratigraphic and a chronostratigraphic correlation and stratigraphic interpretation of two pseudo-wells. After Bashore et al. (1994). (Reprinted with permission of AAPG, whose permission is required for further use.)

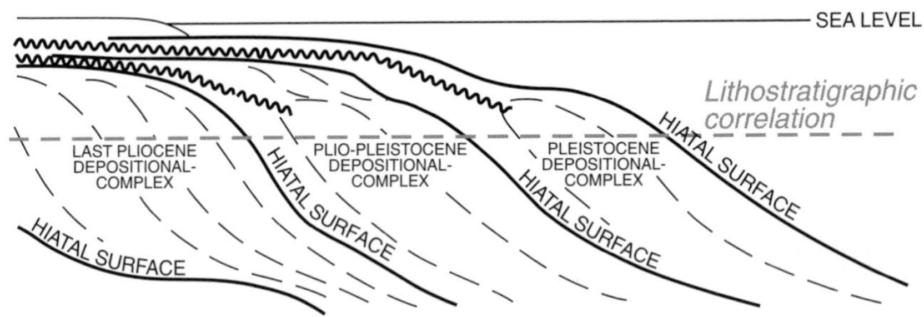

Fig. 11.2. Comparison of a lithostratigraphic and a chronostratigraphic correlation for a series of progradational depositional complexes. Hiatal surfaces represent time intervals with no significant deposition. After Frasier (1974). (Reprinted with permission of AAPG, whose permission is required for further use.)

these disciplines (Figs. 11.1 and 11.3). Before you initiate a well-log correlation project, it is wise to list the data available to you and to develop a correlation strategy, such as choosing lithostratigraphic versus chronostratigraphic correlation (Mulholland, 1994). Chronostratigraphic correlations are the recommended strategy.

Sequence stratigraphy is a very comprehensive subdiscipline of stratigraphy. The landmark publication: Seismic stratigraphy – Application to hydrocarbon exploration (Payton, 1977) introduced the concept that seismic-reflection records image interpretable depositional sequences (the study of which constitutes seismic stratigraphy). Prior to that AAPG publication, seismic-reflection records were used mainly for subsurface structural interpretation of basin geometry and fill. Many of the concepts presented in Payton (1977) had been developed by earlier workers (e.g., a history of the concepts also can be found in Sloss, 1963), but Payton's collection of papers (1977) had the appeal of applying seismic-reflection records to stratigraphic interpretation. Thus, seismic stratigraphy preceded the discipline of sequence stratigraphy.

In 1988, another landmark volume was published by the Society for Sedimentary Geology (SEPM), titled "Sea Level Changes – An Integrated Approach" (Wilgus et al., 1988). That volume introduced many of the modern concepts of sequence stratigraphy. It was soon followed by Van Wagoner et al.'s (1990) "Siliciclastic Sequence Stratigraphy in Well Logs, Cores, and Outcrops", which not only emphasized the rock framework of sequence stratigraphy but also demonstrated that sequence stratigraphy could be applied at a high-resolution, subseismic scale (a history of the development of sequence stratigraphy is also provided in this publication). These seminal works have been followed by additional key books, such as "Sequence Stratigraphy of Foreland Basin Deposits: Outcrop and Subsurface Examples from the Cretaceous of North America" (Van Wagoner and Bertram, 1995), "Isolated Shallow Marine

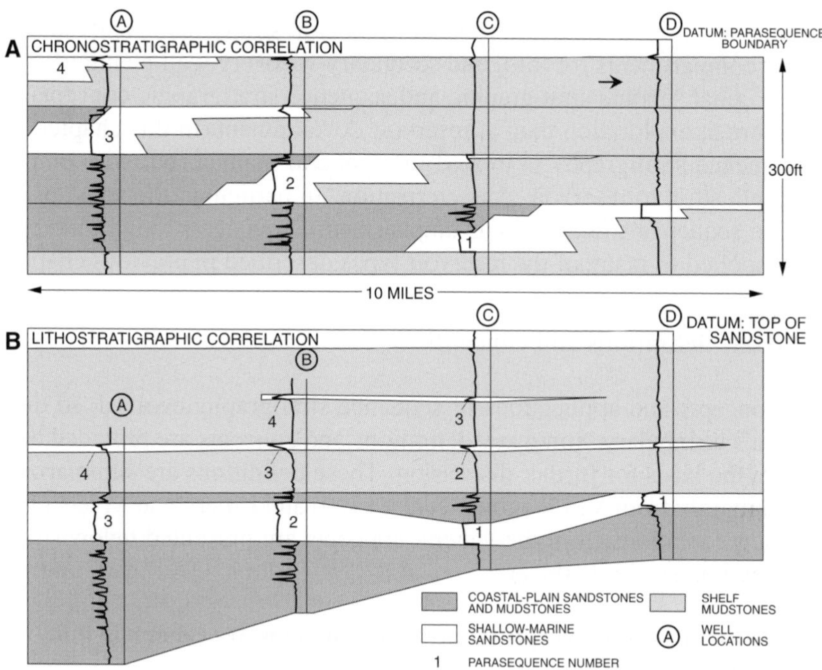

Fig. 11.3. Comparison of well log cross-sections correlated on the basis of (A) chronostratigraphy and (B) lithostratigraphy. The datum for the chronostratigraphic correlation is a parasequence boundary. The datum for the lithostratigraphic correlation is a formation top. The chronostratigraphic correlation indicates deposition of four parasequences, with each subsequently younger parasequence stepping progressively toward the left (landward) and the total defining a retrogradational parasequence set. Based on the chronostratigraphic correlation, sands 1–4 are not connected, which they are according to the lithostratigraphic correlation in (B). After Van Wagoner et al. (1990). (Reprinted with permission of AAPG, whose permission is required for further use.)

Sand Bodies: Sequence Stratigraphic Analysis and Sedimentologic Interpretation" (Bergman and Snedden, 1999), and "Siliciclastic Sequence Stratigraphy – Concepts and Applications" (Posamentier and Allen, 1999). Of course, many papers and other books have now been written on the topic.

Sequence stratigraphy is multidisciplinary in nature. Sequence stratigraphers must be versed in interpretation of seismic-reflection records, well logs, cores, and biostratigraphic and geochemical data. Applications of sequence stratigraphy at the exploration scale include regional correlations of stratal units, and using seismic reflection records or well logs to identify or predict seal, source, trap, and reservoir rocks and their locations in both time (geologic and seismic) and space (within a basin). At the reservoir development scale, applications of sequence stratigraphy include more-accurate fine-scale correlations, and prediction of the location, in time and space, of potential stratigraphic and diagenetic reservoir compartments for infill and secondary-recovery well planning.

The original seismic stratigraphic and sequence stratigraphic concepts were aimed more at exploration than at reservoir development. In this chapter, however, sequence stratigraphy is treated in a general manner, but with emphasis on its application to reservoir characterization. In particular, the value of high-resolution sequence stratigraphy is emphasized. Sequence stratigraphic principles are applied to many of the reservoir types described in previous chapters.

11.2 Basic definitions and concepts

As the concepts and applications of sequence stratigraphy evolved, so did the discipline's vocabulary. Some key definitions and concepts are provided below; they form the basis for further discussion. These definitions are summarized or modified mainly from Van Wagoner et al. (1990) and Emery and Myers (1996). Also, many variations from the general concepts are presented below (for further discussion, see Vail, 1987).

11.2.1 Definitions and concepts related to the ocean water column in time and space

Eustatic sea level: Global sea level as measured between the sea surface at any given time and a fixed datum, normally the center of the Earth (Fig. 11.4). Eustatic sea-level changes occur because of changes in the volume of the ocean basins (by varying ocean-ridge volume) or because of variations in the volume of ocean water (by glaciation and deglaciation).

Relative sea level: The distance or depth between the sea surface and a local moving datum, such as basement or a surface within a submarine sediment accumulation (Fig. 11.4). Relative sea level may change as a result of tectonic subsidence or uplift of a basement datum, subsidence of a datum within the sediment accumulation due to compaction, or vertical eustatic movements of the sea surface. Relative rises in sea level result from subsidence, compaction and/or eustatic sea level rises. Relative falls in sea level result from tectonic uplift and/or eustatic sea level fall.

Water depth: The distance or depth between the sea surface and sea floor (marine sediment surface) at any given time (Fig. 11.4).

Accommodation: The amount of space available in which sediment can accumulate at any point in time. In the vertical sense, it is analogous to water depth.

Relative sea-level cycle/curve: A complete cycle of sea level fall and rise (Fig. 11.5). A falling limb on a sea-level curve occurs during the interval of a fall in relative sea-level. A rising limb of the sea-level curve occurs during the time interval of a rise in relative sea level. The turnaround position occurs during the transition between falling and rising sea level.

Parasequence: A relatively conformable, genetically-related succession of beds or bedsets bounded by marine-flooding surfaces or their correlative surfaces. Most siliciclastic parasequences are progradational in nature, resulting in an upward-shoaling (upward-coarsening and -cleaning) association of shallow marine lithofacies (Figs. 8.10, 11.10, and 11.11). If the rate of sediment supply to a shoreline area exceeds the rate of water deepening as a result of subsidence and/or sea level rise, then sediments will prograde in the basinward

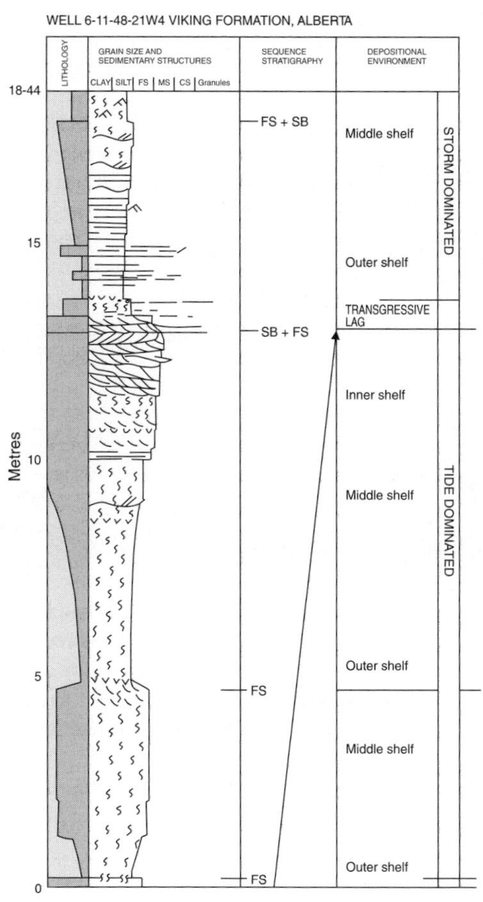

Fig. 11.10. Tide- and storm-dominated parasequences in core, Viking Formation, Alberta, Canada. FS – flooding surface, SB – sequence boundary. At the base of the succession is a flooding surface upon which sit outer shelf strata. The second flooding surface marks a slight increase in water depth from the middle to outer shelf, followed by an upward increase in sandiness of the middle and inner shelf strata. A relatively coarse transgressive lag marks a sequence boundary (beneath), then follows another flooding surface upon which outer shelf strata were deposited, finally followed by middle shelf strata. A fourth flooding surface and possible sequence boundary mark the top of the stratigraphic succession. After Emery and Myers (1996). (Reprinted with permission of Blackwell Science Ltd.)

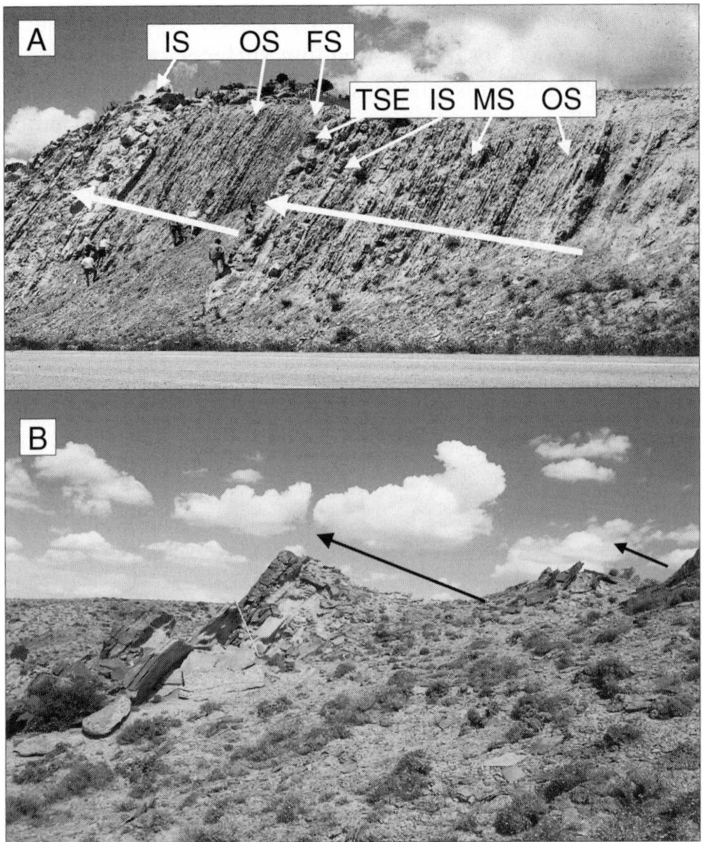

Fig. 11.11. (A) Two shoreface parasequences in Cretaceous strata of Utah. OS – outer shelf, MS – middle shelf, IS – inner shelf, TSE – transgressive surface of erosion, FS – flooding surface. The transgressive surface of erosion is characterized by a relatively coarse-grained lag, abundant burrows, and the presence of shark teeth. Net sandstone content increases from OS to MS and IS. Large arrows point toward coarsening- and thickening-upward strata. (B) Two parasequences, similar to those described in (A) from Cretaceous strata in Wyoming. Arrows point toward coarsening- and thickening-upward trends. The upper surface of these two parasequences contains abundant trace fossils, oyster shells, and wave ripples.

direction (Fig. 11.12, Stage 1). If water depth increases more rapidly than sediment can be supplied, then marine waters will flood landward over the preceding parasequence, forming a condensed section/marine flooding surface that marks the base of a new parasequence (Fig. 11.12, Stage 2). If the rate of sedimentation then exceeds that of relative sea level rise, another progradational parasequence will form, giving rise to two progradational parasequences bounded by a shaly condensed section (which can vertically isolate the two sandstone intervals (Fig. 11.12, Stage 3).

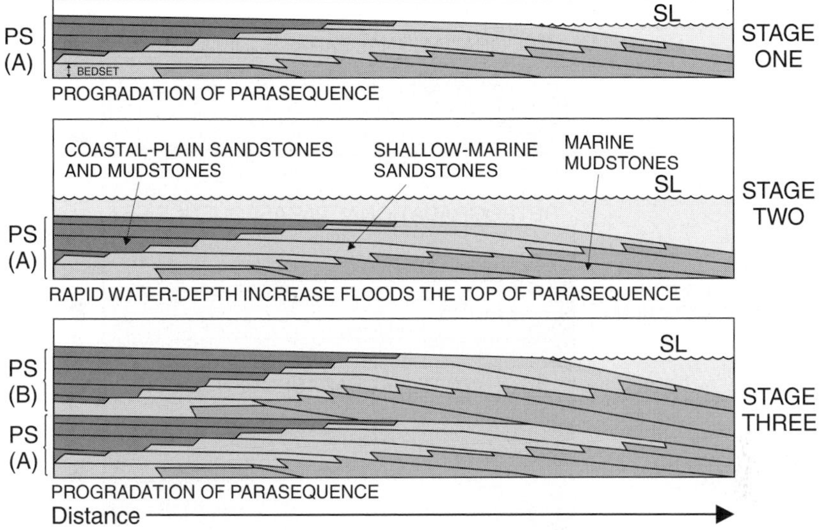

Fig. 11.12. Schematic illustration of the formation of two parasequences, such as those shown in Fig. 11.11. Stage 1 shows the progradation of parasequence A at a time when the rate of deposition exceeds the rate of increase in water depth. Stage 2 shows a rapid water-depth increase, which floods the top of parasequence A, creating a transgressive surface of erosion with chemically precipitated minerals, volcanic ash beds, and/or relatively coarse inorganic or organic (e.g., shark teeth) grains concentrated on the surface. Stage 3 shows deposition of marine shale upon the transgressive surface of erosion, followed by progradation of a second parasequence when the rate of deposition again exceeds the rate of water-depth increase. Bedsets in parasequence B downlap onto the transgressive surface of erosion. After Van Wagoner et al. (1990). (Reprinted with permission of AAPG, whose permission is required for further use.)

Parasequence set: A succession of genetically related parasequences forming a distinctive stacking pattern bounded by major marine-flooding surfaces and their correlative surfaces (Fig. 11.10). Three parasequence sets are presented in Fig. 11.13 to illustrate the variation in stratigraphy that can result from differences between the rate of sediment deposition and the rate of generation of accommodation space. A *progradational parasequence set* forms when a set of individual parasequences are deposited because the rate of deposition exceeds the rate of accommodation. In this instance, each parasequence progrades progressively farther basinward than does the preceding parasequence. This pattern can be recognized on a well log as a set of sandstones that become progressively thicker-bedded and with fewer shale interbeds upward. A *retrogradational parasequence set* forms when the rate of deposition is less than the rate of accommodation. In this instance, each parasequence steps (retrogrades) farther landward than does the preceding parasequence. This pattern can be recognized on a well log as a set of sandstones that become progressively

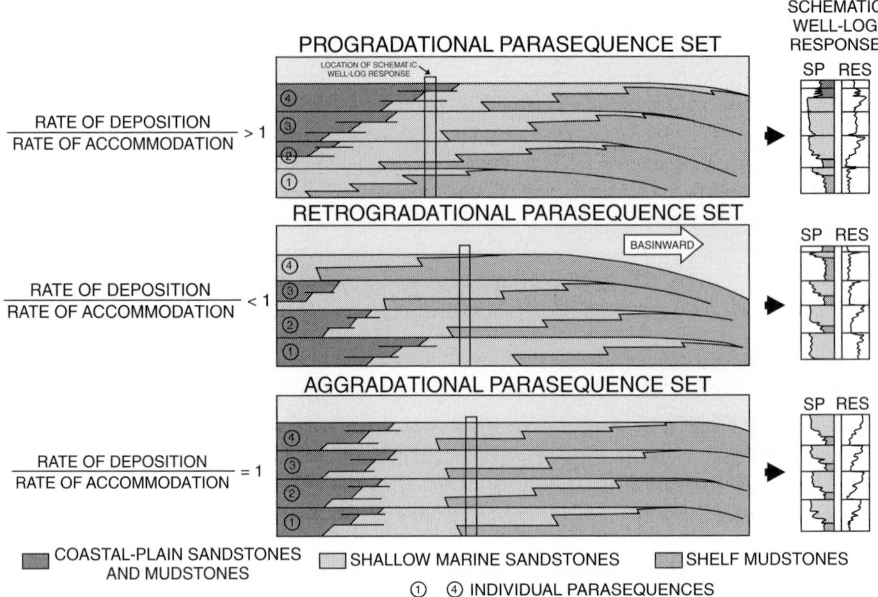

Fig. 11.13. Schematic illustration showing four parasequences stacked to form parasequence sets, and their well log characteristics. For the progradational parasequence set, each parasequence progrades seaward to a greater extent than did the preceding parasequence, giving rise to a coarsening- and thickening-upward well log pattern. Such a parasequence set forms when the rate of deposition exceeds the rate of accommodation over the time interval represented by the parasequence set. The overall progradation is punctuated periodically by a rise in relative sea level, which results in a flooding surface upon which the next younger parasequence is deposited. For the retrogradational parasequence set, each parasequence progrades seaward to a lesser extent than did the preceding parasequence, giving rise to a fining- and thinning-upward well log pattern. Such a parasequence set forms when the rate of deposition is less than the rate of accommodation, over the time interval represented by the parasequence set. The overall retrogradation is punctuated periodically by a stillstand in relative sea level, which results in progradation of the next younger parasequence. For the aggradational parasequence set, each parasequence progrades seaward the same distance as did the preceding parasequence, giving rise to a uniform well log pattern for the four parasequences. Such a parasequence set forms when the rate of deposition is equal to the rate of accommodation, over the time interval represented by the parasequence set. After Van Wagoner et al. (1990). (Reprinted with permission of AAPG, whose permission is required for further use.)

thinner-bedded and more highly interbedded with shale upward. An *aggradational parasequence set* forms when the rate of deposition is approximately equal to the rate of accommodation. In this instance, each parasequence extends basinward about the same distance as does the preceding parasequence. This pattern can be recognized on a well log as a set of sandstones that are of equivalent thickness, separated by shales of equivalent thickness; sedimentary facies are also similar among the sandstones.

Marine flooding surface: A surface separating younger strata (above) from older strata (below), across which there is evidence of an abrupt increase in water depth across that surface. This surface may be accompanied by minor submarine erosion or nondeposition. A marine flooding surface represents the upper surface of a parasequence, above which lie deeper-water strata than were deposited upon the flooding surface. A transgressive lag of coarse grains, shell fragments, shale rip-up clasts and/or chemically precipitated nodules or grains may define the marine flooding surface. A maximum flooding surface (Fig. 11.9) is the surface represented by the maximum landward extent of the shoreline or coastal onlap during a relative sea-level cycle.

Sequence boundary: Subaerial erosion surface (unconformity) or subaerial/submarine surface of nondeposition (disconformity) (Fig. 11.9). Two types of sequence boundaries were defined originally (Van Wagoner et al., 1988). A Type 1 sequence boundary (SB-1 of Fig. 11.9) is defined as a sequence boundary resulting from a relative sea level fall caused by a rate of eustatic fall that exceeds the rate of subsidence. The resulting sea level fall is sufficiently large to move the shoreline to a position at or beneath the position of the offlap break (Fig. 11.7). By contrast, a Type 2 sequence boundary (SB-2 of Fig. 11.9) is defined as a sequence boundary resulting from a less-pronounced relative sea level fall caused by a eustatic fall with a rate slightly less than or equal to the rate of basin subsidence. In this case, the new shoreline does not extend as far seaward as the offlap break.

Depositional system: Three-dimensional assemblages of genetically related lithofacies (Figs. 11.14–11.16).

Systems tract: A three-dimensional assemblage of lithofacies that are genetically linked by sedimentary processes and environments and are bounded by discrete, recognizable surfaces (Fig. 11.9). Systems tracts are arranged laterally and vertically in a predictable manner, within the majority of clastic depositional sequences (Fig. 11.9).

Lowstand systems tract: A systems tract deposited during an interval of relative sea level fall and early sea level rise (Fig. 11.9). There are four component parts to a lowstand systems tract: a mass transport deposit, a basin floor fan, a slope fan (also referred to as a channel-levee complex), and a prograding complex (also referred to as a late lowstand wedge) (Fig. 11.14).

The mass transport deposit forms in deep water as relative sea level first begins to fall from a highstand position. That fall causes the pore pressure in upper slope sediments to alter, resulting in slope instability, large-scale collapse, transport, and then deposition of the collapsed shelf margin strata downslope (Fig. 11.14). Note that Fig. 11.17 also illustrates the same principle with a lacustrine delta, the frontal margin of which has slumped and been eroded due to a lowering of the lake level.

The basin floor fan is deposited during the main falling stage of relative sea level (Fig. 11.14). During that time, shallow marine (shelf and shoreline) areas

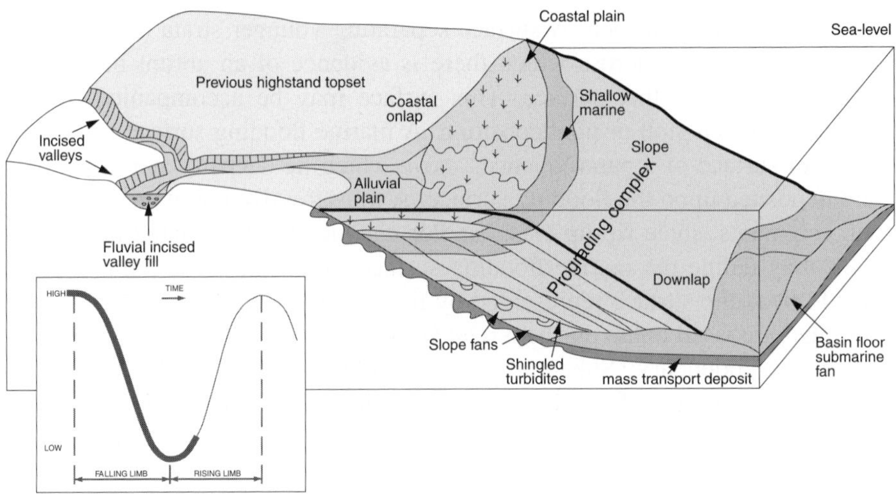

Fig. 11.14. Characteristics of a shelf-to-basin profile during a falling stage of relative sea level. During this time, the shelf area is exposed subaerially, creating a surface of erosion and/or nondeposition (unconformity) and allowing rivers to downcut into preceding highstand deposits, to form incised valleys. Sediments are transported beyond the shelf break and are deposited first in deep water as basin floor submarine fans while sea level falls. Then sediments are deposited as slope fans and prograding complexes when sea level begins to rise. Mass-transport deposits form at the very base of the sequence, because of destabilization of the shelf margin during the early drop in relative sea level. The four types of deposits on the shelf and slope comprise the lowstand systems tract. Modified from Emery and Myers (1996). (Reprinted with permission of Blackwell Science Ltd.)

become subaerially exposed and energetic, sediment-laden rivers flow across them, carving out river valleys called incised valleys (Fig. 11.14; this principle is also demonstrated by the lacustrine delta in Fig. 11.17). The incised valleys act as thoroughfares for the direct transport of sediment over the shelf edge and into deep water (slope and basin) (Fig. 11.14).

Next, during sea-level turnaround and early sea level rise (Fig. 11.5), a smaller volume of generally finer-grained sediment is deposited as the channel-levee complex (Fig. 11.14). Early studies suggested that channels and levees were deposited contemporaneously, but it is now hypothesized that the levees form during the falling stage of sea level, at the same time that the coarser fraction of sediment gravity flows is transported within channels and beyond the levee margins and is deposited as channel-mouth sheets or frontal splays (Fig. 11.18).

Finally, the prograding complex is deposited as sea level continues to rise, creating accommodation on the upper slope (Fig. 11.14). Incised valleys begin to fill in at early turnaround and continue to backfill as rising marine waters encroach on the valleys, forming estuaries (Fig. 11.15). Many incised valleys fill at the base with fluvial deposits and fill stratigraphically higher with estu-

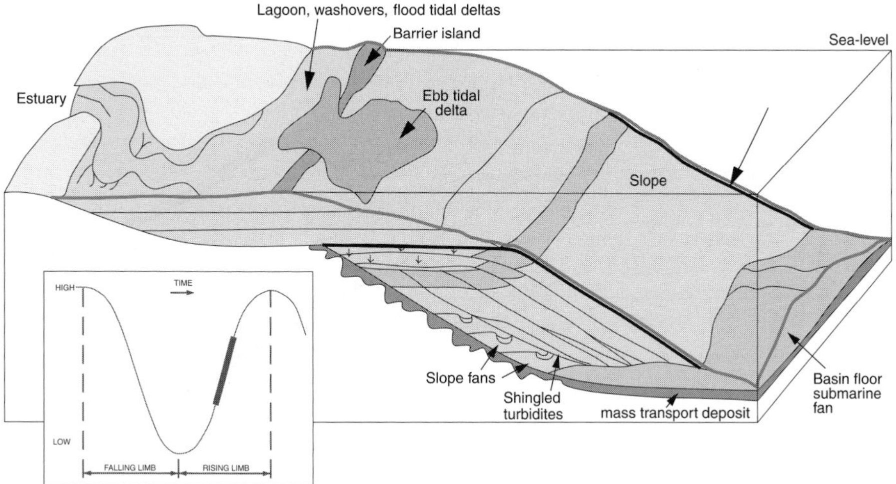

Fig. 11.15. Characteristics of a shelf-to-basin profile during a rapidly rising stage of relative sea level. As sea level rises over the previously exposed shelf edge (Fig. 11.14), sediments are deposited on the shelf, and incised valleys first are filled by basal fluvial deposits, then are filled by estuarine deposits as marine waters encroach into the valleys to form brackish-water estuaries. With a continued rise in sea level, coastal deposits such as barrier islands, lagoon-aeolian-beach-shoreface strata, and ebb- and flood-tidal deltas form. These deposits all form the landward-stepping transgressive systems tract. Mud may be deposited on the shelf and behind (seaward of) the shoreline and nearshore deposits. At the maximum extent of a rise in sea level, a fine-grained, sometimes organic-rich condensed section forms, which is capped by a maximum flooding surface. The maximum flooding surface represents the maximum landward extent of the shoreline and the maximum water depth on the shelf. During this time, very little sediment is deposited in deep water, and the sediment that is deposited there is generally silt and/or clay. Modified from Emery and Myers (1996). (Reprinted with permission of Blackwell Science Ltd.)

arine strata. During much of lowstand time, little sediment is deposited on the shelf, shoreline, and nonmarine environments because those environments generally are exposed subaerially. Commonly, erosion surfaces form during this time (Figs. 11.6 and 11.9).

Transgressive systems tract: A systems tract deposited during an interval of rapid rise in relative sea level (Figs. 11.5 and 11.15). Components of the transgressive systems tract include a variety of shelf and shoreface deposits and a condensed section capped by a maximum flooding surface.

As relative sea level rises above the shelf break, the shelf becomes flooded and shoreface and shelf sediments are deposited. With a continued rapid rise in sea level, the shoreline – and thus, the shoreface strata – migrate in a landward direction, giving rise to a landward-stepping (retrograding) set of shoreface and associated shoreline deposits (Fig. 11.15). At the same time, finer-grained sed-

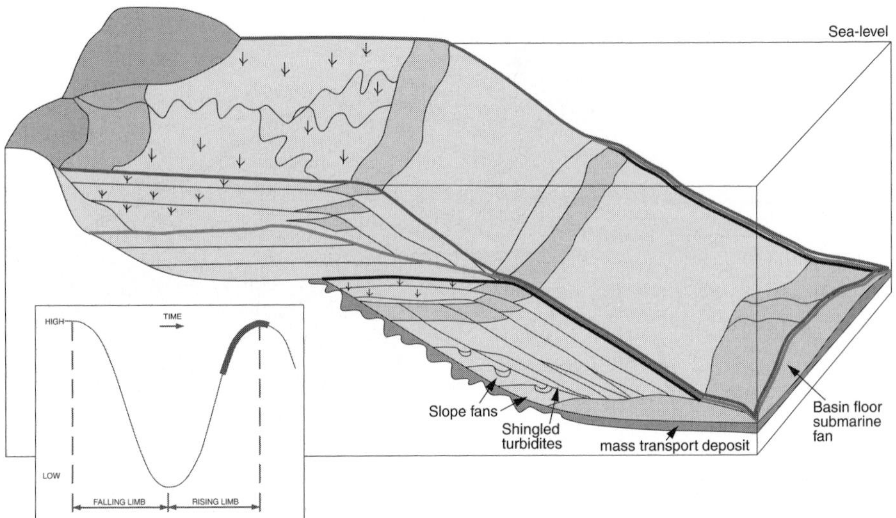

Fig. 11.16. Characteristics of a shelf-to-basin profile during a decrease in rate of sea-level rise. During this time, the rise in sea level increases the amount of accommodation space available in which sediments can be deposited, so sediments prograde seaward. In areas of major river systems, deltas form. In other areas, interdeltaic shoreline strata form. The prograding sediments form downlapping highstand systems tract patterns onto the underlying maximum flooding surface. During this time, very little sediment is deposited in deep water, and the sediment that is deposited there generally is silt and/or clay, although deepwater sands may be deposited in front of major river deltas (i.e., highstand turbidites). Modified from Emery and Myers (1996). (Reprinted with permission of Blackwell Science Ltd.)

iment is deposited on the flooding shelf floor, behind the landward-advancing shoreline.

At the point of maximum rate of sea-level rise (Fig. 11.5), the shoreline reaches its highest landward extent (maximum coastal onlap), and the condensed section is deposited. The condensed section generally is a mud-prone, organic-rich deposit that blankets the shelf, slope, and basin. The condensed section is discussed in more detail below because it is a key deposit that, when identified on seismic reflection records, logs, core and/or outcrop, allows a depositional sequence stratigraphic framework to be established (Loutit et al., 1988).

The maximum flooding surface at the top of the condensed section represents the surface of maximum landward advancement of the shoreline. During transgression, relatively little sediment is deposited in deep water (Fig. 11.15).

Highstand systems tract: A systems tract deposited during continued sea level rise, but when the rate of rise diminishes (Fig. 11.5). At that time, accommodation space increases at a lower rate than the rate at which sediment is supplied, so that sediments prograde basinward, forming broad coastal plains and deltas near major rivers, and interdeltaic shorelines in between (Fig. 11.16; also, Fig. 11.17

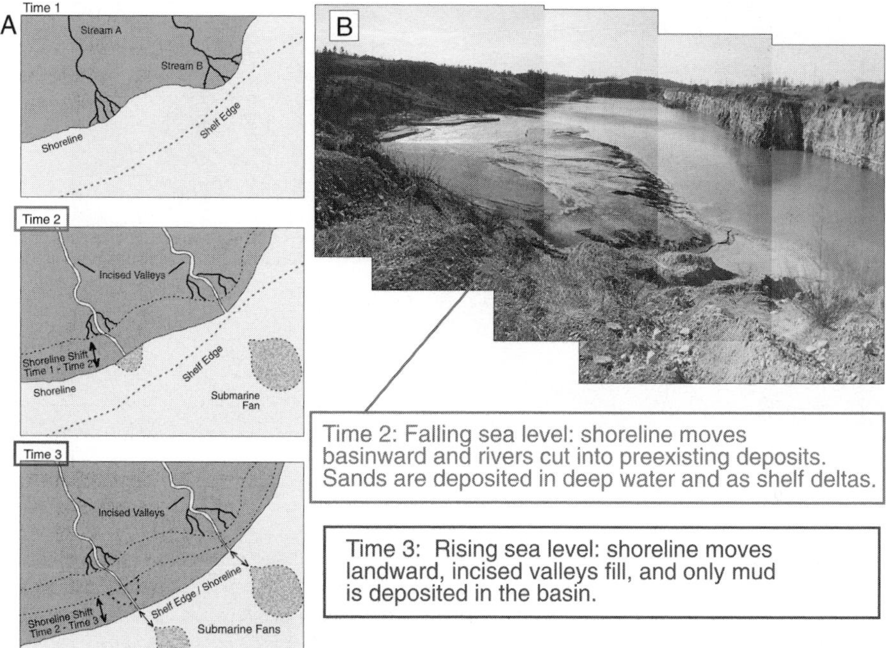
A
Time 1
Stream A
Stream B
Shoreline
Shelf Edge
Time 2
Incised Valleys
Shoreline Shift Time 1 - Time 2
Shoreline
Shelf Edge
Submarine Fan
Time 3
Incised Valleys
Shoreline Shift Time 2 - Time 3
Shelf Edge / Shoreline
Submarine Fans
B
Time 2: Falling sea level: shoreline moves basinward and rivers cut into preexisting deposits. Sands are deposited in deep water and as shelf deltas.
Time 3: Rising sea level: shoreline moves landward, incised valleys fill, and only mud is deposited in the basin.

Fig. 11.17. (A) Schematic diagram illustrating two different size deltas, and the development of submarine fans associated with a relative drop in sea level. Time 1: Highstand in sea level. Time 2: With an initial relative drop of sea level, fluvial incision occurs and sediments bypass across the shelf. A submarine fan develops basinward of the larger delta (Stream B). A shoreline/delta develops basinward of Stream A. Time 3: With a continued lowstand in sea level, a submarine fan develops basinward of Stream A, while the submarine fan basinward of Stream B continues to receive sediment. Modified from Posamentier et al. (1991). (B) Modern river delta within a quarry, showing incisions on the delta surface caused by a combination of fluvial incision and backwasting during a relatively low lake-water level. Any sand entering this system at this time (such as during a storm) will be transported across the exposed surface within the incised valley and deposited at the bottom of the lake. (Reprinted with permission of AAPG, whose permission is required for further use.)

shows an exposed, lacustrine delta that was deposited prior to subaerial exposure, when the lake level was higher). Also during that time, relatively little sediment is deposited in deep water (Fig. 11.16). In deep water, the combined transgressive and highstand deposits may be represented by a thin, but aerially extensive, organic-rich shale.

Shelf margin systems tract: A systems tract deposited during a time of less-pronounced relative sea-level fall (i.e., the rate of eustatic fall is slightly less than or equal to the rate of basin subsidence), such that the shelf becomes partially exposed, but the shoreline does not extend seaward all the way to the offlap break (Fig. 11.7). The resulting deposit, called a shelf margin systems tract, consists of prograding topsets and clinoforms, which become aggradational,

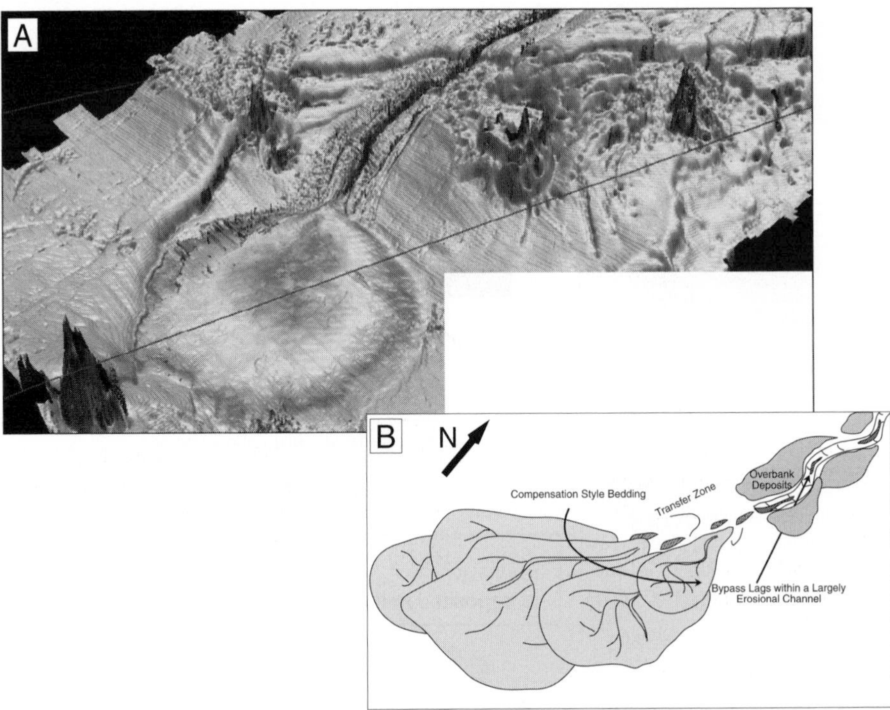

Fig. 11.18. (A) Shallow subsea image, extracted from a 3D seismic reflection volume, showing a lobate sand body (orange/dark-gray and yellow/bright-gray) at the terminus of a leveed-channel. The channel presumably has remained open so that sediment can be transported through it and deposited in this frontal position. After Pirmez et al. (2000), Fig. 8.25. (Reprinted with permission of SEPM, Society for Sedimentary Geology.) (B) Schematic illustration of the same features shown in (A), but interpreted here from an outcrop of the Lewis Shale, Wyoming (Minken, 2004).

and then eventually retrogradational upward (with time). The Type 2 sequence boundary is recognized by a downward shift in coastal onlap, but it does not shift beyond the offlap break (Fig. 11.9). The shelf margin systems tract is recognized most readily on seismic lines and is very difficult, if not impossible, to detect from outcrops, cores, and logs.

11.2.3 Definitions and concepts related to temporal cyclicity of sea-level fluctuations and sediment accumulation, within a chronostratigraphic framework

Cycle (of relative change of sea level): An interval of time during which a relative rise and fall of sea level takes place (Vail et al., 1977). A cycle may be recognized on a local, regional, or global scale. Sea-level cycles are defined

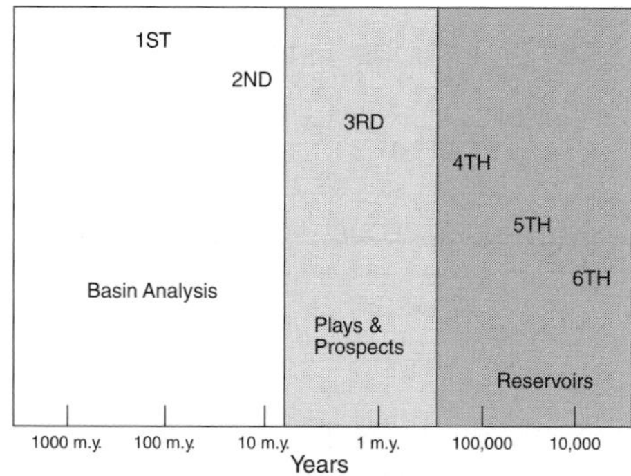

Fig. 11.19. Orders of cyclicity of relative sea level and the level at which various scales of sequence stratigraphic analysis normally are performed. First-order cyclicity occurs on the scale of hundreds of millions of years, second-order cyclicity occurs on the scale of tens of millions of years, third-order cyclicity occurs on the order of a million years, fourth-order cyclicity occurs on the scale of hundreds of thousands of years, fifth-order cyclicity occurs on the scale of tens of thousands of years, and sixth-order cyclicity occurs on the scale of thousands of years. Modified from Mitchum et al. (2002), Fig. 8.25. (Reprinted with permission of SEPM, Society for Sedimentary Geology.)

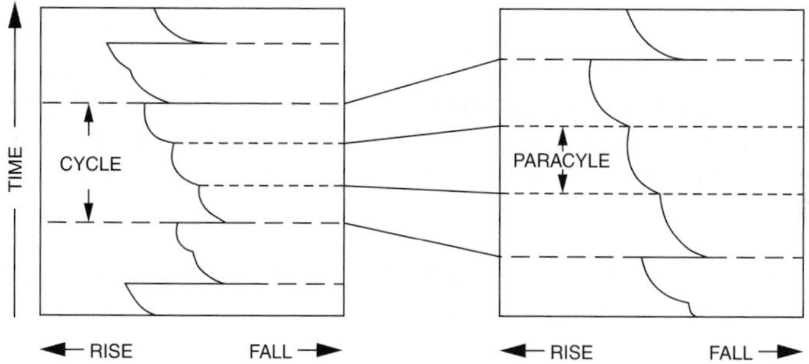

Fig. 11.20. Schematic illustration of cycles and paracycles of relative-sea-level change. Each cycle is bounded by a relatively large drop in relative sea level. Paracycles within a cycle are bounded by early rapid rise, followed by a stillstand in relative sea-level. Sequences form during sea-level cycles, and parasequences form during paracycles. A set of parasequences formed from a set of paracycles is termed a parasequence set. In the right figure, the overall cycle is one of rising relative sea level, which will give rise to a retrogradational parasequence set (Fig. 11.13). At the base of each paracycle, a transgressive surface of erosion will form, upon which will be deposited mainly silt and clay due to a rapid rise in relative sea level. This will be followed by progradation of the parasequence due to the slower rate of rise or to a stillstand of sea level over the remainder of the paracycle time interval (Fig. 11.12). After Vail et al. (1977). (Reprinted with permission of AAPG, whose permission is required for further use.)

according to the duration of the complete cycle (from fall-to-fall or rise-to-rise), as defined in Figs. 11.19 and 11.20.

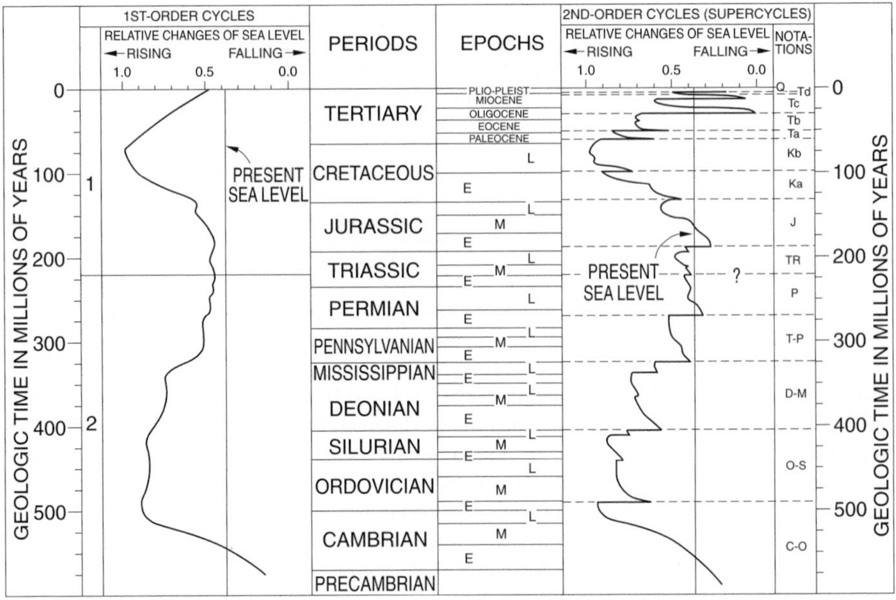

Fig. 11.21. First- and second-order cyclicity in relative sea level. In this chart, a first-order cycle occurs from about middle Cambrian to middle Triassic time, and a second first-order cycle occurs from middle Triassic time to the present. If the cycle is defined on the basis of highstand position, then a first-order cycle occurs between the beginning of Ordovician time and the beginning of Cretaceous time. In the right-hand side of the chart, second-order fall-rise cycles are superimposed upon the first-order cycles. After Vail et al. (1977). (Reprinted with permission of AAPG, whose permission is required for further use.)

Paracycle: A time interval of relative rise and stillstand of sea level, followed by another relative rise, with no significant intervening sea-level fall (Vail et al., 1977) (Fig. 11.20).

First-order cycle: A sea-level cycle with a duration on the order of hundreds of millions of years. During Phanerozoic time (post-Precambrian), there were two such cycles, a 300-m.y. cycle from the Precambrian to the Triassic, and a 225-m.y. cycle from the Triassic to the present (Figs. 11.19 and 11.21).

Second-order cycle: A sea-level cycle generally with a duration on the order of 9–10 million years (Figs. 11.19 and 11.21). Several such cycles are superimposed upon the Phanerozoic first-order cycles shown in Fig. 11.21.

Third-order cycle: A sea-level cycle generally with a duration on the order of 1–5 million years (Fig. 11.19 and 11.22).

Fourth-order cycle: A sea-level cycle with a duration generally on the order of 100,000–250,000 years (Figs. 11.19 and 11.22). Several such cycles are superimposed upon the third-order cycle shown in Fig. 11.22.

Fifth-order cycle: A sea-level cycle generally on the order of tens of thousands of years duration (Fig. 11.19).

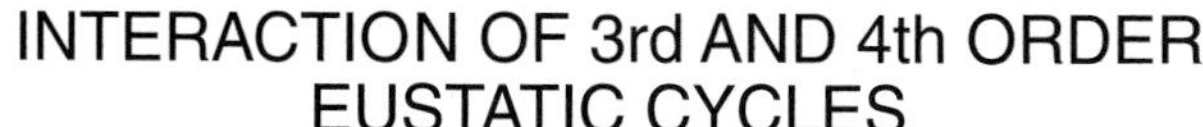

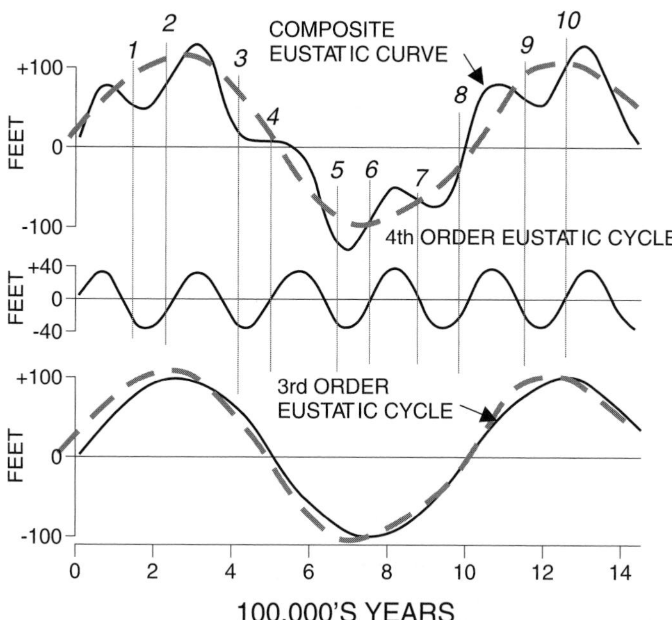

Fig. 11.22. The lower curve is a third-order cycle of about 1 m.y. in duration. The middle curve is a series of fourth-order cycles, each one lasting 100,000 to 200,000 years. The upper curve shows the fourth-order cycles superimposed upon the third-order cycle, and the relative sea-level trends resulting from this superposition of cyclicity. Ten positions of sea level have been marked on the upper curve; depositional style at these 10 times is discussed in the text.

Sixth-order cycle: A sea-level cycle with a duration generally on the order of thousands of years (Fig. 11.19).

Subsidence: the downward settling or sinking of the Earth's surface. The seafloor constantly subsides, but the rate at which it subsides may vary over time as a result of sediment loading and/or tectonic activity. The interaction of subsidence at a constant rate and a composite eustatic sea-level curve composed of third-, fourth-, and fifth-order components over a 1.2-m.y. time interval is shown in Fig. 11.23. A constant subsidence rate of 15 cm/k.y. (0.5 ft/k.y.) has been added to the composite eustatic curve. The two points on the curve labeled SB define a complete third-order relative sea-level cycle, and MFS defines a time during which a third-order maximum flooding surface was emplaced. In this example, the net effect of subsidence and eustasy is a relative rise in sea level.

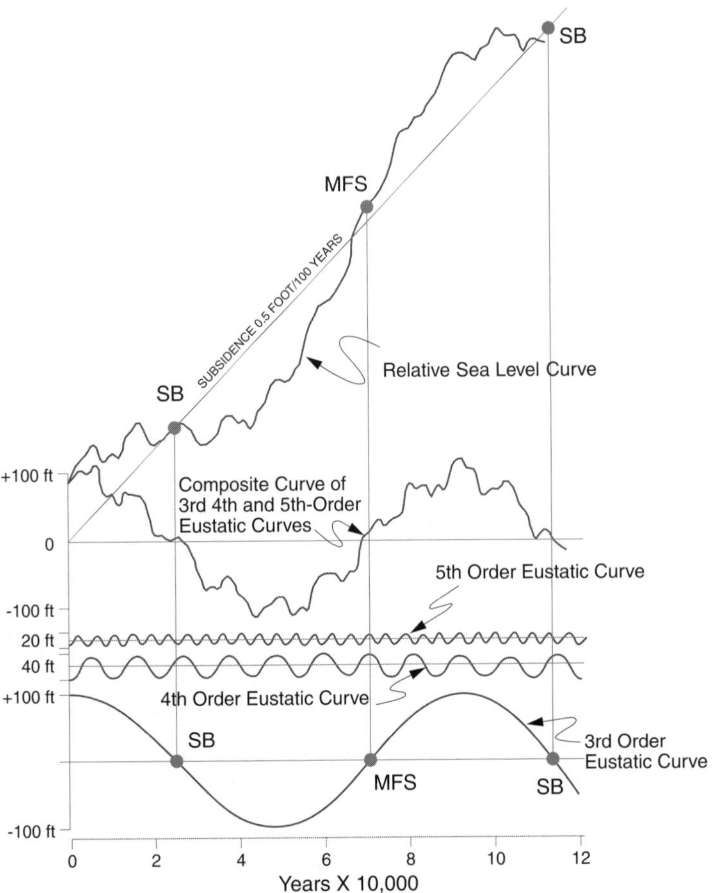

Fig. 11.23. Curves similar to that shown in Fig. 11.22, with the exceptions that fifth-order cyclicity has been added and that a constant rate of basin subsidence has been added to the eustatic curve, giving rise to a relative-sea-level curve that contains both eustatic and tectonic (subsidence) components. The resulting curve shows an overall rise in relative sea level over the 1.2-m.y. time interval represented, with two third-order eustatic sequence boundaries (SB) and one maximum flooding surface (MFS). Smaller scale cycles give rise to relatively minor changes of relative sea level, but these minor changes may have an important effect on overall depositional patterns, particularly at the reservoir scale. After Van Wagoner et al. (1990). (Reprinted with permission of AAPG, whose permission is required for further use.)

11.3 Developing a sequence stratigraphic framework

11.3.1 Identifying a key surface as a starting point

In order to develop a sequence stratigraphic framework, it is necessary first to identify a key stratigraphic surface. The maximum flooding surface and associ-

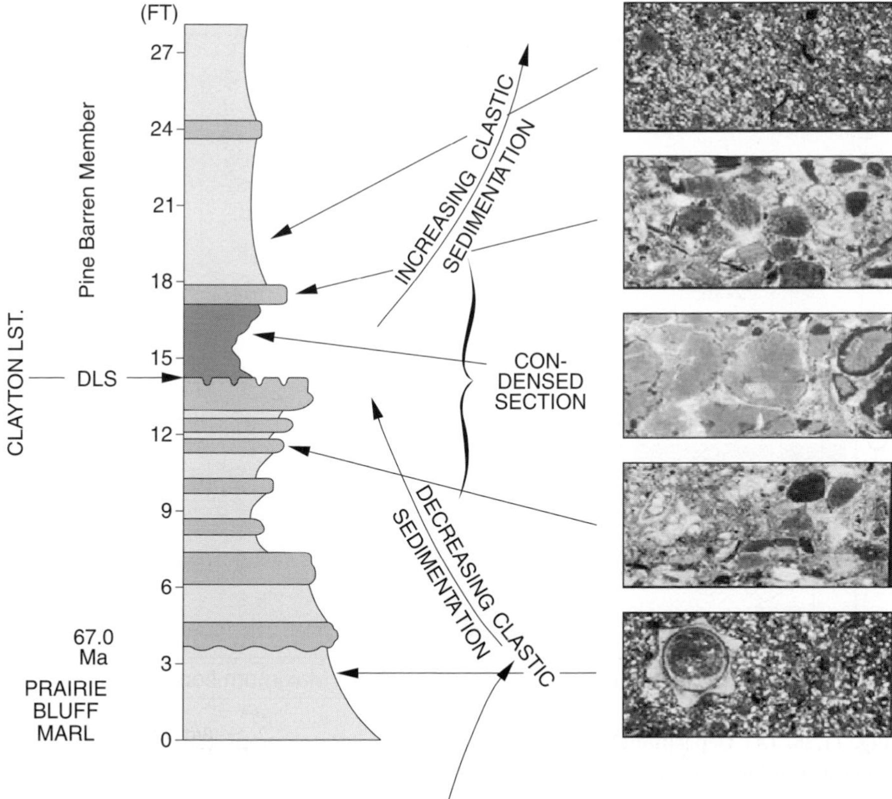

Fig. 11.24. Various lithologic features of a condensed section, including relatively coarser-grained sediments, abundant fossils, and glauconite grains. After Loutit et al. (1988). (Reprinted with permission of SEPM, Society for Sedimentary Geology.)

ated condensed section are perhaps the most readily identifiable components of a depositional stratigraphic sequence, so the available data should be searched for indications of these components (Loutit et al., 1988). Definite characteristics, discussed below, lead to their identification in rock (core or outcrop), seismic reflection records, and well logs.

Because condensed sections represent relatively long intervals of geologic time in which relatively small volumes of fine-grained sediment are deposited over a large marine area, they often are enriched with organic matter and chemically precipitated minerals (e.g., glauconite, siderite, or phosphorite) (Fig. 11.24), and they also exhibit both a high abundance and diversity of microfauna and microflora (discussed in Chapter 4). Thus, on conventional well logs, condensed sections are identified as the interval with the highest gamma-ray count (Figs. 11.25A and 11.25B). On laterally extensive outcrops, condensed sections and their maximum flooding surfaces are recog-

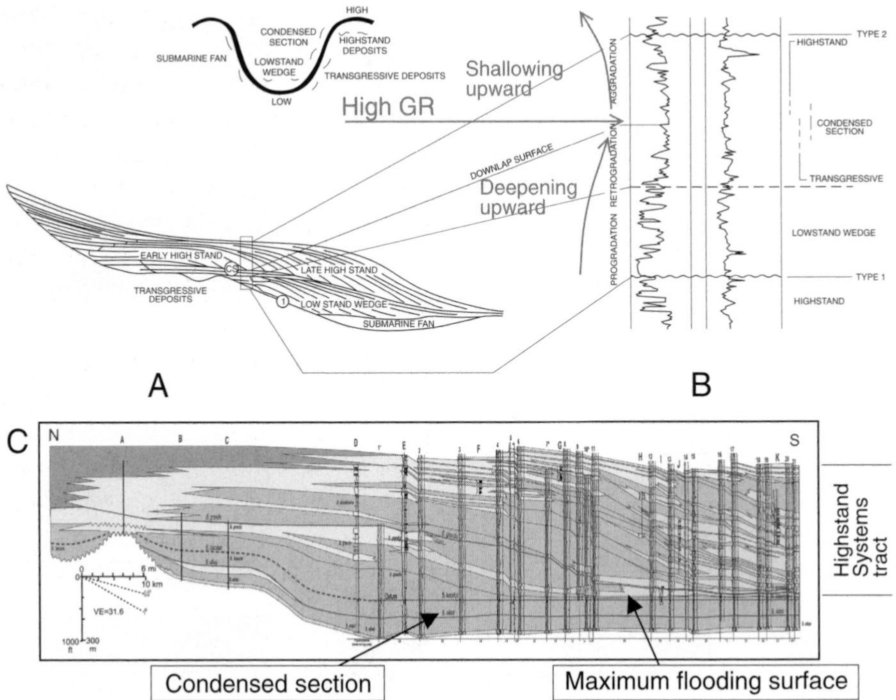

Fig. 11.25. (A) Depositional sequence and relative-sea-level curve. (B) Vertical sequence from a well log, showing the stacking of the various components of the depositional sequence and their well log signature. The base of the sequence is a Type 1 sequence boundary. Lowstand deposits form a fining-upward pattern above that boundary, as relative sea level begins to rise (deepening upward). The transgressive systems tract and its contained condensed section (noted by the maximum gamma-ray-log response, labeled "High GR") form as the rate of sea-level rise increases. The highstand systems tract forms a progradational pattern due to the increased accommodation space during highstand time, and deposition of deltaic and associated strata (shallowing-upward pattern). (A) and (B) are modified from Loutit et al. (1988). (C) Third order transgressive and then progradational pattern of deposition of the Lance Formation–Fox Hills Sandstone–Lewis Shale depositional system in Wyoming (Pyles and Slatt, in press). These figures are compared to illustrate the greater confidence level that we can place on a well-log-based sequence stratigraphic framework developed from a well log cross-section than we can derive from a single conventional well log suite.

nized by overlying highstand systems tract strata that downlap onto that surface (Figs. 11.9 and 11.26A). The same pattern can be observed on seismic reflection records (Fig. 11.26B) and on long well log cross-sections such as the Lance–Fox Hills–Lewis Shale cross-section shown in Fig. 11.25C. A sequence boundary sometimes can also be readily identified in cores, logs, and on seismic. In cores, a sequence boundary may be an erosion surface that forms the base of a thick sandstone bed. On conventional well logs, the surface appears as

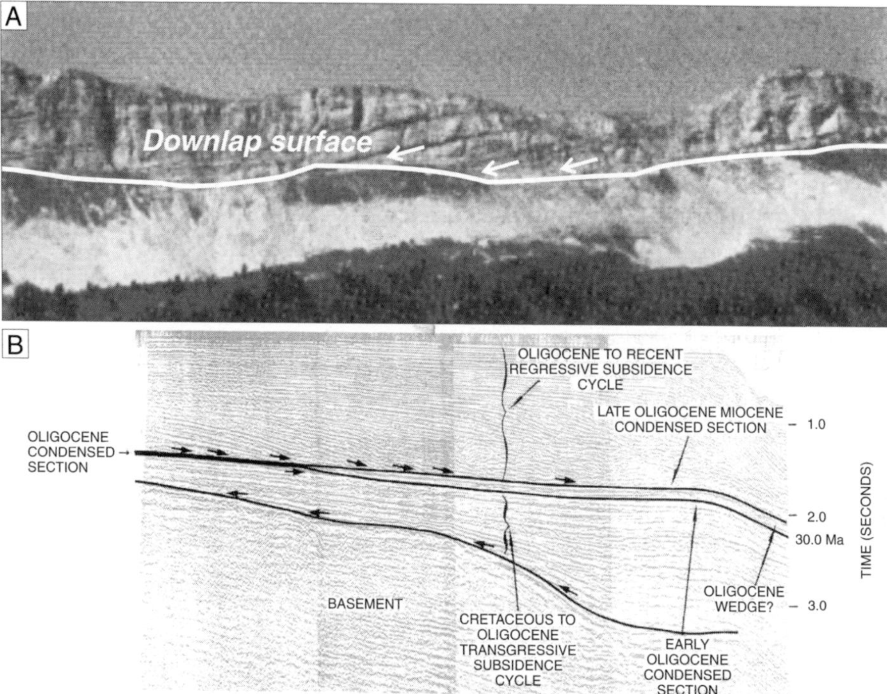

Fig. 11.26. (A) Outcrop in France showing the downlap pattern of a highstand systems tract onto a condensed section. Photo courtesy of McDonaugh (1998). (B) Seismic reflection profile showing the onlap pattern of a lowstand wedge onto underlying basement rocks, and the downlap pattern of highstand strata onto the condensed section. After Loutit et al. (1988). (Reprinted with permission of SEPM, Society for Sedimentary Geology.)

the abrupt contact between finer-grained sediments below, and thick sandstones above (Figs. 11.25B and 11.27B). A sequence boundary is most readily identified on seismic reflection records as a discordance in reflection orientations: specifically, as a seismic reflection that shows an onlapping pattern onto an erosion surface beneath which reflections have a different orientation (Fig. 11.26B)

11.3.2 Identifying and correlating systems tracts

Once the maximum flooding surface or a sequence boundary is identified on a well log, then the related systems tracts can be identified and placed within their predictable sequence stratigraphic position (Fig. 11.25A).

Figure 11.27 compares a chronostratigraphic interpretation with a lithostratigraphic interpretation of a stratigraphic interval from a well log. The lithostratigraphic interpretation defines formations and members. The chronostratigraphic interpretation defines surfaces upon which systems tracts lie.

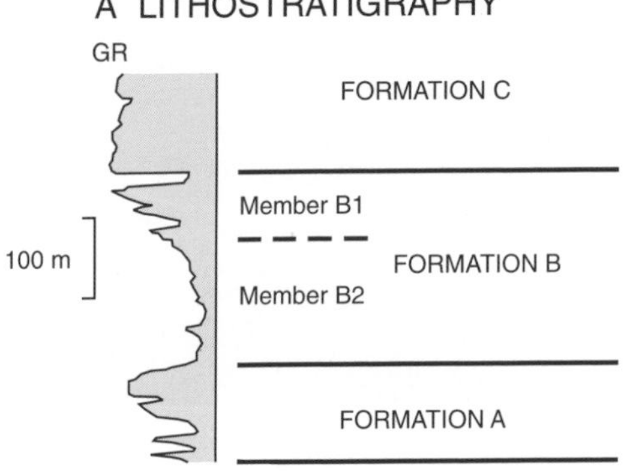

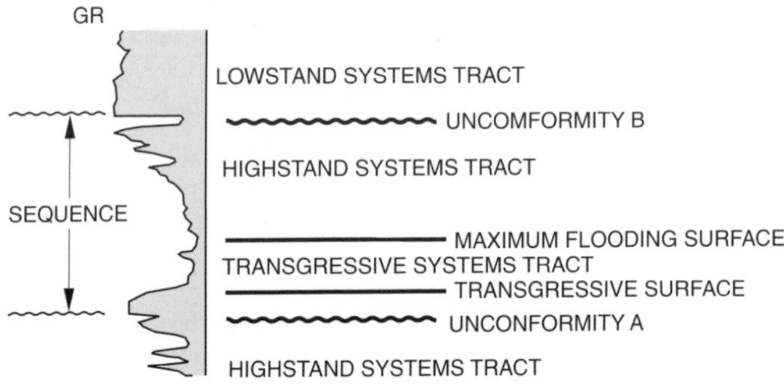

Fig. 11.27. Schematic illustration comparing a well log interpreted on a (A) lithostratigraphic and on a (B) chronostratigraphic basis. Lithostratigraphy defines rock units on the basis of their physical characteristics. Chronostratigraphy, or sequence stratigraphy, defines rock units on the basis of significant surfaces that separate stratigraphic successions. After Posamentier and Allen (1999). (Reprinted with permission of SEPM, Society for Sedimentary Geology.)

Figure 11.28 shows a chronostratigraphic correlation and interpretation for two wells spaced about 1.2 km (2 mi) apart. In this instance, the condensed section was identified as a high-gamma-ray shale, so that the systems tracts above and below could be identified and correlated between the two wells. An outcrop of a similar stratigraphic interval (Fig. 11.28) supports this particular chronostratigraphic interpretation (Slatt et al., in press).

11.3.3 Predicting vertical and lateral distribution of systems tracts and facies (reservoir, hydrocarbon source, and seal rocks)

Within a depositional sequence, various component systems tracts and sedimentary facies are deposited at different times and in different positions along a

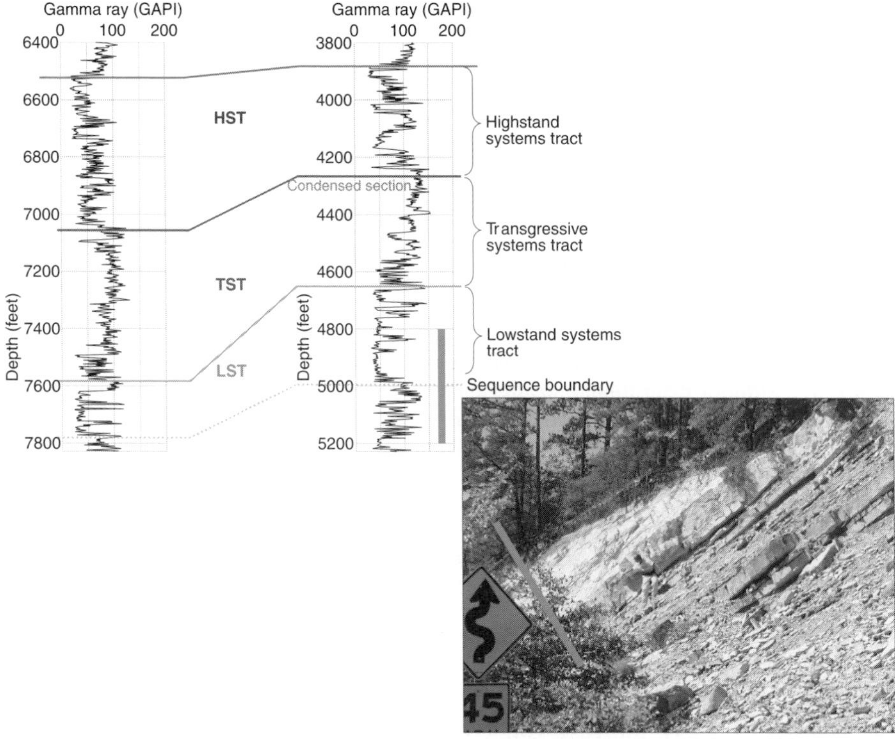

Fig. 11.28. Sequence stratigraphic correlations of two gamma-ray logs of the Pennsylvanian Jackfork Group in eastern Oklahoma, from wells spaced 1.2 km (2 mi) apart. The outcrop shown in the inset is similar to that part of the right-hand well log marked by a vertical orange line. After Slatt et al. (in press).

depositional profile (Fig. 11.9). For example, during the falling stage of relative sea level, most sediments are deposited in deep water, and the exposed continental shelf becomes an erosion surface (i.e., a sequence boundary) (Fig. 11.14). During the rising stage of relative sea level, very little sediment is deposited in deep water, and most sediment is deposited on the shelf (including in incised valleys) and at the shoreline and adjacent onshore coastal plain (Figs. 11.15 and 11.16). Thus, along a depositional profile, the lateral occurrence of different systems tracts and facies can be predicted (i.e., the occurrence of source rock, reservoir rock, and seal rock) (Fig. 11.29).

Also, it is possible to predict the vertical stacking of strata within a depositional sequence at different positions along that depositional profile. For example, at well A (Fig. 11.29), a complete vertical sequence would consist, from the base, upward, of a sequence boundary, a basin floor fan (possibly underlain by mass transport deposit), a slope fan (leveed channel), a late lowstand wedge (prograding complex), a thin transgressive shale, and a thin highstand shale, all capped by a sequence boundary. At well B (Fig. 11.29), the predicted ver-

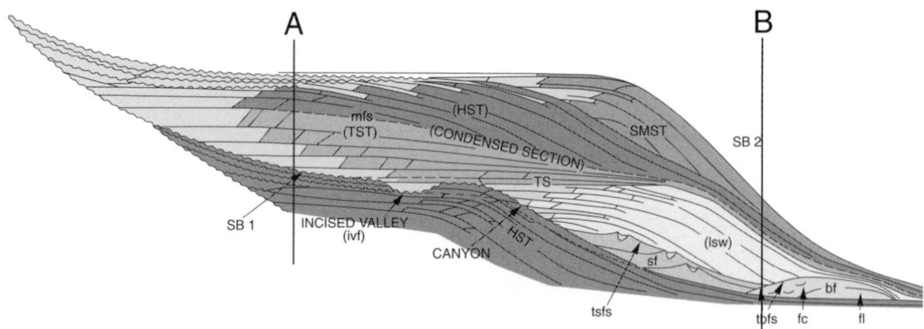

Fig. 11.29. Sequence stratigraphic model showing the positions of sand-prone facies (yellow/gray) within a depositional stratigraphic sequence. See the caption for Fig. 11.9 for definitions of symbols. Figure is modified from Vail (1987). (Reprinted with permission of AAPG, whose permission is required for further use.)

tical sequence would consist, from the base, upward, of a sequence boundary, transgressive shoreface sandstones and shelf mudstones (condensed section), progradational deltaic and coastal plain strata, and possibly fluvial deposits, all capped by a sequence boundary.

The reservoir types associated with this vertical stratigraphy correspond to the deposits and reservoirs discussed in the preceding chapters of this book. At well A (Fig. 11.29), the reservoir types are shoreline and shallow marine deposits (Chapter 8), fluvial deposits, including incised valley fills (Chapter 6), deltaic deposits (Chapter 9), and eolian deposits (Chapter 7). At well B, the reservoir types are deepwater deposits of the lowstand systems tract (Chapter 10).

The discipline of sequence stratigraphy grew from the initial techniques of seismic stratigraphy, which for many years relied heavily on interpretation of 2D seismic lines. Extending a 2D stratigraphic interpretation into the 3rd dimension is fraught with uncertainty. An excellent example of this problem has been presented by Posamentier et al. (1991) (Fig. 11.30). In their example, during a sea-level highstand (Time 1), shoreline sand is trapped only in the largest submarine canyon and is transported into deep water, but during a subsequent sea-level lowstand (Time 2), all three canyons receive sand. A 2D seismic profile shot along the axis of the largest submarine channel would image a submarine fan, thus possibly leading to the erroneous interpretation that the fan is entirely a lowstand deposit.

More frequent use of 3D seismic data has helped geoscientists resolve the issue of 3D interpretation of 2D data, but the problem still remains with 1D well logs, 2D well log cross-sections, and 2D outcrops. Yielding and Apps (1994) provide a graphical explanation for shelf-margin to slope strata in the Gulf of Mexico (Fig. 11.31). From a single well log, coupled with a seismic interpretation, nine depositional sequences were identified. However, only sequences

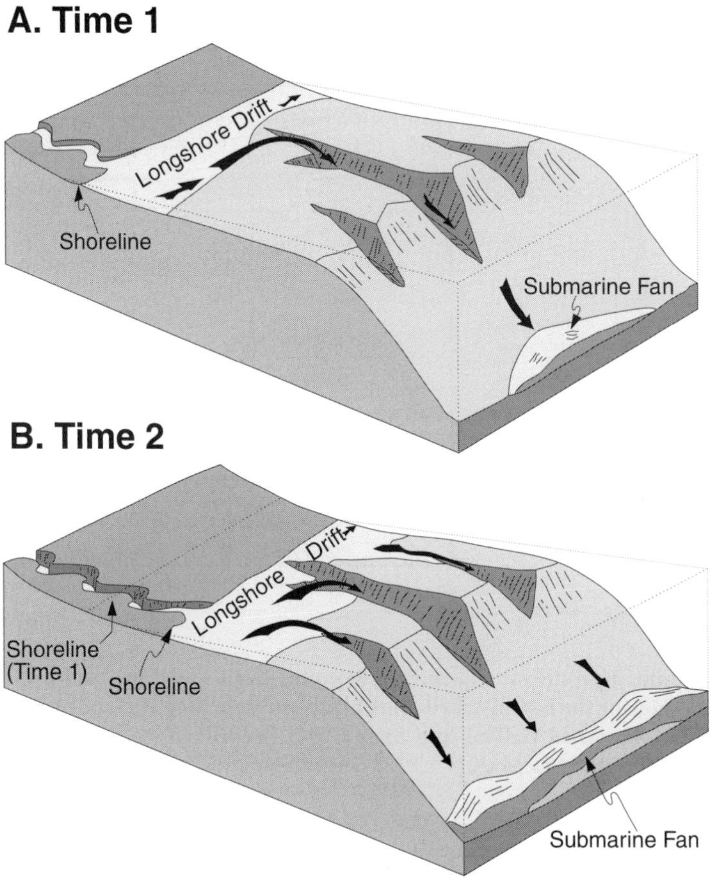

Fig. 11.30. Schematic block diagrams illustrating the effects of changes in sea level on longshore-drift sediment systems. (A) During a relative highstand in sea level (Time 1), only a few large submarine canyons capture longshore-drift sediment. (B) During a relative lowstand in sea level (Time 2), the shoreline and longshore drift have shifted basinward and the volume of sediments delivered to deep water via submarine canyons increases. Modified from Posamentier et al. (1991).

2 and 9 contain a complete vertical stack of components of the lowstand systems tract; the other seven sequences contain only some of these components. This is because of the position of the well relative to the 2D interpretation from the seismic line. The location of the vertical well represents a single point on the seafloor, over which deposition and erosion occurred during a long period of geologic time. Depositional environments and facies will shift laterally in time, relative to the stationary well location. Thus, a complete lowstand systems tract may not be deposited at that location during the complete interval of time.

Nevertheless, it is possible to predict (1) the vertical sequence in a well anywhere along the shelf margin to slope profile and (2) the lateral distribution of

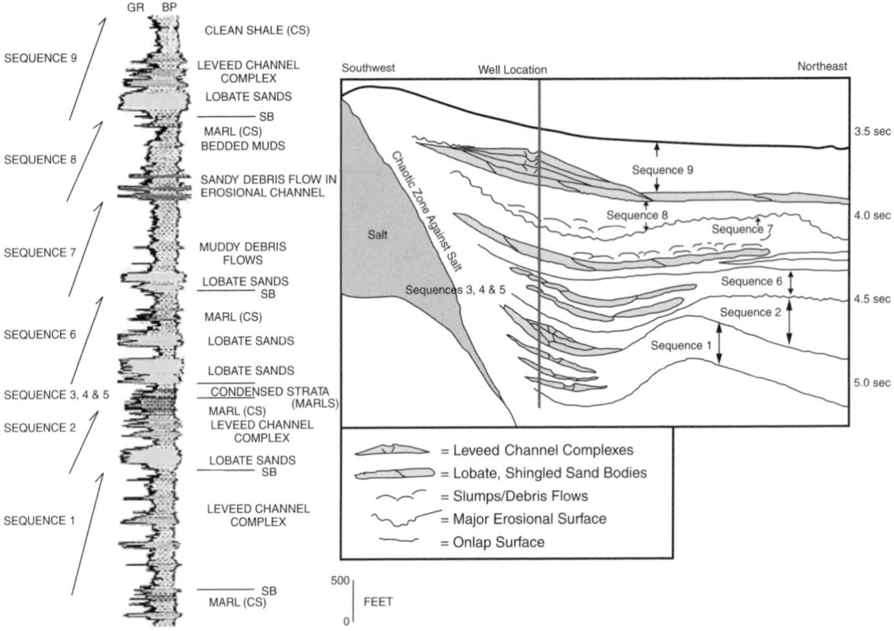

Fig. 11.31. Nine depositional sequences interpreted from a well log, and a 2D schematic illustration showing the lateral distribution of facies comprising the sequences. See text for further discussion. After Yielding and Apps (1994). (Reprinted with permission of SEPM, Society for Sedimentary Geology.)

facies deposited during a single period of time along that profile (Yielding and Apps, 1994).

Figure 11.32 illustrates these two points. In the case of Time 1 (within the time of a "lowstand fan" or a falling stage of relative sea level), the shelf margin and upper slope are characterized by an erosion surface and/or by erosional channel fill, and the middle and lower slopes are characterized by basin floor fan flows (called "lobate turbidites" by Yielding and Apps, 1994) and debris flows (possibly basal mass transport deposits) (Chapter 10). In the case of Time 2 (within the time of a "lowstand wedge" or an early turnaround and rise of relative sea level), the downslope facies tract will consist of shaly progradational deltaics, leveed-channel deposits, and hemipelagic shales. Thus, a well drilled at position A (Fig. 11.32) will consist, from the base, upward, of a sequence boundary overlain by lobes or sheets and possibly mass transport (debris flow) deposits, then topped by a relatively thin interval of shales. A well drilled at position B (Fig. 11.32) will consist, from the base, upward, of a sequence boundary, possibly some erosional channel fill, leveed channel strata, then topped by deltaic deposits.

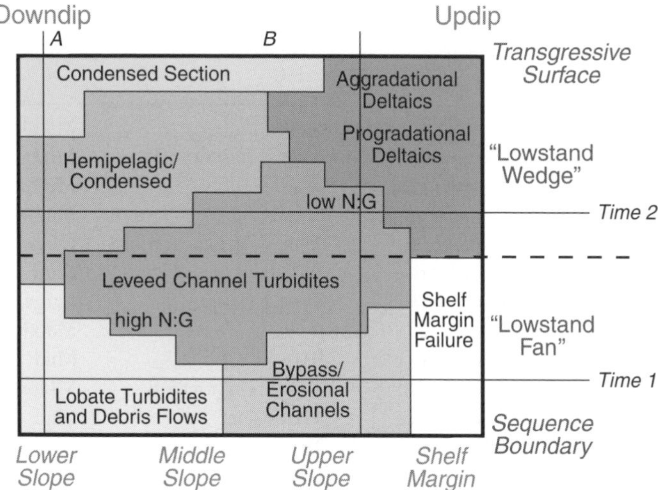

Fig. 11.32. The distribution of sedimentary facies in time and space along a depositional profile from shelf margin to lower slope, Gulf of Mexico. "Lowstand Fan" and "Lowstand Wedge" refer to the geologic times in which those components of a lowstand systems tract are deposited (falling stage and early turnaround in relative sea level). Time 1 and Time 2 refer to specific instances in geologic time and illustrate the various facies that are deposited at those times along the depositional profile. A and B refer to wells that would be located at two positions along the depositional profile, and the various facies that would be deposited throughout Lowstand Fan and Lowstand Wedge time. N:G refers to net:gross sand content. Modified from Yielding and Apps (1994). (Reprinted with permission of SEPM, Society for Sedimentary Geology.)

11.4 High-frequency sequence stratigraphy

11.4.1 General

The term "low-resolution sequence stratigraphy" generally is used to describe stratigraphic cycles at the relatively coarse seismic scale (Posamentier and Weimer, 1993). By contrast, "high-frequency (resolution) sequence stratigraphy" applies to observations at a log, core, and/or outcrop scale, which are of a higher order of frequency (i.e., have a shorter duration of time than third order) (Mitchum and Van Wagoner, 1991). However, advances in seismic resolution (e.g., spectral decomposition, etc.) and the growing use of 3D seismic data for reservoir characterization now allow interpretation of fourth-order cycles from seismic reflection volumes.

Relative sea-level cycles of differing time scales overlap (Figs. 11.20–11.22), which ultimately plays a major role in controlling the style of sedimentation over small and large areas of a basin. This point is illustrated in Fig. 11.21 for the

Table 11.1 Deep-water sediments deposited at different times, expressed along a composite 3rd and 4th order relative-sea-level curve

Position on the sea-level curve	Sea level	Composite curve	Deposition on deep seafloor y
1–2	3rd-order rise		Mainly mud
1	4th-order fall	Type 2 (minor) fall	Mud
2	4th-order rise	Rise	Mud
3–5	3rd-order fall		Mainly sand
3	4th-order fall	Type 1 (major) fall	Large volume of sand
4	4th-order rise	Stillstand	Mud
5	4th-order fall	Type 1 (major) fall	Large volume of sand
6–10	3rd-order rise		Mainly mud
6	4th-order rise	Rise	Mud
7	4th-order fall	Type 2 (minor) fall	Minor sand
8	4th-order rise	Major rise	Mud
9	4th-order fall	Type 2 (minor) fall	Minor sand
10	4th-order rise	Rise	Mud

Table 11.2 Vertical stratigraphy resulting from the composite curve of Fig. 11.22

Top	Mudstone (*Time/position* 10)
	Thin sandstone above a surface of erosion or nondeposition or both (*Time/position* 9)
	Thick mudstone (*Time/position* 8)
	Thin sandstone above a surface of erosion or nondeposition or both (*Time/position* 7)
	Mudstone (*Time/position* 6)
	Thick sandstone above a surface of erosion or nondeposition or both (*Time/position* 5)
	Thin mudstone (*Time/position* 4)
	Thick sandstone above a surface of erosion or nondeposition or both (*Time/position* 3)
Base	Mudstone possibly with a minor surface of erosion/nondeposition (*Time/position* 1–2)

composite third- and fourth-order eustatic curve. Table 11.1 lists the events and deposits associated with 10 temporal positions on the curve. Table 11.2 presents the vertical stratigraphic succession resulting from this composite cyclicity in relative sea levels.

11.4.2 Applications to exploration and reservoir development

Predicting the lateral and vertical distribution of depositional sequences and their component systems tracts and facies plays an important role in exploration for, and development of, sandstone reservoirs. Using Table 11.2 as an example, the main deepwater sandstone (commonly, the exploration target) will occur during the third-order falling stage of relative sea level, and basal and top mudstones may act as seals. At the reservoir scale, the mudstone of Time/position 4 may separate the overlying and underlying sandstones vertically.

In this section, four case studies are presented to illustrate different applications of high-frequency sequence stratigraphy for reservoir characterization. These case studies exemplify the different reservoir types discussed in prior chapters. They illustrate how sequence stratigraphy can be used to better understand reservoir architecture, which in turn can lead to improved well planning and reservoir management.

11.4.2.1 Case Study 1: High-frequency sequence stratigraphy of the Cretaceous-age Lance Formation–Fox Hills Sandstone–Lewis Shale system, Wyoming: identifying a petroleum system for exploration and development

Detailed ammonite zonation and age dating indicate that the Cretaceous-age Lance Formation–Fox Hills Sandstone–Lewis Shale depositional system was deposited during the time interval 69.1–71.3 Ma (Fig. 11.33) (Pyles and Slatt,

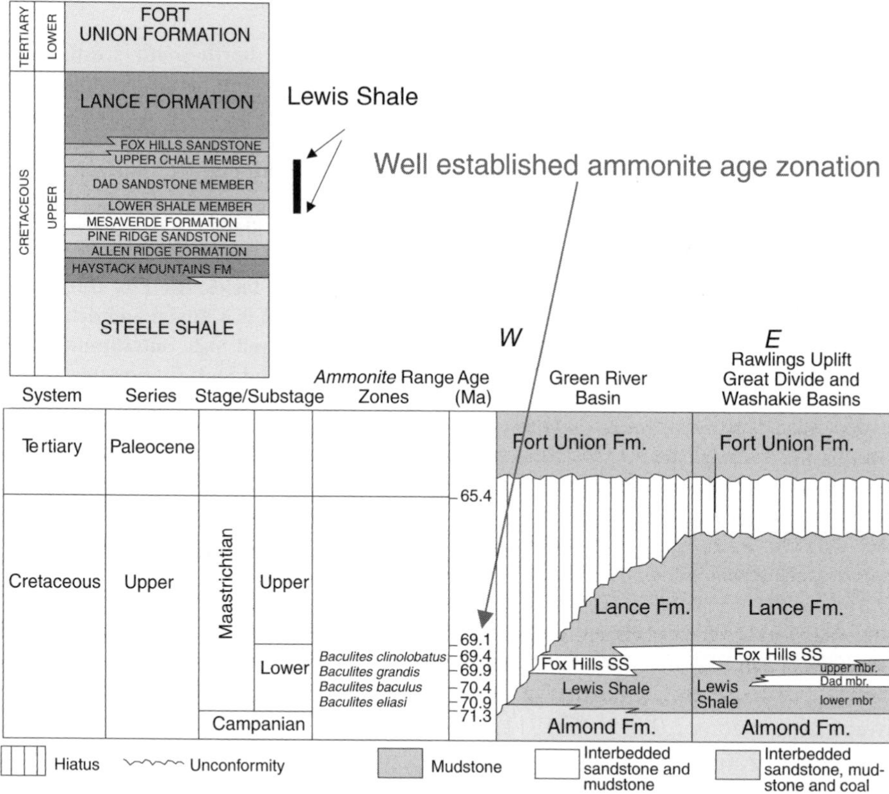

Fig. 11.33. Chronostratigraphic chart of Upper Cretaceous strata in south-central Wyoming. Ages of Baculites (ammonite) zones are shown. Modified from Pyles (2000). Lithostratigraphic column of Upper Cretaceous and lower Tertiary strata is shown in the upper left.

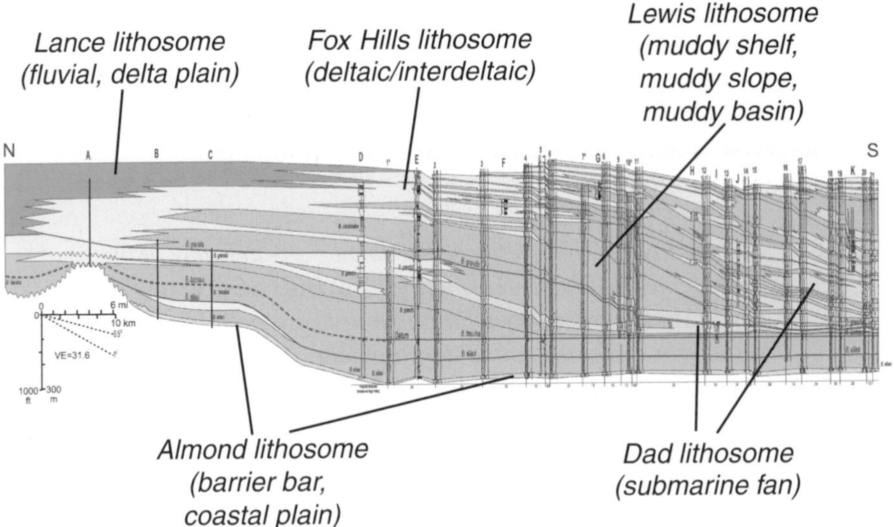

Fig. 11.34. 150-km (90-mi)-long and 900-m (3000-ft)-thick, north–south stratigraphic cross-section of the upper Almond Formation, Lewis Shale, Fox Hills Sandstone, and Lance Formation in the Washakie Basin of Wyoming. The upper surface of the Almond Formation is a transgressive surface of erosion. The lower Lewis Shale (Fig. 11.33) rests horizontally on that surface and is capped by the organic-rich "Asquith marker" horizon (datum used for the cross-section except toward the north). Younger stratigraphic units downlap onto that surface in a progradational pattern. The lower Lewis Shale is a transgressive systems tract capped by the Asquith condensed section and maximum flooding surface. The Dad Sandstone and upper Lewis Shale are basinal and slope facies, the Fox Hills Sandstone is a deltaic/interdeltaic deposit, and the Lance Formation is a fluvial and delta-plain deposit. The cross-section was constructed from subsurface well logs, outcrop-measured stratigraphic sections, and a behind-outcrop well. After Pyles and Slatt (in press).

in press). A 150-km (90-mi)-long by 900-m (3000-ft)-thick stratigraphic section composed of the Lewis Shale, Fox Hills Sandstone, and Lance Formation sits atop the gas-prone Almond Formation (Fig. 11.34). The horizontally bedded lower Lewis Shale overlies the Almond Formation. The Dad Sandstone, upper Lewis Shale, and Fox Hills Sandstone form a downlapping, clinoform pattern of time-equivalent strata onto the lower Lewis Shale. The lower Lewis Shale is capped by an organic-rich, shaly horizon called the "Asquith marker" (Asquith, 1970). Because the time interval over which the Fox Hills–Lewis Shale rocks were deposited is 2.2 m.y. (Fig. 11.33), it is defined as a third-order depositional sequence stratigraphic cycle. The lower Lewis Shale is a third-order transgressive systems tract, and the Asquith marker is its condensed section. The downlap pattern onto the top of the Asquith marker (maximum flooding surface) defines the downlapping strata as a third-order highstand systems tract.

On the basis of this highstand downlap pattern, which is recognized by correlation of time-stratigraphic ammonite biozones, a more-detailed chronostrati-

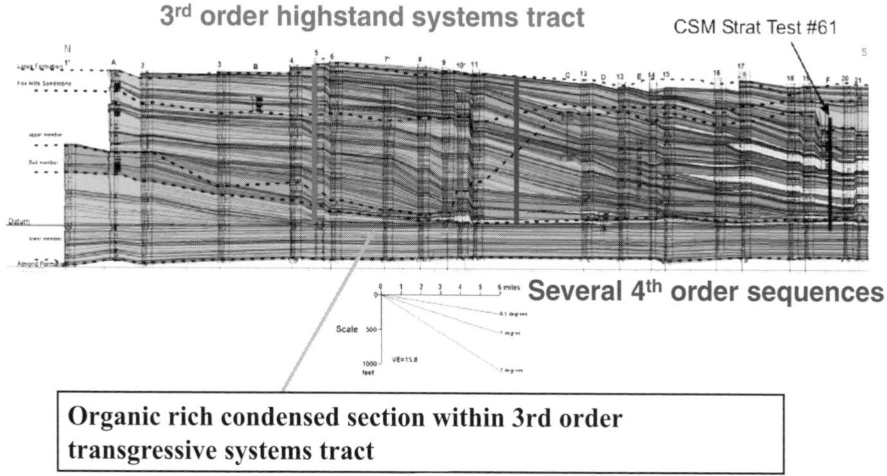

Fig. 11.35. A portion of the high-frequency sequence stratigraphic framework illustrated in Fig. 11.34. The Asquith marker condensed section is shown in green/black horizontal line. Dotted black lines refer to lithostratigraphic (formation) boundaries. The chronostratigraphic downlap surfaces crosscut the lithostratigraphic surfaces. CSM Strat Test #61 is a behind-outcrop well drilled into the Lewis Shale. Modified from Pyles (2000).

graphic framework was developed based on the correlation of bentonite beds (which represent mere instants in geologic time) and shales (Fig. 11.35). Pyles (2000) defined 21 high-frequency fourth-order (21 cycles/2.2 m.y. = 95,000 yrs/cycle) depositional sequences within a 72-km (45-mi) portion of the cross-section of Fig. 11.34, which are superimposed upon the third-order highstand deposits.

The significance of this case study is the fact that fourth-order lowstand basin floor fan sandstones were deposited progressively basinward (toward the south) during the 2.2 m.y. time interval, giving rise to a succession of potentially good gas-reservoir rocks that stack in that direction and are associated with shaly top and side seals. Also, the east–west-trending Cherokee Arch, just to the south of the cross-section, may provide structural trapping of the Dad Sandstones. In addition to the deepwater Dad Sandstone deposits, shoreline and shallow-marine reservoirs of the Fox Hills Sandstone are attractive exploration targets.

For many years, the Lewis Shale was a secondary exploration target that sometimes was tested as it was drilled through on the way to the underlying, gas-prone Almond Formation (Fig. 11.34). Over the last few years, gas exploration within the Lewis Shale has increased dramatically, in part because of the potential for large reservoir sandstones to the south (Figs. 11.34 and 11.35). The sequence stratigraphic framework that was developed for the Lewis Shale defines a complete petroleum system, including source rock (Asquith marker), reservoir rock (Dad Sandstone basin floor fans and leveed channel deposits), seals (shales and bentonites), and traps (structural and stratigraphic pinch-outs).

11.4.2.2 Case Study 2: High-frequency sequence stratigraphy of the Miocene Mt. Messenger Formation, New Zealand: outcrop analog of deepwater reservoirs

The second case study is of the Mt. Messenger Formation in New Zealand, which was deposited in a basin-slope setting during a global period of relative sea-level lowstand and early rise (Fig. 11.36) (Browne and Slatt, 2002). A well-

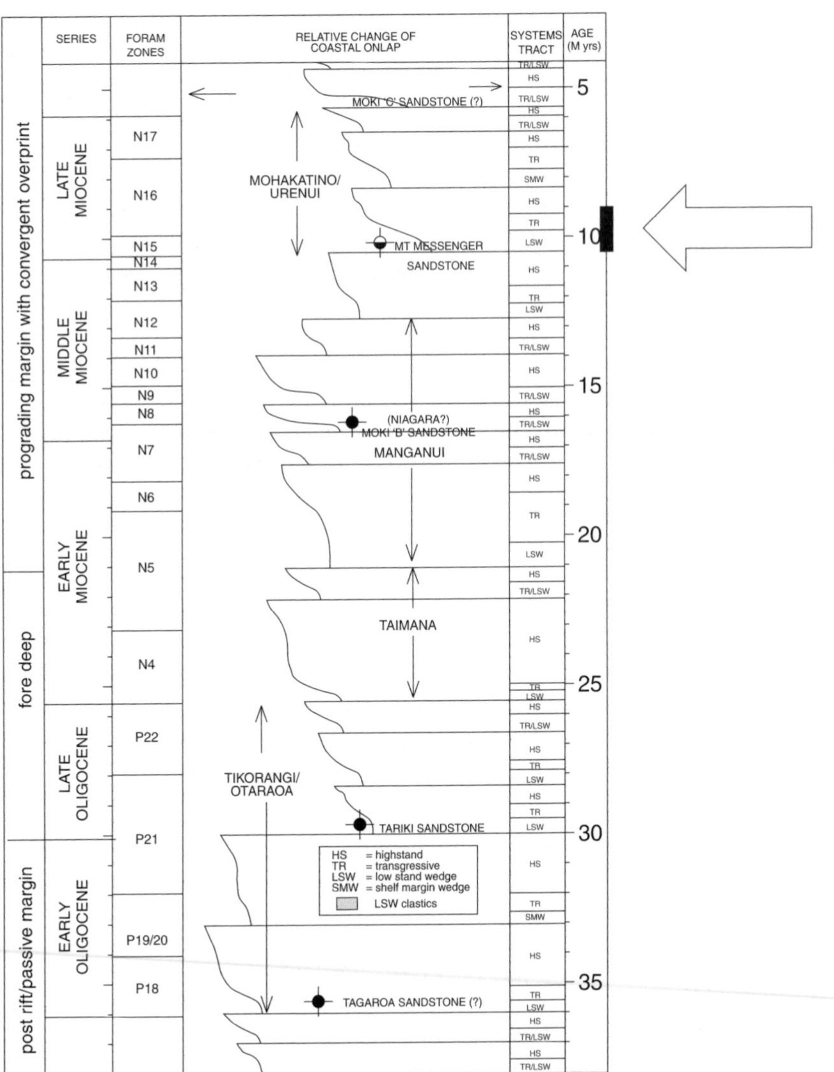

Fig. 11.36. Coastal online chart of the Carnarvon Basin, Australia compared to that of the relative coastal onlap curve of Haq et al. (1987), Fig. 10.30. (Reprinted with permission of Science.)

Fig. 11.37. Outcrop of Miocene Mt. Messenger Formation and Urenui Formation, Taranaki Coast, North Island of New Zealand. The locations of an exploration well and two research wells discussed in the text are shown. The components of a lowstand systems tract (basin floor fan, leveed-channel fill, and prograding complex) occur along the vertical cliffs. Beds dip about 5° toward the lower right. The Mt. Messenger Formation is described by Browne and Slatt (2002).

defined biostratigraphic zonation indicates deposition occurred in water depths greater than 1000–600 m, over a 1-m.y. time interval, thus defining a third-order stratigraphic succession. The beds dip gently along a 10-km-long cliff face (Fig. 11.37), revealing a 600-m (1800-ft)-thick, complete lowstand systems tract (Diridoni, 1996; Browne and Slatt, 2002).

Four individual basin-floor fan sandstones occur along the outcrop face and can be traced inland (in the third dimension) for several kilometers before they pinch out beside slope mudstones (Fig. 11.38). Along the cliff, the four sandstone intervals overlie erosion surfaces cut into underlying mudstones (Fig. 11.39A). The sandstones are interpreted to have been deposited during a third-order fall in relative sea level. Each sandstone unit grades upward into thinner-bedded sandstones interbedded with marly mudstones (Fig. 11.39B), then into thicker-bedded marls (Fig. 11.39C) that contain a high abundance of microfauna. These marls are interpreted as fourth-order transgressive-highstand deposits formed during time intervals in which relatively little clastic sediment was being deposited in the basin. Thus, the four erosionally-based sandstones (Fig. 11.38) are fourth-order lowstand strata deposited during a third-order

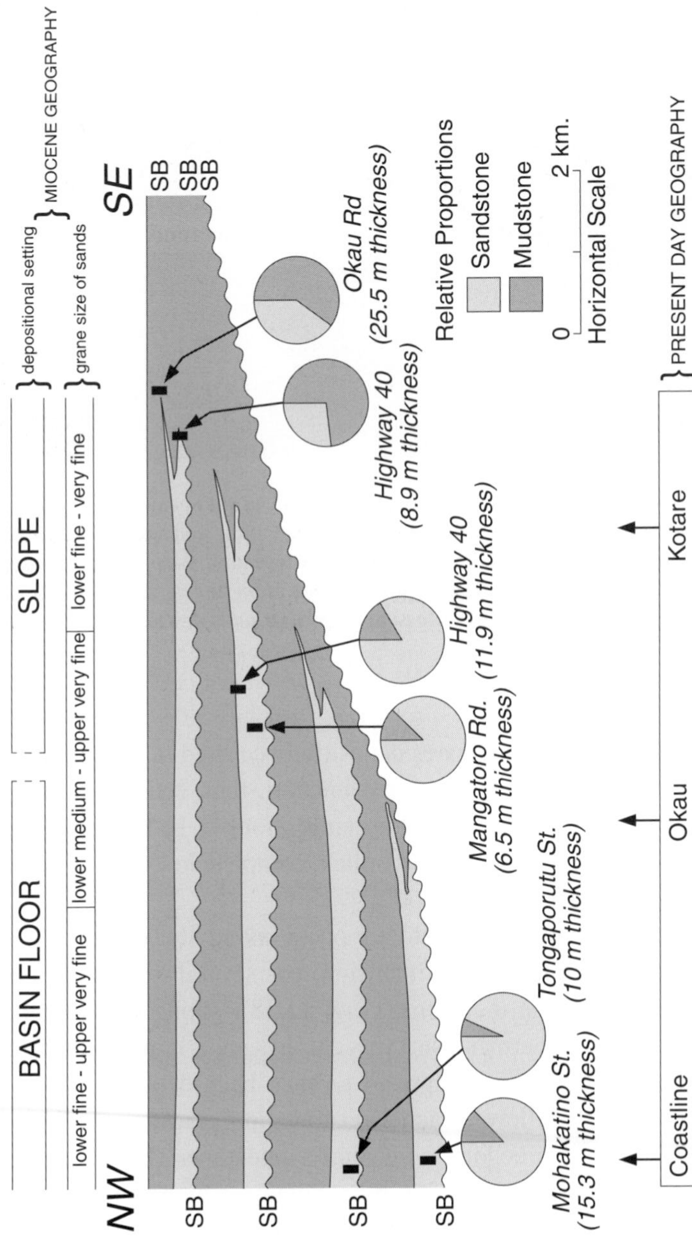

Fig. 11.38. Schematic illustration of a northwest (cliff) to southeast (inland) stratigraphic profile of four basin-floor-fan sandstones (yellow/bright-gray) exposed along the cliff face (Fig. 11.37), and marly mudstones that separate each sandstone interval. SB – high-frequency sequence boundaries at the base of the four basin-floor-fan sandstone intervals. Proportions of sandstone and mudstone are shown by the pie diagrams, and average grain sizes of sandstones along the profile are highlighted. The Mt. Messenger Formation is described by Browne and Slatt (2002). (Reprinted with permission of AAPG, whose permission is required for further use.)

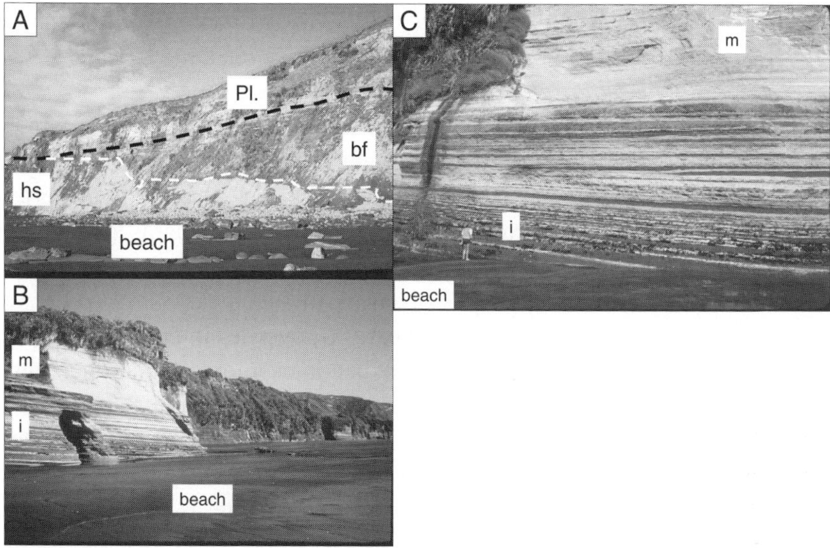

Fig. 11.39. Mt. Messenger Formation outcrop photographs. (A) Basin floor fan (bf) overlying a preceding highstand mudstone (hs). The red/black dashed line marks a fourth-order sequence boundary. Pleistocene fluvial terrace deposits (Pl.) overlie the Mt. Messenger Formation. Base of the terrace deposits is shown by a black dashed line. (B) Interbedded thin-bedded sandstones (dark color) and marly mudstones (i) overlain by marly mudstones (m). The thin sandstone beds are interpreted to be distal basin-floor-fan deposits, and the thick marly mudstone (m) is interpreted to be a fourth-order transgressive-highstand deposit. (C) Close-up of thin interbeds of sandstone (dark color) and marly mudstones (i) grading upward into thick-bedded marly mudstones (m). Note that some of the thin sandstones pinch out laterally. Person for scale.

fall punctuated by at least three composite third- to fourth-order stillstands–highstands. This history probably is analogous to Time/positions 3–5 on the composite eustatic sea level curve of Fig. 11.21.

The significance of this case study is the presence of the marls, which appear to have been deposited over a large area. Widespread deposition of fine-grained, transgressive-highstand marls in deep water is not uncommon in such areas as the deep-water Gulf of Mexico (Fig. 11.31). In an analog subsurface exploration area, such marls are excellent potential regional stratigraphic marker horizons for correlation and interpretation of ages and environments of deposition (Chapter 4). In an analog subsurface reservoir, the marls might act as vertical barriers to communication between sandstones, in addition to being excellent correlation marker beds.

11.4.2.3 Case Study 3: High-frequency sequence stratigraphy of a shoreface parasequence set: Cretaceous-age Terry Sandstone, Hambert–Aristocrat field, Colorado

In Chapter 8, seven stratigraphic intervals were identified as comprising the Terry Sandstone Member of the Pierre Shale in Hambert–Aristocrat field (labeled A–G in Fig. 8.22). The Terry Sandstone was described as a "series of retrogradational shoreface parasequences ...". Here, this parasequence set is described within a sequence stratigraphic framework.

Interval A is shaly and represents open-marine-shelf deposits. To the west, a sharp-based sandstone, representing interval B, overlies interval A (Fig. 8.22). However, toward the east, the sandstone becomes more gradational into an underlying open-marine-shelf shale (compare Figs. 8.23 and 8.24). The two well log patterns follow those of Figs. 8.10 and 8.14, which illustrate lateral facies changes in the seaward direction during a time interval in which one parasequence forms. On the basis of the well log patterns and core description (Slatt, 1997), this parasequence defines an upper shoreface sandstone to the west, grading eastward (paleobasinward) through lower shoreface deposits to offshore shelf deposits (as in Fig. 8.10). A similar lateral well log pattern appears for intervals C–G (Fig. 8.22), although not all well log signatures are identical. Parasequences B–G each are separated by a transgressive marine shale or condensed section/flooding surface. By locating the boundary between a paleolandward, blocky log pattern (upper-middle shoreface) and a paleoseaward, cleaning-upward, gradational log pattern (lower shoreface/offshore shelf) on the cross-section (Fig. 8.22), and by mapping this boundary for intervals B–G from more than 100 well logs in the mapped area, it is possible to demonstrate that the boundary generally steps landward with each successively younger parasequence (Fig. 8.25).

On the basis of the landward step of the parasequence set, the depositional history of the Terry Sandstone Member is illustrated in Fig. 11.40. The parasequences are related temporally to a paracycle curve (Fig. 11.20) that illustrates first a drop in relative sea level, followed by erosion of shelf shales (parasequence A) to yield an unconformity surface, which is then overlain by upper shoreface strata (Figs. 8.9, 8.13, and 8.14). Subsequent parasequences were deposited first by relatively rapid flooding of the shelf and the formation of a condensed section shale, followed by a stillstand of relative sea level and basinward progradation of the shoreface/offshore strata. Each successively younger shoreface parasequence did not extend as far seaward as the preceding parasequence, thereby giving rise to an overall lower-order transgressive systems tract (parasequence set) composed of a series of higher-order parasequences (i.e., a retrogradational parasequence set; Fig. 11.13).

In general terms of reservoir characterization and performance, significant aspects of this depositional history include:

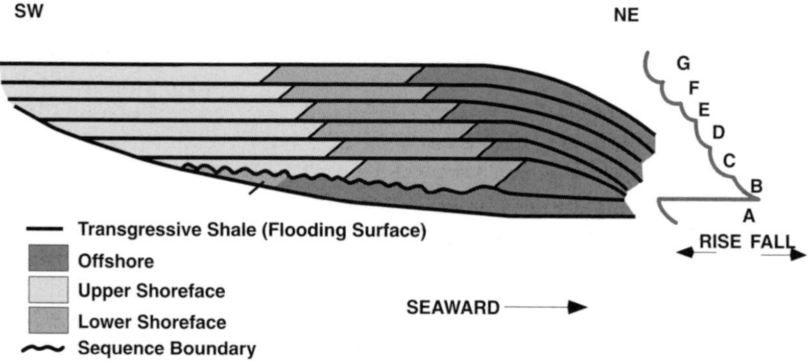

Fig. 11.40. Schematic illustration of the deposition of the Terry Sandstone parasequence set. The base of parasequence B is erosional toward the southwest (paleolandward) and gradational with underlying shelf mudstones toward the northeast (paleoseaward). Each of the succeeding parasequences steps in a landward direction, together forming a retrogradational parasequence set (Fig. 11.1). This parasequence set is a result of a relatively low-frequency (third-order?) cycle of relative sea-level rise, with six higher-frequency (fourth-order?) paracycles of rapid rise (and deposition of a transgressive shale) followed by a relative stillstand of sea level (and progradation of a single shoreface parasequence). After Slatt (1997). (Reprinted with permission of Gulf Coast Section SEPM Foundation.)

- the flooding-surface/condensed-section shales may be laterally extensive and may form vertical barriers between parasequence sandstones;
- within any one parasequence, there will be a basinward decrease in reservoir quality (via a decrease in porosity and permeability) because of the basinward decrease in net sandstone (and probable decrease in sandstone grain size) (compare Figs. 8.23 and 8.24), as discussed in Chapter 5; and
- net sandstone and sandstone reservoir quality of each parasequence within this retrogradational parasequence set will decrease upward, as illustrated in Fig. 11.13 and discussed in Chapter 5.

11.4.2.4 Case Study 4: High-frequency sequence stratigraphy of incised-valley-fill reservoirs: Pennsylvanian Glenn Pool field, Oklahoma, and Southwest Stockholm field, Kansas

In Chapter 6, the Glenn Pool field in Oklahoma was defined as a combination fluvial reservoir with a lower, braided-river facies and an upper meandering-river facies. Reservoir quality is controlled by the facies, with better quality occurring in the braided-river facies (Fig. 6.48). Kerr et al. (1999) and Ye and Kerr (2000) placed the distribution of facies within a sequence stratigraphic framework. The lower braided-river facies is a nonmarine lowstand systems tract. The coarse-grained strata were deposited over a wide geographic area within a major incised valley, with the base being a regional sequence boundary. During early turnaround and rise of relative sea level, the energetics of the fluvial system diminished, and finer-grained and muddier meandering-river/floodplain facies were deposited atop the braided-river facies. Thus, the vertical variation in stratification style, continuity of strata, and reservoir quality are controlled by primary depositional processes during one or more phases of sea-level fluctuation. In this example, the braided-river facies (which comprise a lowstand systems tract) are interconnected and of highest reservoir quality (Fig. 6.48). The meandering-river facies (a transgressive systems tract) was deposited on a lower-energy, muddier floodplain, and thus continuity and connectivity of reservoir facies, in addition to reservoir quality, are not as good. In essence, this reservoir is composed of two quite different architectural elements, each with its own production characteristics.

Also in Chapter 6, Southwest Stockholm field was described as being composed of two incised-valley-fill facies – a low-permeability estuarine facies and a higher-permeability fluvial facies (see Fig. 6.43 and Table 6.3) (Tillman and Pittman, 1993). The valley was incised during a falling stage of relative sea level, and probably filled partially with fluvial sediment during early turnaround and later with estuarine sediment as sea level rose sufficiently to fill the valley with brackish or marine water (Fig. 11.15). The fact that fluvial sediment in part overlies estuarine sediment indicates more than one episode of sea-level fluctuation. As pointed out in Chapter 6, the well-log patterns alone do not reveal any significant difference in net sand of the two facies, yet the permeabilities differ by an order of magnitude. Again, two different architectural elements and reservoir types occur in this field, each with its own probable production characteristics.

11.5 Summary

Sequence stratigraphy is the study of sedimentary rock relationships within a chronostratigraphic framework. The discipline is now mature, having evolved over a 30-year period that began with the concept and applications of seismic stratigraphy presented in a landmark publication by Payton (1977). Sequence stratigraphy provides a powerful tool for predicting the spatial and temporal distribution of reservoir, source, and seal rocks within a stratigraphic interval, at both the exploration and the reservoir scales.

Sequence stratigraphic intervals are deposited in cycles of relative sea-level change (highstand to lowstand and vice versa) over a variety of time scales, from hundreds of millions of years to thousands of years. Superimposition of these cycles in time and space provides predictable stratigraphic patterns from terrestrial to deep-marine environments.

Various criteria have been recognized to identify these stratigraphic patterns and the contained facies. Once a single significant chronostratigraphic surface has been identified and mapped (on seismic lines, in well logs, or in outcrops), then the surface's sequence stratigraphic framework, and the facies contained above and below that surface, can be predicted. Reservoir, hydrocarbon source, and seal rocks also can be predicted in time (geologic and seismic) and space.

High-frequency parasequences – which are composed of laterally continuous shales, and of sandstones that have laterally and vertically variable reservoir quality and continuity – can have a major effect on reservoir performance. Thus, they should be identified as early as possible in the life of a field by sequence stratigraphic analysis.

Four case studies are presented in this chapter to illustrate some applications of sequence stratigraphy to fluvial, shoreface, and deep marine reservoirs discussed in preceding chapters.

Chapter 12

An example of integrated characterization for reservoir development and exploration: Northeast Betara field, Jabung Subbasin, South Sumatra, Indonesia

I Nyoman Suta and Budi Tyas Utomo
PetroChina International, Jakarta, Indonesia

12.1 Introduction

This chapter presents an example of a reservoir characterization using a variety of data, including a 3D seismic survey covering 455 km^2 (175 mi^2), conventional logs from 30 wells, four wells with pressure data (repeat formation tester, RFT and modular dynamic tester, MDT data), conventional and sidewall cores, formation micro-imager (FMI) logs from two wells, and a plethora of reports and data on geochemistry, biostratigraphy, cuttings descriptions, reservoir fluid analysis/PVT, DST data and the like (PetroChina Inc., 1998a–1998d). The field presented here is the Northeast Betara field, in the Jabung block, which is located in the northern side of the South Sumatra Basin, Indonesia (Fig. 12.1). Regional geology and tectonic history of the area have been detailed by Suta (2004).

Northeast Betara field occupies the platform of the basement high to the north of the northeast-trending bounding fault that separates the field from the Gemah field to the south (Fig. 12.2). North-facing normal faults bound the field on the north. The field is bounded by a reverse fault on the west, and on the east by a reactivated normal fault which extends 25 km (15 mi) across the field and another 6 km (4 mi) south into the Gemah field (Fig. 12.1, inset). A similarly trending anticline provides an effective closure for hydrocarbon accumulations in the North Betara and Gemah fields.

The main productive interval in the Jabung Block, including Northeast Betara field, is the Oligocene age lower Talang Akar Formation. The lower Talang Akar is divisible into two major facies: a lower, pebbly, coarse, braided-fluvial

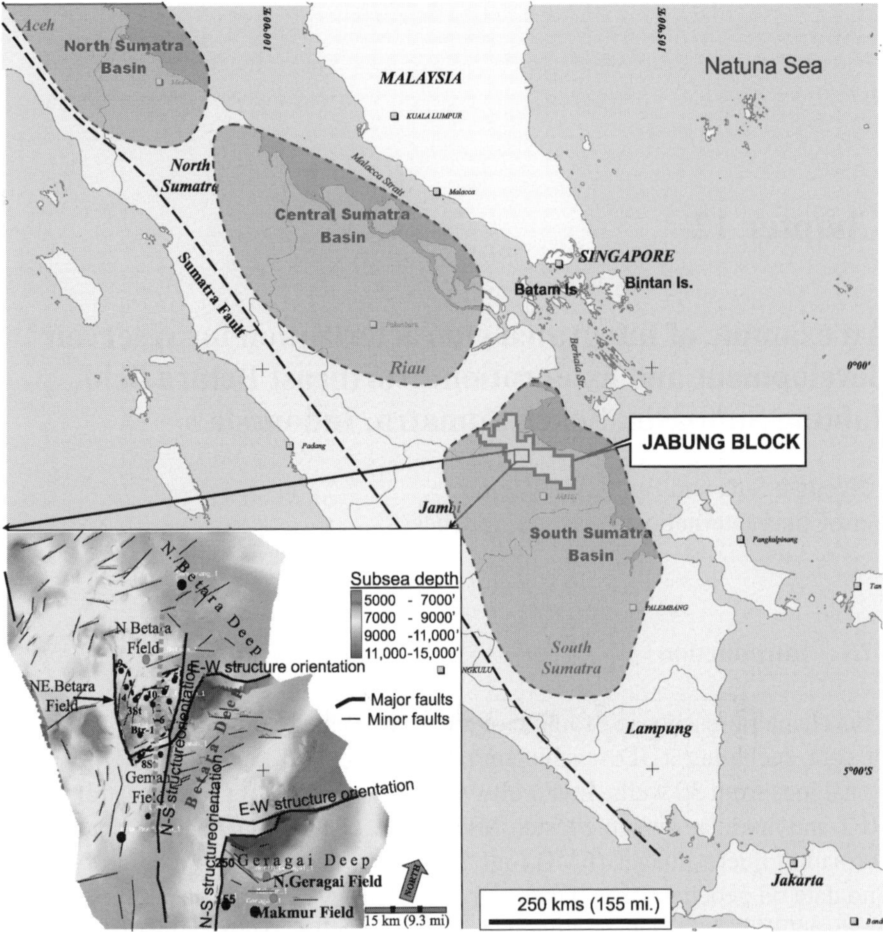

Fig. 12.1. Location map showing Sumatra Island and associated basins. Reservoirs in the Jabung block, including Northeast Betara, the focus of this study, are shown in the inset, which is a 3D depth structure image of the top of basement. Betara complex structures lie adjacent to the Betara deep. After Suta et al. (2000).

sandstone and an upper, finer-grained, sandy meandering-river facies with shale interbeds. The two intervals are separated by a continuous floodplain/marine shale. Hydrocarbon source rocks include intraformational shales and coals of the Talang Akar Formation, and the underlying lacustrine Lahat Formation.

12.2 History of Northeast Betara field

The Northeast Betara field is operated currently by PetroChina International in Indonesia (previously Devon Energy, after the acquisition of Santa Fe Energy

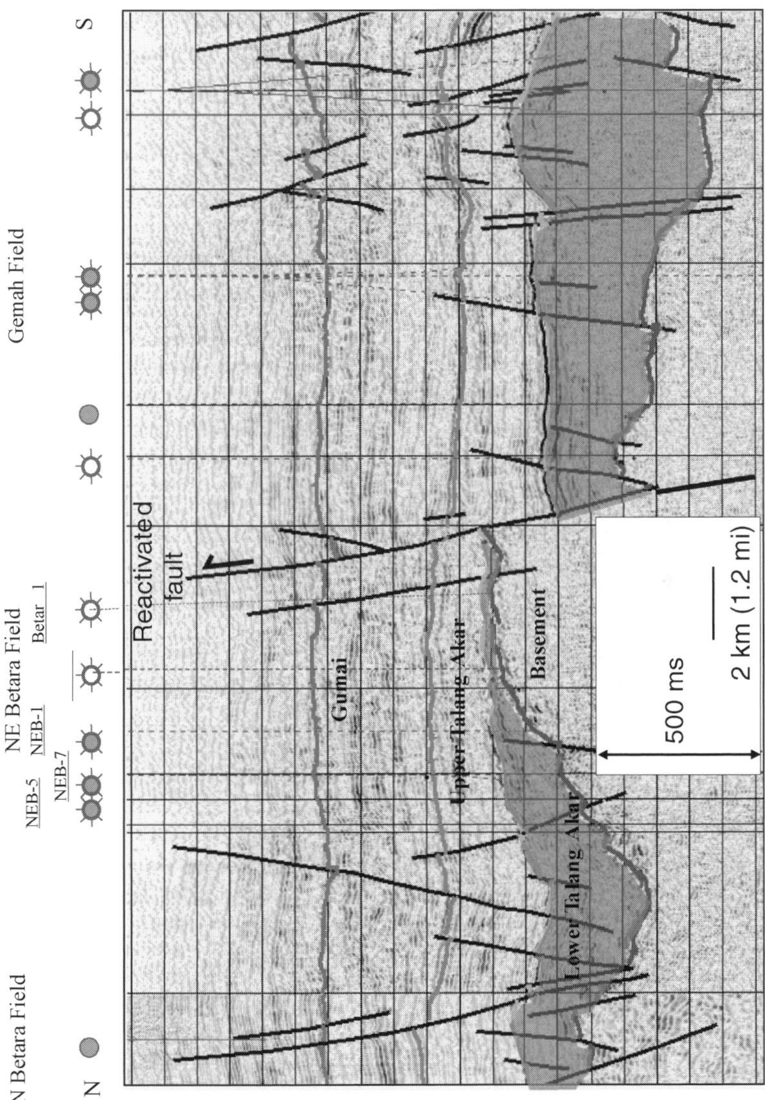

Fig. 12.2. North–south regional seismic section. To the south, Northeast Betara field is bounded by an old reactivated fault trending northeast to southwest. The fault separates the field from Gemah field. To the north it is bounded by a structural low separating the field from North Betara field.

Resources Co.) on behalf of Petronas and Pertamina. At 138 km^2 (34,000 acres), it is the largest field in the Jabung block and lies in a northwesterly trend along with the North Betara field to the north and the Gemah field to the south (Fig. 12.1, inset).

In 1971, the Betara-1 well encountered thin lower Talang Akar sandstones, which had gas shows on the crest of the structure, but that well was abandoned without being tested (Fig. 12.2). In 1995, on the basis of 800 km (500 mi) of new 2D seismic data, the NEB-1 was drilled downdip of the Betara-1, along the flank of the Betara basement high (Fig. 12.2). That well encountered eight individual lower Talang Akar sandstones totaling 81 m (266 ft) thick, 36 m (117 ft) of which were net-hydrocarbon-bearing sandstone. Tests on three different sandstone bodies produced at a combined rate of 18.22 MMCFGPD and 432 BCPD. The NEB-2 delineation well, drilled in 1996 updip to the south, encountered 14 m (42 ft) of net gas pay within two sandstones that tested at 12 MMCFGPD and 207 BCPD. From 1995 to 1999, ten wells were drilled to delineate Northeast Betara field. Five of them encountered an oil rim beneath the gas cap, proving that the Northeast Betara field does not contain just gas, as initially thought. A 3D seismic survey was acquired and processed in 1999–2000 that covered 455 km^2 (175 mi^2) of full fold coverage with a bin size of 20 m (66 ft). Seismic interpretation led to drilling the NEB-11 Hz horizontal well.

12.3 Field characteristics

12.3.1 Total reservoir thickness

The 3D seismic survey was designed to image the lower Talang Akar reservoir at Northeast Betara, Gemah, and North Betara fields (Fig. 12.2). Interpretation and mapping were performed using IESX 3-D and CPS-3 software (Geoframe™ of Schlumberger). Eighteen checkshot surveys from the study and surrounding areas were used to derive a time–depth curve.

The top of basement (at the base of the lower Talang Akar) was easily mapped because of a high impedance contrast between sediment and granite basement (Fig. 12.3A). The top of the formation, which lies at depths of approximately 1300 m (4400 ft) to nearly 1600 m (5400 ft), dips gently northward at about 1–2° in the middle of the field and at 4° at the northern edge (Fig. 12.3B). The lowest elevation is 1800 m (5800 ft), where the top of the formation is a saddle separating Northeast Betara and North Betara fields (Fig. 12.3B). To the south, the beds dip steeply as a result of a series of extensional faults bounding the field and separating it from the Gemah field to the south.

The lower Talang Akar total isopach thickness, mapped as the depth from the top of basement to the top of the formation, is thickest (900 m; more than 3000 ft) in the southeast and in the northern area (1200 m; 4000 ft;) (Fig. 12.4).

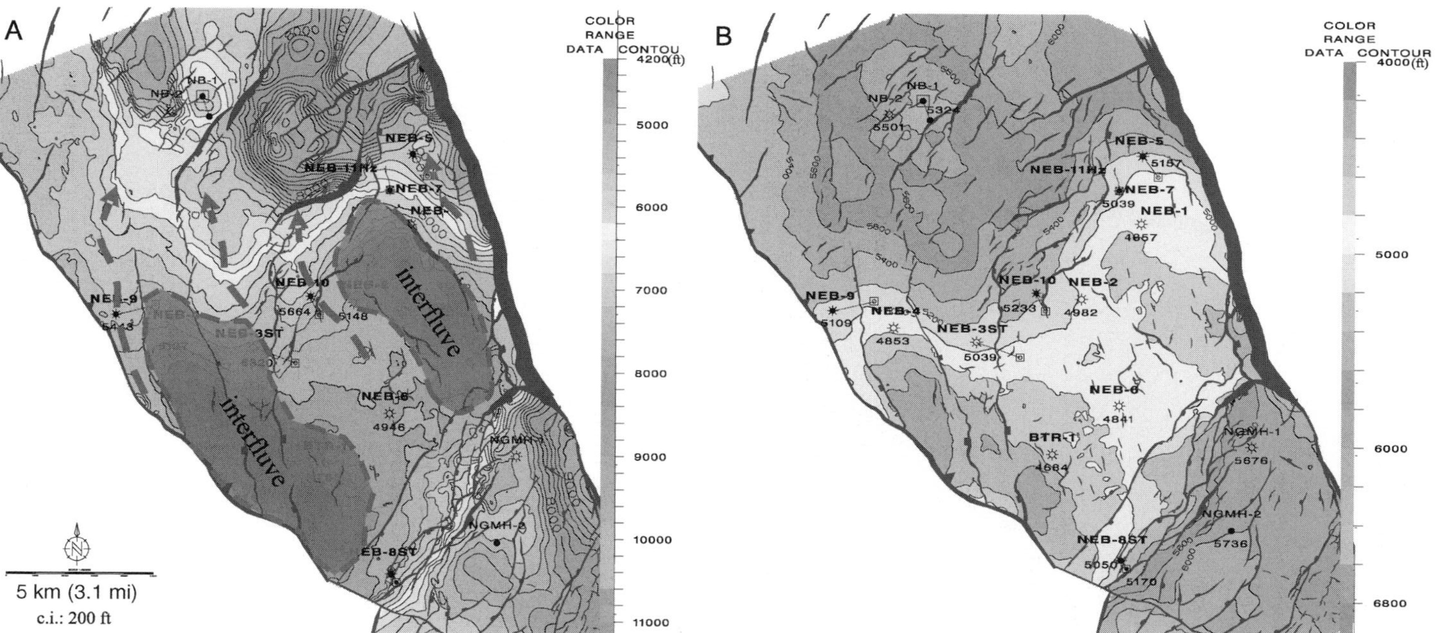

Fig. 12.3. (A) Depth structure map of the base reservoir (top granitic basement). Drainage direction (red*/dark-gray arrows) in the paleolows (valley) where possible prospective sands were deposited. (B) Depth structure map of the top of the lower Talang Akar Formation.

*The indicated color is for a CD which contains all of the figures in color.

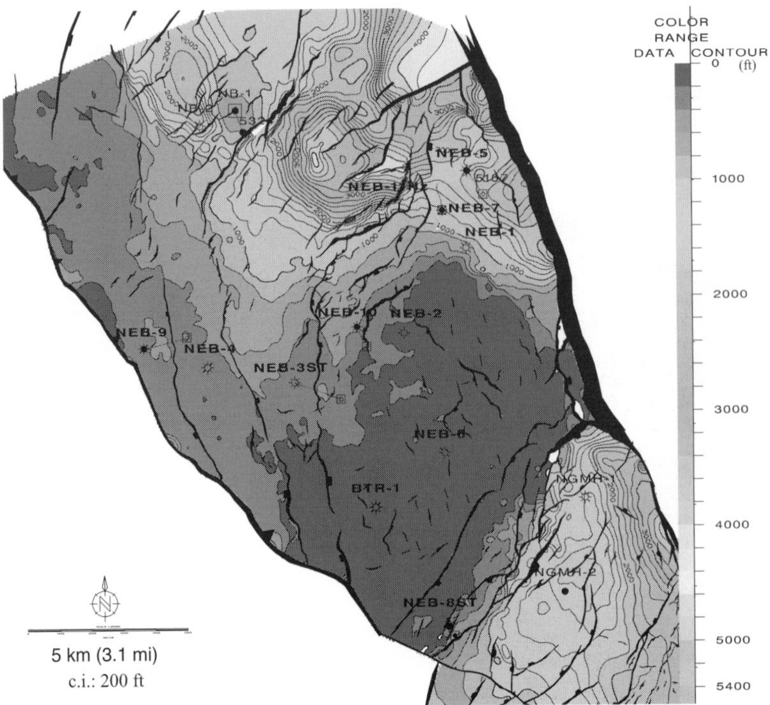

Fig. 12.4. Lower Talang Akar isopach map (total lower Talang Akar sediment thickness).

Gross reservoir thickness varies, from 30 m (100 ft) at the Betara-1 well in the south to 218 m (715 ft) at the NEB-7 well in the northeast. Sediment thickens gradually toward the north-northeast where most oil and gas reserves and wells are concentrated. Reservoir thickness abruptly becomes more than 200 m (600 ft) at the southern edge of the field in well NEB-8/8St, due to the development of a series of old growth faults. A total of 200 m (650 ft) of hydrocarbon column is defined from the top reservoir at the Betara-1 well −1428 m (−4685 ft) to the lowest known oil measured at the NEB-10 well to the north.

12.3.2 Sedimentary facies, properties and distribution

12.3.2.1 Lower braided-river facies

Seismic amplitudes in the lower reservoir are continuous. An isopach map of the lower reservoir interval was constructed by subtraction from seismically mapped lower and upper boundaries (Fig. 12.5A). The isopach map shows a thickness greater than 1100 m (3500 ft) to the north of the NEB-5 well. To the southeast, the sediment exceeds 800 m (2500 ft) in thickness (Fig. 12.5A). No sediment is present over the basement high.

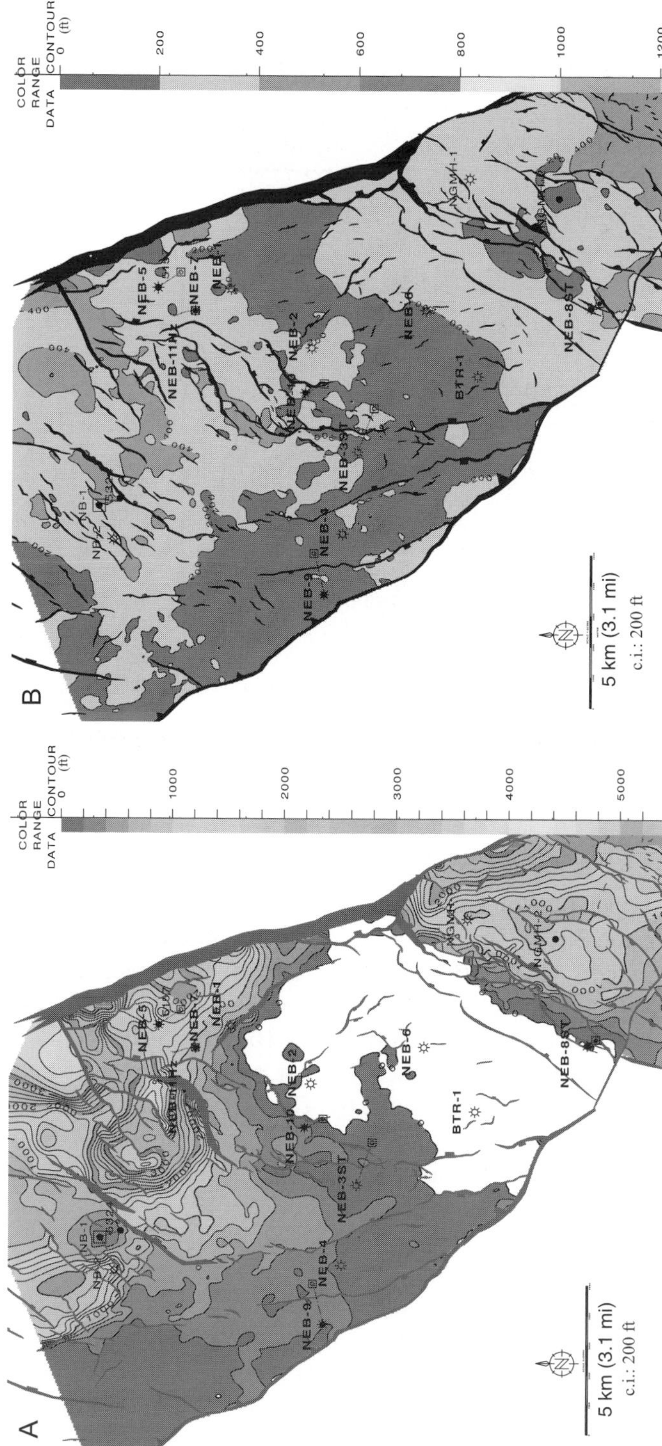

Fig. 12.5. **(A) Lower braided-river reservoir gross isopach map. (B) Upper meandering-river reservoir gross isopach map.**

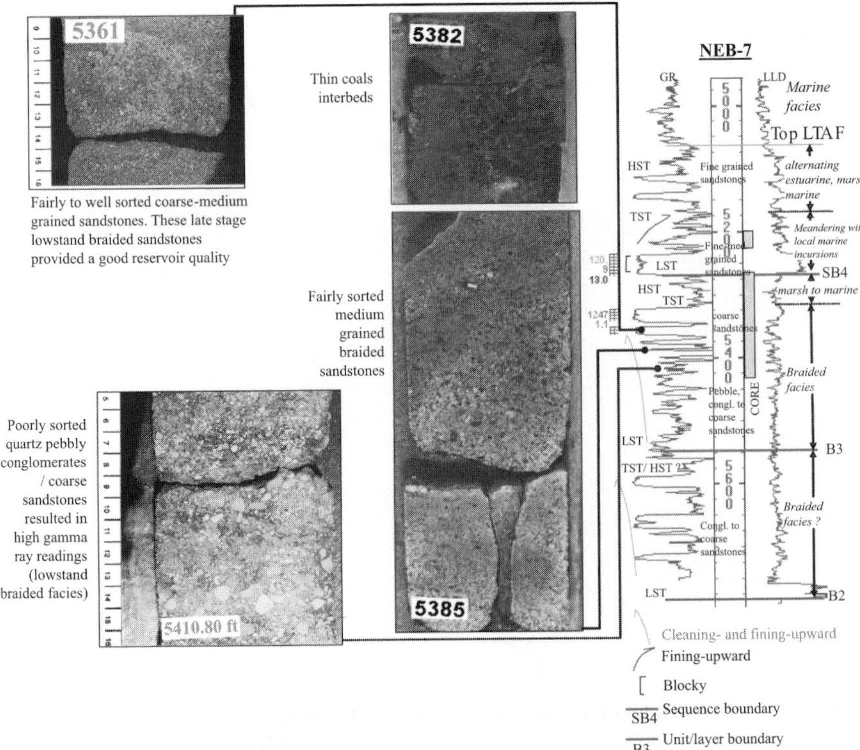

Fig. 12.6. NEB-7 core sections and log from the upper part of the lower braided-river deposit.

The cored upper part of a coarsening- and cleaning-upward braided sandstone facies in the NEB-7 well (Fig. 12.6) is composed of light gray, subangular to subrounded, coarse to very coarse quartz sandstone, with occasional lithic clasts, common granule- and pebble-conglomerates. Pebbly conglomerate to coarse muddy sandstones at a depth of 1649 m (5410 ft) are organized crudely into cleaning-upward sequences that are cross-bedded and slightly parallel-laminated near the top (Fig. 12.6). Thin gray, cross-bedded sandstones at 1641 m (5385 ft) are interbedded with shale and thin coal beds at 1640 m (5382 ft), and are overlain by cleaner and thicker sandstones at 1634 m (5361 ft) (Fig. 12.6). The uppermost interval consists of 8 m (25 ft) of clean, horizontal to cross-bedded sandstones and pebbly sandstones (Fig. 12.7A).

The pebble sandstone of the braided facies only occurs in wells in the northeastern part of the field, possibly due to the presence of a steep basin slope in this area at the time of deposition (Figs. 12.5 and 12.7B). A more gentle basin slope to the south of the field, in Gemah field, resulted in more homogeneous sandstone, with a thick, clean, blocky gamma-ray log response comprising the braided facies.

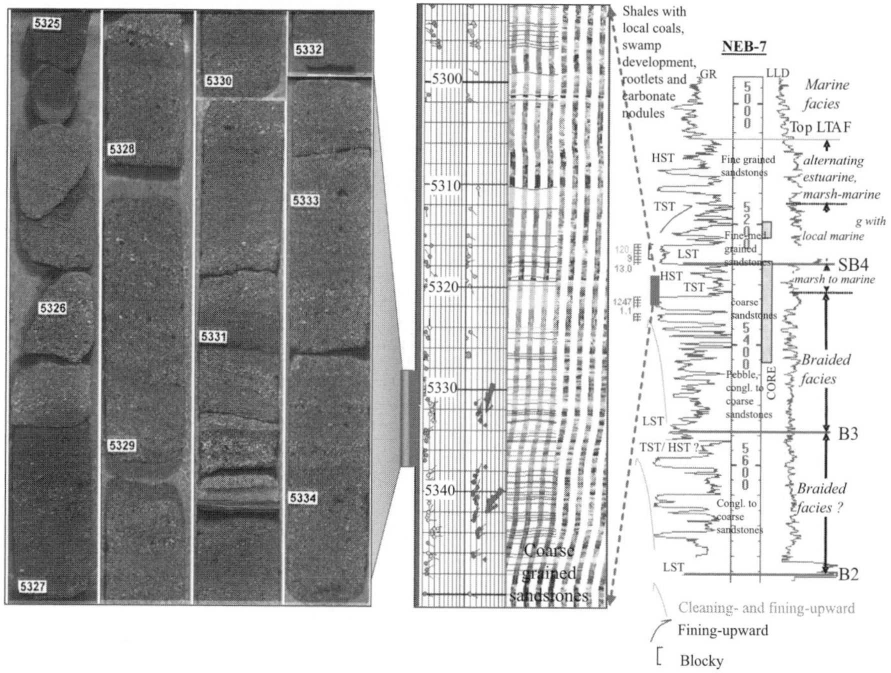

Fig. 12.7A. NEB-7 well logs and core, showing the uppermost coarse to very coarse-grained and well-sorted sandstones with thin coal stringers and shale interbeds. The sandstone interval tested 1250 BOPD in this well.

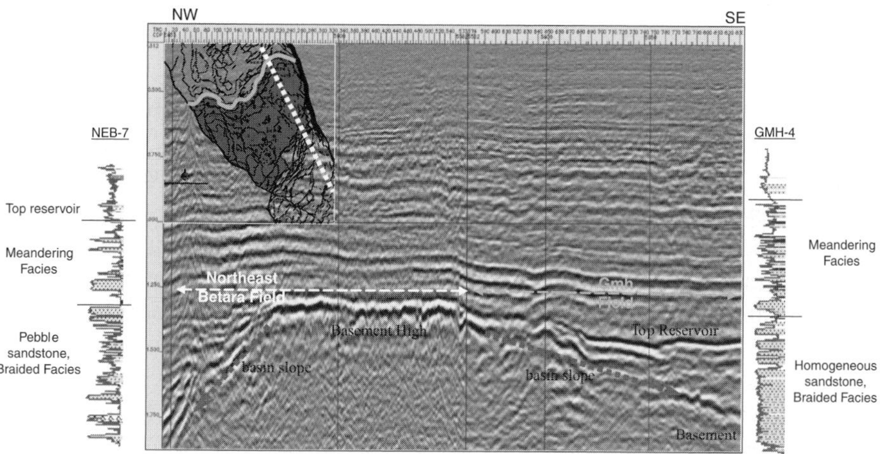

Fig. 12.7B. Seismic Section through Northeast Betara and Gemha (GMH) fields showing a steep basin slope in the northern part of the Northeast Betara Field that provided more pebbly sandstone braided river facies than is found to the south, in Gemha field, where the braided-river facies, deposited on a gentler slope, consists of homogeneous, amalgamated, thick sandstones. (Insert: index map).

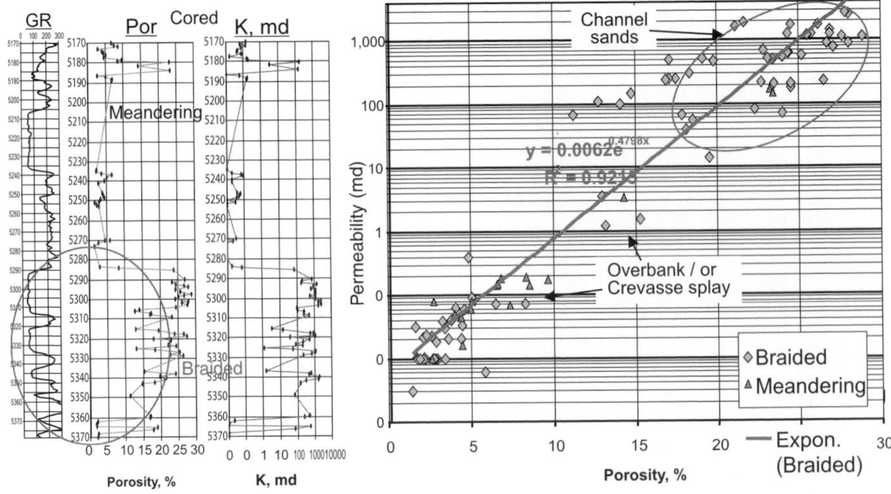

Fig. 12.8. Well logs and core interval in NEB-7. Cored interval shows excellent reservoir quality in the lower reservoir. Limited core in the upper meandering-river facies also exhibits excellent reservoir quality.

The cored interval in well NEB-7 exhibits an average porosity of 23% and permeability as high as 2.6 darcys (Fig. 12.8). This interval tested at 1247 BOPD in NEB-7 well, 1454 BOPD in NEB-9 well, and a total of 10 MMCFGD and 220 BCPD was produced from two test intervals in the NEB-1 well. Log analyses in the upper part of the braided-river facies exhibit an average of 19.6% porosity from NEB-1, -5, -7, -9, and -10 wells.

12.3.2.2 Shale interval

Overlying the lower braided-sandstone interval is a 15 m (50 ft) thick section of shales that, in the uppermost part, contain marine burrows (Fig. 12.9). A borehole-image log from NEB-7 reveals a general structural dip of 1–5° to the west-northwest for sandstones above the shale and 5° to the west-northwest for sandstones beneath the shale (Fig. 12.9). A borehole-image log from NEB-10 reveals a general structural dip of 7° to the west-northwest for sandstones above the shale and 11° west-northwest for sandstones beneath the shale (Fig. 12.10). On the 3D seismic volume, a moderately continuous reflector represents the thick shale that separates the lower from the upper sandstones (Fig. 12.11), suggesting that the shale is a barrier that separates the lower from the upper reservoir intervals.

12.3.2.3 Upper meandering-river facies

The isopach thickness map of the upper reservoir interval, constructed by subtracting the seismically defined lower and upper reservoir facies boundaries, ranges from several meters thick to nearly 150 m (500 ft) in thickness

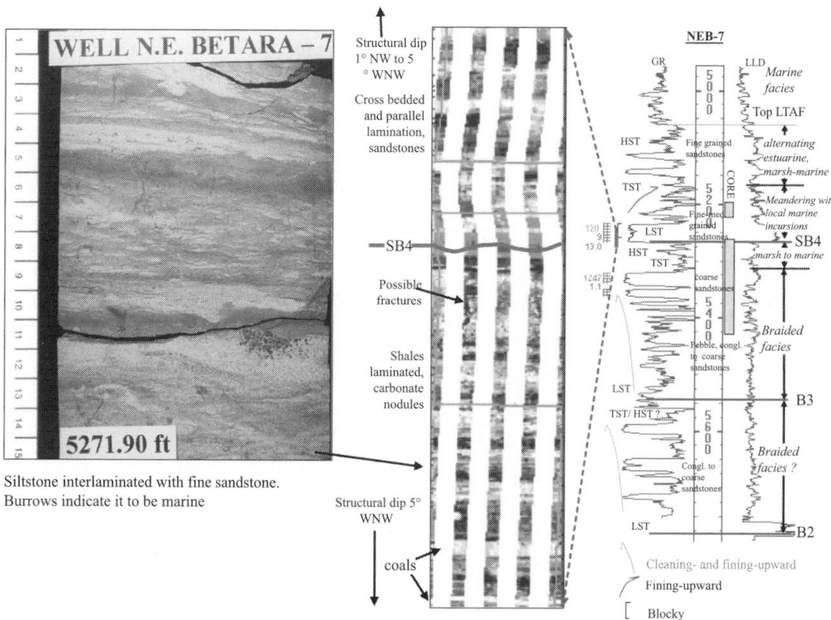

Fig. 12.9. Core photograph of marine, laminated pale gray shales beneath a sequence boundary (SB-4).

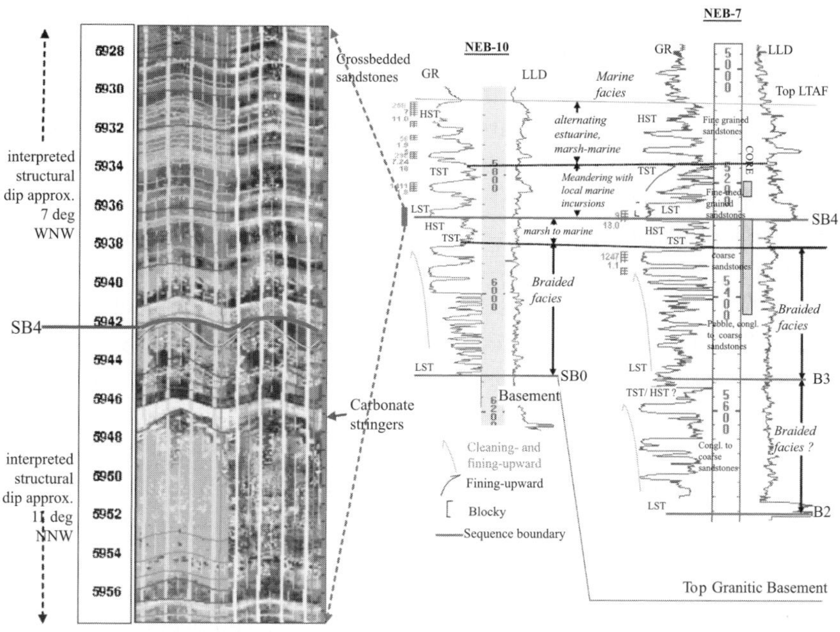

Fig. 12.10. FMI showing interpreted sequence boundary (SB4) in NEB-10, which is correlative with that in the NEB-7 well.

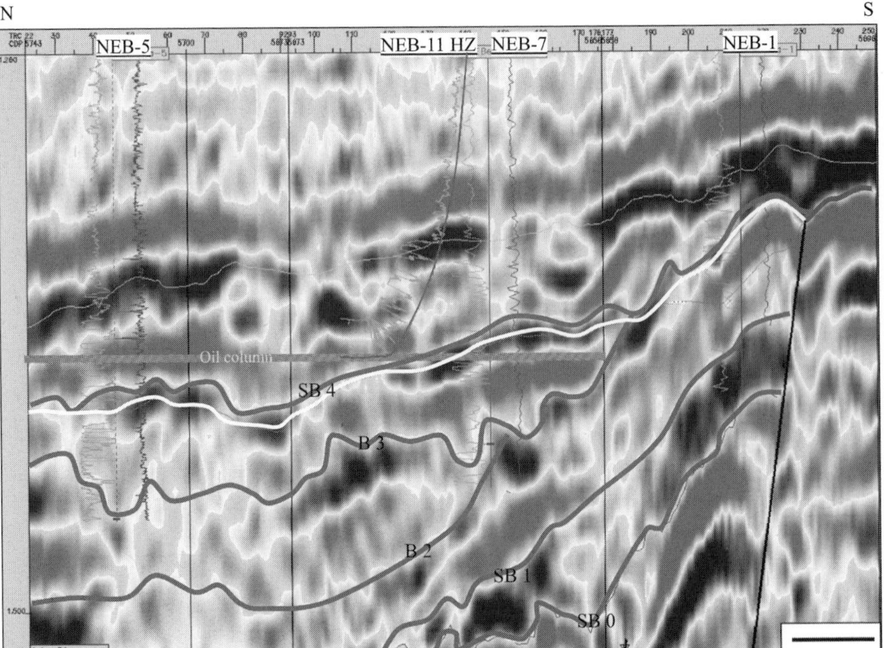

Fig. 12.11. Seismic section through NEB-1, -5, -7, and -11 Hz wells. SB – sequence boundary, B – boundary.

(Fig. 12.5B). The thin reservoir interval is developed in the middle of the field above the crest of the structure around Betara-1 and NEB-6 wells. Slight thickening occurs to the north and south around NEB-5, NEB-11 Hz, and NEB-10 in the north and around NEB-8St in the south (Fig. 12.5B).

The upper meandering-river reservoir facies was cored only in NEB-7, through a relatively thin, muddy–sandy interval, but an FMI log was obtained through the thicker, basal sandstone that comprises a set of stacked, amalgamated beds (Fig. 12.12). The cored interval (Fig. 12.13) is composed of greenish to white, medium- to very-fine-grained, moderately sorted and cemented, texturally mature quartz sandstones with some lithic grains. It grades upward into fining-upward, locally cross-bedded to parallel laminated sandstones as much as 5 m (15 ft) in thickness. The sandstones are interbedded with mudstone/shales that contain roots and thin coal beds, carbonate nodules, and numerous vertical and horizontal burrows.

The cored upper reservoir interval has an average of 19% core porosity and 65 md permeability (Fig. 12.8). Log porosity in the well averages 23%, using a 9% porosity cutoff. This sandstone tested at 120 BCPD and 9 MMCFGD in NEB-7 and tested oil in the NEB-5 and NEB-11 Hz wells.

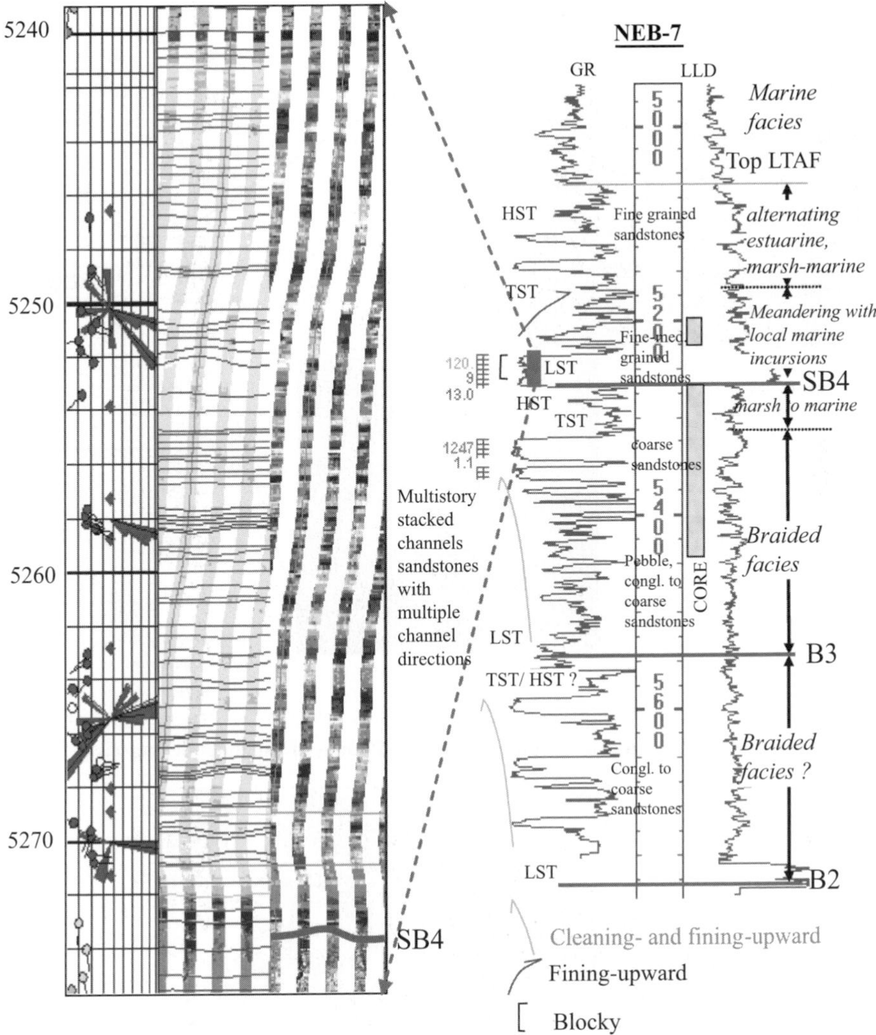

Fig. 12.12. FMI log, showing stacked channel (multistory) sandstones at the base of the meandering facies. Coring was attempted, but core was not recovered over much of the interval. This sandstone interval tested gas in this well and intersected oil in the NEB-11 horizontal well.

RMS amplitudes representing 50 ms in time-thickness below the top of the upper reservoir identified a high sinuosity channel belt in the NEB-5 and NEB-7 areas (Fig. 12.14), further confirming the meandering fluvial nature of the upper reservoir interval. Discontinuities appear on seismic images within the same horizon and are parts of the same channel belt (Fig. 12.15). At this horizon, a 12 m (35 ft) thick oil column was encountered in a sandstone in the NEB-5, -7, and -11 Hz wells. Oil intersected in the NEB-5 and -11 Hz wells is from

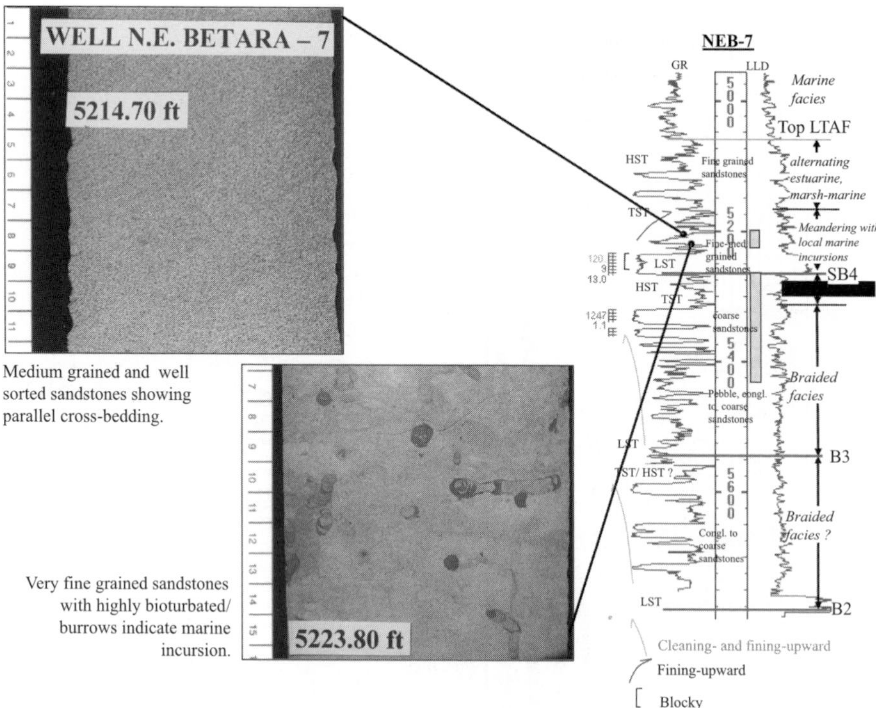

Fig. 12.13. Cored, medium- to fine-grained sandstones of the meandering-river facies.

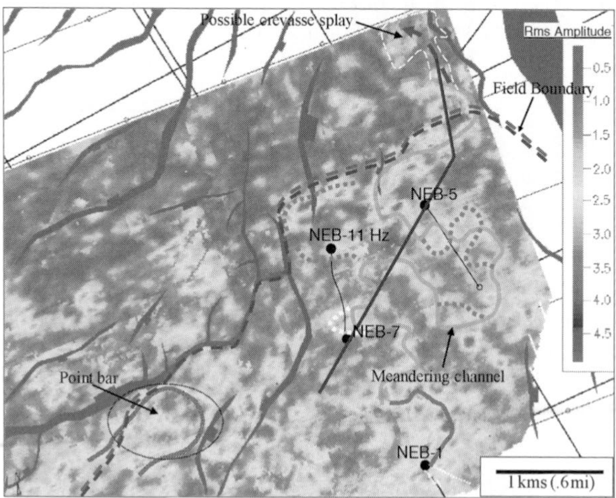

Fig. 12.14. Seismic amplitude showing channel belt development in the upper reservoir of the lower Talang Akar Formation around NEB-1, -5, -7, and -11 Hz wells. The amplitude represents 50 ms below the top of the upper reservoir (top of lower Talang Akar Formation). The channel encountered in NEB-7 well probably is the same genetic unit as that encountered in and around the NEB-5 well (within one compartment). Lateral distribution and orientation of the channels were controlled by faults.

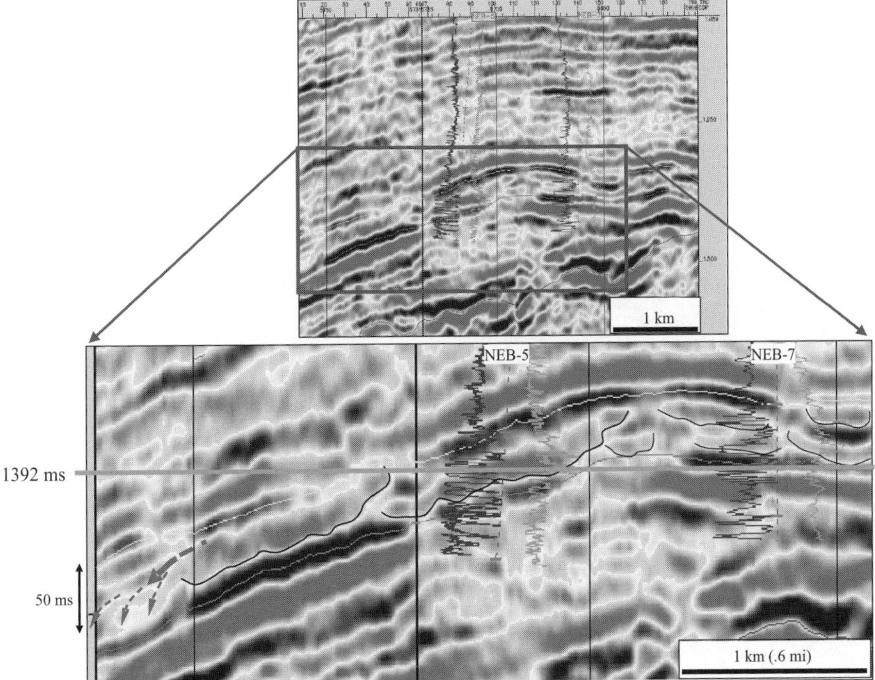

Fig. 12.15. Seismic section across the high-sinuosity channel belt from NEB-7 to NEB-5 wells (Fig. 12.14). The red/dark-gray arrows (lower left) indicate possible crevasse splay deposits, and the green/bright-gray line represents the oil rim at approximately 1392 ms. Black solid lines represent interpreted bases of channels.

the meandering channel sandstones that tested gas in NEB-7. A horizon map at 1392 ms (Fig. 12.16) illustrates the anticipated lateral distribution and complexity of the oil sands and also can be used to reduce the uncertainty in calculating reliable oil reserves.

12.3.3 Reservoir compartments

Figures 12.17 and 12.18 show structural cross-sections across the field. The main fluid contacts are not continuous across the field, which suggests that the reservoir is structurally compartmentalized. Vertical fluid distribution information from drillstem tests (DSTs) (Fig. 12.19) shows that gas occurs opposite oil at the same structural elevations and that gas also occurs downdip from oil, further indicating that the reservoir is compartmentalized. The reservoir can be segmented into seven fault-bounded compartments (Fig. 12.20A) (Saifuddin et al., 2001). For example, gas–oil and oil–water contacts in NEB-9 (compartment-1) and NEB-10 (compartment-4) wells are at different depths from those in NEB-5 and NEB-7 (compartment-5) wells (Figs. 12.18 and 12.20A).

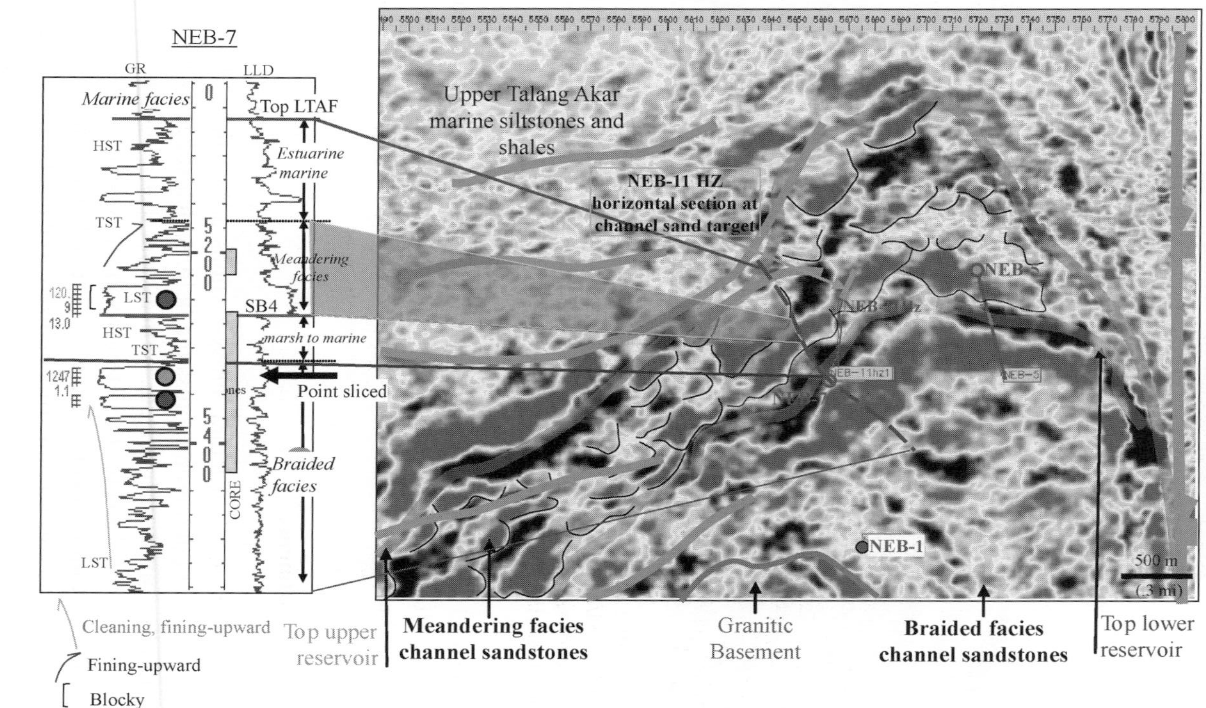

Fig. 12.16. Seismic time slice at 1392 ms two-way traveltime, showing strong amplitude where the oil column was penetrated in the meandering-facies channel sandstone at NEB-5 well and NEB-11 horizontal section, and in braided-facies sandstone in the NEB-7 well. The amplitude "seismic subcrop" feature is correlatible to the Northeast Betara log type. Strong amplitudes might represent channel belt sands in meandering (upper reservoir) and braided (lower reservoir) river facies. The sharp bases of the meandering channel belts are inferred (black thin line). Gray lines are faults.

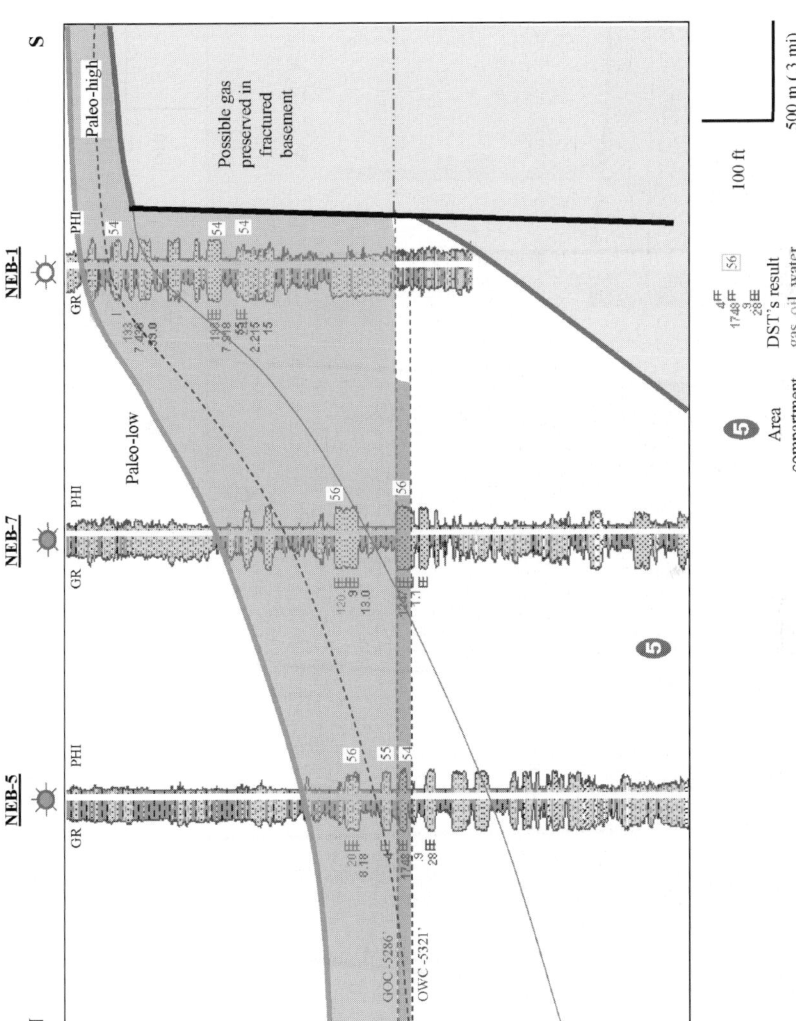

Fig. 12.17. North–south structural cross-section. The wells represent area compartment 5, showing similar fluid contacts and CO_2 content.

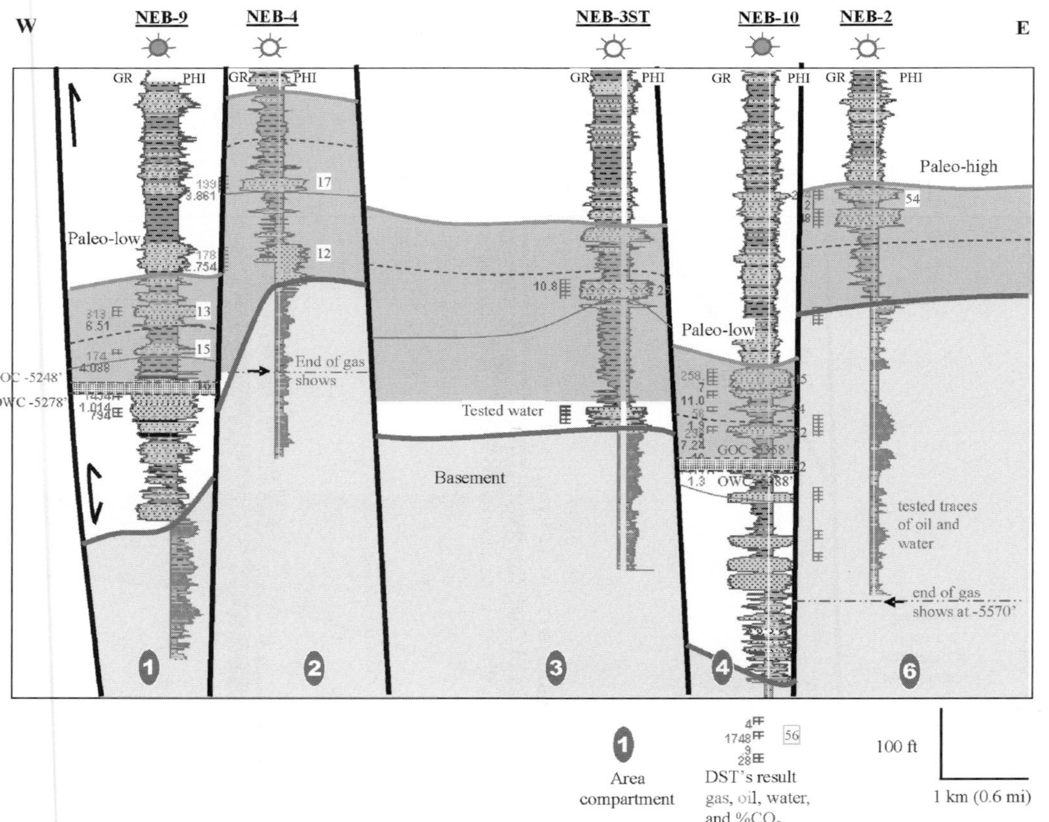

Fig. 12.18. West–east structural cross-section, showing fluid distributions within various reservoir compartments.

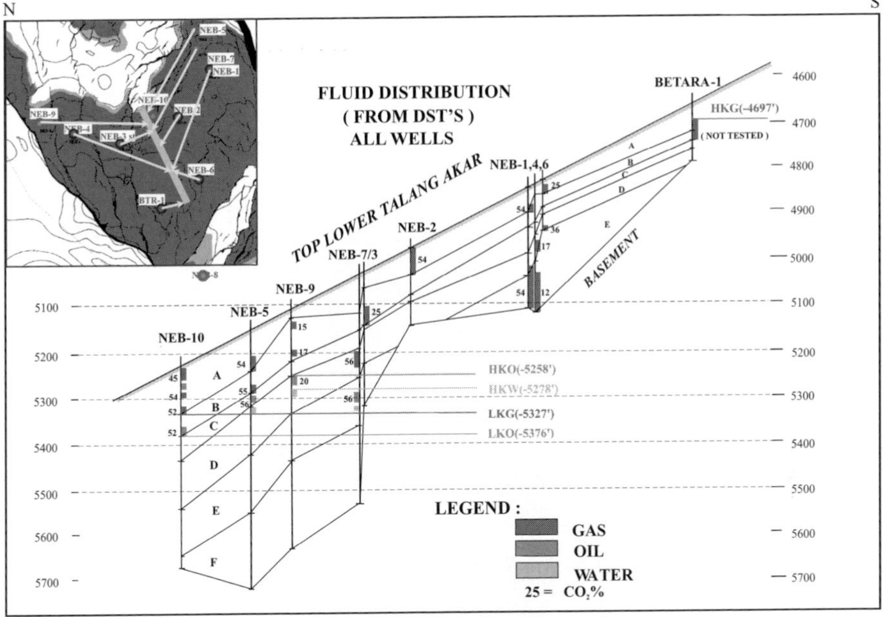

Fig. 12.19. Projected north–south diagrammatic structural well arrangement from the highest reservoir encountered at Betara-1 well to the lowest well at NEB-10. Diagram shows reservoir subdivisions and fluid distribution from DSTs, indicating multiple fluid contacts (PetroChina interpretation prior to 3D seismic acquisition).

The distribution of CO_2 in wells also indicates compartmentalization (Figs. 12.18 and 12.20B). A trend of gradually diminishing CO_2 concentration toward the west suggests that the CO_2 was sourced from the east and migrated to the west. Vertical CO_2 concentration does not vary in the wells.

Compartmentalization is controlled predominantly by structural rifting and later rejuvenation. Extensional basins formed valleys in which sediment was deposited. For example, sandy facies tend to be distributed in the grabens in NEB-1, -5, -7, -9, and -10 wells (Figs. 12.19 and 12.20A). The original north–south block faults bounding compartments 1, 2, 3, 4, and 5 (Fig. 12.20A) have controlled the general pattern of sediment geometry, width, thickness, and orientation during the extensional tectonic phase.

12.4 Depositional model

Prior to this study, workers thought that reservoir sands were deposited throughout the basin (Fig. 12.21A). The present study has shown that block faulting controlled the sandier reservoir development in paleo-low areas (Fig. 12.21B). This model indicates that different sedimentary styles developed between the

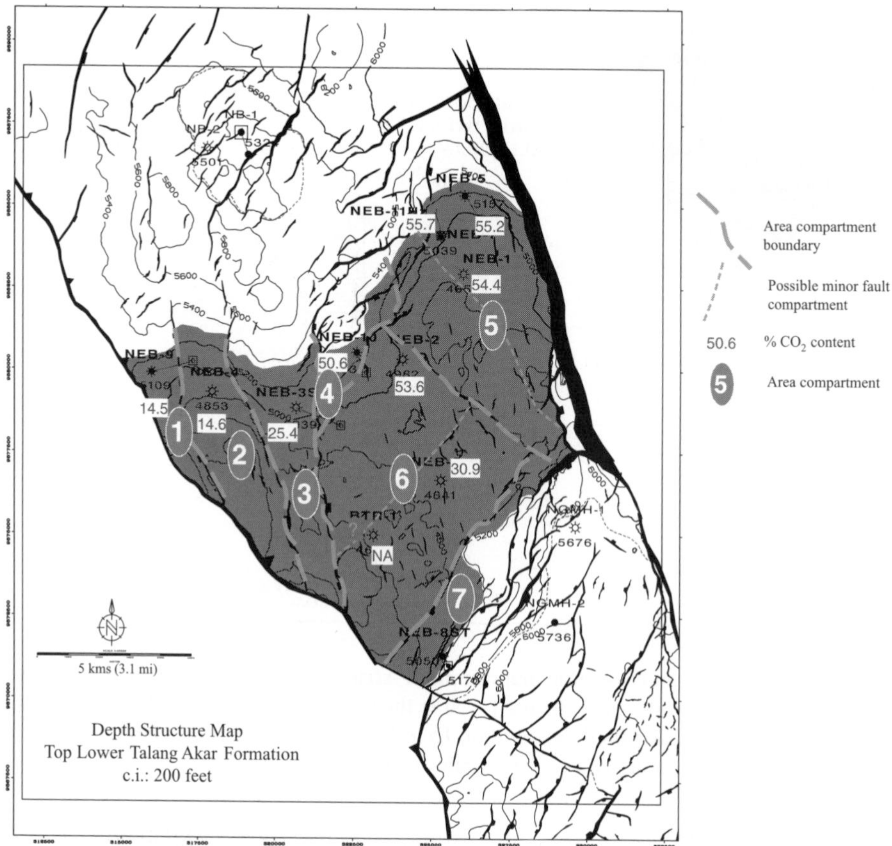

Fig. 12.20A. Top reservoir of the lower Talang Akar Formation, showing the seven area compartments and their CO_2 distribution. Compartments are bounded by faults. Area compartment 6 is divisible into two areas indicated by different CO_2 content.

paleo-lows and interfluves. This model focuses development of the field in sandier paleo-lows.

The model was proven in January 2005 when a gas injection well was drilled to test for the presence of proximal channel sand; instead, the well encountered low porosity/permeable facies with no fluvial channels. The well was then sidetracked to the northeast to target sands within a paleo-low, and encountered thick fluvial channel sandstones that could be used for gas injection.

12.5 Sequence stratigraphy

The 3D seismic volume shows several north-dipping reflectors, which are interpreted to be boundary surfaces (Fig. 12.11). Integration of the seismic, well-log (including FMI), biostratigraphic, and core data led to an improved correlation

CO_2 CONTENT DISTRIBUTION

Summary of CO2 content of seperator gas of PVT data

NORTH EAST BETARA Composition Analysis and CO2 Distribution										
Well	DST	Region 1 Separator Gas(52%CO2)			Region 2 Separator Gas(27%CO2)			Region 3 Separator Gas(16%CO2)		
		%CO2	%C1	%C7+	%CO2	%C1	%C7+	%CO2	%C1	%C7+
NEB-1	1A	54.19	29.35	0.20						
	3	54.37	29.79	0.07						
NEB-2	4	54.52	30.67	0.16						
NEB-3A	2				25.31	51.6	0.11			
NEB-4	1							12.31	61.78	0.11
	2							16.69	58.34	0.13
NEB-5	2	56.37	26.75	0.14						
	4	54.12	31.99	0.12						
NEB-6	2	37.47	37.18	0.61						
NEB-7	2	55.88	24.45	0.27						
	3	55.65	31.88	0.05						
NEB-8ST	1				28.24	45.07	0.27			
NEB-9	2							15.45	50.29	0.33
	4							13.00	63.47	0.05
NEB-10	1	51.05	23.53	0.36						
	2	49.52	22.73	1.58						
	3	52.57	30.93	0.23						
NEB-13		56.26	25.63	0.26						
NEB-20		47.75	27.81	0.27						
NEB-22								18.43	57.54	0.27
NEB-17								21.45	45.17	0.46
Average		**52.29**	28.67	0.332	**26.78**	48.335	0.19	**16.22**	56.10	0.22

NE Betara PVT CO_2 Regions

Ne Betara Field Fluid Region

Region 2-
27%CO2

Region 1-
52%CO2

Region 3-
16%CO2

Region 2-
27%CO2

Fig. 12.20B. Northeast Betara field is divided into three CO_2 regions as shown in the figure. This is done by categorizing the wells into three classes based upon well test data, and using the current fault patterns to define the boundaries among the three regions.

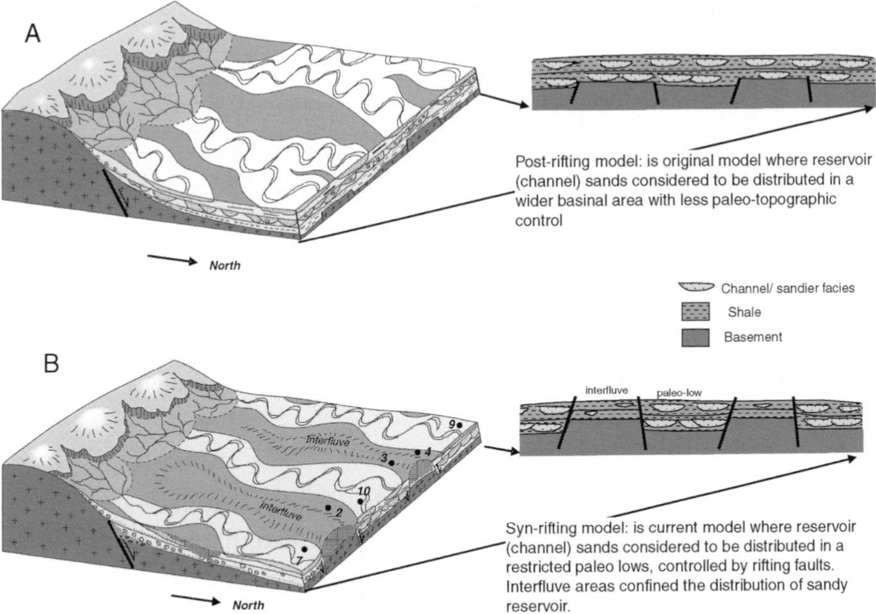
A
Post-rifting model: is original model where reservoir (channel) sands considered to be distributed in a wider basinal area with less paleo-topographic control
North
Channel/ sandier facies
Shale
Basement
B
interfluve
paleo-low
Interfluve
Interfluve
9
4
3
10
2
7
Syn-rifting model: is current model where reservoir (channel) sands considered to be distributed in a restricted paleo lows, controlled by rifting faults. Interfluve areas confined the distribution of sandy reservoir.
North

Fig. 12.21A. Schematic block diagram, showing (A) originally interpreted reservoir LTAF was deposited in postrifting time, and (B) current model indicating that the synrifting reservoir development was controlled by block faulting of paleolow and paleohigh areas.

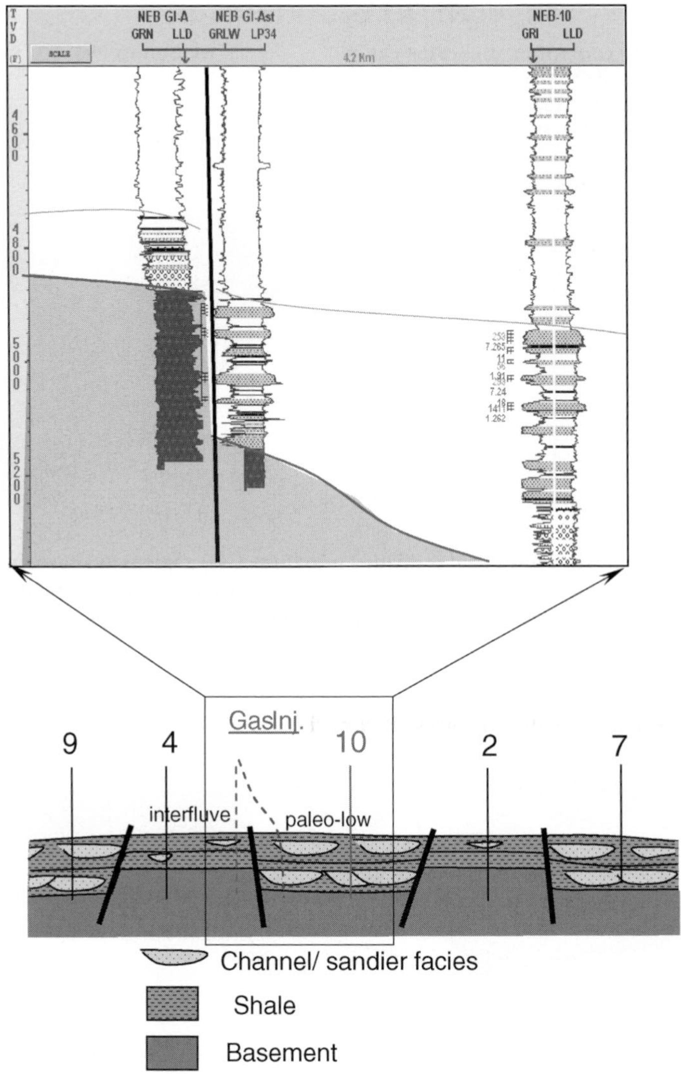

Fig. 12.21B. Section showing a gas injection well which was sidetracked into fluvial sandstones in the paleo-low area.

and interpretation of significant surfaces or boundaries that have stratigraphically compartmentalized the reservoir and distribution of fluid contacts.

From this integration, a high-frequency sequence stratigraphic framework for the reservoir has been developed (Fig. 12.22).

In and around NEB-5, NEB-7, and NEB-10 wells, an unconformity labeled SB0 is clearly visible on FMI logs (Fig. 12.23) and on seismic profiles (Fig. 12.11). This unconformity developed on the top of basement during early

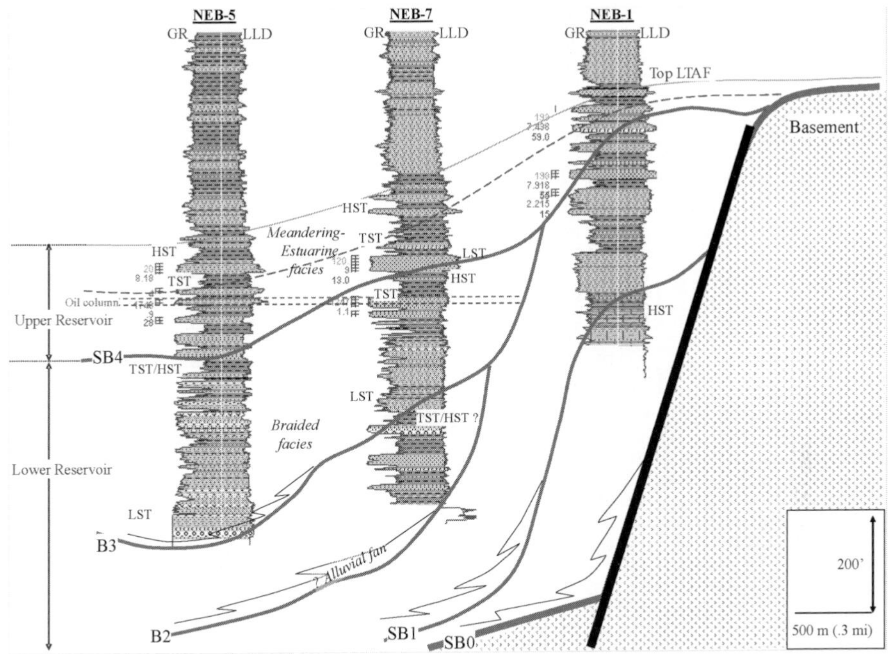

Fig. 12.22. Depositional dip section, showing interpreted sequence boundary succession in the northeastern edge of the Northeast Betara field. Alluvial fans might have developed beneath the braided facies. SB – sequence boundary, LST – lowstand systems tract, TST – transgressive systems tract, HST – highstand systems tract.

basin rifting. The lower braided reservoir facies was deposited on this surface during a sea-level lowstand.

A second significant sequence boundary (SB4) is interpreted to lie at the base of the upper reservoir meandering-channel belt in the NEB-7, -5, -1, and -10 wells (Figs. 12.6, 12.7A, 12.9, 12.10, and 12.23). That sequence boundary marks a major change of depositional style from early transgressive/highstand(?) marine shales to lowstand sandy meandering channel facies when the sea regressed from the area (Figs. 12.7A and 12.23). Coeval braided sands might have been deposited further south, closer to the detrital source, during this time. The subsequent eustatic rise in sea/base level resulted in an overall fining-upward sequence in the upper meandering interval (Fig. 12.5B). A similar sea level/depositional history has been described for a Pleistocene fluvial sequence in the Malay Basin, offshore north of Malaysia (Miall, 2002).

Other, less significant sequence boundaries, labeled SB-1, -2, and -3, were recognized on the seismic sections and well logs (Figs. 12.6, 12.7A, 12.9–12.11, 12.22, and 12.23). A 100-psi difference in pressure below and above boundary SB-3 in the NEB-7 well suggests that it is a significant boundary. In NEB-1,

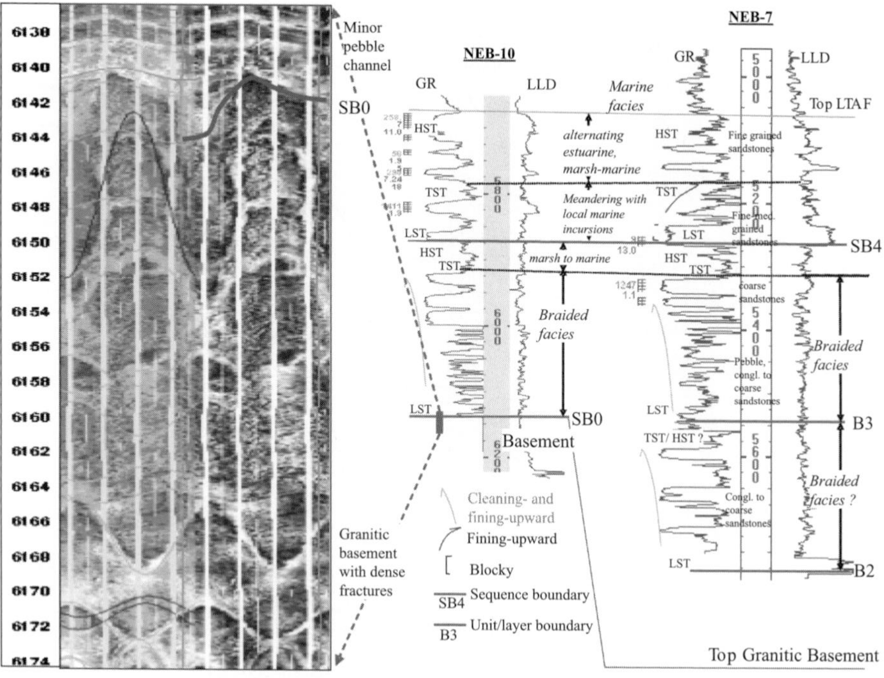

Fig. 12.23. Typical fractured basement shown on the FMI image of the NEB-10 well. The upper part of the section shows an unconformity between granitic basement and the base of the braided channel that represents the main sequence boundary (SB0).

a carbonate interval below the reservoir sands may be another sequence boundary (SB1) (Fig. 12.22).

Figure 12.24 summarizes the sequence development in relation to the proposed relative sea-level or base-level changes through time. Deposition of the lower Talang Akar sandstones was marked by a major sea-level drop during a time of early rifting. Adjacent, emergent basement highs probably were sources of detrital sediments. The paleogeographic features resulted in deposition of a series of coarse-grained, braided sandstones (lower reservoir interval) in confined low areas, resting on SB0. Sea-level/base-level turnaround and rise were marked by deposition of meandering fluvial and estuarine deposits (upper reservoir interval), which then were capped by marine shales of the upper Talang Akar Formation (UTAF) (Fig. 12.24).

A higher-order curve is shown in Fig. 12.24 to explain higher-order sequence development of the reservoir. Sequence boundary SB1 may have formed and separated the carbonate interval (encountered in NEB-1 well) beneath the boundary (high-frequency transgressive/highstand deposit) from the fluvial deposits above the boundary (high-frequency falling stage deposit). During the later stage of rifting, the basin slopes were gentler, so coastal plain and marine

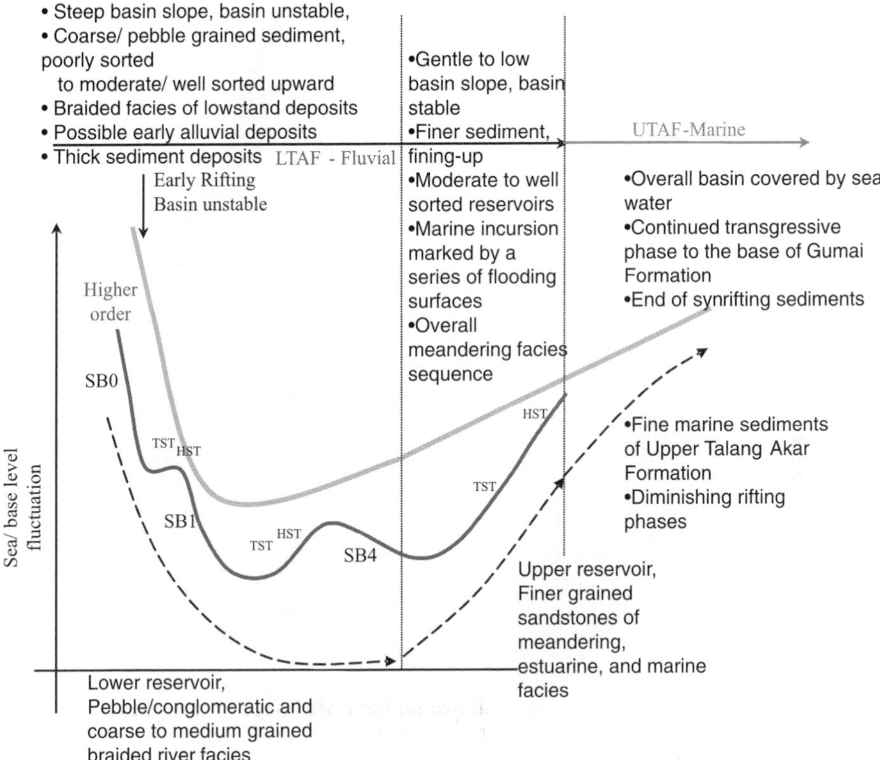

Fig. 12.24. Schematic sea-level/base-level fluctuation in the Northeast Betara field, created on the basis of sequence stratigraphy analyses. Red/black is a modified eustatic curve. Haq et al. (1987), his gray line.

sediments, predominantly shale, were deposited during a high-frequency transgressive/highstand event. This event was followed by a high-frequency drop in sea level to create SB4, which then was followed by deposition of fining-upward meandering-river sands during a rising stage of sea level.

12.6 Field development plan

12.6.1 Field development scenario

Based upon the reservoir characterization, depositional model, sequence stratigraphy, and well test data, a material balance reservoir model was generated to guide early field development. The Northeast Betara field consists of two hydrocarbon reservoirs, an oil reservoir and a gas-condensate reservoir (Fig. 12.25). The volumetric reserves calculation combined with the material balance reser-

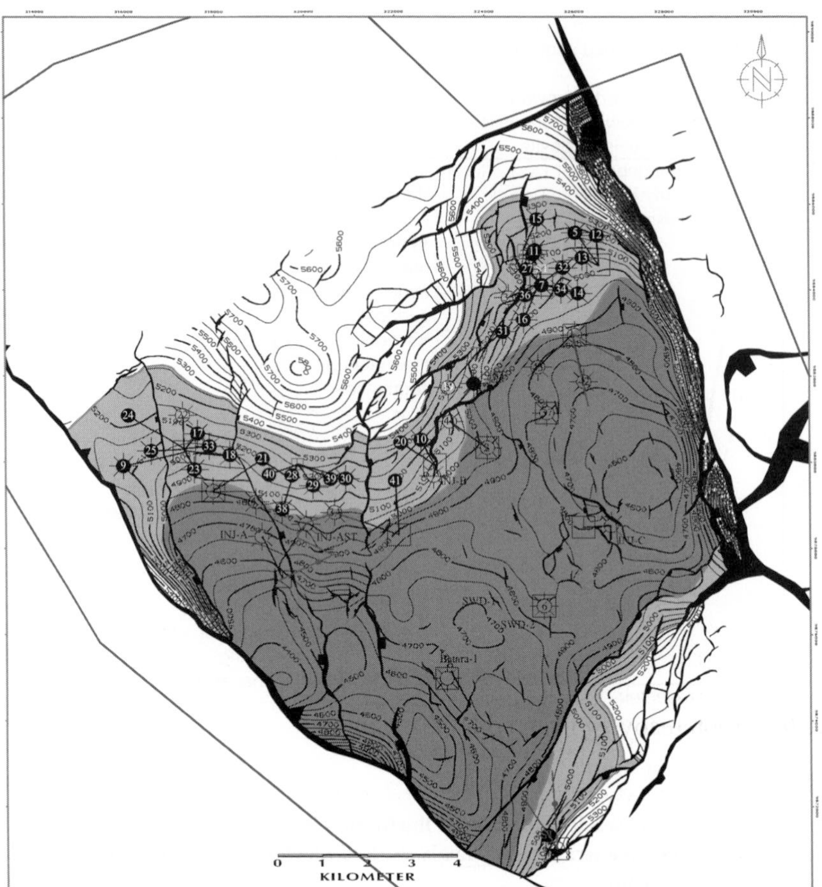

Fig. 12.25. Structurally the oil rim/reservoir in the Northeast Betara field has been subdivided into five areas (blocks) separated by north- to south-trending faults.

voir model indicate the potential oil reserves comprise about 10 to 15% of the potential gas-condensate reserves. Sensitivity analysis using the material balance model to evaluate the impact of gas production start-up to oil recovery is shown in Fig. 12.26. This analysis determined that the oil recovery factor will increase by approximately 7.0% by delaying the gas production schedule for four years. The analysis indicated the oil reservoir drive mechanism was mostly from the gas cap and associated gas, with some limited water drive.

These results led to the recommendation to produce the oil rim reserves in the early stage of development, then to produce the main gas-condensate reservoir. This recommendation was in line with the requirement for Gas Certification, for dealing with the gas buyer, and for building a Gas Processing Plant, all of which require more time compared to oil reservoir development. Other benefits of the

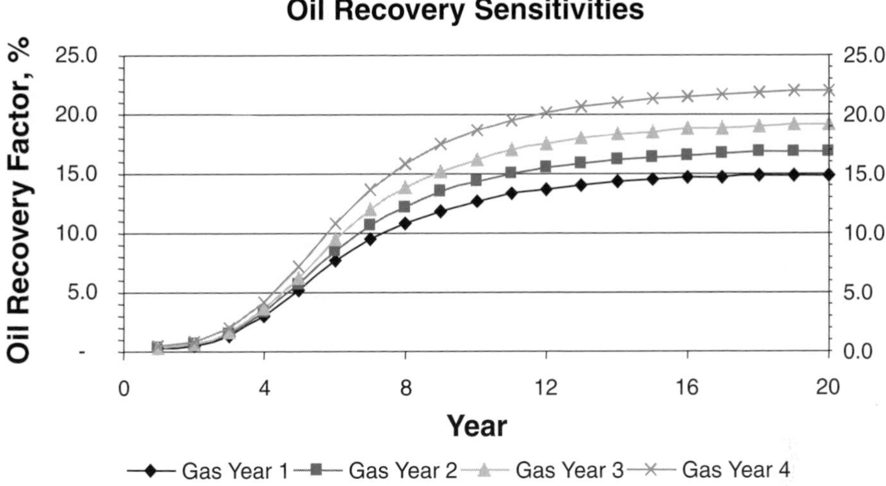

Fig. 12.26. Oil recovery sensitivities based upon the material balance model.

two phases of oil and gas development include additional well data from the early stage of oil development which can be used to generate better geological and reservoir simulation models to improve reservoir management and optimize hydrocarbon recovery from both the oil and gas reservoirs.

12.6.2 Determination of proposed development wells

Both structural and stratigraphic compartments have challenged development of the field. Additional development wells have been, and will continue to be drilled based upon the 3D seismic, coupled with new well and production data. For example, seismic attributes such as acoustic impedance volumes have been applied to the Lower Talang Akar Formation in order to obtain better images of possible oil/gas sands (Fig. 12.27).

12.6.3 Reserves exploitation, challenges and strategy

Major challenges identified for Northeast Betara field development are listed below:

1. The development plan must take into account that the reservoir has a large gas cap, a thin oil rim, and a thin oil column (9–11 m or 30–35 ft) with different gas/oil and oil/water contacts in each area (Figs. 12.18, 12.19, 12.25, and 12.28); fluvial channels are narrow and meandering, with thin sands in some areas (Fig. 12.14).
2. CO_2 content is highly variable, with 55% in the eastern area of the field and 15% in the western area (Fig. 12.20). This variability of CO_2 content is

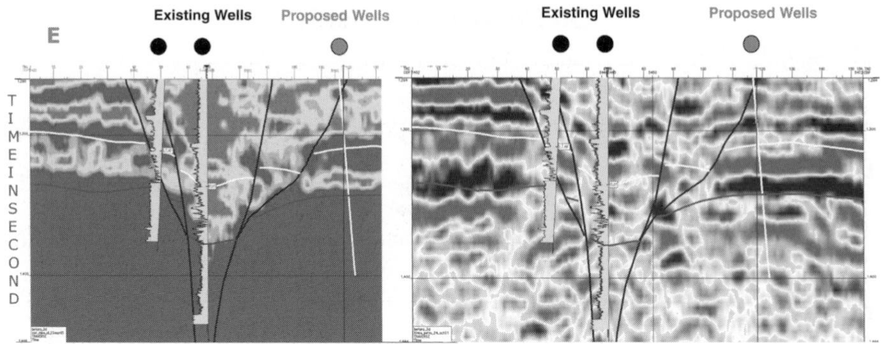

Fig. 12.27. Existing and proposed well locations within this structurally and stratigraphically complex area, as revealed by seismic attribute analysis.

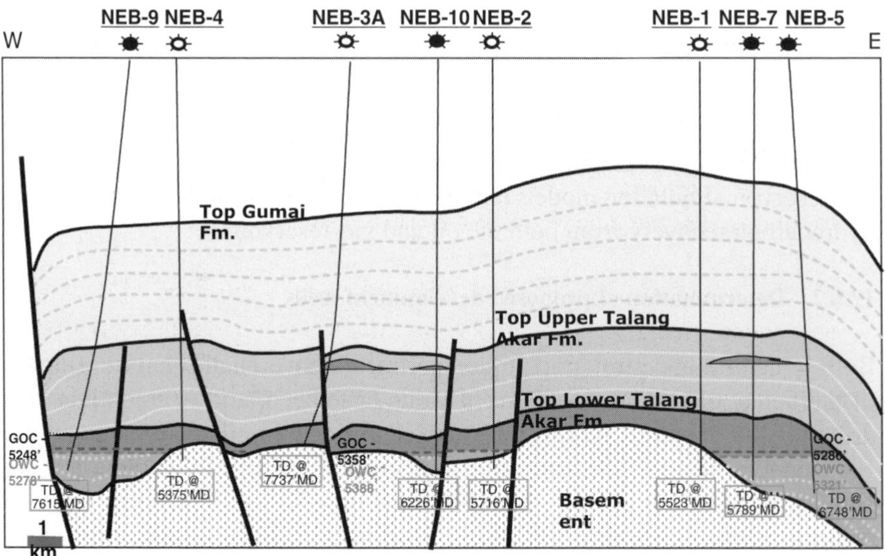

Fig. 12.28. Northeast Betara field showing a large gas cap, a small oil rim, and a thin oil column with different gas–oil (GOC) and oil–water (OWC) contacts in different fault blocks.

very important in gas development, especially for designing the Gas Processing Plant and in making commitments to potential gas buyers. A gas production strategy will be influenced by compartmentalization of CO_2 across the field.

These challenges are the main drivers for generating a strategy which will maximize gas and oil recovery and minimize drilling risk and expense. The following action plan has been recommended to achieve this strategy:

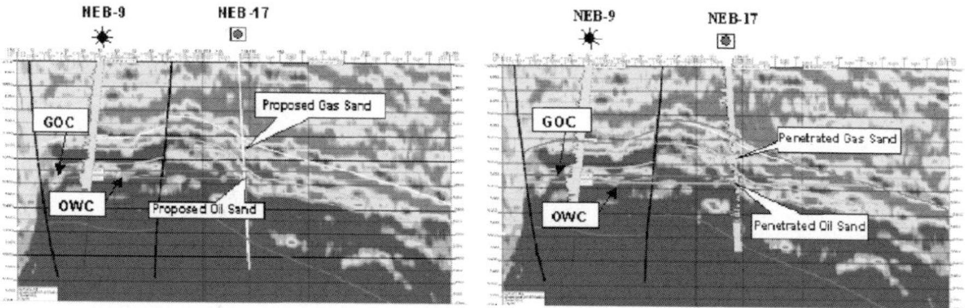

Fig. 12.29. Well development strategy. The left figure is a seismic attribute section in depth showing the proposed NEB-17 well location, designed to penetrate both gas and oil sands. The right figure shows the well successfully penetrated both gas and oil sands, as predicted from the seismic attribute analysis.

1. Select future well locations to penetrate both oil and gas reservoirs. During the first two to four years, the wells should be designated to produce oil. Thereafter, the wells should be converted for gas production. Figure 12.29 shows results of one well which successfully drilled both oil and gas sands on the basis of this strategy.
2. Selectively perforate oil zones to reduce possible water and/or gas coning.
3. Maintain a reasonable oil production rate to manage potential early water/gas encroachment.
4. Install a production string with multi packer and sliding sleeve door to accommodate changing production zones using wireline instead of pulling the drill string and re-running a new string. With this production string design, approximately US$500,000 could be saved for every cycle (Fig. 12.30).
5. Conduct a detailed, multi-disciplinary characterization to optimize well design and select favorable locations in the oil rim. After several years oil production and development, additional well data and production history can be added to generate more comprehensive geological and reservoir simulation models. During the gas development phase, the updated numerical simulation model should be used to (a) provide optimum well spacing, to forecast production of raw gas, oil, condensate, and LPG for gas sales, (b) predict more precisely the CO_2 distribution and content, and (c) determine the field's pressure profile. All of this information will be utilized in the design and construction of production facilities.

The preliminary reservoir simulation was generated on the basis of the existing geological model (Fig. 12.31). This model is being used to propose further development wells and to generate production forecasts (Fig. 12.32). The models are being continually updated as new information is obtained. This workflow is required for reservoir management.

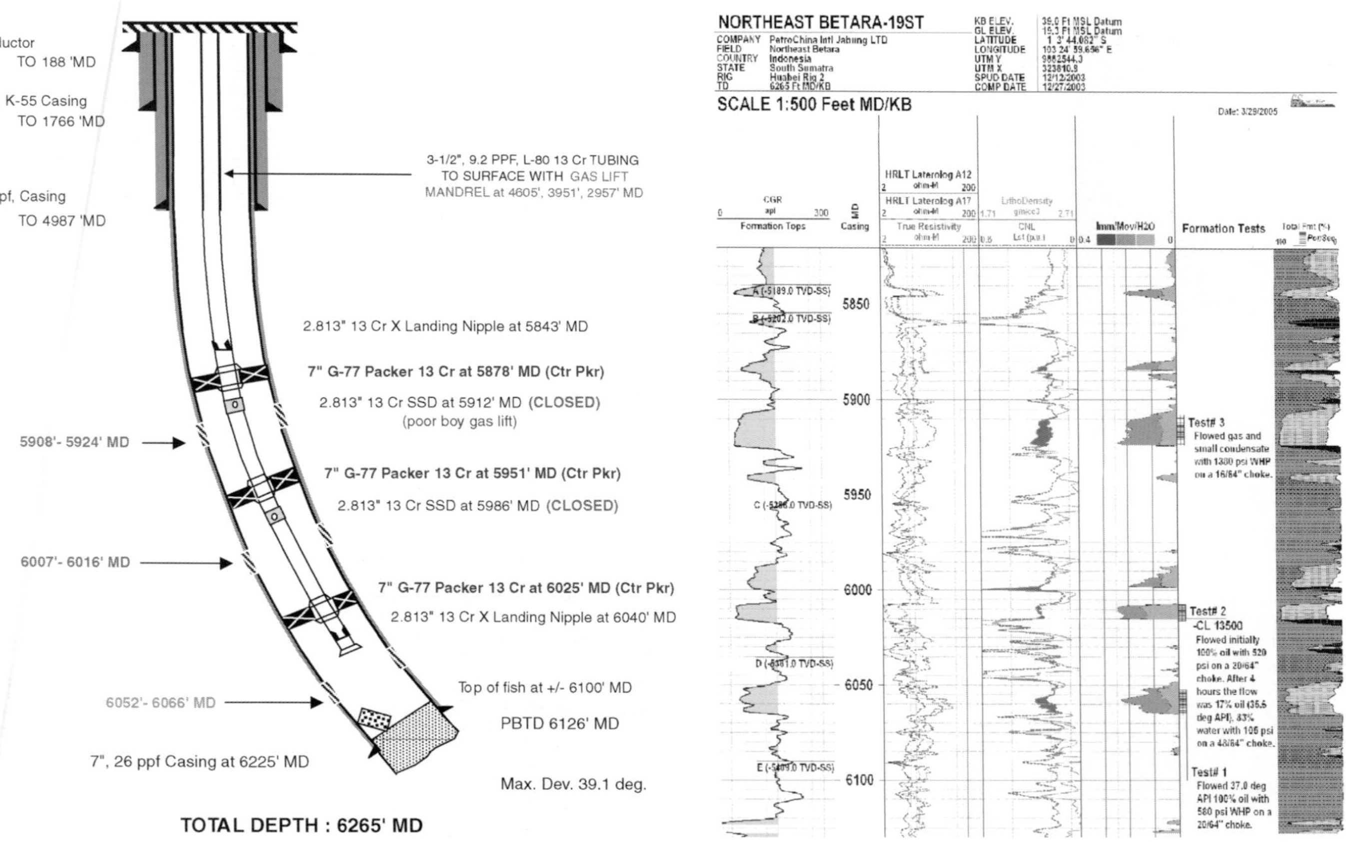

Fig. 12.30. Proposed production string with multi packer and sliding sleeves door.

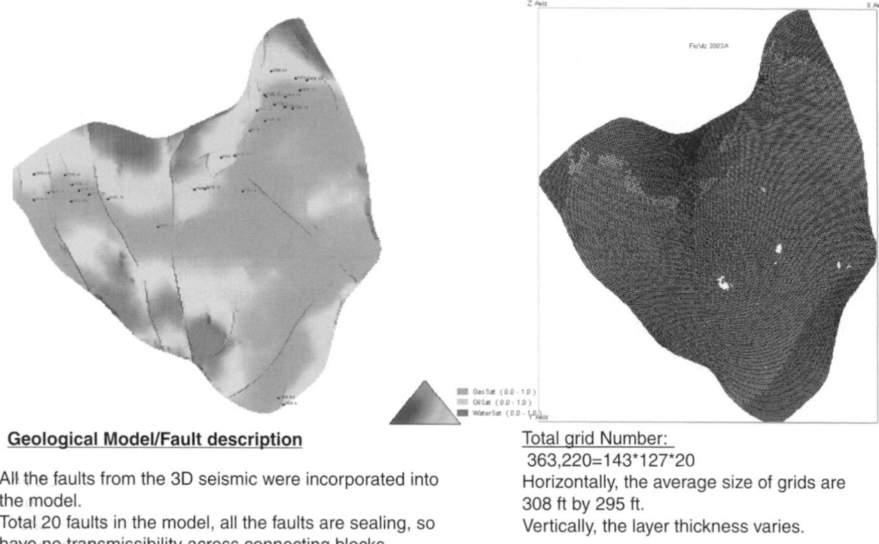

Geological Model/Fault description
All the faults from the 3D seismic were incorporated into the model.
Total 20 faults in the model, all the faults are sealing, so have no transmissibility across connecting blocks.
Total grid Number:
363,220=143*127*20
Horizontally, the average size of grids are 308 ft by 295 ft.
Vertically, the layer thickness varies.

Fig. 12.31. Upscaled geological model, which was imported into a reservoir simulator program.

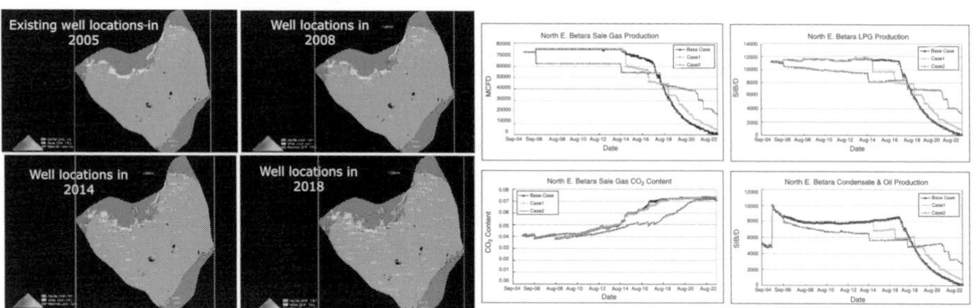

Fig. 12.32. Proposed development wells through the year 2018, and various field production forecasts.

12.7 Conclusions and applications

The lower Talang Akar Formation is the main hydrocarbon-bearing formation in the area. The reservoir is divisible into lower and upper reservoirs, separated by a thick floodplain/marine shale. The lower reservoir is composed of thick, coarse-grained sandstones to conglomeratic and pebble sandstones that were deposited from a series of braided rivers on a steep paleoslope during a falling stage of relative sea/base level. Early lowstand braided-river deposits are coarser grained, more poorly sorted, and of poorer reservoir quality than are the sands deposited during later lowstand. The upper reservoir contains sands that were reworked and recycled from underlying strata. They were deposited as finer-grained, clean, and well-sorted meandering-river sandstones that have good reservoir quality. The base of the upper reservoir exhibits excellent blocky and clean sandstones deposited during a second, but lesser stage of relative sea-level/base-level fall.

The majority of reserves are on the north flank of the reservoir and in the paleo-lows, where a greater volume of sediments and sandstone reservoirs were deposited. The reservoir thickens to the north as a result of normal fault movement during rifting. Thinner reservoir intervals occur on the crest of the basement highs and on paleo-highs and interfluve areas.

The Northeast Betara field is compartmentalized by the following geologic components: (a) block-faulting of basin origin and later structural inversion; (b) limited/noncontinuous lateral geometry of fluvial sandstones; and (c) vertical distribution of facies formed by sequence sets that are characterized by interbedding of different rock types (i.e., pebbly/conglomeratic sandstones, finer sandstones, and shales/mudstones) and by the presence of depositional sequence boundaries. The thick shale that separates the lower braided-reservoir sandstones from the upper meandering-reservoir sandstones is a particularly important barrier to communication between the lower and upper parts of the reservoir.

Structural and stratigraphic compartmentalization controls the distribution of reservoir fluids in Northeast Betara field. This knowledge can lead to improved placement of wells and choice of well types (horizontal, vertical, slant). Recognition of reservoir compartments with varying fluid contacts may constitute an important interwell frontier for future field development. Compartments may leave bypassed reserves, if the compartments are not drilled. Compartmentalization can be identified using the integrated methodology described here. In structurally and stratigraphically compartmentalized reservoirs such as Northeast Betara field, development of an integrated 3D geologic model, numerical reservoir simulation model and production strategy is critical to optimize both oil and gas productions.

In this study, analysis of 3D seismic data contributed greatly to identification of reservoir heterogeneity and compartmentalization. In conjunction with well information, 3D analysis provided information that led to development of a sequence stratigraphic framework. As this study showed, relatively small stratigraphic compartments, as well as faults and sequence boundaries, can be identified and mapped on 3D seismic volumes. Such fine-scale seismic characterization can reduce uncertainty in reservoir assessment and can lead to more accurate calculations of hydrocarbon reserves and better drilling scenarios.

This work is not meant to provide a final or complete reservoir characterization. Crossplots, structural analyses, seismic extraction, and other types of characterizations are introduced as a foundation upon which to plan further work in developing and managing the field and in guiding larger basin studies. It is imperative in any reservoir characterization that reservoir information be continually updated as it is obtained. As part of the development scenario, the reservoir and geological modeling should be dynamic, and periodically updated with new wells and production history, through the life of the field. This workflow will definitely improve production and reservoir performance.

Acknowledgments

The initial phase of this work constituted an M.S. thesis by the senior author at the School of Geology and Geophysics at University of Oklahoma. Later work was performed where the authors are employed at PetroChina, Indonesia. Geologists and geophysicists in the Jabung block exploration and subsurface departments of PetroChina are thanked for their support in providing the data and for their many helpful discussions. The authors greatly acknowledge the permission conceded by BPMIGAS, MIGAS, PetroChina in Indonesia and Partners for allowing publication of this paper.

References

Abbots, I.L. (Ed.), 1991. United Kingdom Oil and Gas Fields, 25 Years Commemorative Volume. Geological Society of London Memoir No. 14, 573 pp.

Abdulah, K., Doud, K., Cook, M., Keller, D., Bellamy, J., Bengtson, M., Jensen, T., Alwin, B., 2004. Reservoir facies within the deepwater sandstones of the Falcon Field – Western Gulf of Mexico. Amer. Assoc. Petrol. Geol. Ann. Mtg. Ext. Abs., April 18–21, Dallas, TX.

Al-Quahtani, M.Y., Ershaghi, I., 1999. Characterization and estimation of permeability correlation structure from performance data. In: R.A. Schatzinger, J.F. Jordan (Eds.), Reservoir Characterization: Recent Advances, Amer. Assoc. Pet. Geol. Memoir 71, pp. 343–358.

Ambrose, W.A., Tyler, N., Parsley, M.J., 1991. Facies heterogeneity, pay-continuity, and infill potential in barrier-island, fluvial, and submarine-fan reservoirs: Examples from the Texas Gulf Coast and Midland Basin. In: A.D. Miall, N. Tyler (Eds.), The Three-Dimensional Facies Architecture of Terrigenous Clastic Sediments and Its Implications for Hydrocarbon Discovery and Recovery, SEPM (Society of Sedimentary Geology) Concepts in Sedimentology and Paleontology 3, pp. 13–21.

Armentrout, J.M., 1991. Paleontologic constraints on depositional modeling: Examples of integration of biostratigraphy and seismic stratigraphy, Pliocene–Pleistocene, Gulf of Mexico. In: P. Weimer and M.H. Link (Eds.), Seismic Facies and Sedimentary Processes of Submarine Fans and Turbidite Systems, Springer-Verlag, New York, pp. 137–170.

Asquith, D.O., 1970. Depositional topography and major marine environments, Late Cretaceous, Wyoming. Amer. Assoc. Petrol. Geol. Bull., 54, 1184–1224.

Asquith, G.B., 1982. Basic Well log Analysis for Geologists. Amer. Assoc. Pet. Geol. Methods In Exploration Series, No. 3, 216 pp.

Atkinson, C.D., McGowen, J.H., Bloch, S., Lundell, L.L., Trumbly, P.N., 1990. Braidplain and deltaic reservoir, Prudhoe Bay Field, Alaska. In: J.H. Barwis, J.G. McPherson, J.R.J. Studlick (Eds.), Sandstone Petroleum Reservoirs, Springer-Verlag, New York, pp. 7–30.

Bahorich, M., Farmer, S., 1995. The coherence cube. The Leading Edge, Oct., 1053–1058.

Balsley, J.K., 1980. Cretaceous wave dominated delta systems, Book Cliffs, east-central Utah. AAPG Continuing Education Course Field Guide.

Bashore, W.M., Araktingi, U.G., Levy, M., Schweller, W.J., 1994. Importance of a geological framework and seismic data integration for reservoir modeling and subsequent fluid-flow predictions. In: J.M. Yarus, R.L. Chambers (Eds.), Stochastic Modeling and Geostatistics, AAPG Computer Applications in Geology 3, pp. 159–176.

Bates, R.L., Jackson, J.A. (Eds.), 1980. Glossary of Geology, 2nd Edition. Amer. Geol. Inst., Virginia, 751 pp.

Beard, D.C., Weyl, P.K., 1973. Influence of texture on porosity and permeability of unconsolidated sand. Amer. Assoc. Petrol. Geol. Bull., 57, 349–369.

Beaubouef, R.T., 2004. Deep-water leveed-channel complexes of the Cerro Toro Formation, Upper Cretaceous, Southern Chile. Amer. Assoc. Petrol. Geol. Bull., 88, 1471–1500.

Beaubouef, R.T., Abreu, V., Van Wagoner, J.C., 2003. Basin 4 of the Brazos–Trinity slope system, western Gulf of Mexico: The terminal portion of a late Pleistocene lowstand systems tract. In:

H.H. Roberts, N.C. Rosen, R.H. Fillon, J.B. Anderson (Eds.), Shelf Margin Deltas and Linked Down Slope Petroleum Systems: Global Significance and Future Exploration Potential, GCS–SEPM Foundation 23rd Annual Bob F. Perkins Research Conference, pp. 182–203.

Beaumont, E.A., 1984. Retrogradational shelf sedimentation in the Lower Cretaceous Viking Formation, central Alberta. In: R.W. Tillman, C.T. Siemers (Eds.), Siliciclastic Shelf Sediments, Soc. Econ. Paleo. and Mineral. (SEPM) Spec. Publ. 34, pp. 163–178.

Bellotti, P., Chiocci, F.L., Milli, S., Tortora, P., Valeri, P., 1994. Sequence stratigraphy and depositional setting of the Tiber Delta: Integration of high-resolution seismic, well logs, and archeological data. J. Sed. Petrol., 64, 416–432.

Berg, R.R., 1975. Depositional environments of Upper Cretaceous Sussex Sandstone, House Creek Field, Wyoming. Amer. Assoc. Petrol. Geol. Bull., 59, 2099–2110.

Berg, R.R., 1986. Reservoir Sandstones. Prentice-Hall, Englewood Cliffs, NJ, 481 pp.

Bergman, K.M., 1994. Shannon Sandstone in Hartzog Draw–Heldt Draw Fields (Cretaceous, Wyoming, USA) reinterpreted as lowstand shoreface deposits. J. Sediment. Res., B64, 184–201.

Bergman, K.M., Snedden, J.W. (Eds.), 1999. Isolated Shallow Marine Sand Bodies: Sequence Stratigraphic Analysis and Sedimentologic Interpretation. Soc. Econ. Paleo. and Mineral. (SEPM) Spec. Publ. 64, 362 pp.

Beydoun, W., Kerdraon, Y., Lefeuvre, F., Lancelin, J.P., 2002. Benefits of a 3D HR survey for Girassol field appraisal and development, Angola. The Leading Edge, 21, 1152–1155.

Bhattacharya, J.P., Giosan, L., 2003. Wave-influenced deltas: Geomorphological implications for facies reconstruction. Sedimentology, 50, 187–210.

Bhattacharya, J.P., Walker, R.W., 1992. Deltas. In: R.G. Walker, N.P. James (Eds.), Facies Models, Geol. Assoc. Canada, pp. 157–177.

Bilinski, P.W., McGee, D.T., Pfeiffer, D.S., Shew, R.D., 1994. Reservoir characterization of the "S" sand, Auger Field, Garden Banks 426, 427, 470, 471. In: R.D. Winn Jr., J.M. Armentrout (Eds.), Turbidites and Associated Deep-Water Facies, SEPM Core Workshop No. 20, pp. 75–93.

Blatt, H., Middleton, G., Murray, R., 1972. Origin Of Sedimentary Rocks. Prentice Hall, Englewood Cliffs, NJ, 634 pp.

Bloch, S., 1991. Empirical prediction of porosity and permeability in sandstones. Amer. Assoc. Petrol. Geol. Bull., 75, 1145–1160.

Blott, J.E., Davis, T.L., Benson, R.D., 1999. Morrow Sandstone reservoir characterization: A 3-D multicomponent seismic success. The Leading Edge, March, 394–397.

Booth, J.R., DuVernay III, A.E., Pfeiffer, D.S., Styzen, M.J., 2000. Sequence stratigraphic framework, depositional models, and stacking patterns of ponded and slope fan systems in the Auger Basin: Central Gulf of Mexico slope. In: P. Weimer, R.M. Slatt, J.L. Coleman, N. Rosen, C.H. Nelson, A.H. Bouma, M. Styzen, D.T. Lawrence (Eds.), Global Deep-Water Reservoirs, GCS–SEPM Foundation 20th Annual Bob F. Perkins Research Conference, pp. 82–103.

Bouma, A.H., 1962. Sedimentology of Some Flysch Deposits: A Graphic Approach to Facies Interpretation. Elsevier, Amsterdam, 168 pp.

Bouma, A.H., 2000. Fine-grained, mud-rich turbidite systems: Model and comparison with coarse-grained, sand-rich systems. In: A.H. Bouma, C.G. Stone (Eds.), Fine-Grained Turbidite Systems, AAPG Memoir 72, SEPM Spec. Publ. 68, pp. 9–19.

Bouma, A.H., Normark, W.R., Barnes, N.E., 1985. COMFAN: Needs and initial results. In: A.H. Bouma, W.R. Normark, N.E. Barnes (Eds.), Submarine Fans and Related Turbidite Systems, Springer-Verlag, New York, pp. 7–11.

Bourke, L., Delfiner, P., Trouiller, J.-C., Fett, T., Grace, M., Luthi, S., Serra, O., Standen, E., 1989. Using formation microscanner images. The Technical Review, 37, 16–40.

Bowen, D.W., Weimer, P., 2003. Regional sequence stratigraphic setting and reservoir geology of Morrow incised-valley sandstones (lower Pennsylvanian), eastern Colorado and western Kansas. Amer. Assoc. Pet. Geol. Bull., 87, 781–815.

Bowen, D.W., Weimer, P., 2004. Reservoir geology of Nicholas and Liverpool Cemetery fields (lower Pennsylvanian), Stanton County, Kansas, and their significance to the regional interpretation of the Morrow Formation incised-valley-fill systems in eastern Colorado and western Kansas. Amer. Assoc. Pet. Geol. Bull., 88, 47–70.

BP, 2000. Statistical Review of World Energy. http://www.bp.com/worldenergy/1999inreview.

Brown, A.R., 1988. Interpretation of Three-Dimensional Seismic Data. Amer. Assoc. Pet. Geol. Memoir 42, 2nd Edition, 253 pp.

Browne, G.H., Slatt, R.M., 2002. Outcrop and behind-outcrop characterization of a late Miocene slope fan system, Mt. Messenger Formation, New Zealand. Amer. Assoc. Pet. Geol. Bull., 86, 841–862.

Carr-Crabaugh, M., Hurley, N.F., Carlson, J., 1996. Interpreting eolian reservoir architecture using borehole images. In: J.A. Pacht, R.E. Sheriff, B.F. Perkins (Eds.), Stratigraphic Analysis: Utilizing Advanced Geophysical, Wireline and Borehole Technology for Petroleum Exploration and Production, GCS–SEPM Foundation 17th Annual Research Conference, pp. 39–50.

Chapin, M.A., Davies, P., Gibson, J.L., Pettingill, H.S., 1994. Reservoir architecture of turbidite sheet sandstones in laterally extensive outcrops, Ross Formation, Western Ireland. In: P. Weimer, A.H. Bouma, B.F. Perkins (Eds.), Submarine Fans and Turbidite Systems, GCS–SEPM Foundation 15th Annual Research Conference, pp. 53–68.

Chapin, M., Terwogt, D., Ketting, J., 2000. From seismic to simulation using new voxel body and geologic modeling techniques, Schiehallion Field, West of Shetlands. The Leading Edge, 19, 408–412.

Ciftci, B.N., Avianatara, A.A., Hurley, N.F., Kerr, D.R., 2004. Outrcop-based three-dimensional modeling of the Tensleep Sandstone at Alkali Creek, Bighorn Basin, Wyoming. In: G.M. Grammer, P.M. Harris, G.P. Eberle (Eds.), Integration of Outcrop and Modern Analogs in Reservoir Modeling, Amer. Assoc. Petrol. Geol. Memoir 80, pp. 235–259.

Clark, J.D., Pickering, K.T., 1996. Submarine Channels: Processes and Architecture. Vallis Press, London, 231 pp.

Clarke, D.D., Phillips, C.C., 2003. Three-dimensional geologic modeling and horizontal drilling bring more oil out of the Wilmington oil field of southern California. In: T.R. Carr, E.P. Mason, C.T. Feazel (Eds.), Horizontal Wells: Focus on the Reservoir, Amer. Assoc. Pet. Geol. Methods in Exploration Series No. 14, pp. 27–48.

Clemenceau, G.R., 1995. Ram/Powell Field, Viosca Knoll 912–956: Deepwater Gulf of Mexico slope fan. In: R.D. Winn Jr., J.M. Armentrout (Eds.), Turbidites and Associated Deep-Water Facies, SEPM Core Workshop No. 20, pp. 55–73.

Clemenceau, G.R., Colbert, J., Edens, D., 2000. Production results from levee-overbank turbidite sands at Ram/Powell Field, deepwater Gulf of Mexico. In: P. Weimer, R.M. Slatt, J.L. Coleman, N. Rosen, C.H. Nelson, A.H. Bouma, M. Styzen, D.T. Lawrence (Eds.), Global Deep-Water Reservoirs, Gulf Coast Section SEPM Foundation Bob F. Perkins 20th Annual Research Conference, pp. 241–251.

Coalson, E.B., Goolsby, S.M., Franklin, M.H., 1994. Subtle seals and fluid-flow barriers in carbonate rocks. In: J.C. Dolson, M.L. Hendricks, W.A. Wescott (Eds.), Unconformity-Related Hydrocarbons in Sedimentary Sequences: Guidebook for Petroleum Exploration and Exploitation in Clastic and Carbonate Sediments, Rocky Mtn. Assoc. Geol., pp. 45–58.

Coates, G.R., Xiao, L., Prammer, M.G., 1999. NMR Logging Principles and Applications. Gulf Publishing Co., Houston, TX, 234 pp.

Coleman, J.M., Roberts, H.H., Murray, S.P., Salama, M., 1981. Morphology and dynamic sedimentology of the eastern Nile shelf. Mar. Geol., 42, 301–326.

Cossey, S.P.J., 1994. Reservoir modeling of deep-water clastic sequences: Mesoscale architectural elements, aspect ratios, and producibility. In: P. Weimer, A.H. Bouma, B.F. Perkins (Eds.), Submarine Fans and Turbidite Systems, GCS – SEPM Foundation 15th Annual Research Conference, pp. 83–93.

Craig, P.A., Bourgeois, T.J., Malik, Z.A., Stroud, T.B., 2003. Planning, evaluation, and performance of horizontal wells at Ram–Powell field, deep-water Gulf of Mexico. In: T.R. Carr, E.P. Mason, C.T. Feazel (Eds.), Horizontal Wells: Focus on the Reservoir, Amer. Assoc. Petrol. Geol. Methods in Exploration Series 14, pp. 95–112.

Davidson, J.P., Reed, W.E., Davis, P.M., 2002. Exploring Earth: An Introduction to Physical Geology, 2nd Edition. Prentice Hall, Englewood Cliffs, NJ, 549 pp.

Deming, D., 2001. Oil: Are we running out? In: M.W. Downey, J.C. Threet, W.A. Morgan (Eds.), Petroleum Provinces of the Twenty-First Century, Amer. Assoc. Pet. Geol. Memoir 74, pp. 45–55.

Diegel, F.A., Karlo, J.F., Schuster, D.C., Shoup, R.C., Tauvers, P.R., 1996. Cenozoic structural evolution and tectonostratigraphic framework of the northern Gulf Coast continental margin. In: M.P.A. Jackson, D.G. Roberts, S. Snelson (Eds.), Salt Tectonics, a Global Perspective, Amer. Assoc. Pet. Geol. Memoir 65, pp. 109–151.

Diridoni, J.L., 1996. Sequence stratigraphic framework of the Miocene Mt. Messenger Formation deep-water clastics, North Taranaki Basin, New Zealand. Unpubl. M.Sc. thesis, Colorado School of Mines, Golden, CO, 165 pp.

Dorn, G.A., Tubman, K.M., Cooke, D., O'Connor, R., 1996. Geophysical reservoir characterization of Pickerill Field, North Sea, using 3-D seismic and well data. In: P. Weimer, T.L. Davis (Eds.), Applications of 3-D Seismic Data To Exploration and Production, Amer. Assoc. Petrol. Geol. Studies in Geology No. 42, pp. 107–121.

Durham, L.S., 2001. The Sneider focus: He knows what he's looking for. Amer. Assoc. Pet. Geol. Explorer, May, 28.

Durham, L.S., 2003. The future looks to be gas fired. Amer. Assoc. Pet. Geol. Explorer, February, 12–14.

Ebanks, W.J. Jr., Scheihing, M.H., Atkinson, C.D., 1992. Flow units for reservoir characterization. In: D. Morton-Thompson, A.M. Woods (Eds.), Development Geology Reference Manual, Amer. Assoc. Petrol. Geol. Methods in Exploration Series No. 10, pp. 282–284.

Edwards, J.D., 2001. Twenty-first century energy: Decline of fossil fuel increase of renewable nonpolluting energy sources. In: M.W. Downey, J.C. Threet, W.A. Morgan (Eds.), Petroleum Provinces of the Twenty-First Century, Amer. Assoc. Pet. Geol. Memoir 74, pp. 21–34.

Ellis, D., 1993. The Rough gas field: Distribution of Permian aeolian and non-aeolian reservoir facies and their impact on field development. In: C.P. North, D.J. Prosser (Eds.), Characterization of Fluvial and Aeolian Reservoirs, Geological Society of London Spec. Publ. 73, pp. 265–278.

Emery, D., Myers, K.J., 1996. Sequence Stratigraphy. Blackwell, Oxford, England, 297 pp.

Favennec, J.P., 2002. Recherche et production du petrole et du gaz-reserves. Couts, contrats, Institut Francais du Petrole, Editions Technip.

Fisher, W.L., Brown, L.F. Jr, 1984. Clastic Depositional Systems – A Genetic Approach to Facies Analysis: Annotated Outline and Bibliography, reprinted and revised. Texas Bureau of Economic Geology, 105 pp.

Fisher, W.L., Brown, L.F., Scott, A.J., McGowen, J.H. (Eds.), 1969. Delta Systems in Exploration for Oil and Gas. Texas Bur. Econ. Geology, 92 pp.

Fisher, W.R., 1991. Future supply potential of U.S. oil and natural gas. The Leading Edge, 11, 15–21.

Fisk, H.N., 1961. Bar finger sands of the Mississippi delta. In: J.A. Peterson, J.C. Osmond (Eds.), Geometry of Sandstone Bodies, Amer. Assoc. Petrol. Geol. Bull., AO55 (Spec. volume), 29–52.

Folk, R.L., 1968. Petrology of Sedimentary Rocks. Hemphill's Book Store, Austin, TX, 170 pp.

Frasier, D.E., 1974. Depositional episodes: Their relationship to the quaternary stratigraphic framework in the northwestern portion of the Gulf Basin. Univ. Texas Bur. Econ. Geol., Geol. Circular 4, 28 pp.

Friedman, G.M., Sanders, J.E., 1978. Principles of Sedimentology. Wiley, New York, 792 pp.

Fugitt, D.S., Herricks, G.J., Wise, M.R., Stelting, C.E., Schweller, W.J., 2000. Production characteristics of sheet and channelized turbidite reservoirs, Garden Banks 191, Gulf of Mexico, U.S.A. In: P. Weimer, R.M. Slatt, J.L. Coleman, N. Rosen, C.H. Nelson, A.H. Bouma, M. Styzen, D.T. Lawrence (Eds.), Global Deep-Water Reservoirs, GCS–SEPM Foundation 20th Annual Bob F. Perkins Research Conference, pp. 389–401.

Galloway, W.E., 1968. Depositional systems of the Lower Wilcox Group, north-central Gulf Coast Basin. Gulf Coast Assoc. Geol. Soc. Trans., 18, 275–289.

Galloway, W.E., 1989. Genetic stratigraphic sequences in basin analysis: Architecture and genesis of flooding surface bounded depositional units. Amer. Assoc. Petrol. Geol. Bull., 73, 125–142.

Galloway, W.E., Hobday, D.K., 1983. Terrigenous Clastic Depositional Systems. Application to Petroleum, Coal, and Uranium Exploration. Springer-Verlag, New York.

Galloway, W.E., Hobday, D.K., 1996. Terrigenous Clastic Depositional Systems, Springer-Verlag, Heidelberg, 489 pp.

Galloway, W.E., Hobday, D.K., Magara, K., 1982. Frio Formation of the Texas Gulf Coast Basin: depositional systems, structural framework, and hydrocarbon origin, migration, distribution, and exploration potential. The University of Texas at Austin, Bur. Econ. Geology Rept., Invest. No. 122, 78 pp.

Gardner, M.H., Borer, J.M., 2000. Submarine channel architecture along a slope to basin profile, Brushy Canyon Formation, West Texas. In: A.H. Bouma, C.G. Stone (Eds.), Fine-Grained Turbidite Systems, AAPG Memoir 72, SEPM Spec. Publ. 68, pp. 195–214.

Garich, A.M., 2004. Porosity types and relation to deepwater sedimentary facies of subsurface Jackfork Group sandstones, Latimer and LeFlore Counties, Oklahoma. Unpubl. M.S. thesis, Univ. Oklahoma, 94 pp.

Gastescu, P., 1992. Danube Delta – Tourist Map. Editura Sport–Turism, Bucuresti.

Gaynor, G.C., Scheihing, M.H., 1988. Shelf depositional environments and reservoir characteristics of the Kuparuk River Formation (Lower Cretaceous), Kuparuk Field, North Slope, Alaska. In: Giant Oil and Gas Fields: A Core Workshop, Soc. Econ. Paleo. and Mineral. (SEPM) Core Workshop No. 12, pp. 333–390.

Gaynor, G.C., Swift, D.J.P., 1988. Shannon Sandstone depositional model: Sand ridge dynamics on the Campanian western interior shelf. J. Sed. Petrol., 58, 868–880.

Gratton, P.J.F., 2004. Changing along with the times: President's column. Amer. Assoc. Pet. Geol. Explorer, July, 3.

Greaves, R.J., Fulp, T.J., 1988. Three-dimensional seismic monitoring of an enhanced oil recovery process. In: A.R. Brown (Ed.), Interpretation of Three-Dimensional Seismic Data, AAPG Memoir 42, 2nd Edition, pp. 198–211.

Green, C., Slatt, R.M., 1992. Complex braided stream depositional model for the Murdoch Field, Block 44/22 U.K. southern North Sea, braided rivers: Form, process and economic applications. Abstract, The Geological Society, London, May 6–7.

Grier, S.P., Marschall, D.M., 1992. Reservoir quality. In: D. Morton-Thompson, A.M. Woods (Eds.), Development Geology Reference Manual, Amer. Assoc. Petrol. Geol. Methods in Exploration Series No. 10, pp. 275–277.

Gunter, G.W., Finneran, J.M., Hartmann, D.J., Miller, J.D., 1997. Early determination of reservoir flow units using an integrated petrophysical method. In: Proc. Soc. Petrol. Engn., Ann. Tech. Conf. and Exhibit., No. SPE-38679, pp. 373–380.

Halderson, H.H., Damsleth, E., 1993. Challenges in reservoir characterization. Amer. Assoc. Pet. Geol. Bull., 77, 541–551.

Hamilton, D.S., Tyler, N., Tyler, R., Raeuchle, S.K., Holtz, M.H., Yeh, J., Uzcategui, M., Jimenez, T., Salazar, A., Cova, C.E., Barbato, R., Rusic, A., 2002. Reactivation of mature oil fields through advanced reservoir characterization: A case history of the Budare field, Venezuela. Amer. Assoc. Petrol. Geol. Bull., 86, 7, 1237–1262.

Handford, C.R., Loucks, R.G., 1993. Carbonate depositional sequence and systems tracts – Responses of carbonate platforms to relative sea-level changes. In: R.G. Loucks, J.F. Sarg (Eds.), Carbonate Sequence Stratigraphy: Recent Developments and Applications, Amer. Assoc. Pet. Geol. Memoir 57, pp. 3–42.

Haq, B.U., Hardenbol, J., Vail, P.R., 1987. Chronology of fluctuating sea levels since the Triassic (250 million years ago to present). Science, 235, 1156–1167.

Hardage, R.A., Levey, R.A., Pendleton, V., Simmons, J., Edson, R., 1996. 3-D seismic imaging and interpretation of fluvially deposited thin-bed reservoirs. In: P. Weimer, T.L. Davis (Eds.), Applications of 3-D Seismic Data to Exploration and Production, AAPG Studies in Geology No. 42/SEG, Geophysical Developments Series No. 5, pp. 27–34.

Hart, B.S., Plint, A.G., 1994. Tectonic influence on deposition and erosion in a ramp setting: Upper Cretaceous Cardium Formation, Alberta Foreland Basin. Amer. Assoc. Petrol. Geol. Bull., 77, 2092–2107.

Hartanto, K., Widianto, E., Safrizal, S., 1991. Hydrocarbon prospect related to the local unconformities of the Kuang area, South Sumatra Basin. In: Proceedings of the Indonesian Petroleum Association, 20th Annual Convention, Jakarta, pp. 17–35.

He, W., Anderson, R.N., Xu, L., Boulanger, A., Meadow, B., Neal, R., 1996. 4D seismic monitoring grows as production tool. Oil & Gas Journal, May 20, 41–46.

Heezen, B.C., 1956. Corrientes de turbidez del Rio Magdalena. Boletin de la Sociedad Geografica de Colombia, 51–52, 135–143.

Heezen, B.C., Ewing, M.H., 1952. Turbidity currents and submarine slumps, and the 1929 Grand Banks earthquake. Amer. J. Sci., 250, 849–873.

Heezen, B.C., Menzies, R.J., Schneider, E.D., Ewing, M.H., Grainelli, N.C.L., 1964. Congo submarine canyon. Amer. Assoc. Petrol. Geol. Bull., 48, 1126–1149.

Heymans, M.J., 1998. Evaluating reservoir compartmentalization by correlating laboratory and field data. In: R.M. Slatt (Ed.), Compartmentalized Reservoirs in Rocky Mountain Basins, Rocky Mtn. Assoc. Geol., pp. 207–218.

Hobson, J.P., Fowler, M.L., Beaumont, E.A., 1982. Depositional and statistical exploration models, Upper Cretaceous offshore sandstone complex, Sussex Member, House Creek field, Wyoming. Amer. Assoc. Petrol. Geol. Bull., 66, 689–707.

Holman, W.E., Robertson, S.S., 1994. Field development, depositional model, and production performance of the turbiditic 'J' sands at prospect Bullwinkle, Green Canyon 65 Field, outer shelf, Gulf of Mexico. In: P. Weimer, A.H. Bouma, B.F. Perkins (Eds.), Submarine Fans and Turbidite Systems, GCS–SEPM Foundation 15th Annual Research Conference, pp. 139–150.

Hunter, R.E., 1977. Basic types of stratification in small eolian dunes. Sedimentology, 24, 361–387.

Hurley, N.F., 1994. Recognition of faults, unconformities, and sequence boundaries using cumulative dip plots. AAPG Bull., 78, 1173–1185.

Hurley, N.F., Aviantara, A.A., Kerr, D.R., 2003. Structural and stratigraphic compartments in a horizontal well drilled in the eolian Tensleep Sandstone, Byron Field, Wyoming. In: T.R. Carr,

H.H. Roberts, B. Van Hoorn (Eds.), Petroleum Systems of Deepwater Basins: Global and Gulf of Mexico Experience, GCS–SEPM Foundation 21st Annual Bob F. Perkins Research Conference, pp. 1–22.

Pfeiffer, D.S., Mitchell, B.T., Yevi, G.Y., 2000. Mensa: Shell's Mississippi Canyon Block 731 Field – An integrated field study. In: P. Weimer, R.M. Slatt, J.L. Coleman, N. Rosen, C.H. Nelson, A.H. Bouma, M. Styzen, D.T. Lawrence (Eds.), Global Deep-Water Reservoirs, GCS–SEPM Foundation 20th Annual Bob F. Perkins Research Conference, pp. 756–775.

Phillips, S., 1987. Dipmeter interpretation of turbidite-channel reservoir sandstones, Indian Draw Field, New Mexico. In: R.W. Tillman, K.J. Weber (Eds.), Reservoir Sedimentology, SEPM Spec. Publ. 40, pp. 113–128.

Piper, D.J.W., Normark, W.R., 1983. Turbidite depositional patterns and flow characteristics, Navy Submarine Fan, California Borderland. Sedimentology, 30, 681–694.

Piper, D.J.W., Normark, W.R., 2001. Sandy Fans – From Amazon to Hueneme and beyond. Amer. Assoc. Petrol. Geol. Bull., 85, 1407–1438.

Pirmez, C., Beaubouef, R.T., Friedmann, S.J., Mohrig, D.C., 2000. Equilibrium profile and base level in submarine channels: Examples from Late Pleistocene systems and implications for the architecture of deepwater reservoirs. In: P. Weimer, R.M. Slatt, A.H. Bouma, D.T. Lawrence, J. Coleman Jr., M. Styzen, H. Nelson (Eds.), Deepwater Reservoirs of the World, GCS–SEPM Foundation 20th Annual Bob F. Perkins Research Conference, Houston, Dec. 3–6, pp. 782–805.

Pittman, E.D., 1992. Relationship of porosity and permeability to various parameters derived from mercury injection-capillary pressure curves for sandstone. Amer. Assoc. Petrol. Geol. Bull., 76, 191–198.

Porter, K.W., Weimer, R.J., 1982. Diagenetic sequence related to structural history and petroleum accumulation: Spindle Field, Colorado. Amer. Assoc. Petrol. Geol. Bull., 66, 2543–2560.

Posamentier, H.W., 2001. Lowstand alluvial bypass systems: Incised vs. unincised. Amer. Assoc. Pet. Geol. Bull., 85, 1771–1793.

Posamentier, H.W., Allen, G.P., 1999. Siliciclastic sequence stratigraphy – concepts and applications. Soc. Econ. Paleo. and Mineral. (SEPM) Concepts in Sedimentology and Paleontology 7, 210 pp.

Posamentier, H.W., Allen, G.P., James, D.P., Tesson, M., 1992. Forced regressions in a sequence stratigraphic framework: Concepts, examples, and exploration significance. Amer. Assoc. Petrol. Geol. Bull., 76, 1687–1709.

Posamentier, H.W., Erskine, R.D., Mitchum, R.M. Jr., 1991. Models for submarine-fan deposits within a sequence-stratigraphic framework. In: P. Weimer, M.H. Link (Eds.), Seismic Facies and Sedimentary Processes of Submarine Fans and Turbidite Systems, Frontiers in Sedimentary Geology, Springer-Verlag, New York, pp. 127–136.

Posamentier, H.W., Weimer, P., 1993. Siliciclastic sequence stratigraphy and petroleum geology – Where to from here? Amer. Assoc. Petrol. Geol. Bull., 77, 731–742.

Potter, P.E., Scheidegger, A.E., 1966. Bed thickness and grain size: Graded beds. Sedimentology, 7, 233–240.

Prosser, D.J., Maskall, R., 1993. Permeability variation within Aeolian sandstones: A case study using core cut sub-parallel to slipface bedding, The Auk Field, central North Sea. In: C.P. North, D.J. Prosser (Eds.), Characterization of Fluvial and Aeolian Reservoirs, Geological Society of London Special Publ. 73, pp. 377–398.

Pyles, D.R., 2000. A high-frequency sequence stratigraphic framework for the Lewis Shale and Fox Hills Sandstone, Great Divide and Washakie Basins, Wyoming. Unpubl. M.S. thesis, Colorado School of Mines, Golden, CO, 212 pp.

Pyles, D.R., Slatt, R.M., 2000. A high-frequency sequence stratigraphic framework for shallow through deep-water deposits of the Lewis Shale and Fox Hills Sandstone, Great Divide and

Washakie Basins, Wyoming. In: P. Weimer, R.M. Slatt, J.L. Coleman, N. Rosen, C.H. Nelson, A.H. Bouma, M. Styzen, D.T. Lawrence (Eds.), Global Deep-Water Reservoirs, Gulf Coast Section – SEPM Foundation Bob F. Perkins 20th Annual Research Conference, pp. 836–857.

Pyles, D.R., Slatt, R.M. Stratigraphic evolution of the Upper Cretaceous Lewis Shale, southern Wyoming: Applications to understanding shelf to base-of-slope changes in stratigraphic architecture of mud-dominated, progradational depositional systems. In: T. Nelsen, R. Shew, G. Steffens, J. Studlick, Atlas of Deepwater Outcrops, Amer. Assoc. Petrol. Geol. Studies in Geology, vol. 56, in press.

Reading, H.G., 1986. Sedimentary Environments and Facies. Blackwell Scientific, 117 pp.

Reading, H.G., Richards, M., 1994. Turbidite systems in deep-water basin margins classified by grain size and feeder system. Amer. Assoc. Petrol. Geol. Bull., 78, 792–822.

Reedy, G.K., Pepper, C.F., 1996. Analysis of finely laminated deep marine turbidites: Integration of core and log data yields a novel interpretation model. In: Proc. Soc. Petrol. Engn. (SPE) Ann. Tech. Conf. and Exhibit, CO, pp. 119–127.

Richards, M., Bowman, M., Reading, H., 1998. Submarine-fan systems I: Characterization and stratigraphic prediction. Mar. Petrol. Geol., 15, 687–717.

Roberts, M.T., Compani, B., 1996. Miocene example of a meandering submarine channel-levee system from 3-D seismic reflection data, Gulf of Mexico Basin. In: J.A. Pacht, R.E. Sheriff, B.F. Perkins (Eds.), Stratigraphic Analysis Utilizing Advanced Geophysical, Wireline and Borehole Technology for Petroleum Exploration and Production, GCS–SEPM 17th Annual Research Conference, Houston, pp. 241–254.

Robinson, J.W., McCabe, P.J., 1997. Sandstone-body and shale-body dimensions in a braided fluvial system: Salt Wash Sandstone Member (Morrison Formation), Garfield County, Utah. Amer. Assoc. Pet. Geol. Bull., 81, 1267–1291.

Romero, G., 2004. Stratigraphy and composition of turbidite deposits, Jackfork Group, Lynn Mountain syncline, Pushmataha and LeFlore Counties, Oklahoma. M.S. thesis, University of Oklahoma, 238 pp.

Ross, J.G., 1997. The Philosophy of Reserve Estimation. SPE Paper 37960.

Rossen, C., Sickafoose, D.K., 1994. 3-D seismic expression and architecture of deep-water reservoirs at Ram/Powell field, Viosca Knoll Block 912, Gulf of Mexico. In: P. Weimer, A.H. Bouma, B.F. Perkins (Eds.), Submarine Fans and Turbidite Systems, GCS–SEPM Foundation 15th Annual Research Conference, pp. 309–310.

Saifuddin, F., Soeryowibowo, M., Suta, I.N., Chandra, B., 2001. Acoustic impedance as a tool to identify reservoir targets: A case of the NE Betara-11 Horizontal well, Jabung block, South Sumatra. In: Proceedings of the Indonesian Petroleum Association, 28th Annual Convention, Jakarta, poster exhibition.

Saller, A.H., Noah, J.T., Schneider, R., Ruzuar, A.P., 2003. Lowstand deltas and a basin-floor fan, Pleistocene, offshore east Kalimantan, Indonesia. In: H.H. Roberts, N.C. Rosen, R.H. Fillon, J.B. Anderson (Eds.), GCS–SEPM Bob F. Perkins 23rd Annual Research Conference, pp. 421–440.

Sanchez, M.E.N., 2003. Integrating seismic inversion, spectral decomposition and other seismic attributes in the Furrial Area, Venezuela and Stratton Field, Texas. Unpubl. M.S. thesis, Univ. Oklahoma, 134 pp.

Sanchez, M.E., 2004. Integrating seismic inversion, spectral decomposition and other seismic attributes in the Furrial area, Venezuela and Stratton Field, Texas. Unpubl. M.S. thesis, Univ. Oklahoma, 134 pp.

Sarg, J.F., 1988. Carbonate sequence stratigraphy. SEPM Special Publication No. 42, 155–181.

Scruton, P.C., 1960. Delta building and the deltaic sequence. In: F.P. Shepard, F.B. Phleger, T.H. Van Andel (Eds.), Recent Sediments, Northwest Gulf of Mexico, Amer. Assoc. Petrol. Geol., Tulsa, 394 pp.

Shaffer, B.L., 1990. The nature and significance of condensed sections in the Gulf Coast late Neogene sequence stratigraphy. In: G. Kinsland, T. Cagle (Eds.), Transactions 40th Ann. Meet. Gulf Coast Assoc. Geological Societies, pp. 767–776.

Shew, R.D., 1997. Deepwater Gulf of Mexico Core Workshop. Short Course #1: GCAGS Convention, New Orleans, LA.

Shew, R.D., Rollins, D.R., Tiller, G.M., Hackbarth, C.J., White, C.D., 1994. Characterization and modeling of thin-bedded turbidite deposits from Gulf of Mexico using detailed subsurface and analog data. In: P. Weimer, A.H. Bouma, B.F. Perkins (Eds.), Submarine Fans and Turbidite Systems, GCS–SEPM Foundation 15th Annual Research Conference, pp. 327–334.

Shiralkar, G.S., Volz, R.F., Stephenson, R.E., Valle, M.J., Hird, K.B., 1996. Parallel computing alters approaches, raises integration challenges in reservoir modeling. Oil & Gas Journal, May 20, 48–56.

Shirley, M.L. (Ed.), 1966. Deltas and Their Geologic Framework. Houston Geol. Soc., 252 pp.

Siemers, C.T., Ristow, J.H., 1986. Marine-shelf bar sand/channelized sand shingled couplet, Terry Sandstone member of Pierre Shale, Denver Basin, Colorado. In: T.F. Moslow, E.G. Rhodes (Eds.), Modern and Ancient Shelf Clastics: A Core Workshop, Soc. Econ. Paleo. and Mineral. (SEPM) Core Workshop No. 9, Atlanta, June 15, pp. 269–324.

Sippel, M.A., 1996. Integration of 3-D seismic to define functional reservoir compartments and improve waterflood recovery in a Cretaceous reservoir, Denver Basin. In: S. Longacre, B. Katz, R. Slatt, M. Bowman (Eds.), AAPG/EAGE Research Symposium, Compartmentalized Reservoirs: Their Detection, Characterization, and Management, The Woodlands, TX.

Slatt, R.M., 1984. Continental shelf topography: Key to understanding the distribution of shelf sand ridge deposits from the Cretaceous Western Interior Seaway. Amer. Assoc. Petrol. Geol. Bull., 68, 1107–1120.

Slatt, R.M., 1997. Sequence stratigraphy, sedimentology and reservoir characteristics of the Upper Cretaceous Terry sandstone, Hambert-Aristocrat Field, Denver Basin, Colorado. In: K.W. Shanley, B.F. Perkins (Eds.), Shallow Marine and Nonmarine Reservoirs, GCS–SEPM Foundation 18th Annual Research Conference, pp. 289–302.

Slatt, R.M., 1998. Foreword: Compartmentalized reservoirs – The exception or the rule? In: R.M. Slatt (Ed.), Compartmentalized Reservoirs in Rocky Mountain Basins, Rocky Mtn. Assoc. Geol., pp. v–vii.

Slatt, R.M., Browne, G.N., Davis, R.J., Clemenceau, G.R., Colbert, J.R., Young, R.A., Anxionna, N., Spang, R.J., 1998. Outcrop–behind outcrop characterization of thin bedded turbidites for improved understanding of analog reservoirs: New Zealand and Gulf of Mexico. Soc. Petrol. Engn. Ann. Mtg., New Orleans, SPE Paper 49563, 845–853.

Slatt, R.M., Hopkins, G.L., 1991. Scaling geologic reservoir description to engineering needs. J. Petrol. Tech., 202–210.

Slatt, R.M., Jordan, D.W., Davis, R.J., 1994. Interpreting formation microscanner log images of Gulf of Mexico Pliocene turbidites by comparison with Pennsylvanian turbidite outcrops, Arkansas. In: P. Weimer, A.H. Bouma, B.F. Perkins (Eds.), Submarine Fans and Turbidite Systems, GCS–SEPM Foundation 15th Annual Research Conference, pp. 335–348.

Slatt, R.M., Mark, S., 2004. Geologic knowledge key to reservoir characterization. Amer. Oil and Gas Reporter, 111–113.

Slatt, R.M., Omatsola, B., Garich-Faust, A.M., Romero, G.A. Potential stratigraphic reservoirs in the Jackfork Group, southeastern Oklahoma. In: T. Nelsen, R. Shew, G. Steffens, J. Studlick, Atlas of Deepwater Outcrops, Amer. Assoc. Petrol. Geol. Studies in Geology, vol. 56, in press.

Slatt, R.M., Phillips, S., Boak, J.M., Lagoe, M.B., 1993. Scales of geologic heterogeneity of a deep water sand giant oil field, Long Beach Unit, Wilmington field, California. In: E.G. Rhodes, T.F. Moslow (Eds.), Frontiers in Sedimentary Geology, Marine Clastic Reservoirs, Examples and Analogs, Springer-Verlag, New York, pp. 263–292.

Slatt, R.M., Stone, C.G., Weimer, P., 2000. Characterization of slope and basin facies tracts, Lower Pennsylvanian Jackfork Group, Arkansas, with applications to deepwater (turbidite) reservoir management. In: P. Weimer, R.M. Slatt, J.L. Coleman, N. Rosen, C.H. Nelson, A.H. Bouma, M. Styzen, D.T. Lawrence (Eds.), Global Deep-Water Reservoirs, GCS-SEPM Foundation 20th Annual Bob F. Perkins Research Conference, pp. 940–980.

Slatt, R.M., Thomasson, M.R., Romig Jr., P.R., Pasternack, E.S., Boulanger, A., Anderson, R.N., Nelson Jr., H.R., 1996. Visualization technology for the oil and gas industry: Today and tomorrow. Amer. Assoc. Pet. Geol. Bull., 80, 453–459.

Slatt R.M., Weimer, P., 1999. Petroleum geology of turbidite depositional systems: Part II, Subseismic scale reservoir characteristics. The Leading Edge, May, 562–567.

Sloss, L.L., 1963. Sequences in the cratonic interior of North America. Geol. Soc. Amer. Bull., 74, 93–114.

Snedden, J.W., Bergman, K.M., 1999. Isolated shallow marine sand bodies: Deposits for all interpretations. In: K.M. Bergman, J.W. Snedden (Eds.), Isolated Shallow Marine Sand Bodies: Sequence Stratigraphic Analysis and Sedimentologic Interpretation, Soc. Econ. Paleo. and Mineral. (SEPM) Spec. Publ. 64, pp.1–11.

Sneider, R.M., 1987. Practical Petrophysics for Exploration and Development. Amer. Assoc. Petrol. Geol. Short Course Lect. Notes, variously paginated.

Sneider, R., 1999. Teams usually win competition. Amer. Assoc. Pet. Geol. Explorer, December, 24–27.

Sprague, A.R., Sullivan, M.D., Campion, K.M., Jensen, G.N., Goulding, F.J., Sickafoose, D.K., Jennette, D.C., 2002. The physical stratigraphy of deep-water stratal a hierarchical approach to the analysis of genetically related stratigraphic elements for improved reservoir prediction. Amer. Assoc. Petrol. Geol. Ann. Convention Abstracts, Houston, TX, pp. 10–13.

Srivastava, R.M., 1994. An overview of stochastic methods for reservoir characterization. In: J.M. Yarus, R.L. Chambers (Eds.), Stochastic Modeling and Geostatistics, Amer. Assoc. Pet. Geol. Computer Applications in Geology no. 3, pp. 3–16.

Stelting, C.E., Bouma, A.H., Stone, C.G., 2000. Fine-grained turbidite systems: Overview. In: A.H. Bouma, C.G. Stone (Eds.), Fine-Grained Turbidite Systems, Amer. Assoc. Petrol. Geol. Memoir 72, SEPM Spec. Publ. 68, pp. 1–8.

Stelting, C.E., Pickering, K.T., Bouma, A.H., Coleman, J.M., Cremer, M., Droz, L., Meyer-Wright, A.A., Normark, W.R., O'Connell, S., Stow, D.A.V., 1985. DSDP Leg 96 Shipboard Scientists, Drilling results on the middle Mississippi Fan. In: A.H. Bouma, W.R. Normark, N.E. Barnes (Eds.), Submarine Fans and Related Turbidite Systems, Springer-Verlag, New York, pp. 275–282.

Stephen, K.D., Clark, J.D., Gardiner, A.R., 2001. Outcrop-based stochastic modeling of turbidite amalgamation and its effects on hydrocarbon recovery. Petroleum Geoscience, 7, 163–172.

Sujanto, F.X., 1997. Substantial contribution of petroleum systems to increase exploration success in Indonesia. In: Proceedings of an International Conference on Petroleum Systems of SE Asia & Australia, Indonesian Petroleum Association, Jakarta, pp. 1–14.

Sullivan, M., Jensen, G., Goulding, F., Jennette, D., Foreman, L., Stern, D., 2000. Architectural analysis of deep-water outcrops: Implications for exploration and production of the Diana sub-basin, western Gulf of Mexico. In: P. Weimer, R.M. Slatt, J.L. Coleman, N. Rosen, C.H. Nelson, A.H. Bouma, M. Styzen, D.T. Lawrence (Eds.), Global Deep-Water Reservoirs, GCS–SEPM Foundation 20th Annual Bob F. Perkins Research Conference, pp. 1010–1031.

Sullivan, M.D., Van Wagoner, J.C., Jennette, D.C., Foster, M.E., Stuart, R.M., Lovell, R.W., Pemberton, S.G., 1997. High resolution sequence stratigraphy and architecture of the Shannon Sandstone, Hartzog Draw Field, Wyoming: Implications for reservoir management. In: K.W. Shanley, B.F. Perkins (Eds.), Shallow Marine and Nonmarine Reservoirs: Sequence Stratigraphy, Reservoir Architecture and Production Characteristics, Gulf Coast Sec.

Soc. Econ. Paleon. and Mineral. (SEPM) Foundation 18th Annual Res. Conf., Houston, pp. 331–344.

Suta, I.N., 2004. Reservoir characterization of a lower Talang Akar reservoir, Northeast (NE) Betara Field, Jabung Sub-Basin, South Sumatra, Indonesia. M.Sc. thesis, Univ. of Oklahoma, 74 pp.

Suta, I.N., desAutels, D., Gresko, M., 2000. NE Betara field, South Sumatra, Indonesia: Stratigraphic architecture of a lower Talang Akar reservoir. Presentation at the Annual Amer. Assoc. Petrol. Geol. International Convention, Bali.

Suta, I.N., Xiaoguang, L., 2005. Complex stratigraphic and structural evolution of Jabung Sub-basin and its hydrocarbon accumulation; Case study from Lower Talang Akar Reservoir, South Sumatra Basin, Indonesia. In: Proceedings of International Petroleum Technology Conference, Doha – Qatar, IPTC 10094.

Taylor, G., 1995. Seeing inspires believing: 'Coherence cube' seeks a fuller view of seismic. AAPG Explorer, September, 1 and 18.

Tillman, L.E., 1989. Sedimentary facies and reservoir characteristics of the Nugget Sandstone (Jurassic), Painter Reservoir Field, Uinta County, Wyoming. In: E.B. Coalson, S.S. Kaplan, C.W. Keighin, C.A. Oglesby, J.W. Robinson (Eds.), Petrogenesis and Petrophysics of Selected Sandstone Reservoir of the Rocky Mountain Region, Rocky Mtn. Assoc. Geologists, pp. 97–108.

Tillman, R.W., Martinsen, R.S., 1984. The Shannon shelf-ridge sandstone complex, Salt Creek Anticline area, Powder River Basin, Wyoming. In: R.W. Tillman, C.T. Siemers (Eds.), Siliciclastic Shelf Sedimentation, Soc. Econ. Paleo. and Mineral. (SEPM) Spec. Publ. 34, pp. 85–142.

Tillman, R.W., Martinsen, R.S., 1987. Sedimentologic model and production characteristics of Hartzog Draw Field, Wyoming: A Shannon Sandstone shelf-ridge sandstone. In: R.W. Tillman, K.J. Weber (Eds.), Reservoir Sedimentology, Soc. Econ. Paleo. and Mineral. (SEPM) Spec. Publ. 40, pp. 15–112.

Tillman, R.W., Pittman, E.D., 1993. Reservoir heterogeneity in valley-fill sandstone reservoirs, southwest Stockholm Field, Kansas. In: W. Linville (Ed.), Reservoir Characterization III, Proc. 3rd Inter. Reservoir Char. Tech. Conf., Tulsa, pp. 51–105.

Tillman, R.W., Siemers, C.T. (Eds.), 1984. Siliciclastic Shelf Sedimentation. Soc. Econ. Paleo. and Mineral. (SEPM) Spec. Publ. 34, 268 pp.

Turner, J.R., Conger, S.J., 1981. Environment of deposition and reservoir properties of the Woodbine Sandstone at Kurten Field, Brazos County Texas. Gulf Coast Assoc. Geological Societies Transactions, 31, 213–232.

Tye, R.S., Bhattacharya, J.P., Lorsong, J.A., Sindelar, S.T., Knock, D.G., Puls, D.D., Levinson, R.A., 1999. Geology and stratigraphy of fluvio–deltaic deposits in the Ivishak Formation: Applications for development of Prudhoe Bay Field, Alaska. Amer. Assoc. Pet. Geol. Bull., 83, 1588–1623.

Tye, R.S., Hickey, J.J., 2001. Permeability characterization of distributary mouth bar sandstones in Prudhoe Bay field, Alaska: How horizontal cores reduce risk in developing deltaic reservoirs. Amer. Assoc. Petrol. Geol. Bull., 85, 459–475.

US Geological Survey, 2000. World Petroleum Assessment. US Geological Survey Digital Data Series No. 60.

Vail, P.R., 1987. Seismic stratigraphy interpretation using sequence stratigraphy. Part 1: Seismic stratigraphy interpretation procedure. In: A.W. Bally (Ed.), Atlas of Seismic Stratigraphy, Amer. Assoc. Petrol. Geol. Studies in Geology 27, 1, pp. 1–10.

Vail, P.R., Mitchum, R.M. Jr., Thompson III, S., 1977. Seismic stratigraphy and global changes of sea level, Part 3: Relative changes of sea level from coastal onlap. In: C.E. Payton (Ed.),

Seismic Stratigraphy – Applications to Hydrocarbon Exploration, Amer. Assoc. Petrol. Geol. Memoir 26, pp. 63–81.

Van Wagoner, J.C., Bertram, G.T. (Eds.), 1995. Sequence Stratigraphy of Foreland Basin Deposits. Amer. Assoc. Petrol. Geol. Memoir 64, 490 pp.

Van Wagoner, J.C., Mitchum, R.M., Campion, K.M., Rahmanian, V.D., 1990. Siliciclastic Sequence Stratigraphy in Well Logs, Cores, and Outcrops, Amer. Assoc. Petrol. Geol. Methods in Exploration Series 7, 55 pp.

Van Wagoner, J.C., Posamentier, H.W., Mitchum, R.M., Vail, P.R., Sarg, J.F., Loutit, T.S., Hardenbol, J., 1988. An overview of sequence stratigraphy and key definitions. In: C.K. Wilgus, B.S. Hastings, C.G.St.C. Kendall, H. Posamentier, C.A. Ross, J.C. Van Wagoner (Eds.), Sea-Level Changes – An Integrated Approach, Soc. Econ. Paleo. and Mineral. (SEPM) Spec. Publ. 42, pp. 39–45.

Vavra, C.L., Kaldi, J.G., Sneider, R.M., 1992. Geologic applications of capillary pressure: A review. Amer. Assoc. Petrol. Geol. Bull., 76, 840–850.

Walker, R.G., 1978. Deep-water sandstone facies and ancient submarine fans: Models for exploration for stratigraphic traps. Amer. Assoc. Petrol. Geol. Bull., 62, 932–966.

Walker, R.G., 1980. Facies Models. Geoscience Canada Reprint Series 1, 211 pp.

Walker, R.G., 1992. Facies, facies models, and modern stratigraphic concepts. In: R.G. Walker, N.P. James (Eds.), Facies Models: Response to Sea Level Change, Geol. Assoc. Can., pp. 1–14.

Walker, R.G., Bergman, K.M., 1993. Shannon Sandstone in Wyoming: A shelf-ridge complex reinterpreted as lowstand shoreface deposits. J. Sed. Petrol., 63, 839–851.

Walker, R.G., Eyles, C.H., 1991. Topography and significance of a basin-wide sequence-bounding erosion surface in the Cretaceous Cardium Formation, Alberta, Canada. J. Sed. Petrol., 61, 473–496.

Weber, K.J., 1987. Computation of initial well productivities in aeolian sandstone on the basis of a geological model, Leman Gas Field, U.K. In: R.W. Tillman, K.J. Weber (Eds.), Reservoir Sedimentology, Soc. Econ. Paleon. and Mineral. (SEPM) Spec. Publ. No. 40, pp. 333–354.

Weimer, P., Crews, J.R., Crow, R.S., Varnai, P., 1998. Atlas of the petroleum fields and discoveries in northern Green Canyon, Ewing Bank, and southern Ship Shoal and South Timbalier area (offshore Louisiana), northern Gulf of Mexico. Amer. Assoc. Pet. Geol. Bull., 82, 878–917.

Weimer, R.J., 1992. Developments in sequence stratigraphy: Foreland and cratonic basins. Amer. Assoc. Pet. Geol. Bull., 76, 965–982.

Wilgus, C.K., Hastings, B.S., Kendall, C.G.St.C., Posamentier, H., Ross, C.A., Van Wagoner, J.C., 1988. Sea-Level Changes – An Integrated Approach. Soc. Econ. Paleo. and Mineral. (SEPM) Spec. Publ. 42, 407 pp.

Witton-Barnes, E.M., Hurley, N.F., Slatt, R.M., 2000. Outcrop characterization and subsurface criteria for differentiation of sheet and channel-fill strata: Example from the Cretaceous Lewis Shale, Wyoming. In: P. Weimer, R.M. Slatt, J.L. Coleman, N. Rosen, C.H. Nelson, A.H. Bouma, M. Styzen, D.T. Lawerence (Eds.), Global Deep-Water Reservoirs, Gulf Coast Section, SEPM Foundation Bob F. Perkins 20th Annual Research Conference, pp. 1087–1104.

Ye, L., Kerr, D.R., 2000. Sequence stratigraphy of the Middle Pennsylvanian Bartlesville Sandstone, northeastern Oklahoma: A case of an underfilled incised valley. Amer. Assoc. Pet. Geol Bull., 84, 1185–1204.

Yielding, C., Apps, G., 1994. Spatial and temporal variations in the facies associations of depositional sequences on the slope: Examples from the Miocene–Pleistocene of the Gulf of Mexico. In: P. Weimer, A.H. Bouma, B.F. Perkins (Eds.), Submarine Fans and Turbidite Systems, GCS–SEPM Foundation 15th Annual Research Conference, pp. 425–437.

Index